Biologically-Inspired Systems

Volume 20

Series Editor

Stanislav N. Gorb, Department of Functional Morphology and Biomechanics, Zoological Institute, Kiel University, Kiel, Germany

Motto: Structure and function of biological systems as inspiration for technical developments

Throughout evolution, nature has constantly been called upon to act as an engineer in solving technical problems. Organisms have evolved an immense variety of shapes and structures from macro down to the nanoscale. Zoologists and botanists have collected a huge amount of information about the structure and functions of biological materials and systems. This information can be also utilized to mimic biological solutions in further technical developments. The most important feature of the evolution of biological systems is multiple origins of similar solutions in different lineages of living organisms. These examples should be the best candidates for biomimetics. This book series will deal with topics related to structure and function in biological systems and show how knowledge from biology can be used for technical developments in engineering and materials science. It is intended to accelerate interdisciplinary research on biological functional systems and to promote technical developments. Documenting of the advances in the field will be important for fellow scientists, students, public officials, and for the public in general. Each of the books in this series is expected to provide a comprehensive, authoritative synthesis of the topic.

Vincent Bels • Stanislav N. Gorb • Roi Gurka
Editors

Biomimicry – A Functional Approach

Springer

Editors

Vincent Bels
Muséum National d'Histoire Naturelle
Sorbonne Université, Institut de
Systématique Evolution Biodiversité
Paris, France

Stanislav N. Gorb
Department of Functional Morphology
and Biomechanics, Zoological Institute
Kiel University
Kiel, Germany

Roi Gurka
Department of Physics and Engineering
Science
Coastal Carolina University
Conway, SC, USA

ISSN 2211-0593 ISSN 2211-0607 (electronic)
Biologically-Inspired Systems
ISBN 978-3-032-11948-3 ISBN 978-3-032-11949-0 (eBook)
https://doi.org/10.1007/978-3-032-11949-0

This Springer imprint is published by the registered company Springer Nature Switzerland AG
The registered company address is: Gewerbestrasse 11, 6330 Cham, Switzerland

Preface

The present volume is devoted to the study and application of biomimicry at the intersection of biology, engineering, and design. On the basis of several case studies, this book aims to reflect the growing maturity of the field and its capacity to connect diverse areas of knowledge through a shared concern for efficiency, resilience, and sustainability. Rooted in diverse biological observations and analyses, these studies are oriented toward practical applications, demonstrating how concepts derived from natural systems can inform technological innovation and vice versa.

This work is addressed to researchers, students, and practitioners across a wide range of disciplines who share an interest in the principles that underlie nature's designs. It is offered as an invitation to engage with studies integrating biological approaches with biomimetics as a mode of inquiry that is at once analytical and creative, empirical and visionary perspective.

We would thank all the contributors of this book and all colleagues who agreed to provide comments on the chapters, particularly Dr. Filipe Natalio, Dr. Katherine Nagel, Dr. Aaron Bauer, Dr. Norbert Hoeller, Dr. Frank Fish, and Dr. Christine Rollard.

Paris, France — Vincent Bels
Kiel, Germany — Stanislav N. Gorb
Conway, SC, USA — Roi Gurka

Contents

Chapter 1
Introduction: From Functional Morphology to Biomimetics

Vincent Bels, Roi Gurka, and Stanislav N. Gorb

Abstract In this chapter, we explore the evolutionary and functional significance of various case studies associated with biomimicry in ten relevant chapters with various aims. By the "form–function" relationships over a wide diversity of organisms, the chapters highlight how organisms' structures, performances, and behaviors relate to their survival strategies, design, and how these are utilized in building the diversity of applications in biomimicry. Comparative perspectives between organisms and their potential applications emphasize how artificial systems adopt structural and functional integration, interacting with their dynamic environments. The work underscores the importance of bridging functional morphology, ethology, and biomimicry to generate insights into both biological evolution and the design of resilient biomimicry.

Keywords Adaptation · Form-function · Evolution · Biomimicry

1.1 A Context

Over the past 50 years, conceptual approach and empirical biologically inspired designs have shifted from studying structure–function relationships of organisms

V. Bels (✉)
Muséum national d'Histoire naturelle, Sorbonne Université, Institut de Systématique Evolution Biodiversité, UMR 7205 CNRS/MNHN/EPHE/UA, Station de Biologie Marine de Dinard, Dinard, France
e-mail: vincent.bels@mnhn.fr

R. Gurka
Department of Physics and Engineering Science, Coastal Carolina University, Conway, SC, USA
e-mail: rgurka@coastal.edu

S. N. Gorb
Department of Functional Morphology and Biomechanics, Zoological Institute, Kiel University, Kiel, Germany
e-mail: sgorb@zoologie.uni-kiel.de

V. Bels et al. (eds.), *Biomimicry – A Functional Approach*, Biologically-Inspired Systems 20, https://doi.org/10.1007/978-3-032-11949-0_1

toward a broader and more integrated framework including implementation of this knowledge into engineering sciences: bionics and biomimicry. This research direction has rapidly adopted a multidisciplinary approach, integrating numerous characteristics of animal traits from morphology to behavior to designing artificial systems following similar functional principles (Vincent 2006; Wahl 2006; Armour and Vincent 2006; Coelho and Versos 2011; Jiang et al. 2024). This extended the concept beyond engineering sciences to sustainability and systems, emulating the nature principles considered as a model, a measure, and a mentor for innovation (Nachtigall 1988, 1997; Vincent 1988; Benyus 1997; Bar-Cohen 2006). It is well known that the shape (form at various levels) and functioning of various model animals have long inspired numerous engineering disciplines (Sane et al. 2016). In the literature of the last decade, hundreds of examples of form (structure)–function relationships have been reported in diverse animal lineages studied under various environmental conditions. These examples are often converged according to their basic physical principles. For example, the streamlined forms of dolphins, sharks, penguins and even marine turtles (Bang et al. 2016) have influenced hull design and drag-reduction coatings for improving movements of swimming and flying engineering systems (i.e., Bechert et al. 2000; Jaar 2025). The segmented limbs of arthropods and their lightweight yet strong exoskeletons have informed robotic actuators and prosthetic designs (Cornejo et al. 2024). Other examples associate functional and biomechanical studies related to the fluid dynamics of swimming: exploring how fish and aquatic tetrapods generate thrust and control movement in a large number of organisms from a highly diverse radiation of the Vertebrate Bauplan. The results of these studies provide detailed information about numerous locomotory principles (i.e., body undulation and fin movements) for the development of bioinspired underwater robots (Gordon et al. 2000; Lauder 2005; Fish 2006; Lauder and Tytell 2006; Triantafyllou et al. 2016, 2020). Various modes of terrestrial (i.e., hexapodal, quadrupedal, and bipedal) and aerial locomotion are probably within the most well-known examples of animal behavior with various *Baupläne* for the potential development of applications in engineering based on the studies of mechanisms through integrated biological and physical approaches. For example, the biomechanics of attachment and movement in amphibians, lizards (i.e., geckos, anoles, skinks), and various insects have revealed sophisticated strategies for understanding mechanisms of adhesion, friction control, and locomotion that have inspired developments in robotics (i.e., Gorb 2008; Crawford 2016; Higham et al. 2019; Langowski et al. 2020; Li et al. 2021). Advancing our understanding of jumping mechanisms in both vertebrates and insects increased our knowledge on the biological role of this locomotor mode at various levels (i.e., stability, energy) in take-off and landing phases that can be applied in the development of jumping robots (i.e., Paskins 2007; Chang et al. 2019).

Numerous examples show that biomimicry utilize holistic approaches emphasizing that the traits studied from various perspectives primarily related to forms (structures at various levels) and functions including behaviors of the organisms. Probably a key evolution of biomimicry is in its strong integration with biological approaches favoring understanding of organisms' diversities and their lives to explain evolutionary and ecological mechanisms under the process of natural selection as pointed by Darwin and Wallace in the nineteenth century (Wallace 1858; Darwin 1871). Potentially, engineering sciences can use this information to develop human applications at various levels. The main goal of this book content is to provide a series of case studies in highly different Baupläne, in such an approach highlighted by Gravish and Lauder (2018) who conclude: "*The aim of robotics-inspired biology research is to 'close the loop', and use both physical models and reactive robotic systems to generate new hypotheses for biological investigation.*" Such approach permits a more complete understanding of the organisms than examining separate traits. By this way, biomimicry/biomimetics/bionics embodies integrated research approach to organisms in their phylogenetic relationships, in order to grasp as much as possible, their responses to their proximal and distal environmental constraints and help to conceive the optimal strategies that organisms use to survive within their environments. This design can enable the resilient innovation and development of engineering systems and processes that align with ecological sustainability.

Two well-known examples illustrate such holistic approach showing the loop between biomimicry and fundamental biological approach (i.e., ethology, behavioral ecology, and evolution) as possible comprehension of the organisms and their use for sustainable human activities. The first example concerns the dry adhesion enabled by a complex micro–nano hierarchical structure, where setae branch into spatulae, maximizing surface contact without chemical bonding. The knowledge of subdigital lamellae in *Gekko* and *Anolis* and numerous other lizards (i.e., Chamaeleonidae) (Spinner et al. 2013) has evolved over decades, beginning with morphological observations at the beginning of the twentieth century. Later studies have demonstrated fine-scale functional variation related to environmental influences across families and species (see, e.g., Ernst and Ruibal 1966; Nachtigall 1974; Russell 2002; Russell and Johnson 2007; Garner and Russell 2021; Riedel et al. 2024). Classical comparative biological studies have revealed striking convergences in toe pad structure in lizards (i.e., geckos and anoles; Losos 1990; Irschick et al. 1996; Russell and Garner 2023). Engineers have replicated this structure using microfabrication and flexible polymers, resulting in synthetic adhesives with directional, self-cleaning, and reversible properties. Initially studied in the field of biology, the remarkable ability of these lizards to adhere to vertical and inverted surfaces was associated to van der-Waals forces through millions of microscopic spatulae on their toe pads (i.e., Autumn et al. 2002; Griffing et al. 2022, 2024). Beyond adhesion, biomimicry expanded to explore how geckos self-clean their adhesive pads, resist wear over thousands of steps, and manage precise detachment (Gardner and Russell, 2025 [this volume, Chap. 4]). These insights enhanced the development of reversible adhesives (Gorb et al. 2007), wall-climbing robots, and even biomedical patches that require secure yet residue-free attachment (Kim et al. 2009), as well as provide the understanding of adaptive evolutionary mechanisms of these lizards to proximal (i.e., tree surface) environmental constraints. The actual transition to integrating biomimicry and biology entails a deeper consideration of how geckos interact with their environment as adaptive, resilient organisms. Furthermore, this form–function relationship coupled with such an integrated biology–biomimicry approach can enhance the functional potential in the adaptive responses of the two groups of organisms (*Anolis*—Iguanidae and *Gekko*—Gekkonidae) during their convergent radiation. Finally, this sheds light on the ability of organisms to respond to environmental variability in terms of biology, and to the building of principally novel material classes using an engineering approach.

The study of shark skin has long served as a biological model in surface engineering, with early experimental biology approaches focused on drag reduction inspired by dermal denticles: microscopic, ribbed scales aligned along the body surface (Reif and Dinkelacker 1982). These structures presumably reduce turbulence and demonstrate hydrodynamic resistance, prompting synthetic imitations in swimwear and aircraft coatings. Over the last 10 years, studies of shark skin transitioned from a mechanical model in biology to a classical reference in biomimicry (Zehra and Mobin 2024). In such transitional approach including biomimicry and ecology, shark skin antifouling properties gained attention for entirely different applications (i.e., medical and maritime surfaces, Chien et al. 2019). Notably, the computational exploration also facilitates a more precisely defined morphospace, important for new potential sustainable applications. For example, machine learning enhances adaptive surface control for explaining shark radiation, but also biofilm resistance and antifouling performance, extending the function of shark skin mimicry beyond hydrodynamics to long-term surface maintenance in marine engineering.

Hundreds of other examples can be used to demonstrate that biomimicry has increasingly drawn upon integration of numerous aspects of animal biology for innovations in both sciences: (i) establishing evolutionary and adaptive mechanisms and (ii) offering insights into adaptable, efficient, and sustainable design.

Historically, the "form–function" paradigm which is a central concept in biology, rooted in the anatomical work of Georges Cuvier (1835), has become one major holistic approach combining biology and biomimicry. This paradigm is based on the relationship between a biological structure ("form") in terms of properties ("function") associated with a "biological role," but not always directly reflecting this role; although Cuvier (1835) proposed that the structure of organisms could be inferred from function due to their integration into a coherent whole. Amundson and Lauder (1994) distinguish two major concepts of biological function: (i) the "selected-effect" account and the "biological role" account. Whereas "selected-effect" influences the evolutionary history and adaptive design, the "biological role" functions are defined by the immediate contribution of a trait of one organism. These functions are indispensable for the study of forms in comparative anatomy and function in functional morphology, since historical selection pressures are often impossible to reconstruct. Ultimately, these two concepts are complementary rather than mutually exclusive, each offering distinct explanatory contributions to

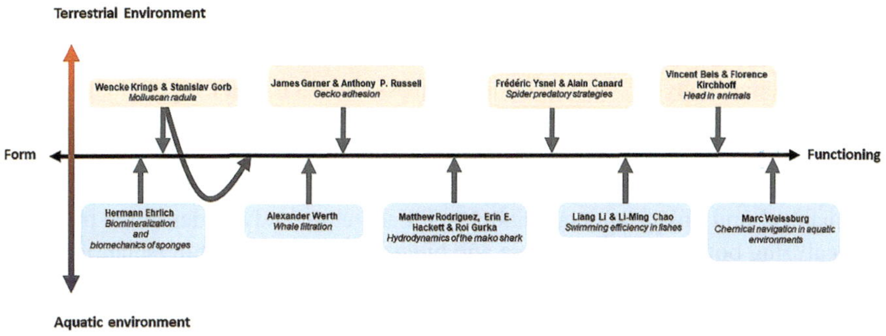

Fig. 1.1 All the chapters using animals from different lineages living in terrestrial and aquatic habitats embrace the holistic approach of *"form–function"* necessary for the development in biomimicry (@Vincent Bels)

evolutionary biology. The "form–function" paradigm assumes that morphological and behavioral traits associated to determined performances evolve under selective pressures to fulfil specific ecological role along the animal evolution. This paradigm becomes formalized by integrating the anatomical structure and biomechanical performance, as well as the behavioral context and neuromotor control into the analysis of organismal relationships with their environment (Lauder 1991; Lauder and Thomason 1995).

The chapters of this book explore the relationships between "form" and "functions" in diverse lineages of organisms through various case studies (Fig. 1.1). Beyond structure or form, animal functioning has become an equally important focus in biomimicry as demonstrated in all of the chapters of this volume. A major advancement lies in combining morphology and behavior into integrated system understanding with applications having been made in biomimicry during the last 30 years. It is clear that the integration of artificial intelligence (AI) in integrated approach Biology–Biomimicry significantly enhances the capacity to analyze, replicate, and optimize natural forms and functions by exploring with novel paradigms the evolutionary processes, the ecological constraints (i.e., constraints produced by human activities), the neuro–morphological spaces of each type of organisms, etc. (Kim et al. 2021). Recent discourse in biomimicry includes researchers advocating for "biologizing" design considering animals as co-designers or informants in the innovation process, not merely as resource models. Animal-inspired systems are now leveraged to address global challenges in sustainability, including energy efficiency and climate-responsiveness (Yanlic and Coskun 2024). Thus, the ethics of AI in biomimicry requires strong attention because it can obscure or hide the real relationship between natural models and technological outcomes, potentially deriving from misrepresentations of biological systems. Hence, ethical biomimicry must ensure that innovation for human life does not disconnect from biodiversity and its evolutionary and ecological roots.

1.2 A Content

The collective body of work presented across these chapters highlights the fundamental interdependence of biological form and function. From the microarchitectures of sponges and radulae to the structural diversity of gecko adhesion and baleen filtration, we observe how nature's forms evolve to cope with specific mechanical and environmental challenges. Each case study underscores the value of analyzing both physical structure and biological function, not in isolation, but as co-evolved systems to provide a coherent framework for engineering design inspired by evolution.

Hermann Ehrlich presents an in-depth overview of pioneering contributions to the fields of biomineralization and sponge biomechanics, with a particular focus on the siliceous skeletons of marine and freshwater sponges (Fig. 1.2). This chapter bridges biology, materials science, and nanotechnology, addressing the evolutionary significance of sponge biomineralization. Through comparative analyses of different sponge species, the author elucidates the diversity of biomineralization pathways. His multidisciplinary approach, combining biochemistry, crystallography, and electron microscopy, uncovers complex mineral–organic interfaces that influence both material structure and physiological properties. These findings have broader implications for understanding the role of mineralized tissues in living organisms and for designing biomimetic materials. By decoding nature's strategies for precise, energy-efficient material synthesis, this chapter demonstrates groundwork for applications ranging from bioactive scaffolds and biosilica-based drug delivery systems to sustainable technologies for water filtration and environmental remediation.

Fig. 1.2 Illustration of the chapter by Herman Ehrlich on biomineralization and sponge biomechanics (@ Herman Ehrlich)

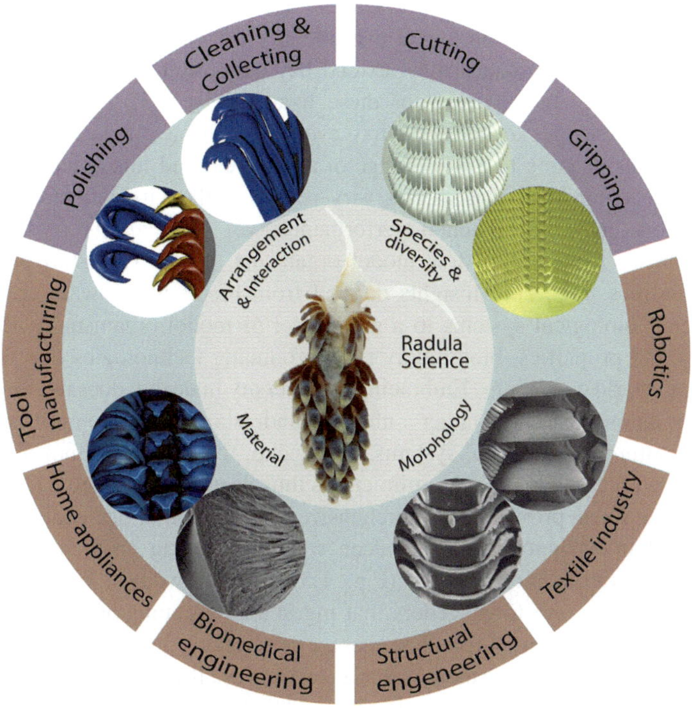

Fig. 1.3 Illustration of the chapter by Wencke Krings and Stanislav Gorb on the comprehensive exploration of the molluscan radula for examining its biomechanical properties (@Wencke Krings)

Wencke Krings and Stanislav Gorb provide a comprehensive exploration of the molluscan radula and examine its biomechanical properties and relevance for biomimetic applications, by focusing on the radular structure–function relationships, mechanical performance, and its potential as a model for engineered materials and devices (Fig. 1.3). Reflecting evolutionary adaptation to various feeding strategies and ecological niches, the teeth of various gastropods show significant variation in shape, size, and mechanical properties. The authors discuss how hardness and stiffness contribute to radular function, emphasizing the trade-offs between wear resistance and flexibility. They highlight how spatial gradients in mechanical properties within individual teeth, a combination of hard tooth tips and more elastic bases, enhance durability and functional efficiency of the radular system. Finally, the authors contextualize their findings within the broader framework of biomimetics. The radula is not merely a biological curiosity but a functional prototype for developing wear-resistant tools, flexible abrasives, and adaptive micro-robotic elements.

The chapter by James Garner and Anthony P. Russell evaluates the progress and limitations of gecko-inspired adhesion research over the past 25 years, suggesting a shift toward biodiversity-aware approaches to improve the design and performance of gecko-based synthetic adhesives (GSAs). They present a comprehensive synthesis of biological, functional, and evolutionary dimensions of gecko adhesion; emphasizing that biomimetics must extend beyond simplified models of a single species to embrace the diversity of these lizards (Fig. 1.3). A central topic of the chapter is that gecko adhesion has evolved multiple times across distantly related lineages, leading to a wide range of anatomical and functional adaptations. The authors demonstrate that species vary widely in their toe pad structure, setal configuration, and environmental performance, yet the majority of biomimetic research draws from only a few model organisms, primarily *Gekko gecko*. The chapter outlines three core insights derived from biodiversity-aware approaches. By reducing biological systems to a narrow set of model organisms with limited structures and properties, biomimicry risks producing technologies that fail under certain real-world conditions. Embracing biodiversity in design does not imply complexity for its own sake, but rather a nuanced understanding of the multifunctional, adaptable nature of biological systems. The gecko, for long, a symbol of adhesion excellence with application in biomimicry, is thus deeply reconsidered (Fig. 1.4).

Alexander Werth provides a comprehensive examination of the feeding strategies and biomechanical innovations of baleen whales, focusing on their potential for bioinspired filtration technologies (Fig. 1.5). Baleen whales, members of the Mysticeti suborder, are filter feeders that ingest vast amounts of water containing prey, such as krill and small fish. The chapter investigates the anatomy and function of baleen plates, exploring their material properties, flow dynamics, and role in facilitating cross-flow filtration that is a method that contrasts with traditional dead-end filtration in engineered systems. Often facing flow resistance and fouling, baleen enables high-throughput filtration through a dynamic, self-cleaning, and anisotropic matrix. Three distinct modes of feeding are identified among baleen whales: skim feeding (e.g., right whales), engulfment feeding (e.g., rorquals), and suction feeding (e.g., gray whales). Each mode involves unique morphological adaptations and flow regimes, but all share the underlying principle of efficient, selective filtration driven by hydrodynamics and anatomical specialization. Importantly, the chapter connects these biological insights to engineering through the concept of functional abstraction. The author proposes that rather than mimicking the exact structure of baleen, engineers should replicate its functional principles, such as maintaining unidirectional flow, managing variable particle sizes, and reducing energy input through passive mechanics. Applications envisioned include industrial water purification, microplastic removal, and air filtration systems designed for resilience and energy efficiency.

Matthew Rodrigez, Erin E. Hackett, and Roi Gurka present an in-depth analysis of the mako shark's (*Isurus* spp.) lunate caudal fin from a hydrodynamic and biomimetic perspective (Fig. 1.6). The focus lies on the unique morphological adaptations of mako shark to move very fast with minimal energy expenditure, that is, high hydrodynamic performances. The chapter discusses locomotion using body–caudal–fin oscillatory motions and the unicity of shortfin mako as an hydromechanically efficient swimmer and relate this to its morphological features, among its symmetrical lunate caudal fin coupled with the specific mode

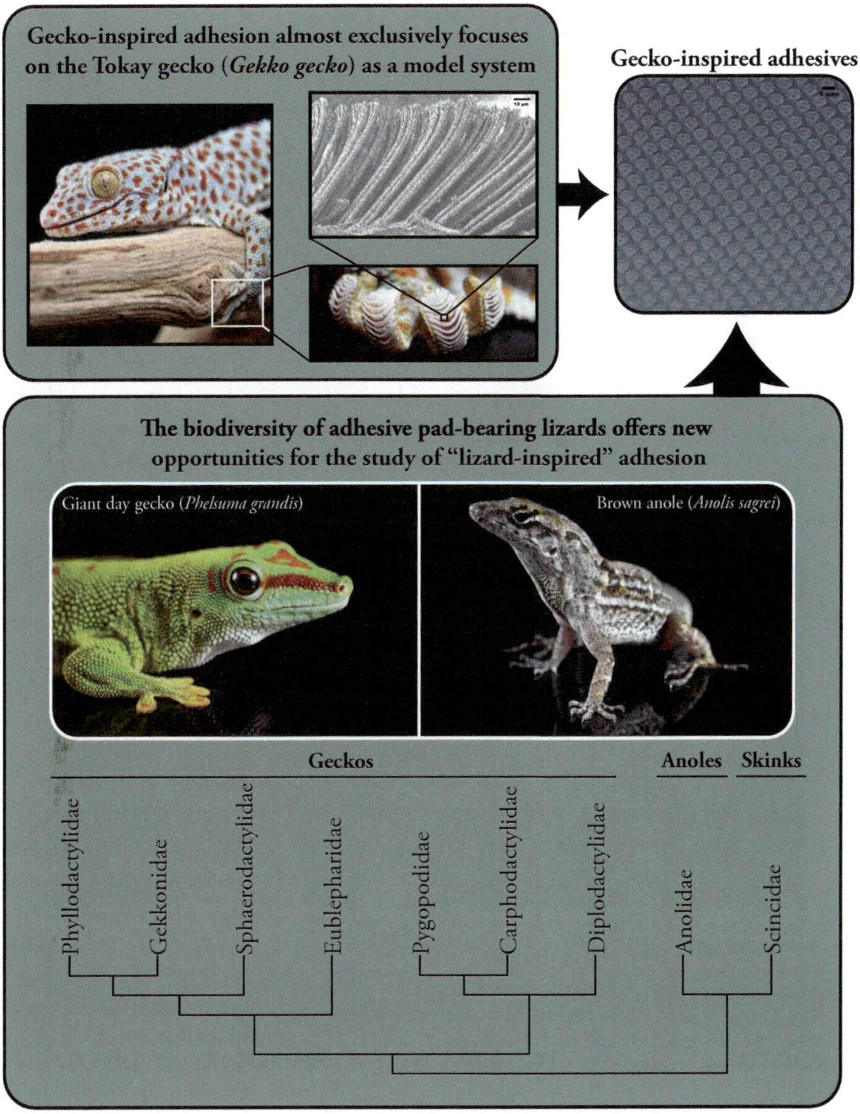

Fig. 1.4 Illustration of the chapter by James Garner and Anthony P. Russell on the functional and evolutionary aspects of gecko adhesion (@ James Garner and Anthony P. Russell)

of locomotion: thunniform, which is common among large fish, such as Bluefin tuna and some of the Lamnidae sharks as well as blue whale. This comparative study investigates two caudal fin morphologies in the shortfin mako shark. Using PIV analysis, the research demonstrates that lunate tails paired with thunniform body undulations produce narrower and more energetic wakes than heterocercal fins. This morphology supports high-speed, long-distance swimming with improved

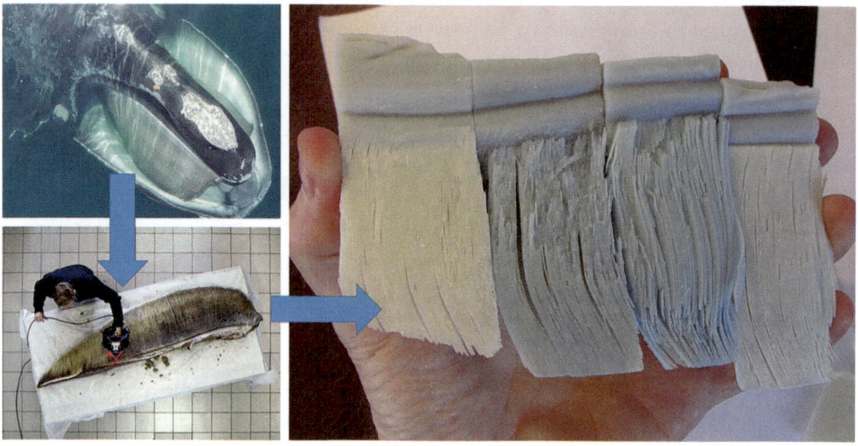

Fig. 1.5 Illustration of the chapter by Alexander Werth that explores the characteristics of the baleen whales (@Alexander Werth)

Fig. 1.6 Illustration of the chapter by Matthew Rodrigez, Erin E. Hackett, and Roi Gurka showing the skar model used in the study

hydromechanical efficiency. The chapter provides foundational data for robotic fin design and underscores the performance trade-offs between thrust and efficiency that differ by fin structure.

Liang Li and Li-Ming Chao explore multiple strategies that fish use to optimize swimming efficiency, including streamlined morphologies, drag-reducing textures, and collective behavior in schools (Fig. 1.7). Their work combines computational fluid dynamics (CFD), experimental modeling, and biological observations. The review delineates passive strategies (e.g., denticles, body shape) and active control mechanisms (e.g., vortex phase matching, proprioception). The authors discuss how these mechanisms scale from individual swimmers to schooling groups.

Frédéric Ysnel and Alain Canard explore how the predation strategies of spiders (order Araneae) offer a rich foundation for biomimetic innovation (Fig. 1.8).

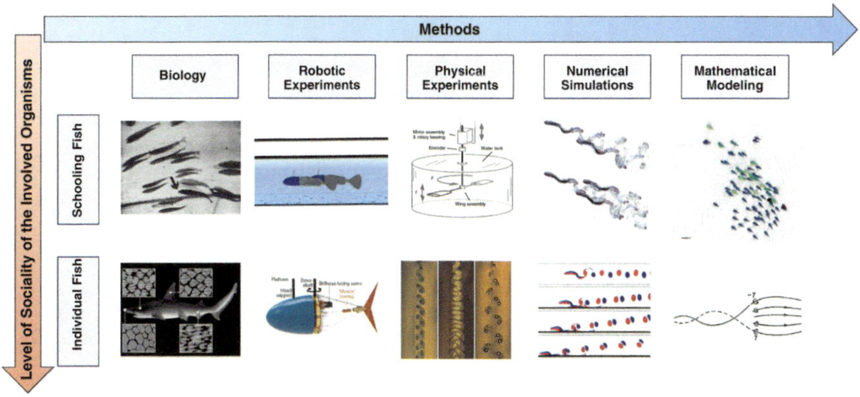

Fig. 1.7 Illustration of the chapter by Liang Li and Li-Ming Chao that explores multiple strategies that fish use to optimize swimming efficiency (@Liand Li). At one end, biological studies capture authentic behaviors of live fish, providing the highest realism but offering limited opportunity for manipulation. Moving along the spectrum, robotic experiments retain some biological features while allowing greater control of parameters, and physical experiments simplify biological details further yet permit extensive manipulation of flow and structure. Numerical simulations provide highly flexible environments for testing hypotheses, though at the cost of reduced biological fidelity, while mathematical modeling represents the most abstract approach, maximizing analytical control and predictive power but incorporating minimal biological detail

Fig. 1.8 Illustration of the chapter by Frédéric Ysnel and Alain Canard showing the hydrophobic "microhairs" on their appendages in *Dolomedes fimbriatus* (@ Alain Canard)

As ubiquitous predators in terrestrial ecosystems, spiders have evolved numerous specialized tactics for survival and hunting, demonstrating great potential toward biomimetic transfer from their vision, locomotion, mechanoreception, deceptive signaling, and silk production. The authors describe the structure and function of the visual system. Wandering spiders and ambush predators exploit terrain features for stealth and maneuverability. This form of movement is efficient, low-energy and adaptable; traits that could be potentially harnessed in legged robotics

Fig. 1.9 Illustration of the chapter by Marc Weissburg showing the multidisciplinary exploration of how aquatic organisms successfully locate chemical sources in turbulent environments. (**a**) Blue crab. (**b**) Crab drag plume (@ Marc Weissburg)

and prosthetics. The authors examine how spiders use deceptive signals to lure prey. Importantly, such deception is often finely tuned to the sensory capacities of the target species; providing a model for highly specific signal-target systems in engineering. The chapter concludes by advocating a systems-level approach to biomimicry, in which spider behavior and physiology are viewed as integrated systems rather than isolated traits.

Marc Weissburg presents a multidisciplinary exploration of how aquatic organisms successfully locate chemical sources in turbulent environments, and how these strategies can inform the design of autonomous systems and sensors (Fig. 1.9). The chapter focuses on aquatic animals that use complex behaviors that account for the highly variable and discontinuous nature of chemical plumes in water. The behaviors of blue crabs, lobsters, and fish serve as case studies, illustrating how movement patterns, sensor morphology, and temporal sampling strategies improve tracking success. A central theme of the chapter is the "active sampling," in which animals dynamically control their position and sensory apparatus to maximize information gain. Successful navigation depends on the coupling between locomotion and sensing, often mediated by feedback mechanisms and internal decision rules. Based on this approach, the inspiration from biological systems for artificial plume tracking technologies is largely discussed. Drawing from robotic applications, specific design principles derived from nature are suggested. Finally,

Fig. 1.10 Illustration of the chapter by Vincent Bels and Florence Kirchhoff that opens the discussion on the flexibility of the head in animals and robots (@Florence Kirchhoff)

the author reviews experimental robotic platforms that mimic these biological principles, including mobile robots equipped with chemical sensors and behavioral algorithms modeled on animal movement.

Vincent Bels and Florence Kirchhoff present a conceptual and functional framework for understanding and designing sensorimotor integration systems in bio-inspired robotics (Fig. 1.10). By examining how biological heads coordinate sensory input and motor output to control adaptive movement, the chapter provides a transdisciplinary basis for discussing the potential role of a head in robots. The core argument is that in bilaterian evolution, the head has emerged as a central structure that aggregates sensory organs, integrates environmental data, and directs locomotion via neural and musculoskeletal coordination. Unlike conventional robotics that often separate perception and action, the head system employs a closed-loop architecture allowing real-time sensorimotor feedback. Behavioral experiments and control demonstrate that the functioning of the head is of major biological role in goal-oriented tasks, in adapting to changing environments, and in integrating multisensory cues to refine its responses. These responses are not hardcoded but emerge from task-dependent coordination between sensor modalities and motor effectors.

To conclude this volume (Chap. 11), challenging the "form–function" coupling in nature and in biomimicry, Julian Vincent explores the conceptual and functional parallels between biology and engineering, emphasizing the role of analogical reasoning as a bridge between natural and technological systems. In this chapter, Julian Vincent articulates how biological systems, though evolved under constraints vastly different from those encountered in human design, offer a wealth of solutions adaptable to engineering problems and proposes that the core advantage of biomimicry lies not entirely in copying nature. A central claim is that many biological trade-offs—between strength and flexibility, efficiency and redundancy, or speed and accuracy—can be systematically analyzed to inspire innovative solutions for human survival. The chapter moves beyond surface-level biomimicry to advocate for a deeper understanding of biological "contradictions," which resemble engineering design challenges. The author suggests that these apparent paradoxes are reconcilable through methods that uncover the underlying strategies embedded in nature—strategies. In exploring these analogies, the author categorizes various approaches to biomimicry, distinguishing between form-based (mimicking biological shapes), process-based (replicating biological functions), and system-based (adopting ecological principles) strategies. He critiques the overreliance on superficial form-mimicry in commercial products and urges a move toward more systemic and functional interpretations of biology.

By embracing the complexity, adaptability, and ingenuity of biological systems, engineers can unlock new design paradigms that align with ecological principles and long-term sustainability goals on the basis of a greater true interdisciplinary collaboration between biologists, engineers, and materials scientists.

References

Amundson R, Lauder GV (1994) Function without purpose: the uses of causal role function in evolutionary biology. Biol Philos 9(4):443–469. https://doi.org/10.1007/BF00850375

Armour RH, Vincent JFV (2006) Rolling in nature and robotics: a review. J Bionic Eng 3:195–208. Springer Science and Business Media LLC

Autumn K et al (2002) Evidence for van der Waals adhesion in gecko setae. Proc Natl Acad Sci USA 99:12252–12256

Bang K, Kim J, Lee S-I, Choi H (2016) Hydrodynamic role of longitudinal dorsal ridges in a leatherback turtle swimming. Sci Rep 6:34283. Nature Publishing Group UK, London

Bar-Cohen Y (2006) Biomimetics: biologically inspired technologies. CRC Press, Boca Raton

Bechert DW, Bruse M, Hage W, Meyer R (2000) Fluid mechanics of biological surfaces and their technological application. Naturwissenschaften 87:157–171. Springer Science and Business Media LLC

Benyus JM (1997) Biomimicry: innovation inspired by nature. HarperCollins, New York

Chang B, Myeong J, Virot E, Clanet C, Kim H-Y, Jung S (2019) Jumping dynamics of aquatic animals. J R Soc Interface 16:20190014. The Royal Society

Chien S et al (2019) Gecko-inspired adhesive systems for robotics. Bioinspir Biomim 14:031001. https://doi.org/10.1088/1748-3190/ab0e33

Coelho DA, Versos CAM (2011) A comparative analysis of six bionic design methods. Int J Des Eng 4:114. Inderscience Publishers

Cornejo J, Sierra-Garcia JE, Gomez-Gil FJ, Weitzenfeld A, Acevedo FE, Escalante I, Recuero E, Wehrtmann IS (2024) Bio-inspired design of hard-bodied mobile robots based on arthropod morphologies: a 10-year systematic review and bibliometric analysis. Bioinspir Biomim 19(5). https://doi.org/10.1088/1748-3190/ad5778. IOP Publishing

Crawford NA (2016) The biomechanics of tree frog adhesion under challenging conditions. PhD Thesis, University of Glasgow

Cuvier G (1835) Leçons d'anatomie comparée. Fortin, Masson, Paris

Darwin C (1871) The descent of man and selection in relation to sex. John Murray, London

Ernst V, Ruibal R (1966) The structure and development of the digital lamellae of lizards. J Morphol 120:233–265. Wiley

Fish FE (2006) Limits of nature and advances of technology: what does biomimetics have to offer to aquatic robots? Appl Bionics Biomech 3(1):49–60

Garner AM, Russell AP (2021) Revisiting the classification of squamate adhesive setae: historical, morphological and functional perspectives. R Soc Open Sci 8(2):202039

Gorb SN et al (2007) Biological fibrillar adhesives: functional principles and biomimetic applications. Annu Rev Mater Res 37:415–443

Gorb SN (2008) Biological attachment devices: exploring nature's diversity for biomimetics. Philos Trans R Soc A Math Phys Eng Sci 366:1557–1574

Gordon MS, Hove JR, Webb PW, Weihs D (2000) Boxfishes as unusually well-controlled autonomous underwater vehicles. Physiol Biochem Zool 73(6):663–671

Gravish N, Lauder GV (2018) Robotics-inspired biology. J Exp Biol 221:jeb138438

Griffing AH, Gamble T, Cohn MJ, Sanger TJ (2022) Convergent developmental patterns underlie the repeated evolution of adhesive toe pads among lizards. Biol J Linn Soc 135:518–532. Oxford University Press UK

Griffing AH, Gamble T, Behere A, Higham TE, Keller GM, Resener J, Sanger TJ (2024) Developmental patterns underlying variation in form and function exhibited by house gecko toe pads. Integr Comp Biol 64:1494–1504. Oxford University Press

Higham TE, Russell AP, Niewiarowski PH, Wright A, Speck T (2019) The ecomechanics of gecko adhesion: natural surface topography, evolution, and biomimetics. Integr Comp Biol 59:148–167. Oxford University Press

Irschick DJ et al (1996) A comparative analysis of ecological and behavioral correlates of locomotor performance in lizards. Ecol Monogr 66:387–421

Jaar SG (2025) Nature-inspired innovation: evaluating biomimicry in hull design for enhanced engineering performance. The National High School Journal of Science (NHSJS) – NHSJS Reports.

Jiang M, Deng W, Lin H (2024) Sustainability through biomimicry: a comprehensive review of bionic design applications. Biomimetics 9:507. MDPI

Kim S et al (2009) Smooth vertical surface climbing with directional adhesion. IEEE Trans Robot 24:65–74

Kim S et al (2021) Directional dry adhesives for climbing robots. Adv Mater 33:2007345

Langowski JK, Dodou D, van Assenbergh P, van Leeuwen JL (2020) Design of tree-frog-inspired adhesives. Integr Comp Biol 60:906–918. Oxford University Press

Lauder GV (1991) Functional design of the feeding mechanism in lower vertebrates: a perspective on evolution. Am Zool 31:1–19

Lauder GV (2005) Biomechanics and behavior of fish swimming. In: Shadwick RE, Lauder GV (eds) Fish physiology, vol 23. Academic Press, pp 23–58

Lauder GV, Thomason JJ (1995) On the inference of function from structure. In: Thomason JJ (ed) Functional morphology in vertebrate paleontology. Cambridge University Press, pp 1–18

Lauder GV, Tytell ED (2006) Hydrodynamics of undulatory propulsion. Fish Physiol 23:425–468

Li M, Shi L, Wang X (2021) Physical mechanisms behind the wet adhesion: from amphibian toe-pad to biomimetics. Colloids Surf B: Biointerfaces 199:111531. Elsevier

Losos JB (1990) The evolution of form and function: morphology and locomotor performance in Caribbean *Anolis* lizards. Evolution 44:1189–1203

Nachtigall W (1974) Biological mechanisms of attachment. Springer-Verlag, Berlin, Heidelberg, New York

Nachtigall W (1988) Bionik — Grundlagen und Beispiele für Ingenieure und Naturwissenschaftler. Carl Hanser, München

Nachtigall W (1997) Bionik: Grundlagen und Beispiele für Ingenieure und Naturwissenschaftler, 2nd edn. Springer, Berlin

Paskins KE (2007) The mechanics and energy economy of animal jumping and landing applied to autonomous robots. University of Bath, Bath

Reif W-E, Dinkelacker A (1982) Functional morphology of squamate toe pads. Zoomorphology 99:37–51

Riedel J, Eisele K, Gabelaia M, Higham TE, Wu J, Do QH, Nguyen TQ, Meneses CG, Brown RM, Ziegler T, Grismer LL, Russell AP, Rödder D (2024) Ecologically-related variation of digit morphology in Cyrtodactylus (Gekkota, Squamata) reveals repeated origins of incipient adhesive toepads. Funct Ecol 38:1630–1648. Wiley

Russell AP (2002) Integrative functional morphology of the gekkotan adhesive system (Reptilia: Gekkota). Integr Comp Biol 42:1154–1163. Oxford University Press

Russell AP, Garner AM (2023) Solutions to a sticky problem: convergence of the adhesive systems of geckos and anoles. In: Bels VL, Russell AP (eds) Convergent evolution: animal form and function. Springer, Cham, pp 221–255

Russell AP, Johnson MK (2007) Mechanical properties of gecko setae. J R Soc Interface 4:1071–1079

Sane SP et al (2016) Bioinspiration and biomimicry: what can engineers learn from biologists? J Appl Sci Eng 19:1–6

Spinner M, Westhoff G, Gorb SN (2013) Subdigital and subcaudal microornamentation in chamaeleonidae—a comparative study. J Morphol 274:713–723

Triantafyllou MS, Weymouth GD, Miao J (2016) Biomimetic survival hydrodynamics and flow sensing. Annu Rev Fluid Mech 48(1):1–24

Triantafyllou MS, Winey N, Trakht Y, Elhassid R, Yoerger D (2020) Biomimetic design of dorsal fins for AUVs to enhance maneuverability. Bioinspir Biomim 15(3):035003

Vincent J (1988) Where biology leads, industry can follow. Biologist 35(5):274–276

Vincent JF (2006) The materials revolution. J Bionic Eng 3:217–234

Wahl DC (2006) Bionics vs. biomimicry: from control of nature to sustainable participation in nature. In: Design and nature III: comparing design in nature with science and engineering, vol 87. WIT Press, Southampton, pp 289–298

Wallace AR (1858) On the tendency of varieties to depart indefinitely from the original type. J Proc Linn Soc Zool 3:53–62

Yanlic BS, Coskun A (2024) Understanding the needs of nonhuman stakeholders in design process: an overview of and reflection on methods. arXiv. http://arxiv.org/abs/2407.14750. Accessed 26 July 2025

Zehra S, Mobin M (2024) Biomimetic coatings: an introduction. In: Smart biomimetic coatings, pp 1–17. https://doi.org/10.1016/B978-0-443-21738-8.00001-1

Chapter 2
Architectured Biosilica in Sponges a Unique Source for Bioinspired Design

Hermann Ehrlich

Abstract Sponges (phylum Porifera), which arose at the dawn of the evolution of multicellular organisms, are a unique storehouse of the biosilica-based bioarchitectures represented by their skeletal formations. The diversity of size (from micro- to meter-long dimensions), shape (up to 100 morphotypes), porosity (from nano- to macro-scale), and hierarchical ornamentation of their interconnected siliceous networks remain to be in a focus of modern materials science and biomimetics. Principles of templating and scaffolding activities of polysaccharide- and protein-based biosynthesis of highly structured biosilica in sponges will be critically analysed in this chapter. Special attention is paid to fundamental role of actin in patterning of sophisticated, branched geometries of poriferan biosilica. Also discussed are prospects for using fundamental knowledge about the hierarchical architecture of biosilicates in sponges to create new functional and sustainable composites.

Keywords Biosilica · Sponges · Spicule · Porifera · Nanomaterials · Biocomposites · Bioinspired design · 3D scaffolds

2.1 Introduction: Sponge as a "Living Pump" and a "Living Filter"

Our current understanding, as engineers and bio-inspired material science experts, is that sponges, body plan and function in aquatic habitats makes them "living pumps", originating more than 800 Myr ago (Turner 2021). The relation between form and function in both ancient and modern groups of sponges remains stable

H. Ehrlich (✉)
Adam Mickiewicz University in Poznan, Center for Advanced Technology, Poznan, Poland

Poznan University of Technology, Faculty of Chemical Technology, Institute of Chemical Technology, Poznan, Poland
e-mail: hermann.ehrlich@amu.edu.pl

V. Bels et al. (eds.), *Biomimicry – A Functional Approach*, Biologically-Inspired Systems 20, https://doi.org/10.1007/978-3-032-11949-0_2

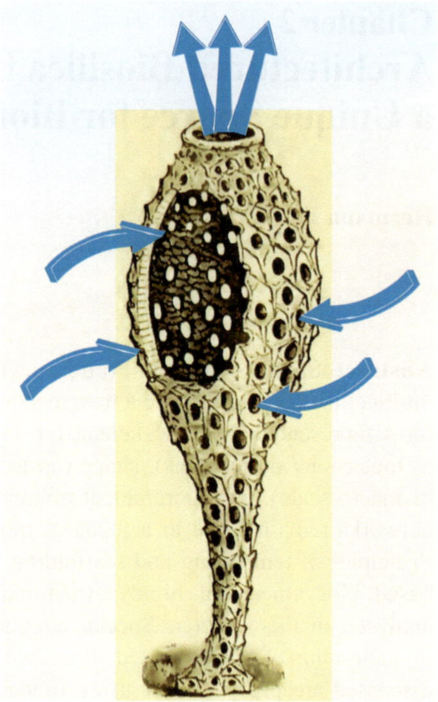

Fig. 2.1 Schematic view of the cross-section of "classical sponge" originally represented by Minchin (1900) with respect to the incurrent and excurrent openings (arrows) which take part in the entry and exit of water from the body of the sponge. The body encloses a central cavity called spongocoel. Water enters the spongocoel through the ostia (small pores) and passes out through the osculum (large pores)

(Suarez and Leys 2022). In the classical understanding of sponges, these organisms must be strongly attached to the substratum and have a mechanically robust and simultaneously porous frame of biocomposite nature, allowing them to obtain food by filtering water from the environment. Moreover, the habitat of sponges may include the deep-sea as well as Arctic and Antarctic regions of the world's oceans.

The simplified diagram in Fig. 2.1 shows how water containing various ingredients enters the body of the sponge, and then filtered, with a radically altered chemical composition, leaves it.

Probably one of the best descriptions of body form and water filtration in sponges that are understandable to non-spongologists can be found in Edward Minchin's classic work from 1900. Thus, he wrote: "*The typical sponge form is that of a hollow vase or sac, attached by its base to some object. At its upper extremity is a conspicuous opening, termed the osculum, and the wall is perforated by numerous minute apertures, the pores. During life water enters by the pores, and passes, either directly or after a more or less tortuous course along a system of canals in the body wall, into the central space or gastral cavity, to emerge by the osculum*" (Minchin 1900).

Marine sponges pump impressive amounts of water (100–1200 mL h^{-1} g^{-1}) (Orani et al. 2018) to satisfy their physiological needs for organic and inorganic compounds. Thus, they effectively remove from seawater both particulate organic matter (POM) dissolved organic matter (DOM) (Morganti et al. 2021; Ribes et al.

2023). Also, Silicon (Si) and calcium (Ca) uptake (Maldonado et al. 2012; Marron et al. 2016) is crucial for formation and development of poriferan skeletons, structural constructs fundamental to the very existence of these organisms. These skeletons play an important role as scaffolding structures involved in the formation of the so-called *aquiferous systems* (for details see Reiswig 1975; Langenbruch 1983; Weissenfels 1992; Leys et al. 2011; Goldstein et al. 2024) a central anatomical and functional structure (Lavrov and Ereskovsky 2022) mentioned by Minchin as "*a system of canals in the body wall*" (Minchin 1900). Up until today, the "sponge pump" is well-documented in the literature with respect to anatomical, structural, and functional peculiarities (i.e. Reiswig 1971; Larsen and Riisgård 1994; Riisgård and Larsen 1995; Leys et al. 2011). However, not all sponges have filtering abilities. For example, in highly specialized carnivorous demosponges (Vacelet and Boury-Esnault 1995), the absence of an aquiferous system is well documented (Riesgo et al. 2007).

Functional morphology of sponges (Schönberg 2021) as "living pumps" depends on diverse environmental or anthropogenic factors that may select for certain growth forms of these aquatic invertebrates (Bidder 1923; Bell and Barnes 2000). As listed by Christine Schönberg in her fundamental paper, such factors include "*Hydrodynamics (flow structure and strength, mixing); Sedimentation (turbidity, sediment deposition, abrasion); Nutrients (gradients, anthropogenic eutrophication); Physical damage (storms, human activity through fisheries or construction); Substrate type (hard, sandy, muddy bottom); Biological interaction (predation, competition, symbiosis); Illumination (photosynthetic ability, UV damage); pH (ocean acidification); Temperature (climate, season, climate change, heat events); Bathymetry (a product of pressure, hydrodynamics, illumination, sedimentation); Salinity (gradients, freshwater influx); as well as Oxygenation (saturation state)*" (Schönberg 2021).

In order to get an idea of the diversity of body shapes of both marine and freshwater sponges, it is recommended that you familiarize yourself with the materials provided in two well-recognized publications, called "*Atlas of Sponge Morphology*" (De Vos et al. 1991) and "*Thesaurus of sponge morphology*" (Boury-Esnault and Rützler 1997), respectively. The known poriferan species are amazingly numerous and diverse. Currently (2025), there are 9704 valid Porifera species: Class Calcarea—827; Class Demospongiae—8032; Class Hexactinellida—708; Class Homoscleromorpha—136 (de Voogd et al. 2025). Fundamental to the scientific description of each species of sponge is assessment of the morphological parameters of each organism including whether they contain a biomineralized skeleton. Without doubt, the progress in morphological analysis of sponges depends on development of modern imaging techniques (Woźnica et al. 2023).

Considering the specifics of this chapter is devoted to biosilica-containing sponges, as an example of the diversity of body shapes of sponges, I would like to present selected representatives of the class of glass sponges (Hexactinellida) (Fig. 2.2), the description of which stretches far back to the nineteenth (Owen 1841; Claus 1868; Kent 1870; Schulze 1880, 1885, 1886, 1887, 1899; Marshall 1875; Bowerbank 1875, 1876; Zittel 1877; Ijima 1895) and early twentieth century

Fig. 2.2 Body forms of selected representatives of glass sponges (Hexactinellida) as museum exhibits initially dried on air: (**a**) Euplectellidae (aff. *Regadrella*); (**b**) Pheronematidae, *Pheronema* aff. *Carpenter*; (**c**) Rossellidae; (**d**) Dactylocalycidae, *Iphiteon panicea.* (Images courtesy Konstantin Tabachnick)

(for overview see Chun 1900; Ijima 1901, 1902, 1903, 1927; Schulze 1900, 1904; Kirkpatrick 1907; Woodland 1907; von Lendenfeld 1915; Dendy and Burton 1926; Burton 1928; Okada 1928).

The description of these sponges, morphological features continues actively today (Beaulieu 2001; Tabachnick and Reiswig 2002; Janussen et al. 2004; Tabachnick and Collins 2008; Reiswig and Kelly 2018; Federwisch et al. 2020; Gong et al. 2023). Some of them (i.e. *Euplectella* sp. and *Monorhaphis* sp.) have been recognized even as "*textbook sponges*".

Despite the clearly visible massive siliceous skeletons of dried glass sponges (Fig. 2.2), representatives of this class also belong to "living pumps". Also in the photographs (Fig. 2.2a, c, and d), the porous structure of the walls of their bodies is apparent, making it possible to characterize these 3D constructs as an example of a "living filter", too (Fig. 2.3).

Sponges adapted to feed on a sparse suspension of viruses, bacteria, and phytoplankton (Maldonado et al. 2010; Riisgård and Larsen 2022) using also selective filtration of ultraplankton (Yahel et al. 2006), nano-, and microplankton (Varamogianni-Mamatsi et al. 2021). Being suspension feeders (Riisgård and Larsen 2022), they use special flagellated cells called choanocytes (Pozdnyakov et al. 2018) to actively filter a volume of water equivalent to many times their body volume, each hour (Leys et al. 2011). As recently suggested by Ogawa and co-workers, the architecture of poriferan choanocyte chambers maximizes mechanical pumping efficiency (Ogawa et al. 2024). Also, the architecture of flagellated chambers of the canal system within the sponge body wall force the flagella-driven flow through their collar filter, due to the formation of a "hydrodynamic gasket" above the collar (Asadzadeh et al. 2020). Moreover, marine glass sponges, due to specificity of their siliceous skeletons and choanocyte chambers (for details see Falcucci et al. 2024), rely on the passive conversion of horizonal flow to vertical flow of surrounding water (Vogel and Bretz 1972; Vogel 1977).

Giant biosilica-based reefs composed of glass sponges are an impressive example of the biological role of these water filtration systems in nature. Hexactinellids represented by such frame builders are heavily silicified hexactinosidan species; *Aphrocallistes vastus*, *Farrea occa*, and *Heterochone (Chonelasma) calyx* formed reefs, which occur in depths of 30–200 m, are up to 25 m tall, and discontinuously cover, for example, an area of about 1000 km^2 in Hecate Strait and Queen Charlotte Sound, British Columbia (Boutillier et al. 2013); 80% of *A. vastus* sponge dry weight is composed of biogenic silica (Chu et al. 2011). Such sponge reefs are particularly dense communities of filtration units, which remove seven times more carbon than can be provided by vertical flux alone (Kahn et al. 2018). These habitats have tremendous filtration capacity (465–47,300 L/m^2 water per day) (Stevenson et al. 2020). According to Sally Leys group, a dense population of reef hexactinellids can process water at 83,000 $L\ s^{-1}$, removing carbon at a rate of 0.96 $g\ C\ m^{-2}\ d^{-1}$ (Chu and Leys 2010). Notably, the mechanical integrity of the sponges skeleton is vital to reef formation, which are built at temperatures ranging between 6 and 12 °C (Tabachnick et al. 2017), through larval sponges settling atop the fused dead skeletons of previous generations (Stevenson et al. 2020). The resulting biosilica-containing sponge reefs that grow in the open ocean, represent unique large-scale biosilica-containing "living filters" in the open ocean, which can withstand large ocean swells associated with winter storms due to their framework-like architecture.

In connection with the above, an intriguing question arises: what will happen to the body of the sponge if it interacts with dispersed fine bottom sediments, triggered, for example, by a storm? Recently, it has been reported that deep-sea glass sponge *Vazella pourtalesii* was able to engage in self-cleaning by removing sediment with contractive behavior (Grinyó et al. 2024). The sediment under study

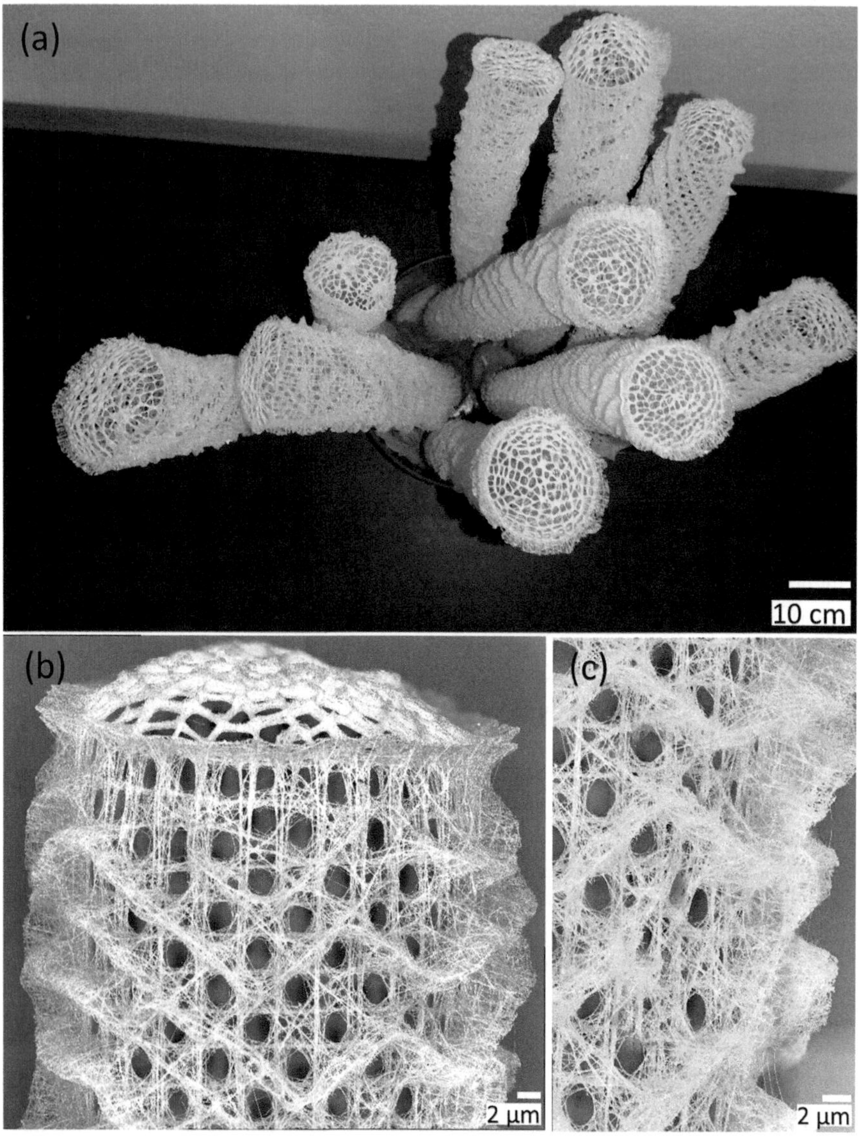

Fig. 2.3 Dried *Euplectella aspergillum* glass sponges (**a**) represent characteristic example of the sponge body as a "living filter" having a broad diversity of highly ordered micropores (**b**, **c**). The species is studied since 1841 (see Owen 1841; Schulze 1880)

was cleared from the sponge surface within 72 h, which is highly relevant for its filtering capacity. It could be suggested that this ability is based on recently discovered tissue-wide organization of a contractile module in sponges composed of actin, striated-muscle myosin II, and transgelin (Colgren and Nichols 2022). Thus, these marine invertebrates with long evolutionary history are capable of coordinated whole-body contractions that purge debris from internal aquiferous system. The biomimetic potential of self-cleaning phenomenon observed in sponges as "living filters" should be unambiguously investigated in future research.

From the standpoint of materials science and engineering, hierarchically organized structures observed within the filtering system of sponges (Fig. 2.3) can be viewed as an example of graded/gradient porous biological materials (Miao and Sun 2009), where the internal pore surfaces are modified with organic, inorganic, as well as biological (i.e. cellular) coatings. The study of this kind of structural formations motivates the development of biomimetic approaches to solve problems, that are similar in the technical aspect, for example the design of effective filter-based tools including ultralight and compressible inorganic sponges with hierarchical porosity (for details see Wang et al. 2020).

2.2 Structural Diversity of Poriferan Biosilica

Biosilica is the main structural component of skeletal formations in representatives of the following classes of sponges: Hexactinellida, Demospongiae, and Homoscleromorpha. While glass sponges and homoscleromorphs belong exclusively to the sea, some representatives of Demospongia live in freshwater habitats. This demonstrates that the process of biosilicification in sponges, similar to such biosilicifiers as diatoms, can be carried out in a range of aquatic environments having different salt concentrations and pH. However, unlike diatoms, which can synthesize their frustules up to 98 °C (Ehrlich 2019), skeletal formations in sponges are formed in the temperature range from −1.9 °C (Hentschel 1929; Tabachnick et al. 2017) to +30 °C (Chin et al. 2020). Below, I have taken the liberty of presenting a short analysis of the structural diversity of skeletal siliceous formations in sponges in terms of their sizes, shapes, and ornamentation, which can provide useful information for biomimetics and bioinspired materials science.

2.2.1 Structural Diversity in Size

Knowledge on sponge dimensions (Brendstedt 1929), especially of the biosilica-based types, is crucial for better understanding of both the structural diversity and biological functions (Bell 2008) of these invertebrates, as well as the manner in which these organisms have become highly specialized to live and survive under diverse environmental conditions since the Precambrian.

A wide range of sizes is observed among these organisms. One of the thinnest sponges *Myceliospongia*, measures several millimetres thick with a diameter of 25 cm, but this sponge does not synthesize any biomineral structures (Vacelet et al. 2002). A focus on the smallest biosilicifier sponges leads us to the smallest known hexactinellids, *Oopsacas minuta*, habituating in a Mediterranean cave on the outskirts of Marseille, which have a body height from 4 to 12 mm (see for details Fig. 2.4). In some demosponges, which synthesize biosilica-based microscleres called spicules, the body size ranges from 3.5 to 14 mm (Fig. 2.4). The sizes of representatives from class Homoscleromorpha, in which siliceous spicules were also found, fluctuate in approximately the same range (Fig. 2.4).

In contrast, what are the largest sizes recorded by researchers for biosilica-forming sponges? Already mentioned above are the of gigantic formations of hexactinellids; forming reefs that are impressively large and structurally sound (see also Fig. 2.5). However, individual hexactinellids can also be striking in size. The largest known glass sponge is *Laguinelline* sp., and the longest is *Monoraphis chuni*, whose spicule can reach up to 3 m in height (Fig. 2.5).

2.2.2 Structural Diversity in Shape and Ornamentation

The structural diversity of biosilica-based formations in sponges called spicules (both mega-and microscleres) (Uriz et al. 2003; Łukowiak et al. 2023) is rivalled only with that of silicoflagellates (Deflandre 1950; McCartney et al. 2014), radiolarians (Haeckel 1887; Biard 2022), and diatoms (Bahls 2018; Fu et al. 2023). It includes more than 80 and 46 morphotypes in Demospongiae and Hexactinellida, respectively, network connectivity (monaxons, triaxons, tetraxons), as well as sophisticated superficial ornamentation (Ehrlich et al. 2022, 2024).

Typical examples of hexactinellids network connectivity are represented in Fig. 2.6. The SEM images presented here help to show that the microporosity of such skeletons also determines the microporosity of the body walls of the glass sponges themselves, which correspondingly function as "living filtration systems". The principles of bioarchitecture regarding the unique biosilica-based honeycomb-like skeleton of the deep-sea glass sponge *Aphrocallistes beatrix* (Fig. 2.6a) have been discussed recently by Voronkina and co-workers (2023). Inspired by such constructs, diverse models (Zimmermann et al. 2024), including 3D printing, using PLA-, resin-, and synthetic-glass have been prepared based on microtomography-based 3D reconstruction (see for details Voronkina et al. 2023).

However, compared to this species of hexactinellids, the limit of the most complex biosilica architecture at the micrometer- and millimeter scale is perhaps the species *Euplectella* (Fig. 2.3), an example of "*hexaradiate spiculed*" sponges (Kent 1870). Using nineteenth century scientific terminology, it possesses a "*siliceo-fibrous skeleton*" (Bowerbank 1869) made by "*trifurcate hexaradiate stellate spicules*" (Owen 1841; Schulze 1880). The discoverer of the hexactinellid species known as *Euplectella aspergillum*, Richard Owen, reported the "*exclusive beauty and regularity of the texture of the walls of the cone*" (Owen 1841). Already in those days, using an optical microscope, researchers were able to discover the sophisticated geometry of the skeleton of this sponge species, which amazed them. The drawings presented in a number of publications perfectly depict the principles

Class	Species Name	Size	Presence of Biosilica	Reference
Hexactinellida	*Oopsacas minuta*	4 mm in length	+	Bakran-Petricioli et al. (2007)
	*Leucopsacus orthodocus*	11 mm total length, 6.5 mm in diameter, stalk 4 mm long	+	Tabachnick (2002a)
	Vitrollula fertile	12–15 mm long; 6–8.5 mm in diameter	+	Tabachnick (2002b)
	Cyrtaulon sigsbeei	Up to 4.5 cm tall by3 cm wide	+	Reiswig (2002)
Demospongiae	*Lycopodina hypogea*	stalk 3.7-14 mm long, 0. 1 - 0. 1 8 mm in diameter	+	Vacelet and Boury-Esnault (1996)
	Cryptosyringa membranophila	4.5 x 3 mm	+	Rützler et al. (2014)
	Tentorium levantinum	3.5 mm by 2.2 mm in size	+	Ilan et al. (2003)
Homoscleromorpha	*Plakina finispinata*	13 mm wide by 1–3 mm thick	+	Lage et al. (2018)
	Plakinastrella pseudolopha	1.5 cm high by 0.5 cm in diameter	+	Lage et al. (2018)
	Plakortis ruetzleri	0.5–2.0 cm wide by 3–5 mm thick	+	Lage et al. (2018)
	Plakina crypta	1–2 cm^2 by 1–5 mm thick	+	Muricy et al. (1998)

Fig. 2.4 Selected biosilica-synthesizing sponges from three classes that have the smallest known body sizes. The image represents the smallest glass sponge, *Oopsacas minuta*. (Image courtesy Alexander Ereskovsky)

Class	Species Name	Size	Presence of Biosilica	Reference
Hexactinellida	*Monorhaphis* sp.	Up to 3 m long; up to 3 cm in diameter	+	Wang et al. 2009
	*Aphrocallistes vastus*	The colony measured 3.4 m in length, 1.1 m in height and 0.5 m in width	+	Austin et al. 2007
	Aspidoscopulia sp.	Up to 2 m in height and 70 cm in width	+	Tabachnick et al., 2011; Wysokowski et al., 2018;
	Lanuginelline sp.	over 3.5 m in length, 2.0 m in width and 1.5 m inheight	+	Wagner & Kelley, 2016; Dohrmann et al., 2023;
	Anoxycalyx (Scolymastra) Joubini	Up to 2 m in height	+	Dayton et al., 2013
	Hyalonema sieboldi	Up to 1 m in height	+	Ehrlich et al., 2010
	Hexactinosidan reefs	discontinuously mantle about 1000 km^2 of seafloor; cconsist of interconnected circularmounds and ridges up to 200 m in diameter; up to 25 m in height	+	Conway et al., 1991; 2001; Krautter et al., 2006; Chu & Leys, 2010; Stone et al., 2013; Han et al., 2022;

Fig. 2.5 Selected biosilica-synthesizing sponges from two classes that have the largest known body sizes. The image represents one of the largest demosponge, *Biemna* sp., from our collection. The body of this sponge contains millions of siliceous spicules, which are responsible for the integrity of its skeleton

Demospongiae	*Xestospongia muta*	Up to 1.8 m in height and 1m in diameter	+	Humann & DeLoach 2002; McMurray et al. 2008;
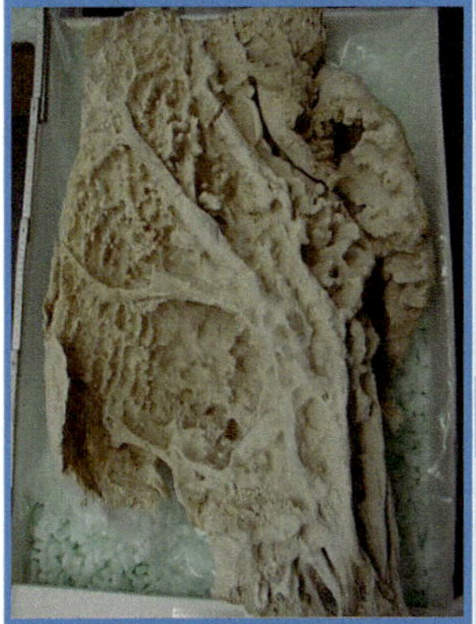	*Biemna* sp.	Up to 1.2 m in height and 70 cm in diameter	+	Ehrlich et al., 2024
	Ianthella basta	Up to 2.5 m in height and diameter	Silica-chitin composite	Ehrlich et al., 2010; Rohde & Schupp, 2012; Kertmen et al., 2021;
	Aplysina archeri	Up to 1.5 m long tubes	Silica-chitin composite	Wysokowski et al., 2020
	Verongula gigantea	Up to 1.5 m height and 70 cm in diameter	Silica-chitin composite	Ehrlich et al., 2010; Wysokowski et al., 2018

Fig. 2.5 (continued)

of the structural organization of biosilica-based fibers, for example, multi-layering and existence of axial filaments within axial channels of interconnected spicules (Fig. 2.7). Up until today, *E. aspergillum* (Fig. 2.8) has been the most studied hexactinellid, involving various directions in modern biomimetics (Sundar et al. 2003; Aizenberg et al. 2004; Weaver et al. 2007; Ehrlich et al. 2024).

Also, hexactinellid microslerae, which are not fused to the siliceous fibers of the main frame-forming skeleton, are striking in their structural diversity (see for examples, Schulze 1896, 1904; Ijima 1901, 1903). At this point, it is rational to cite the opinion of the English zoologist Arthur Dendy which is as follows: "*microscleres are much more enigmatical, for they usually occur scattered promiscuously throughout the soft tissues, while they exhibit the most surprising variety of exquisitely symmetrical forms, the meaning of which it is impossible to interpret in terms of the physiological requirements of the sponge*" (Dendy 1926). Structural peculiarities of hexactinellids microscleres are represented in Fig. 2.9.

Representatives of the class Demospongiae, being the most numerous in terms of species among sponges, also form a huge number of morphologically diverse spicules (Carter 1849; Hinde 1887–1893), such as megascleres (Van Soest et al. 2010, 2012), microscleres (Łukowiak 2020; Łukowiak et al. 2022a, b), and gemmuloscleres (Pinheiro et al. 2015; Manconi et al. 2016), which have traditionally been used in taxonomic studies of both fossilised (Pisera 2006; Drozdov et al. 2022; Łukowiak et al. 2022a, b; Chen et al. 2023) and modern species (Jones 1984; Santín et al. 2021). According to the opinion of Hartman, "*within a single sponge*

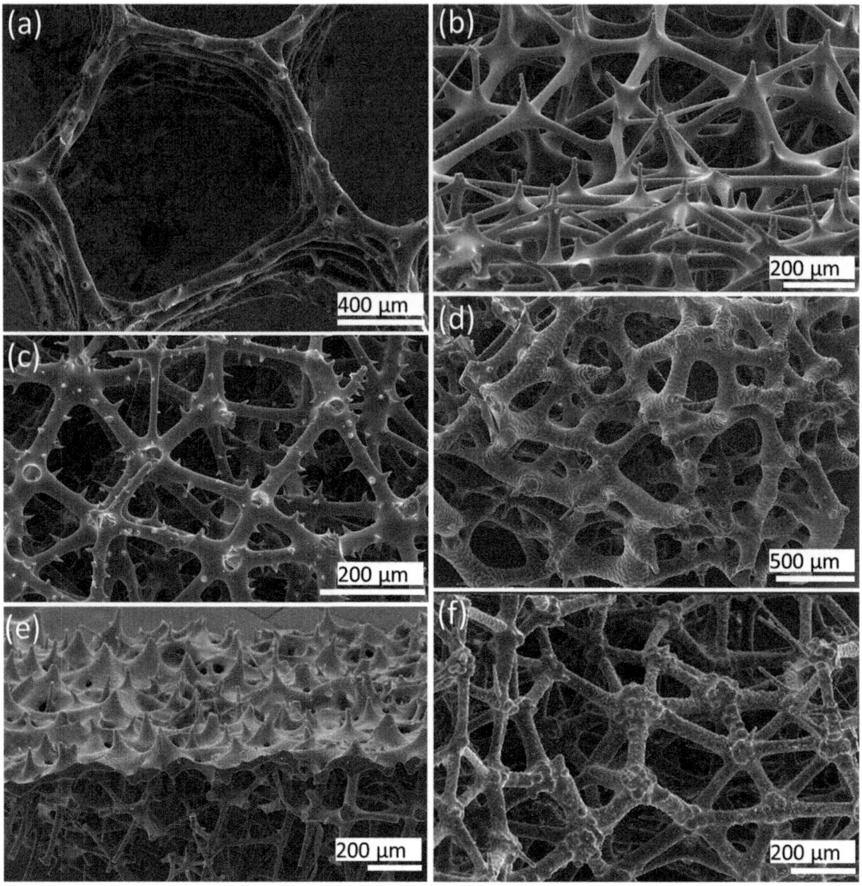

Fig. 2.6 Micromorphology of glass sponges skeletons as "living filters". SEM imagery: (**a**) honeycomb architecture of skeleton of *Aphrocallistes beatrix* glass sponge (for details see Voronkina et al. 2023); (**b–f**) characteristic network-like triaxonic (cubic)- and hexactinic-fused skeletal constructs of diverse hexactinellids. (Images courtesy Kontantin Tabachnick and Andrzej Pisera)

species from one to eight or more kinds of spicules may occur" (Hartman 1981). Spicules are also formed in the larvae of both marine and freshwater sponges. Their morphological characteristics are given in several articles (Evans 1899; Okada 1928; Schröder 1936; Bergquist and Sinclair 1968; Leys 2003).

Hand-made drawings of such structures have been represented in numerous classical publications (for overview see Donati 1753; Carter 1849; Thoulet 1884; Gray 1868; Sollas 1888; Noll 1888; von Lendenfeld 1894; Bütschli 1901; Vosmaer 1902; Vosmaer and Wijsman 1905; Minchin 1909; Schulze 1923, 1925; Dendy 1925). In recent time, the morphology of microscleres of demosponges origin has been traditionally investigated using SEM (Uriz et al. 2003; Schoeppler et al. 2017).

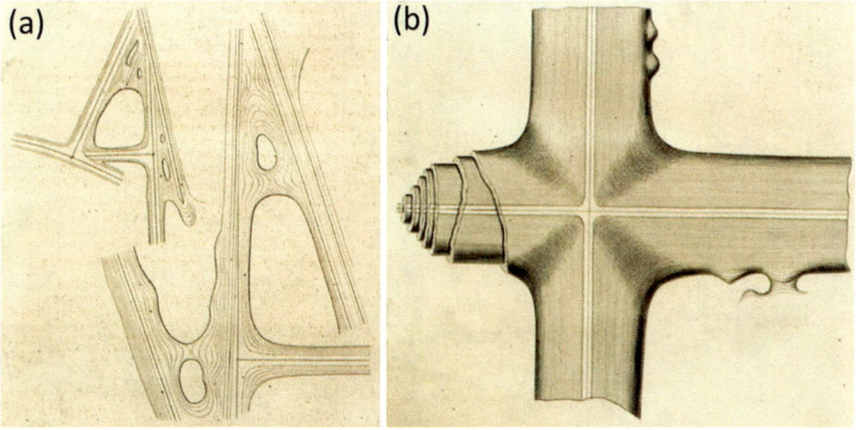

Fig. 2.7 Drawings of *E. aspergillum glass* sponge skeletal fragments prepared by Carl Claus already in 1868 clearly represent their complex geometry (**a**) as well as the presence of axial filaments within axial channels of multi-layered glassy construct (**b**). (Images adapted from Claus (1868))

Several examples are represented in Fig. 2.10. Methods for measuring of sponge spicules have recently been reported (Rieuwpassa et al. 2023). It is suggested that the smallest microsclere to be found in demosponges remain to be the 2.5–10 μm large spherules in the representatives of the genus *Caminus* or *Caminella* (Cárdenas et al. 2018) as well as bipocilla that is 6 μm in size, which occurs in *Lophon proximum* demosponge (Desqueyroux-Faúndez and Van Soest 1996). Intriguingly, the rate of spicule formation in freshwater sponge *Ephydatia muelleri*, both under in vitro and in vivo conditions, was measured as 1–10 μm/h (Imsiecke et al. 1995).

Following the classic paper by Gualtherus Vosmaer entitled "*On the shape of some siliceous spicules of sponges*" was published in 1902 (Vosmaer 1902), results of microscopic studies including SEM with respect to the forms and shapes of a broad diversity of poriferan siliceous spicules, have been reported many times (Hartman 1981; Uriz et al. 2003; Karcz et al. 2015; Wysokowski et al. 2018; Łukowiak 2020; Rasbold et al. 2023). The fundamental source of information about the structural features of the spicule formations listed above is undoubtedly the recent work by Magdalena Lukowiak and co-workers entitled "*The terminology of sponge spicules*" (Łukowiak et al. 2022b). Numerous SEM images presented in this publication demonstrate the high morphological variability of spicules and their geometrically complex ornamentation, which remains uniquely sophisticated from the human viewpoint.

Similarities in the structural organization of spicules with demospongia are also found in the biosilica-based microscleres of homoscleromorphs (Fig. 2.11). It is to note that some species which belong to class Homoscleromorpha possess very small microscleres with respect to size. For example, irregular spheres of 2–15 μm in diameter and microstrongyles of 4–10/2–4 μm as well as microrhabds of 2–5.5–

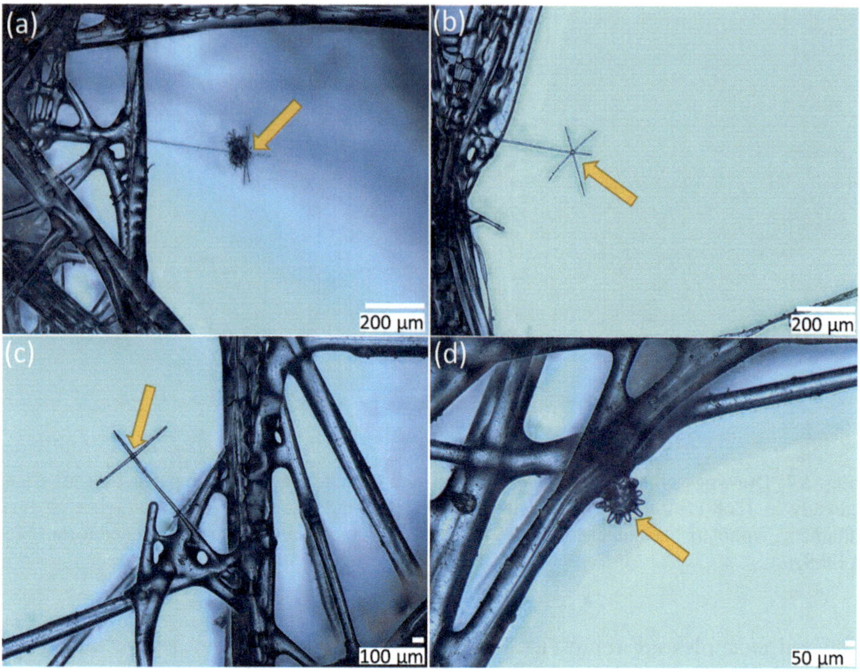

Fig. 2.8 Digital microscopy images of selected fragments of *E. aspergillum* glass sponge skeleton represent diversity of their micro-ornamentation (arrows): dermal hexactine (**a**) with several microscleres including a floricome; dermal hexactins (**b**, **c**); and club-like microsclere (**d**) fused to rigid siliceous skeletal framework

8/0.5–1.0 µm in size have been reported in *Plakortis quasiamphiaster* and *Plakortis hooperi* species, respectively (Muricy 2011). Also, siliceous spheres of 1–20 µm in diameter have been observed in *Plakortis spinalis* homoscleromorph (Domingos et al. 2013).

Along with megascleres and microscleres, demosponges, mainly freshwater species, form biosilica-based microstructures called gemmoscleres (or gemmuloscleres), which are found in the so-called *gemmules* are present (for details see Simpson and Gilbert 1974) (Fig. 2.12). These egg-like structures represent some kind of asexual spherical cyst that contains stem cells that are initially in a state of metabolic arrest known as diapause (Loomis 2010). In other words, gemmules, which are in the state of quiescence (Schill et al. 2006), represent resistant stages of sponge development (Manconi and Pronzato 2007). It is suggested (Gost et al. 2023) that the ability of sponges to form resistance gemmules represents some kind of defense mechanism against environmental adversity. Gemmules have also been found in diverse marine sponges (see for overview Fell 1974, 2005; Jetton et al. 1987).

Gemmule coats contain oriented gemmoscleres embedded within them (Simpson et al. 1985). In such demosponge species as *Corvoheteromeyenia ezcurra*,

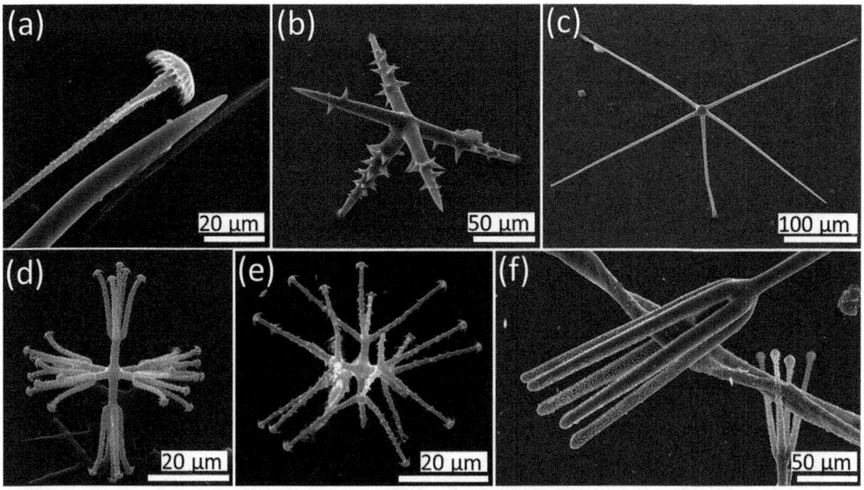

Fig. 2.9 SEM imagery of selected microscleres of glass sponges origin: (**a**) clavule; (**b**) spiny pentactin; (**c**) pentactin; (**d**) aspidodiscohexaster; (**e**) discohexaster; (**f**) two scopules. (Images courtesy Andrzej Pisera and Konstantin Tabachnick)

microsized gemmuloscleres are striking in their structural diversity in the form of oxeas, birotules, pseudobirotules, strongyles, botryoidal, and club-like radially to tangentially arranged formations (Pinheiro et al. 2015). Similar highly sophisticated gemmuloscleres have been visualized using SEM in representatives of the genus *Heterorotula* (Manconi et al. 2022). Figure 2.13 represents SEM imagery of sombrero-like gemmuloscleres localized on the surface of *Drulia* sp. demosponge gemmule.

Exceptionally preserved tri-layered theca of fossilized gemmules embedded with the siliceous gemmosclere-like structures has been shown by Du and co-workers (2015). These may be the oldest reported gemmules that have been found in the Doushantuo Formation in the Yangtze Gorges area (South China), that were deposited between 635 and 580 Ma (Du et al. 2015).

2.3 Function of Poriferan Biosilica

One might get the idea that the diversity of specular structures described above also determine the variety of functions that they perform in the corresponding sponge species. However, the functional load of sponge spicules can be described in only a few directions. An analysis of the available literature allows us to highlight the following examples (Fig. 2.14).

In principle, megascleres, microscleres, and gemmuloscleres serve to create various kinds of robust framework matrices that do not allow the cell mass to exist

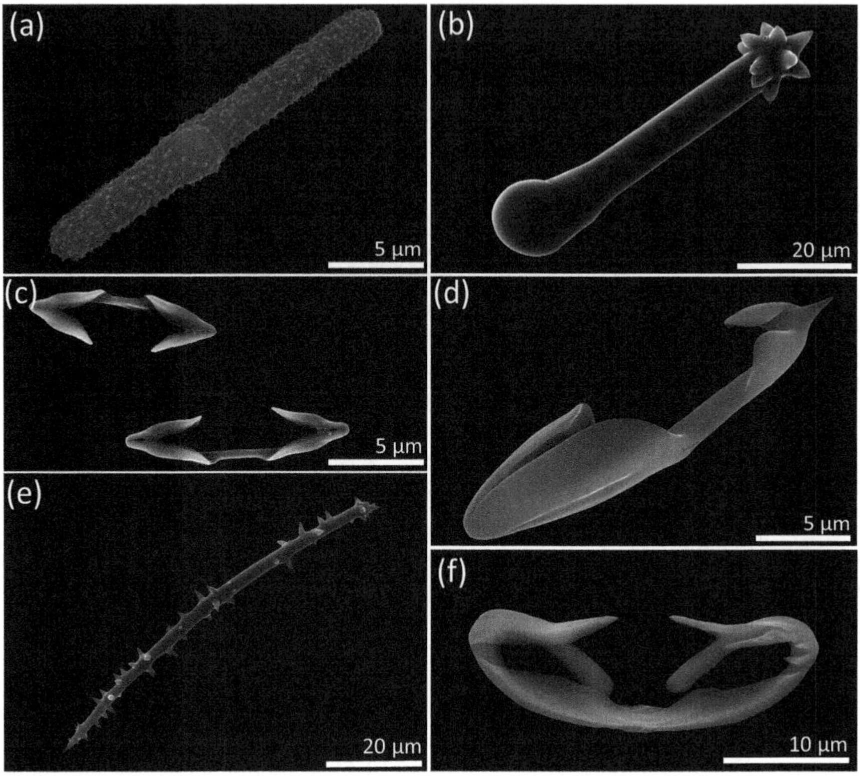

Fig. 2.10 SEM imagery of selected Demospongiae spicules: (**a**) microacanthostrongyle (*Suberites lutkenii*); (**b**) tylostyle (*Hymeraphia stellifera*); (**c**) palmate isochele (*Isodictya palmata*); (**d**) palmate anisochela (*Lophon piceum*); (**e**) acanthostyle (*Crellomima imparidens*); (**f**) palmate chelae (*Artemisina arcigera*). (Images courtesy Alexander Ereskovsky)

in the form of a gel-like amoeboid structure, but instead to actively participate in skeletogenesis and the formation of mechanically stable sponge bodies with porous walls. The protective functions of spicules also prevail; whether from the effects of destructive storms or from attempts by other organisms to use the sponge as a food source. An extremely specialized function is the participation of hook-like spicules in the retention of those organisms (i.e. small crustaceans) that are used by carnivorous marine sponges (Fig. 2.15) as food. The scientifically proven participation of sponge siliceous spicules in the transmission of light (Aizenberg et al. 2004) is still a poorly studied phenomenon in vivo, although it served as an impetus for a number of biomimetic studies in the field of biophotonics.

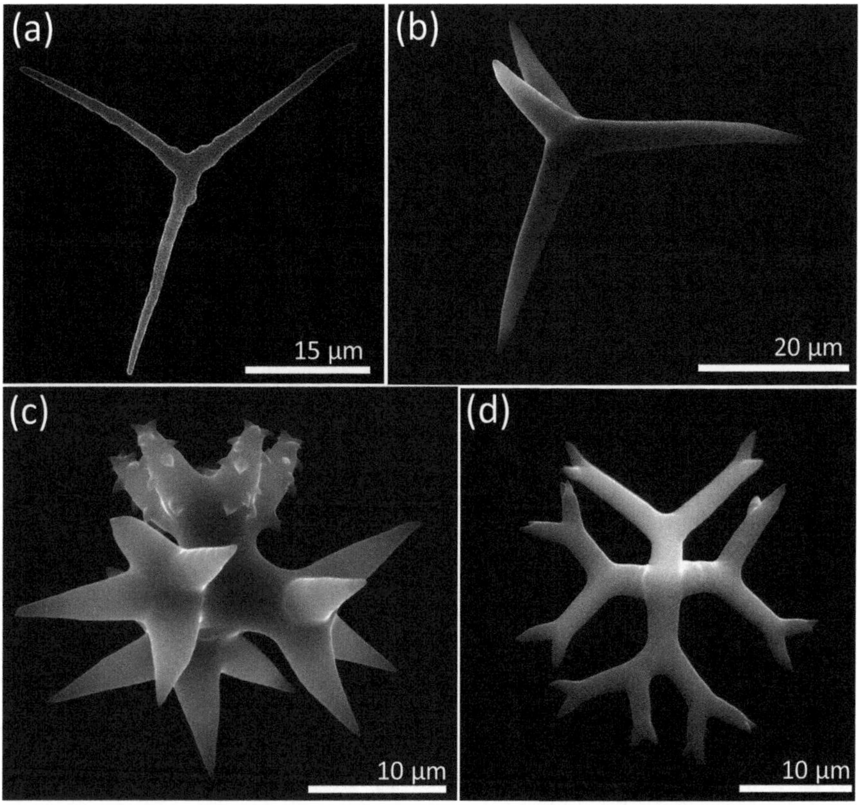

Fig. 2.11 SEM imagery of selected Homoscleromorpha spicules: (**a**) triod (*Plakina crypta*); (**b**) calthrop (*Corticium diamantense*); (**c**) candelabra (*C. diamantense*); (**d**) tetralophous calthrop (*Plakina jamaicensis*). (Images courtesy Alexander Ereskovsky)

2.4 Chemistry and Biochemistry of Poriferan Biosilica

The scientific terminology "biosilica" generally implies a composite material in which there is an inorganic component of silicon-based composition with precipitation mediated by a living organism, and an organic component of biological origin. The formation of the biosilica itself always takes place only in one of the biological systems, be it at the prokaryotic (bacteria) or eukaryotic (protists, diatoms, sponges, plants) level. Biosilicification in fish is also known (Ehrlich 2019).

An analysis of the literature regarding the inorganic component of biosilica allows us to draw the following conclusions. Fundamental differences in structural disorder and bonding environments between biogenic and geologically formed hydrous silicas (Opal-A, hyalite, geyserite) have recently been reported by Farfan and co-workers (2022). Although that sponge biosilica only melt above 2000 °C (Wang et al. 2010), its formation occurs in the range of temperatures between −1.9

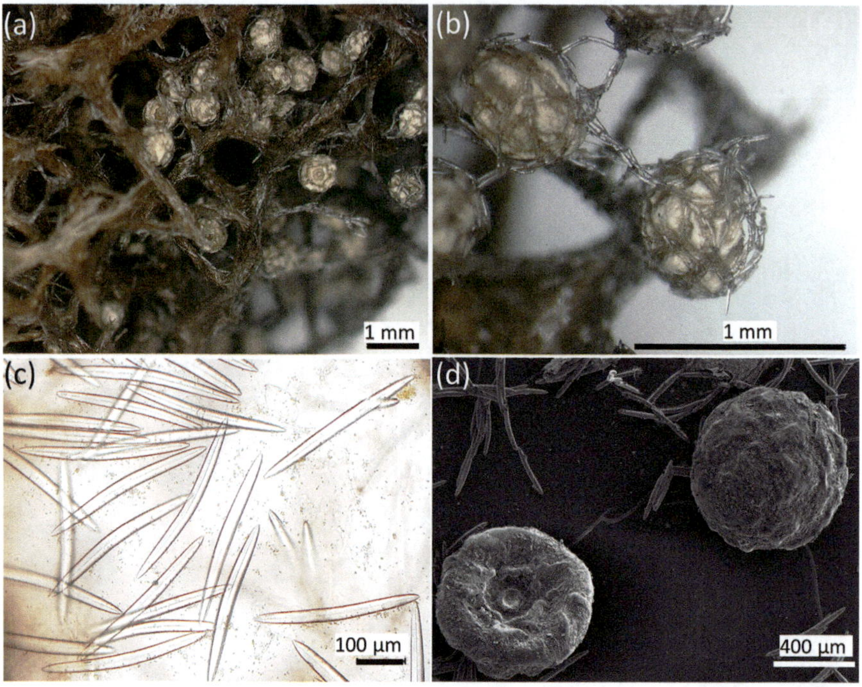

Fig. 2.12 Amazonian fresh-water demosponge *Drulia* sp. produces egg-like gemmules (**a**) which become to be arrested (**b**) within organic skeleton using glassy microscleres (**c**). SEM image (**d**) represents both microscleres and gemmules that are tightly covered with sombrero-like gemmuloscleres (see also Fig. 2.13)

and +30 °C that drastically differ from high temperature levels known for similar silicas of geological origin.

The history of the discovery of biosilica in sponges, briefly summarized (see for details Vosmaer and Wijsman 1905), is as follows:

1826—Grant found that spicules contain silica;
1841—Bowerbank showed that also some organic matter is also present within spicules;
1884—Thoulet suggested spicula as pure, organic-free opal;
1885—Sollas likewise finds that the silica resembles opal.

To form a biosilica with the general chemical formula, ($SiO_2 \cdot nH_2O$), a source of silicon itself is required in the sponge marine as well as freshwater environment. The silicon circle in worldwide aqueous environments is well described with respect to formation and dissolution of poriferan biosilica (see for overview Jørgensen 1944; Tréguer and Pondaven 2000; Maldonado et al. 2022). It is hypothesized by Marron and co-workers that Si transport in sponges is of ancient origin. Corresponding Si transporters "*may have arisen initially to prevent Si toxicity in*

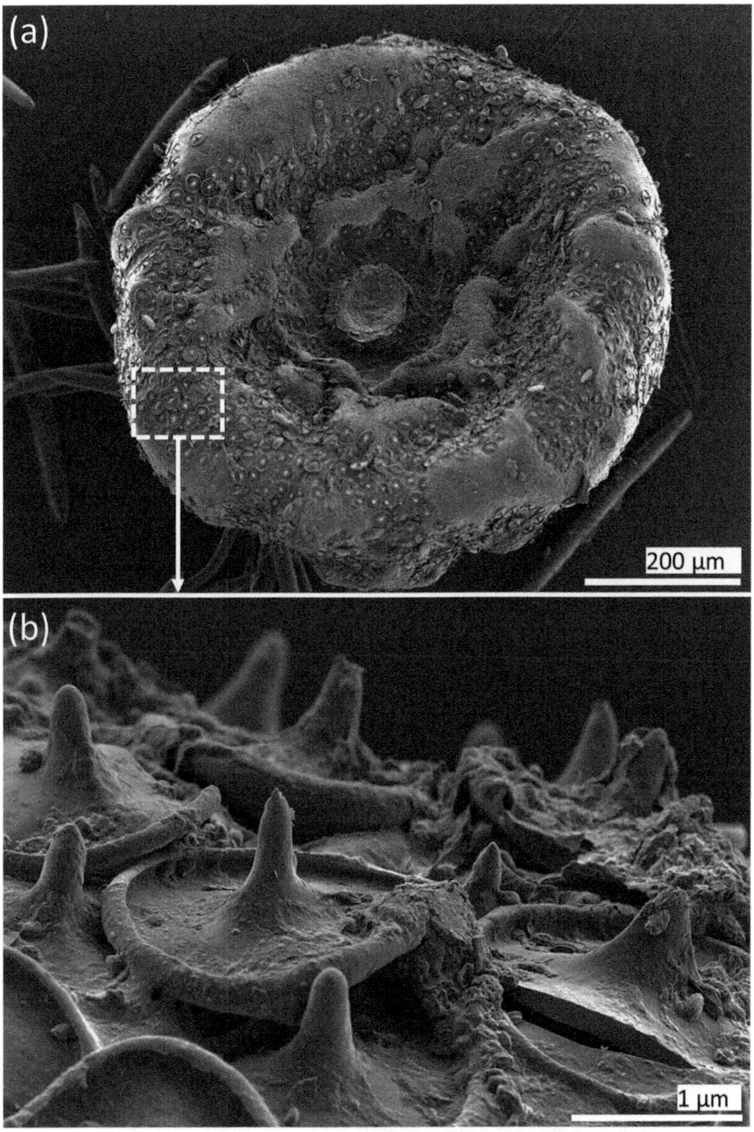

Fig. 2.13 SEM imagery of the surface of the gemmule (**a**) found in *Drulia* sp. demosponge. Gemmuloscleres with characteristic sombrero-like micro-architecture (**b**) tightly cover it and built in this way with corresponding protective layer made of structured biosilica

the high Si Precambrian oceans, with subsequent biologically induced reductions in Si concentrations of Phanerozoic seas" (Marron et al. 2016). Some marine demosponges consume up to 0.19 mmol Si m^{-2} daily and about 6.4×10^6 mol Si yearly (Lopez-Acosta et al. 2018). Siliceous structures in sponges are elaborated

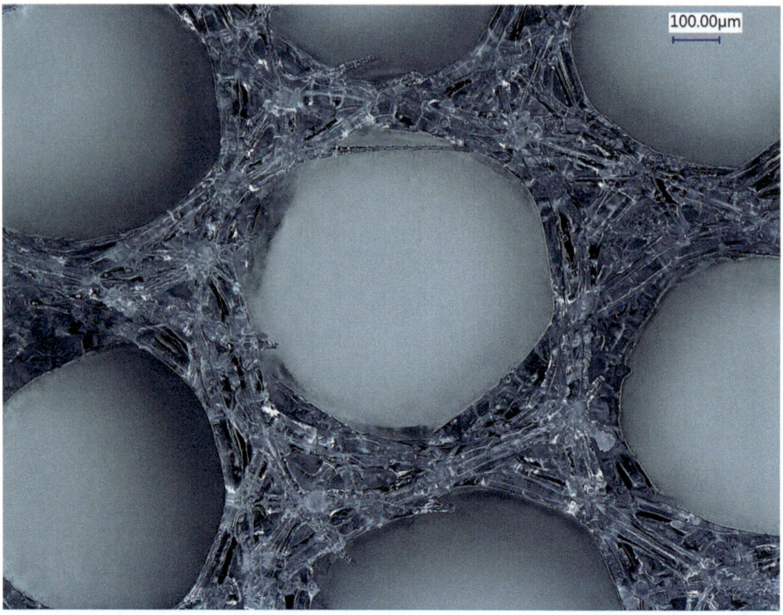

Function	References
Mechanical support for the soft parts of the sponge body	Chen et al., 2022; Woznica et al., 2022
Structural support for the three-dimensional filtration system of the sponges	Bart et al., 2019
Agglomerates from spicules are involved in the generation of skeletal pores	Voronkina et al., 2023
Formation of the armor due to aggregation of microscleres that is resist against wave action	Hill & Hill, 2002
Spicules mechanically protect the sponges against predators	Hill & Hill, 2002; Hill et al., 2005; Ferguson & Davis, 2008
Attachment to the hard (rocky) as well as to the soft (sandy) substrates	Beaulieu, 2001; Ehrlich et al., 2010; Wysokowski et al., 2018.
Spicules are involved in "crawling locomotion" of selected demosponges species	Bond & Harris, 1988; Morganti et al., 2021
Spicules can help larvae stay buoyant while in the plankton or reach the bottom at settlement	Uriz et al., 2003
Sponge siliceous spicules in living specimens transmit light into deeper tissue regions	Brümmer et al., 2008; Müller et al., 2013
Hook-like spicules are involved in prey capture in the carnivorous sponges	Vacelet & Duport , 2004
The spicules of *Terpio hoshinota* demosponge are arranged to support both encrusting basal tissue and pioneering tissue to colonize corals	Hirose et al., 2024
Pneumatic cellular layer of gemmules is reinforced with spicules	Pinheiro et al.,2015
Gemmuloscleres help gemmules to be attached on the feathers of birds and scales of fish	Carballo et al., 2024

Fig. 2.14 The overview on function of poriferan biosilica. The digital microscopy image represents characteristic skeletal fragment of *A. beatrix* that has been formed by this glass sponge

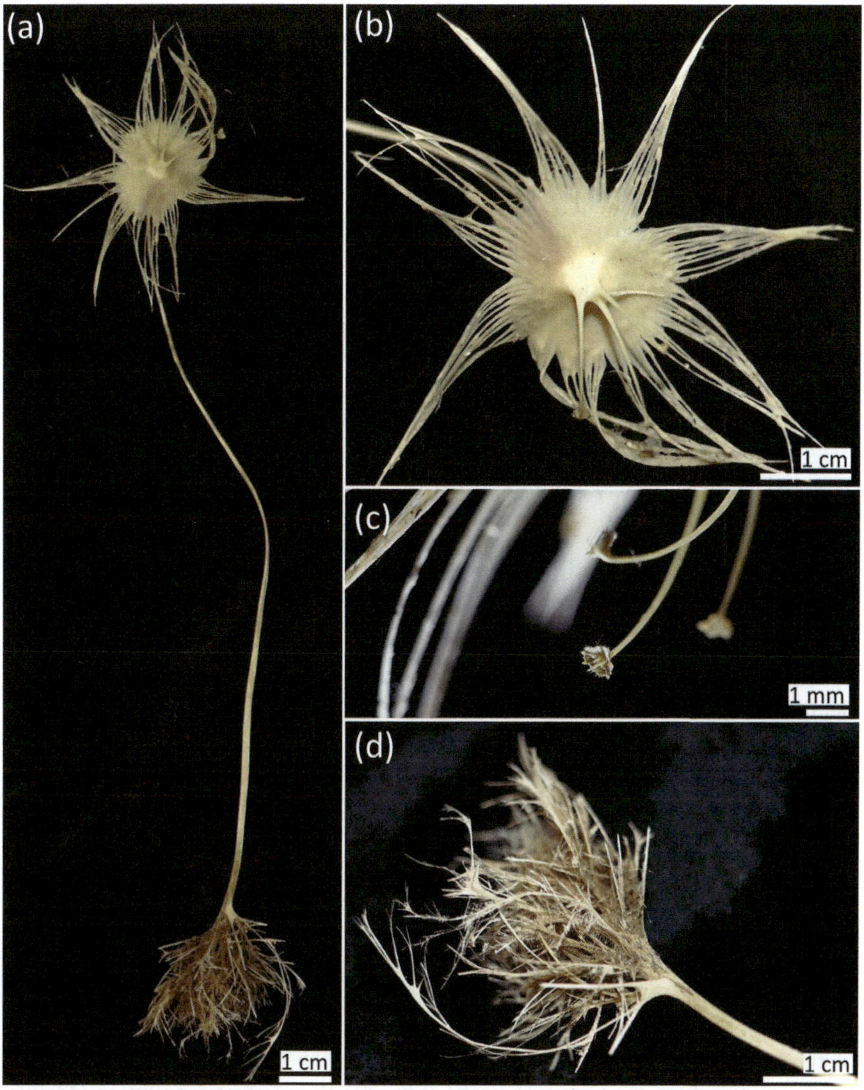

Fig. 2.15 Carnivorous Cladorhizid sponges (**a**) possess numerous filaments having morphology similar to flower petals (**b**, **c**) that protrude from the head (**b**) and contain hook-like microscleres, which are located on their surface. Representatives of small free-living polychaete worms as well as diverse small crustaceans remain to be usual prey organisms (for details see Vacelet and Duport 2004). These sponges are anchored in sandy marine sediment by spicule-containing shoots (**d**) that resemble the root system of plants

by corresponding cells, which possess some mechanism to take up the silicic acid ($Si(OH)_4$) dissolved in water and polymerize it into nanoorganized biogenic silica (Maldonado et al. 2011). The condensation degree of nanoparticulated biosilica (ca. 50 nm in size) of spicule origin is very similar (0.93) to that of synthetic silica colloids (Masse et al. 2016).

The water–silica relationship was the focus of numerous studies on sponge spicules. Earlier in 1905, Vosmaer and Wijsman, proposed two versions of chemical formula for sponge silica: $2(SiO_2) + H_2O$ and $5(SiO_2) + H_2O$ (Vosmaer and Wijsman 1905; Bidder 1925). Later, Minchin in 1910 reported SiO_2/H_2O proportions in a range from 2 to 5 (Minchin 1910), Vinogradov, in his fundamental work entitled "*Elementary Composition of Porifera*" (Vinogradov 1953) wrote that biosilica from tetraxonid demosponges have a ratio of 5/1, however that from hexactinellids have a ratio of 4/1 (Vinogradov 1953). In 2011, using photoemission spectroscopy of club-formed spicule of *Caulophacus* sp. glass sponge, we have shown the presence of two kinds of silicon oxides: SiO_2 (minor) and $SiO_{1.2}$ (dominated) (Ehrlich et al. 2011). Hydrated nature of silica within spicules from *E. aspergillum* glass sponge has been recently studied in relationship with their micromechanical properties (Morankar et al. 2023).

According to Sandford, the spicule chemical composition both for demosponges and hexactinellids was 85.2% SiO_2, 12.3% water, and 2.5% other elements (mainly S, Al, K, Ca, and Na) (Sandford 2003). Similar analytical results have been recently reported for biosilica of selected sponges where Ca, Fe, Cl, K, Zn have been identified additionally to Si (Drozdov et al. 2021). The composition (atomic %) of Suberites domuncula demosponge spicules had been measured by ICP-AES was found to be: O 66.1, Si 32.8, Al 0.006, Mg 0.004, K 0.026, Fe 0.002, Na 1.07, Ca 0.004 (Mugnaioli et al. 2009). Results on chemical content of spicules isolated from two selected species of demosponges and two of hexactinellids obtained by Masse and co-workers using elemental analysis, showed similar amounts of Si (28–34 wt%) (Masse et al. 2016). However, in demosponges, the content of C (6.6–6.8 wt%), H (1.39–1.61 wt%), as well as N (1.23–1.55 wt%) was higher than that in representatives of glass sponges, C (2.35–2.94 wt%), H (0.80–0.88 wt%), and N (0.14–0.26 wt%) (Masse et al. 2016).

It cannot be ruled out that the elemental composition of sponge spicules depends on the specifics of their habitats in terms of both the geological features of the substrates on which they grow (Ferretti et al. 2019) and abiotic pollution (Mayzel et al. 2014). Some of marine demosponges habituating in environments with high concentration of As and Ba due to activity of sponge-associated bacteria (Keren et al. 2017). Others, like the demosponge *Theonella conica*, are known through hyperaccumulation of molybdenum (Shoham et al. 2024), or formation of "armoured", iron-rich siliceous spicules due to activity of submarine hydrothermal fluid vents located near to the volcanic islands of the Aeolian Arc (Tyrrhenian Sea, Italy) (Ferretti et al. 2019).

Thus, poriferan biosilica is recognized in the literature as example of X-ray amorphous, hydrated silica. What about findings of diverse crystalline phases within sponge spicules? Previously, crystallographic theory of the origin of spicules has been proposed and discussed by Geo P. Bidder in his work published in *Nature* (Bidder 1925). Other publications appeared periodically regarding the discovery of crystalline or paracrystalline structures in sponge spicules (Wilkinson and Garrone 1980; Mugnaioli et al. 2009). For example, crystalline smectite-like alumosilicate with the average atomic composition (atomic %) of O 63.5; Si 21.2; Al 11.5; Mg 1.9; K 0.9; Fe 0.9 has been identified using high resolution electron microscopy (HRTEM) and electron diffraction within primordial spicules of *Suberites domuncula* demosponge (Mugnaioli et al. 2009). In the same sponge species, pseudohexagonal nanoparticles between 5 and 15 nm in size, displaying lattice parameters corresponding to hematite (Fe_2O_3) and mixed iron oxide phases typically attributed to ferritin, have been reported in 2014 by Natalio and co-workers (2014). It cannot be ruled out that a form of silicate such as cristobalite could ever be found inside sponge spicules. Cristobalite is a polymorph of SiO_2 in which the silica tetrahedra are packed in a three-layer structure and remains to be a high-temperature modification of quartz (Gutiérrez-Castorena and Effland 2010). As recently reported (Fuchs et al. 2014, 2015) proteinaceous silicatein-containing filaments isolated from *Negombata magnifica* demosponge can induce direct crystallization of biosilica to cristobalite at relatively low temperatures.

The presence of calcium carbonates within siliceous spicules of selected extant and extinct sponges remains to be an intriguing topic. For example, spicules of Cambrian sponge *Eiffelia globosa* have a bilayered structure suggestive of a silica–carbonate biocomposite (Botting and Butterfield 2005). The first record of a crystalline calcite as biomineral, other than silica that the hexactinellid sponges, for example, *Caulophacus* sp. are capable to synthetise, has been reported by us in 2011 (Ehrlich et al. 2011). The nanocrystals of calcite have been undoubtably detected within club-like formations on the ends of siliceous spicules. Calcite reinforced silica–silica joints in the complex skeleton of this sponge, which strongly resembles the design of the Eiffel Tower (Ehrlich 2019, 2024), are crucial for its integrity and mechanical robustness. Spinose regions that are adjacent with calcite nanotuned spicules are then joined by biosilica-made cross-links, leading to hierarchically structured, unusually strong cross-spicule linkages. Similar spine-like microstructures have also been found in skeleton of *E. aspergillus* glass sponge (see for SEM images Wysokowski et al. 2018; Ehrlich 2019, 2024) that as *Caulophacus* sp. belongs to the order Lyssacinosida. Despite the enormous popularity of this sponge as a topic of study, which is known as Venus' flower basket among researchers (Aizenberg et al. 2005; Weaver et al. 2007; Monn et al. 2015; Chen et al. 2022; Morankar et al. 2022, 2023), this particular phenomenon is often either ignored or simply goes unnoticed.

Thus, the examples of occurrence of inorganic crystalline phases within sponge biosilica are more likely to be exceptional cases rather than typical, in contrast to crystalline phases of organic origin.

At this point, let me move on to a brief analysis of the biochemistry of spicules as well as siliceous skeletal frameworks in sponges especially due to the exitance of structural polysaccharides and proteins within these formations, which have characteristic crystalline features. Considering the bioorganic components in the siliceous skeletal structures of sponges, I think it is logical to provide generally accepted information about their origin. Probably the simplest explanation can be found in the work of Arthur Dendy from 1921, who wrote the following:

> Two kinds of cells—initial cells and silicoblasts—are concerned in spicule formation; the former cells secrete the organic material (spiculin) which forms the axial thread or protorhabd around which the silicoblasts collect and deposit silica. A growing spicule may come to be completely enveloped by a silicoblast, which has accordingly been regarded by nearly all observers as the mother-cell in which the spicule originates. (Dendy 1921)

Nowadays, so-called *axial thread* (Bütschli 1901) is termed as *axial filament*, and the rather exotic term *spiculin* involves several proteins at once. We can speak about diverse "*silica-associated proteins*" (Shimizu et al. 2024). The axial filaments possess diameter between 0.1 and 2 μm (Tang et al. 2019) and usually are not immured in biosilica, but instead are found in a somewhat free-floating state inside the so-called *axial channel* (Duncan 1881) (Fig. 2.16), which is filled with some liquid of unknown chemical composition due to the lack of technical capabilities to carry out appropriate bioanalytical studies within such small volumes. It is well-recognized that the axial filament strictly repeats the morphological features of the spicules themselves in which it was found. For example, in star-shaped silica spicules (asters) of *Stelletta grubii* demosponge, the filament is shown for the first time to be radial, however in the four-rayed spicules (triaenes), it is four-rayed (see for details Simpson et al. 1985). Also, the symmetry of the axial filament has been suggested to be expressed and developed prior to the deposition of non-organized biosilica on the outer surfaces of the filament (Reiswig 1971).

If the protein origin of the axial filament in sponges did not raise much doubt, then the identification of the protein that forms its basis has turned out to be a rather difficult undertaking until now. In the seventies of the last century, it was possible to clearly determine that the axial filament does not contain collagen. For example, in the axial filament isolated from spicules of *Hymeniacedon sinapium* demosponge, no detectible hydroxyproline or hydroxylysine have been observed (Shore 1972).

A real breakthrough in deciphering the nature of the axial filament was made in Dan Morse's group in 1998. We can say that the publication of their article entitled "*Silicatein alpha: cathepsin L-like protein in sponge biosilica*" (Shimizu et al. 1998) began an entire era of monopoly recognition of silicateins as the main protein components of this intra-spicule structure (Cha et al. 1999; Weaver and Morse 2003; Murr and Morse 2005; Brutchey and Morse 2008). Cathepsins are the most abundant lysosomal proteases and play a crucial role in intracellular protein degradation (Yadati et al. 2020). Silicateins with molecular weight of 24–27 kDa have been recognized as catalytic triad hydrolases (Shimizu and Morse 2018). The silicatein version of the nature of axial filaments in sponges as a universal one has been actively promoted by Werner Müller's group for many years (Müller et al. 2007a, b, c). Silicatein has been identified not only in spicules of demosponges like *Suberites domuncula* and *Geodia barretti*, but also in such glass sponge species as *Monorhaphis chuni* (Müller et al. 2008a) and *Crateromorpha meyeri* (Müller et al. 2008a). In 2013, silicateins were even hailed as a novel paradigm in bioorganic

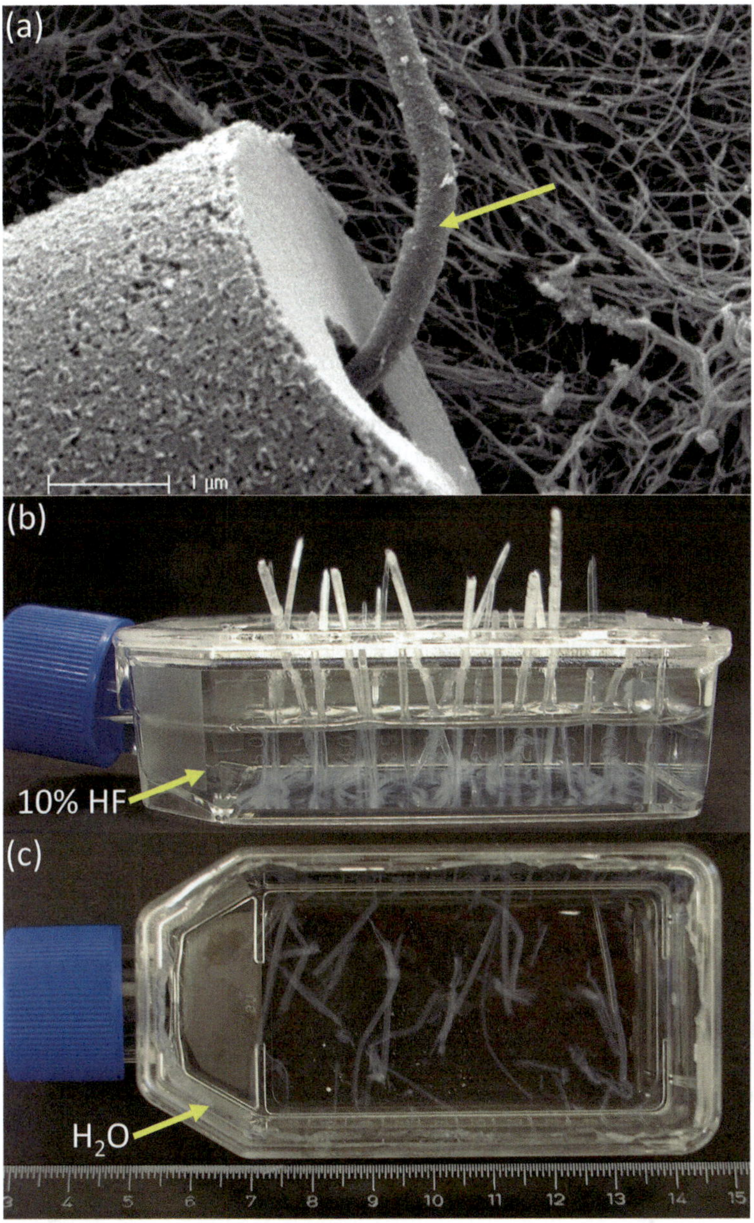

Fig. 2.16 SEM image of the axial filament (arrow) within the spicule of *S. domuncula* marine demosponge (**a**) Axial filaments can be isolated from anchoring spicules of *Corynonema populipherum* using gentle desilicification in HF (**b**) in large amounts (**c**)

chemistry (Müller et al. 2013). To put it diplomatically, "silicatein fever" has also spread to areas such as the synthesis of metal oxides, metal phosphates, polylactides, and polymeric materials composed of organic metal compounds from their corresponding precursors (Tahir et al. 2005; Shimizu and Morse 2018). The apogee can be considered recent publications on the participation of silicatein on control of the $CaCO_3$ crystal morphology with respect to synthesis of calcite (Gao and Nakamura 2022; Gao et al. 2024). Recently, it was shown that recombinantly obtained silicateins did not affect the crystalline structure of the alkaline earth metal (Ca, Ba, Sr) carbonate, but did modify the crystalline habit of corresponding biomorphs (Sánchez-Puig et al. 2021). The selectivity of silicateins (Egedeuzu et al. 2025) in relation to a wide variety of molecular substrates somewhat undermines their uniqueness in the formation of the poriferan axial filament and the glassy spicules themselves. Also, returning to the topic of axial filaments of sponges, their metal oxide or calcium carbonate nature has not yet been discovered.

In the scientific community, an equally intriguing situation has developed around the phenomenon of discovering the crystalline properties of organic compounds in the axial filaments of both demosponges and hexactinellids. Such features have been reported in numerous publications that started with fiber diffraction studies of spicules from diverse marine sponges (Croce et al. 2003, 2004) and continued with investigations using electron diffraction and other modern high-resolution X-ray techniques (Werner et al. 2017; Schoeppler et al. 2017; Zlotnikov et al. 2014; Görlich et al. 2020). Lattice parameters of mesoscopic hexagonal proteinaceous crystal-like structure have been measured with $a = (9.88 \pm 0.06)$ nm in the basal plane and $c = (10.83 \pm 0.02)$ nm, co-aligned with the filament's main axis for *Monorhaphis chuni* glass sponge axial filament (Zlotnikov et al. 2014). In *T. aurantium* demosponge spicules perfect proteinaceous hexagonal superstructure with lattice parameters of $a = 5.95 \pm 0.01$ nm and $c = 11.89 \pm 0.01$ nm, with the c-axis oriented parallel to the long axis of the spicule have been reported by the Igor Zlotnikov's group in 2020 (Görlich et al. 2020).

It should be noted that the physicists did a wonderful job and the data they obtained does not raise any doubt. However, none of the above-mentioned works identified silicateins at all. The researchers simply accepted as truth the presence of silicateins in the axial filaments of all types of sponges they studied as truth. As it turned out later (Ehrlich et al. 2022), this was their historical and unfortunate mistake.

2.5 Principles of Biosilicification and Biosilica Pattering Using Organic Templates

There is no doubt that the process of spiculogenesis in sponges, regulated by certain genes and cells, is based on a biochemical form of biomineralization called biosilicification. Silicic acid and organic molecules are essential participants in this process. Modern view on mechanisms of biosilicification including sponges has been recently represented by Patricia Dove in the impressive review entitled "*Silica–Biomacromolecule Interactions: Toward a Mechanistic Understanding of Silicification*" (McCutchin et al. 2025). The involvement of a number of biomolecules, from the structural aminopolysaccharide chitin to various polyamines, peptides, and proteins, has been either proposed or experimentally proven. For example, only from spicules of marine demosponge *Petrosia ficiformis*, 21 proteins, including cathepsins, silicatein beta, proteins with strong homologies with enzymes like sulphatases, glycosidases, lipid-related proteins, phosphatases, and some others with unknown function have been isolated (Pozzolini et al. 2022). It is impossible to exclude the participation of a large number of biomacromolecules with different chemical properties in biosilcification. In 2009, it was shown (Wallace et al. 2009) that cooperative interactions between oppositely charged surface species play important roles in directing the onset of silica nucleation. Wallace and co-workers reported that amine-terminated surfaces do not promote silica nucleation in vitro, whereas carboxyl and hybrid NH(3)(+)/COO(−) substrates are active for silica deposition as well as that rate of silica nucleation is approximately 18x faster on the hybrid substrates than on carboxylated surfaces (Wallace et al. 2009).

Among templates for biosilicification in sponges, the following should be highlighted:

(a) *Chitin*

This nanofibrillar polysaccharide belongs to universal templates for biomineralization (Ehrlich 2010). Its presence has been unambiguously shown in siliceous spicules of glass sponges like *Rossella fibulata*, *Farrea occa* (Ehrlich et al. 2007), and *S. hawaiicus* (Ehrlich et al. 2016). However, the direct participation of chitin in biosilicification in sponges remain to be a mystery. In the recent paper by Godigamuwa and co-workers, creation of silica on chitin in ambient conditions using silicatein fused with a chitin-binding domain has been reported for the first time (Godigamuwa et al. 2021).

(b) *Polyamines*

Long-chain polyamines (LCPAs), or polypeptides modified with LCPAs are known to be involved in biosilicification in diatoms (Kröger et al. 2000). Matsunaga and co-workers have identified the LCPAs from the marine sponge *Axinyssa aculeata* and represented some evidence that sponge-derived LCPAs can deposit silica and that the LCPA derivatives are associated with spicules (Matsunaga et al. 2007).

(c) *Lysozyme*

Lysozyme as a positively charged protein appears to induce silicate polymerization through electrostatic interactions. It has been found in lysosome (Radons et al. 1994). It is suggested that lysozyme, together with other lysosome enzymes, could be involved in biosilica machinery in sponges (Pozzolini et al. 2022). Formation of silica–protein composited due to lysozyme induced silicification in vitro is well-reported in literature (van den Heuvel et al. 2018; Bruno et al. 2022; Macchiagodena et al. 2024).

(d) *Silicateins*

Silicatein, a silica-synthesizing, catalytic triad hydrolase (Shimizu and Morse 2018), was discovered in the silica spicules comprising the skeletons of certain marine sponges (Shimizu et al. 1998; Cha et al. 1999; Wang et al. 2012). Its sequence similarity is closest to that of the mammalian cathepsin L, a catalytic triad hydrolase and protease (Shimizu and Morse 2018). The presence of silicateins in spicules has been reported in such demosponges as *Latrunculia oparinae* (Povarova et al. 2015; Shkryl et al. 2016), *Tethya aurantia* and *Suberitis domuncula* (Müller et al. 2005), *Haliclona oculata*, *H. simulans*, and *H. indistincta* (Aguilar-Camacho et al. 2019), *Petrosia ficiformis* (Armirotti et al. 2009); *Lubomirskia baicalensis* (Kaluzhnaya et al. 2007), *Ephydatia fluviatilis* (Mohri et al. 2008). Thus, silicateins were only found in species belonging to the class Demospongiae but excluding subclasses Keratosa and Verongimorpha (Aguilar-Camacho et al. 2019). The publications regarding the detection of silicateins in two species of glass sponges *Crateromorpha meyeri* (Müller et al. 2008a) and *Monorhaphis chuni* (Müller et al. 2008b) clearly contain an error. According to the modern view (Riesgo et al. 2015), hexactinellids did not express silicatein genes principally. At the same time, some species of non-spiculate marine demospongia possess silicatein genes (Kozhemyako et al. 2009).

(e) *Cathepsins*

Cathepsins are a type of lysosomal globulin hydrolase and are crucial for many physiological processes (Riesgo et al. 2015; Jiang et al. 2023). Similar to silicateins, they are involved in efficient silica synthesis in vitro, for example, of tetra(glycerol)orthosilicate (Povarova et al. 2018). Also, silicatein alpha-cathepsin L chimeras with the ability to condense silica from solution have been generated previously (Fairhead et al. 2008).

(f) *Glassin*

Glassin is a histidine-, aspartic acid-, threonine-, and proline-rich protein isolated from glass sponge *Euplectella* sp. in 2015 (Shimizu et al. 2015). It directs silica polycondensation at room temperature and neutral pH and guiding extracellular deposition of peripheral silica.

(g) *Hexaxilin and Perisilin*

Both proteins have been recently isolated from spicules of hexactinellid sponges *Euplectella curvistellata* and *Vazella pourtalesii*. Hexaxilin forms the axial filament to intracellularly pattern the main symmetry of the skeletal parts, while perisilin appears to operate in their thickening, similar to glassin extracellular deposition of peripheral silica in these sponges (Shimizu et al. 2024). Unfortunately, the presence of F-actin as the main structural component of axial filaments of these hexactinellids has been overlooked.

(h) *Collagen*

Unique hydroxylated fibrillar collagen that contains an unusual [Gly–3Hyp–4Hyp] motif has been isolated in 2010 from anchoring spicules from the metre-long stalk of the glass rope sponge *Hyalonema sieboldi* (Ehrlich et al. 2010). It was suggested that this highly specific motif is predisposed for silica precipitation and provides a novel template for biosilicification in nature, never reported before.

(i) *Peptide*

Although no specific peptides have been isolated from sponge spicules, many synthetic peptides have been proposed by researchers as examples of bioinspiration. Consequently, inspired by biosilicification of glass sponges, diverse catalytic peptides, which formed silica structures under selected conditions, have been designed (for overview see Naik et al. 2003; Galloway et al. 2018; Sola-Rabada et al. 2018; Strunge et al. 2022).

Thus, an analysis of the literature allows us to conclude that there are a number of biomacromolecules that can or do participate in the process of biosilicification in sponges. Logically, it was previously assumed that these compounds are also involved in biosilica pattering. For example, silicatein is traditionally thought to be the key molecule determining the morphology of spicules (Mohri et al. 2008; Müller et al. 2013; Schoeppler et al. 2017). However, the "perfectionism in glass" (Ehrlich 2019), which we never cease to admire when studying the diversity of morphological forms in biosilica-forming sponges, turns out to have a different participant in this process than expected. In this case, I would like to quote Jenie L. Wulff's wonderful phrase that "*sponges are masters of intrigue*" (Wulff 2016). And the intrigue is that, as it turned out and was experimentally proven (Ehrlich et al. 2022, 2024; Voronkina et al. 2023, 2025) in a huge number of sponges from three classes (Ehrlich 2022), only F-actin performs the function of biosilica pattering.

Previously, a number of inhibitor studies and direct microscopic observations have shown that actin plays an important role in shaping microscale silica structure already in unicellular organisms (i.e. diatoms) (see for details: Schmid 1980; Blank and Sullivan 1983; Cohn et al. 1989). Tesson and Hildebrand (2010) using the different species showed that actin filaments play a substantial role in the formation of meso- and micro-scale silica structure in diatom frustule formation. Actin was visualized in diatoms *Cyclotella cryptica* (Van de Meene and Pickett-Heaps 2004), identifying an actin ring that defined the full extent of the valves and another concentration of actin filaments associated with the growing front of silica deposition. However, since actin has traditionally been perceived as a strictly intracellular protein, no one has ever looked for it and has not been predicted in such structures as spicules, or skeletal frameworks of sponges.

Our approach of F-actin isolation from poriferan biosilica is based on the gentle application of a drop of 10% HF onto the surface of a spicule, or fragment of network-like skeleton placed on an inclined plastic surface (Ehrlich et al. 2022). This acid dissolves silica leaving the organic matrix intact. Based on the results of our careful and year-long research including SEM and TEM observations, we did not find any negative effect of non-buffered HF on the organization or integrity of actin in our samples. Previously, similar results were obtained on diverse biological systems by other authors. For example, concentrated HF was used for dissolution of silica in the diatom *Thalassiosira pseudonana*, where researchers also found actin (Frigeri et al. 2006).

The use of HF is a traditional and proven method in spongology for dissolving siliceous spicules. This method has proven itself both in cytology, developmental biology, molecular biology, biochemistry, etc., of sponges. As shown in numerous investigations, HF does not have negative effects on cells, extracellular matrix, or the internal organic structures of spicules (Drum 1968; Shore 1972; Palmer 1930; Weaver et al. 2003; Sabella et al. 2007).

Identification of isolated axial filaments have been made using diverse methods including application of traditional and highly specific indicators of F-actin such as iFluorTM 594-Phalloidin, iFluorTM 488-Phalloidin, as well as iFluorTM 350-Phalloidin (Ehrlich et al. 2022, 2024; Voronkina et al. 2025) (Fig. 2.17). In our studies, we took the choice to confirm the presence of actin within silicified skeletal structures of selected sponges species habituating both in cold (−1.9 to 4 °C) (*Aphrocallistes beatrix*; *Asconema setubalense*; *Caulophacus arcticus*; *Euplectella subarea*; *Farrea* sp.; *Hyalonema (Corynonema) populiferum*; *Hyalonema* sp.; *Malacosaccus* sp.; *Monorhaphis chuni*; *Rossella Antarctica*; *Walteria flemmingi*; *Cladorhiza corona*) and warm (20–24 °C) (*Spongilla lacustris*; *Petrosia ficiformis*; *Geodia cydonium*) environments. Accordingly, the wide temperature range covered by us, at which both marine (including carnivorous *C. corona*) and freshwater species of sponges develop various forms of biosilicates, allows us to draw the conclusion that the presence of actin in poriferan biosilica is universal (Ehrlich et al. 2024).

Since axial filaments inside the spicules are found unbound to the mineralized phase, we assume that F-actin here plays no direct role in biosilicification. The occurrence of axial filaments within axial channels of spicules being not embedded into the silica is a well-recognized phenomenon observed using TEM (see for overview Uriz et al. 2003; Fig. 7 in Müller et al. 2006) as well as SEM (Figs. 11, 12 in Wysokowski et al. 2018).

The absence of hydroxyproline (3-Hyp and 4-Hyp) residues responsible for the attachment of silicic acid to the protein molecule, as is the case in collagen (Ehrlich et al. 2010), further excludes actin as a direct template for silicification. Until today, over 150 actin-binding and actin-associated proteins have been identified that influence cross-linking, localization, polymerization dynamics, and organization of actin (Winder and Ayscough 2005; Gao and Nakamura 2022). Consequently, it cannot be ruled out that fibrillar actin may act as a template for another low (glassin, hexaxilin, perisilin, silicateins), or high molecular weight biomolecules (hydroxylated collagen), which have been still recognized as possible players in the biosilicification of sponges. Let us remind ourselves once again that none of the above-mentioned proteins, unlike actin, can bifurcate or branch, not to mention assume its radial form.

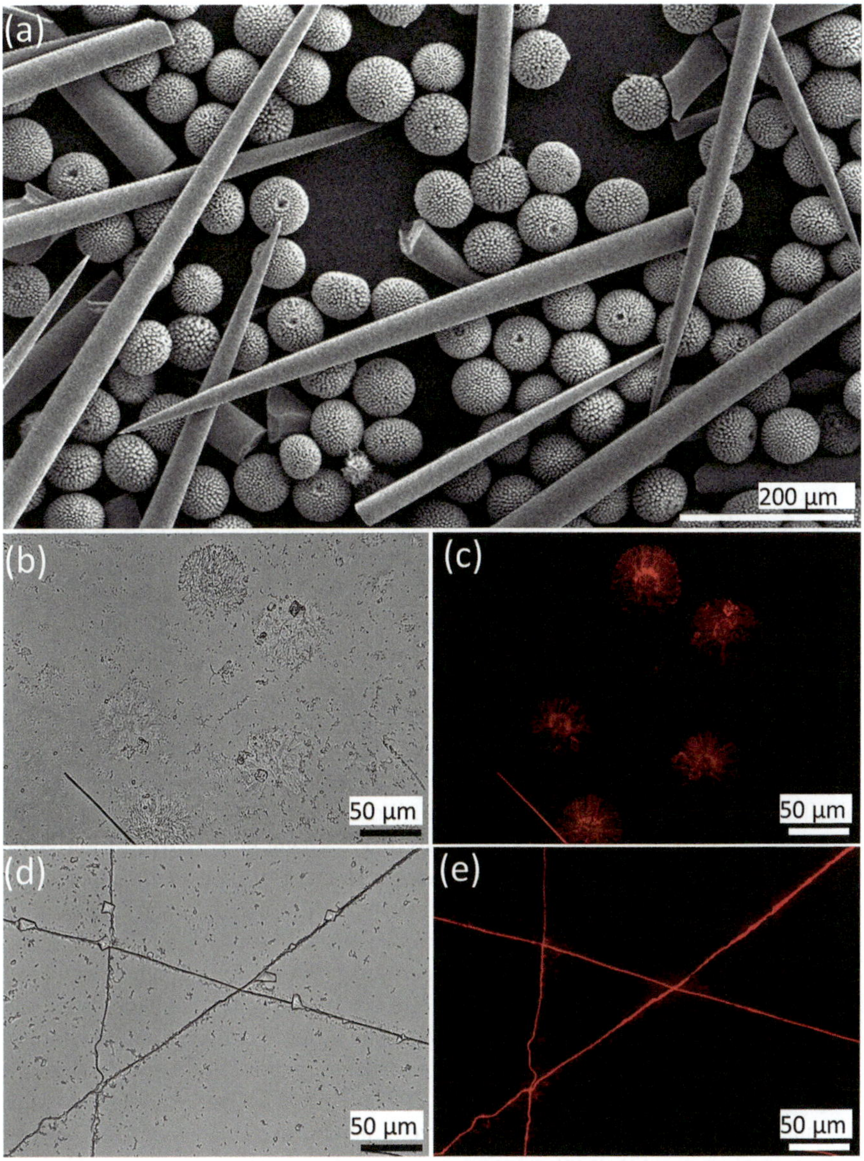

Fig. 2.17 Understanding the morphogenesis of *Geodia* asters shape remains to be a challenging task since 1989 (Simpson 1989). SEM image (**a**) of *Geodia cydonium* demosponge siliceous spicules represents microscleres in the form of ball-shaped sterrasters and megascleres known as oxeas. Their gentle desilicification using 10% HF led to obtaining of F-actins (silactins) (Ehrlich et al. 2024) which resemble the size and shape of original siliceous structures (**b**, **d**). Fluorescence microscopy of phalloidin-stained axial filaments from sterrasters (**c**) and oxeas (**e**) of *G. cydonium* sponge confirms their actin nature (for details see Voronkina et al. 2025)

From the standpoint of biomimetics, "*the prospect of using actin filaments to create new silicate-containing materials and biomaterials with sophisticated three-dimensional architecture can be suggested as extremely interesting precisely also in model experiments with native fibrillar actin in the presence of precursors for biomineralization under laboratory conditions*" (Ehrlich et al. 2022).

2.6 Biomimetic Potential of Sponges Biosilica

2.6.1 Brief Introduction into Biomechanics of Spicules

Studies on the structural biology of glass sponges spicules largely determine the understanding of the various effects found in the investigations of their biomechanics. In this case, we are not talking about axial filaments, but about the composite nature of the biosilica layer located between the axial channel of the spicule and its surface (Fig. 2.16a). In simplified terms, we are dealing with a biosilic-based tube whose walls have a sandwich structure with alternating layers of organic and inorganic phases. This phenomenon was known to researchers as early as the nineteenth century (see for details Minchin 1909), but was described in more detail in Schmidt's work published in 1926 and entitled "*On the nature of lamination and the mutual behaviour of organic and inorganic substances in siliceous sponge needles*" (Schmidt 1926). Lamination is common for spicules to be found in both demosponges and hexactinellids regardless of their size (Weaver et al. 2010).

Effects of laminate architecture on diverse mechanical properties of poriferan biosilica is well-investigated (see for detail Woesz et al. 2006; Miserez et al. 2008; Monn et al. 2015, 2020; Monn and Kesari 2017; Zhang et al. 2016; Monn 2018; Fernandes et al. 2021). Thus, laminated anchor spicules of *Euplectella aspergillum* have been studied as natural models of strength (Monn et al. 2015). According to Michel Monn and co-workers, "*the unique internal geometry that maximizes a beam's strength correlates well with the geometry observed in the native spicules*" (Monn et al. 2015). Mesoscale elastic properties of *Euplectella* spicules are well-described (Zhang et al. 2016). Also, it was shown that the layered architecture of this sponge spicules profoundly affects the fracture process by inducing multiple toughening mechanisms (Morankar et al. 2022). He and co-workers suggested that the unique cell-after-cell and layer-by-layer fracture mechanisms of glass sponge lattice structure provide its interior and entire integrity during compression (He et al. 2022). Nowadays, it is recognized that interplay of strength, flexibility, and sponge biosilica's nano- and microstructure is principal for future functional biomimicry (Meng et al. 2025).

2.6.2 Bionic Design

In my personal opinion, the fashionable passion for searching for strictly bio-inspired motifs in classical architectural solutions may be only partially correct. Despite the structural similarities between the glass sponge's skeleton (i.e. *Euplectella*, or *Caulophacus*) and the Eiffel Tower, it is unlikely that the architect who designed and built it from 1887 to 1889 was under the influence of this biological construct. In modern time, however, sophisticated sponges skeletons can play important role in so-called *Bio-Inspired Intelligent Architecture* (Chayaamor-Heil 2023; Chayaamor-Heil et al. 2023; AlAli et al. 2023). The trend of "sustainability through biomimicry" (Jiang et al. 2024) has been gaining momentum especially in the last few years (Verbrugghe et al. 2023).

Hexactinellids like *Euplectella* species possess a skeleton that represents typical porous lightweight structure (Fig. 2.3) with high toughness, high stability, and high strength. Based on these morphological features, Li and co-workers carried out the finite-element analyses and fabricated corresponding constructs using stereolithography 3D-printing technology (Li et al. 2020). Recently, a new bionic design of thin-walled bilinear tubes with excellent crashworthiness inspired by glass sponge skeletons has been proposed by Liu and co-workers (2024). Finally, a surrogate model is established to perform multi-objective optimization on the optimal bilinear configuration bionic glass sponge tube selected by the improved complex proportional assessment synthesis method (Liu et al. 2024).

A separate area can be distinguished by creation of honeycomb cellular structures inspired by the corresponding skeletons of glass sponges (see for details, Voronkina et al. 2023). As reported by Wang et al. (2022), bio-inspired honeycomb column, thin-walled structures, were fabricated by additive manufacturing selective laser melting using AlSi10Mg Al-Si alloy.

2.6.3 Bioinspired Photonic Devices

Biological materials of poriferan biosilica origin represent well-organized nano-scaled structures, which are characterized by a low refractive index (i.e. 1.4386–1.4452, Sundar et al. 2003; Kulchin et al. 2007, 2009) and an extremely low absorption coefficient in the visible range (Mcheik et al. 2018). Numerous studies have shown that the hexactinellids spicules can guide light thanks to total internal reflection (Aizenberg et al. 2004). In 2016 we have observed supercontinuum generation (SG) in the anchoring needle-like spicules of marine deep-sea glass sponge *S. hawaiicus* (Ehrlich et al. 2016). As highlighted in *Nature Photonics*, "*whereas the fabrication of commercial glass fibres requires very high temperatures, hexactinellid sponges such as S. hawaiicus produce the spicules responsible for their anchorage at temperatures close to 4 °C. The authors speculate that a better understanding of the dispersive and nonlinear properties of these natural optical fibres could facilitate the synthesis of new light-guiding materials*" (Donati 2016).

According to Mehmet Sarikaya "*sponge spicules offer bioinspired lessons for potential biomimetic design of optical fibers with long-term durability*" (Sarikaya et al. 2001). Thus, a biomimetic molecular approach to print optical fibers with participation of silicateins has been developed in Werner Müller's group. The experiments revealed that after enzymatic silicatein-guided reaction with an ortho-silicate substrate, a specific biosilica mantle is formed that can act as a light waveguide (Müller et al. 2014). Also, it was shown that recombinant silicatein-α can direct the formation of optical waveguides in-vitro through soft microlithography (Polini et al. 2012). Recently, it was reported that silicatein-expressing *Escherichia coli* bacteria can self-assemble a shell of polysilicate "bioglass" around themselves (Sidor et al. 2024). Such polysilicate-encapsulated bacteria can focus light into intense nanojets and has the potential to revolutionize the development of multiple optical and photonic technologies.

2.6.4 Ceramic and Silicone Sponges

Biosilica-producing sponges, whose bodies possess hierarchical pores and channels originally determined by skeletal structures, represent a biomimetic model for development of morphologically similar ceramic sponges with a high potential of practical applications. For example, inspired by leuconoid sponge, Xu and co-workers designed a biomimetic macroporous ceramic composite sponge for high-precision logic delivery of molecules and cells regulated by mechanical stimulus (Xu et al. 2017). Facile method for large-scale fabrication of highly compressible, temperature resistant SiO_2-Al_2O_3 composite ceramic sponges by blow spinning and subsequent calcination has been proposed by Jia and co-workers (2020). According to these authors, "*the anisotropic lamellar ceramic sponges exhibit high compression fatigue resistance, strain-independent zero Poisson's ratio, robust fire resistance, temperature-invariant compression resilience from −196 to 1000 °C, and excellent thermal insulation with a thermal conductivity as low as 0.034 W m^{-1} K^{-1}. In addition, the lamellar structure also endows the ceramic sponges with excellent sound absorption properties*" (Jia et al. 2020). Fabrication of multifunctional ceramic foams that mimic marine sponges for sound absorption, waterproofing, and antibacterial applications has been recently reported also by Zhang and co-workers (2024).

The production of silicone sponges is generally a recognized technology, many of which are patented (i.e. Kim 2006; Hirabayashi and Koike 2008). Often, silicone sponges can hardly be distinguished from natural bath sponges, known also as commercial sponges (Jesionowski et al. 2018). Silicone sponges produced using such silicone are very flexible and have excellent resistance to heat and cold, excellent air permeability, and elasticity. For example, the strong, compressible, bendable, and stretchable (SCBS) silicone sponges (Li et al. 2019) were fabricated by solvent-controlled hydrolytic polycondensation of silanes in the presence of co-solvents of water, followed by being dried at ambient pressure at 60 °C. The SCBS silicone sponges are. The SCBS silicone sponges show high stability in a wide range of temperature (−196 to 300 °C), are superhydrophobic/superoleophilic, and highly compressible (90%), bendable (20 mm), and stretchable (50%) (Li et al. 2019).

Durable silicone sponges with superhydrophobic and magnetic properties have been used for hydrocarbons and oil absorption from water. The absorption of the developed sponge is in the range of 700–1700% depending on the organic substance (Li et al. 2014).

2.6.5 Dermabrasive Materials

The presence of pointed ends of the poriferan siliceous spicules determines their natural use by sponges as mechanical protection against predators. However, these same spicules, being a durable material, have found application in traditional methods of medicine and in modern dermatology. When rubbed into human skin, the spicules break through the corresponding surface layer and can be used as dermabrasive materials for therapeutic purposes. For example, the WO2008043071A2 patent entitled "*Use of spongilla spicules as a dermabrasion device or resurfacing modality*" is proof of this (Villani 2008).

Diverse spicules-based therapeutic compositions have been recently reported worldwide. Spicules can be applied topically as novel microneedles to overcome skin barrier by creating plenty of new microchannels in stratum corneum (Liang et al. 2020). Thus, Zhang and co-workers used exosomes derived from human umbilical cord mesenchymal stem cells in combination with marine sponge *Haliclona* sp. spicules for treatment of rejuvenating photoaged mouse skin (Zhang et al. 2020). The same spicules combined with cationic flexible liposomes have been used to increase the delivery of small interfering RNA into viable skin cells in vitro and in vivo (Liang et al. 2020). Also, the fabrication of mesoporous silicon dioxide coated *Haliclona* sp. spicules to enhance the delivery of the insoluble photosensitizer protoporphyrin IX (PpIX) into deep skin layers and mediate photodynamic therapy for metastatic melanoma in mice has been reported (Liang et al. 2024). Topical delivery of PpIX using spicules completely eradicated the primary melanoma in mice in 10 days without recurrence or metastasis over 60 days (Liang et al. 2024).

Spicule-based topical delivery systems offer promising strategies for delivering therapeutic peptides also via a transdermal or dermal route (Zhang et al. 2021). For example, in the case of insulin, spicules dramatically increased the transdermal flux of this peptide (457.0 ± 32.3 ng/cm^2/h) compared to its passive penetration (5.0 ± 2.2 ng/cm^2/h) in vitro (Zhang et al. 2021).

2.6.6 Tissue Engineering

Scaffolding strategies are an integral part of tissue engineering (Tsurkan et al. 2020). Biosilica-based skeletons of sponges are ready-to-use matrices of such a structure. It is not surprising that the creation of biomimetic models of silicate-containing scaffolds is logically included in modern tissue engineering. Potential use of sponge biosilica for tissue engineering applications has been already recognized (see for details Granito et al. 2017; Gabbai-Armelin et al. 2019; Dos Santos Jorge Sousa et al. 2023; Müller et al. 2024). It was reported (Rahmanisa et al. 2022) that 3D biosilica scaffolds from marine sponges *Melophlus sarasinorum* and *Xestospongia testudinaria* are biocompatible for cell growth and differentiation of human Wharton's jelly mesenchymal stem cells in bone tissue engineering. Biosilica from marine demosponge *Geodia barretti* has been effectively used for the development of 3D-printed scaffolds for bone tissue regeneration applications (Colares 2021). Choi and co-workers described application of marine-sponge-derived spicule microparticles (SPMs) with unique structural and compositional features for bone tissue engineering (Choi et al. 2024). In vitro assessment demonstrated dose-dependent increases in osteogenic gene expression and bone morphogenetic protein-2 protein in response to SPMs. Rapid vascularized bone regeneration has been demonstrated due to the osteogenic activity of SPM-reinforced injectable hydrogel that was evaluated in a mouse calvaria defect model (Choi et al. 2024).

The biosilica-containing skeletal structures of sponges have served as a powerful motivating model for the creation of 3D composite materials using non-biological silicates and non-sponge biopolymers (Aslanbay Güler et al. 2024). As an example, multiscale design and synthesis of biomimetic gradient protein/biosilica composites for interfacial tissue engineering has been proposed by the group of David Kaplan (Guo et al. 2017). It was shown that the gradient silicified silk/R5 peptide composites promoted and regulated osteogenic differentiation of human mesenchymal stem cells in an osteoinductive environment in vitro (Guo et al. 2017). Bioactive sponge-inspired 3D scaffolds have been created with a composite of type I collagen sponges, porous poly(ε-caprolactone) scaffolds, platelet-rich plasma, BMP2-loaded nanoporous silicon enclosure microparticles, mineralizing peptide amphiphiles, and mesenchymal stem cells (Murphy et al. 2011). Recently, *Euplectella* glass sponge-inspired "brick-and-mortar" scaffold has been created using oxidized bacterial cellulose, bacterial cellulose, and polydopamine (Meng et al. 2025). It presents a promising approach for vascular tissue engineering and regenerative medicine.

Another direction of bio-inspired materials science that has developed is the 3D printing of sponge-like bioglass-containing scaffolds for tissue engineering (Wang et al. 2014; Baino and Fiume 2020) (Fig. 2.18). Polley and co-workers represented an approach for 3D printing of piezoelectric and bioactive barium titanate-bioactive glass scaffolds for bone tissue engineering. The 3D-printed ceramic-crystallised glass composite scaffolds show both suitable mechanical strength and bioactive behaviour (Polley et al. 2023).

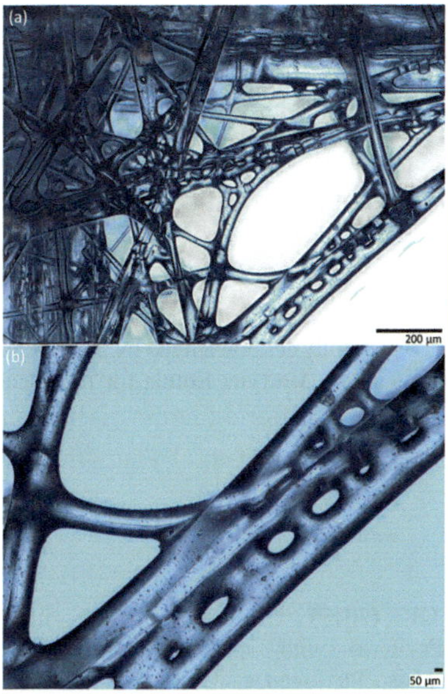

Biomimetic Directions
Mimicking biosintering
Bioinspired selective laser melting
Bio-based generative honeycomb design
Multifunctional design
Bioinspired architecture: next generation of high-performance buildings, skyscrapers, and bridges
Biomimetics of lightweight structural biomaterials
Deep-sea sponge-inspired tubular metamaterials
Crashworthiness design of a sponge-inspired multicell tube
Architectured ceramic fibers and damage evolution
Computational modeling of spicule-inspired nested structures
Artificial intelligence-enhanced bioinspiration
Analytic modeling, finite element modeling, and experimental validation
3D printing of sponge spicule-inspired flexible bioceramic-based scaffolds
Bioinspired stereolithography
Bioinspired models for tissue engineering scaffolding

Fig. 2.18 An overview on biomimetic directions formed in recent years as a result of deep study of biosilica-based skeletal structures of poriferan origin (for details see Ehrlich et al. 2024). The digital light microscopy image (**a, b**) represents the fragment of hierarchically constructed skeleton of *E. aspergillum* glass sponge as example of Eiffel Tower structural motif made by nature

2.7 Conclusion

The biological purpose of biosilica-based structures in sponges is strictly determined by the lifestyle of these ancient aquatic organisms, and, first of all, the need to obtain food by filtering water from the surrounding environment. Accordingly, glass sponges' skeletons must be porous and at the same time mechanically strong. Or, as in the case of demospongia and homoscleromorpha, microscleral formations made of biosilica contribute to the creation of elastic skeletons by contracting and holding organic spongin fibers. In both cases, nature has not spared ingenuity in creating a huge variety of biosilica structures, sometimes with an architecture difficult to imagine for humans at the microstructural level. According to the modern view, diverse low-molecular compounds with appropriate enzymatic activity, in association with such non-organized templates as collagens, silactins, spongin, and chitin contribute to the formation of complex geometry of hierarchically organized poriferan skeletons.

The collection of information on the morphological diversity of micro- and macrosclerotic formations in silica-synthesizing sponges, which began in the sixteenth century, continues even today, although dependent on progress in microscopy and modern visualization methods as well. It should be understood that of the thousands of species of sponges that currently exist, only a small proportion of these organisms have been studied, although this was already enough to create a whole series of biomimetic directions of a functional architectural order.

The issues of molecular mechanisms of formation of biosilica structures in sponges remain largely open and hotly debated in the scientific community. It is possible that the ambitiousness of scientific projects in this direction will allow us to obtain true knowledge about the principles of this kind of fascinating bioarchitecture in glass.

Acknowledgements The financial supports provided by National Science Centre, Poland under MAESTRO-12 programme (no. 2020/38/A/ST5/00151) are gratefully acknowledged by the author. I express special gratitude to Dr. Alona Voronkina, Anita Kubiak, Martyna Kotula for technical assistance in preparing the manuscript.

References

Aguilar-Camacho JM, Doonan L, McCormack GP (2019) Evolution of the main skeleton-forming genes in sponges (phylum Porifera) with special focus on the marine Haplosclerida (class Demospongiae). Mol Phylogenet Evol 131:245–253. https://doi.org/10.1016/j.ympev.2018.11.015

Aizenberg J, Sundar VC, Yablon AD, Weaver JC, Chen G (2004) Biological glass fibers: correlation between optical and structural properties. Proc Natl Acad Sci USA 101:3358–3363. https://doi.org/10.1073/pnas.0307843101

Aizenberg J, Weaver JC, Thanawala MS, Sundar VC, Morse DE, Fratzl P (2005) Skeleton of *Euplectella* sp.: structural hierarchy from the nanoscale to the macroscale. Science 309(5732):275–278. https://doi.org/10.1126/science.1112255

AlAli M, Mattar Y, Alzaim MA, Beheiry S (2023) Applications of biomimicry in architecture, construction and civil engineering. Biomimetics 8:202. https://doi.org/10.3390/biomimetics8020202

Armirotti A, Damonte G, Pozzolini M, Mussino F, Cerrano C, Salis A, Benatti U, Giovine M (2009) Primary structure and post-translational modifications of silicatein beta from the marine sponge *Petrosia ficiformis* (Poiret, 1789). J Proteome Res 8(8):3995–4004. https://doi.org/10.1021/pr900342y

Asadzadeh SS, Kiørboe T, Larsen PS, Leys SP, Yahel G, Walther JH (2020) Hydrodynamics of sponge pumps and evolution of the sponge body plan. elife 9:e61012. https://doi.org/10.7554/eLife.61012

Aslanbay Güler B, Morçimen ZG, Taşdemir Ş, Demirel Z, Turunç E, Şendemir A, Imamoglu E (2024) Design of chemobrionic and biochemobrionic scaffolds for bone tissue engineering. Sci Rep 14:13764. https://doi.org/10.1038/s41598-024-63171-z

Austin WC, Conway KW, Barrie JV, Krautter M (2007) Growth and morphology of a reef-forming glass sponge, *Aphrocallistes vastus* (Hexactinellida), and implications for recovery from widespread trawl damage. Por Res Biodivers Innov Sustain:139–145

Bahls L (2018) Evidence from checklists for a Holarctic (circumboreal) kingdom of diatoms. PhytoKeys 108:17–24

Baino F, Fiume E (2020) 3D printing of hierarchical scaffolds based on mesoporous bioactive glasses (MBGs)—fundamentals and applications. Materials 13(7):1688. https://doi.org/10.3390/ma13071688

Bakran-Petricioli T, Vacelet J, Zibrowius H, Petricioli D, Chevaldonné P, Raa T (2007) New data on the distribution of the "deep-sea" sponges *Asbestopluma hypogea* and *Oopsacas minuta* in the Mediterranean Sea. Mar Ecol 28(Suppl. 1):10–23. https://doi.org/10.1111/j.1439-0485.2007.00179.x

Bart MC, de Vet SJ, de Bakker DM, Alexander BE, van Oevelen D, van Loon EE, van Loon JJWA, de Goeij JM (2019) Spiculous skeleton formation in the freshwater sponge *Ephydatia fluviatilis* under hypergravity conditions. PeerJ 6:e6055. https://doi.org/10.7717/peerj.6055

Beaulieu SE (2001) Life on glass houses: sponge stalk communities in the deep sea. Mar Biol 138(4):803–817. https://doi.org/10.1007/s002270000500

Bell JJ (2008) The functional roles of marine sponges. Estuar Coast Shelf Sci 79(3):341–353. https://doi.org/10.1016/j.ecss.2008.05.002

Bell JJ, Barnes DKA (2000) A sponge diversity centre within a marine "island". Hydrobiologia 440(1):55–64. https://doi.org/10.1023/A:1004145011422

Bergquist PR, Sinclair ME (1968) The morphology and behaviour of larvae of some intertidal sponges. N Z J Mar Freshw Res 2(1):426–437. https://doi.org/10.1080/00288330.1968.9515247

Biard T (2022) Diversity and ecology of Radiolaria in modern oceans. Environ Microbiol 24(5):2179–2200. https://doi.org/10.1111/1462-2920.16004

Bidder G (1925) The origin of sponge-spicules. Nature 115:298–299

Bidder GP (1923) The relation of the form of a sponge to its currents. J Cell Sci s2-67(266):293–324. https://doi.org/10.1242/jcs.s2-67.266.293

Blank GS, Sullivan CW (1983) Diatom mineralization of silicic acid VI. The effects of microtubule inhibitors on silicic acid metabolism in *Navicula saprophila*. J Phycol 19:39–44

Bond C, Harris AK (1988) Locomotion of sponges and its physical mechanism. J Exp Zool 246(3):271–284

Botting JP, Butterfield NJ (2005) Reconstructing early sponge relationships by using the Burgess Shale fossil *Eiffelia globosa*, Walcott. Proc Natl Acad Sci USA 102(5):1554–1559. https://doi.org/10.1073/pnas.0405867102

Boury-Esnault N, Rützler K (1997) Thesaurus of sponge morphology. Smithsonian Institution Press, Washington, DC

Boutillier J, Masson D, Fain I, Conway K, Lintern G, O'Malley M, Davies S, Mahaux P, Olsen N, Nguyen H, Rutherford K (2013) The extent and nature of exposure to fishery induced remobilized sediment on the Hecate Strait and Queen Charlotte Sound glass sponge reef. Canadian Science Advisory Secretariat research document 1919-5044, 2013/075, 84 p

Bowerbank JS (1869) A monograph of the siliceo-fibrous sponges. Part I. Proc Zool Soc Lond 1869(1):66–100, pls III–VI

Bowerbank JS (1875) A monograph of the siliceo-fibrous sponges. Proc Zool Soc Lond 18:272–281

Bowerbank JS (1876) A monograph of the siliceo-fibrous sponges. Part VI. Proc Zool Soc Lond 1876:535–540

Brendstedt W (1929) Die Schwämme der Nord- und Ostsee. In: Grimpe G, Wagler E (eds) Die Tierwelt der Nord- und Ostsee, vol 11b. Akademische Verlagsgesellschaft, Leipzig, pp 93–180

Brümmer F, Pfannkuchen M, Baltz AG, Hauser T, Thiel V (2008) Light inside sponges. J Exp Mar Biol Ecol 367(2):61–64. https://doi.org/10.1016/j.jembe.2008.06.036

Bruno F, Gigli L, Ferraro G, Cavallo A, Michaelis VK, Goobes G, Fratini E, Ravera E (2022) Lysozyme is sterically trapped within the silica cage in bioinspired silica-lysozyme composites: a multi-technique understanding of elusive protein-material interactions. Langmuir 38(26):8030–8037. https://doi.org/10.1021/acs.langmuir.2c00836

Brutchey RL, Morse DE (2008) Silicatein and the translation of its molecular mechanism of biosilicification into low temperature nanomaterial synthesis. Chem Rev 108(11):4915–4934

Burton M (1928) Report on Some deep-sea sponges from the Indian Museum collected by the R. I. M. S. "Investigator". Part II. Tetraxonida (concluded) and Euceratosa. Rec Zool Surv India 30(1):110–138. https://doi.org/10.26515/rzsi/v30/i1/1928/163204

Bütschli O (1901) Einige Beobachtungen über Kiesel- und Kalknadeln von Spongien. Z Wiss Zool 69:235–286

Carballo JL, Cruz-Barraza JA, García-Gómez JC (2024) A revision of dispersal strategies in freshwater sponges: the journey of the Ponto-Caspian sponge *Rosulaspongilla rhadinaea* (Porifera: Spongillidae), a new alien species for Europe. Limnol Rev 24:577–592

Carter HJ (1849) Notes introductory to the study and classification of the Spongida. Ann Mag Nat Hist 4:81–100

Cárdenas P, Vacelet J, Chevaldonné P et al (2018) From marine caves to the deep sea, a new look at Caminella (Demospongiae, Geodiidae) in the Atlanto-Mediterranean region. Zootaxa 4466(1):174–196

Cha JN, Shimizu K, Zhou Y, Christiansen SC, Chmelka BF, Stucky GD, Morse DE (1999) Silicatein filaments and subunits from a marine sponge direct the polymerization of silica and silicones *in vitro*. Proc Natl Acad Sci USA 96(2):361–365. https://doi.org/10.1073/pnas.96.2.361

Chayaamor-Heil N (2023) From bioinspiration to biomimicry in architecture: opportunities and challenges. Encyclopedia 3:202–223. https://doi.org/10.3390/encyclopedia3010014

Chayaamor-Heil N, Perricone V, Gruber P, Guéna F (2023) Bioinspired, biobased and living material designs: a review of recent research in architecture and construction. Bioinspir Biomim 18:041001. https://doi.org/10.1088/1748-3190/acd82e

Chen H, Jia Z, Li L (2022) Lightweight lattice-based skeleton of the sponge *Euplectella aspergillum*: on the multifunctional design. J Mech Behav Biomed Mater 135:105448

Chen A, Porras L, Ma H, Hou X, Liu J, Wörheide G (2023) A new sponge genus from the Chengjiang biota with an intriguing combination of skeletal characters. Paläontol Z 97(3):443–450. https://doi.org/10.1007/s12542-023-00652-y

Chin YY, Prince J, Kendrick G, Abdul Wahab MA (2020) Sponges in shallow tropical and temperate reefs are important habitats for marine invertebrate biodiversity. Mar Biol 167(11):164. https://doi.org/10.1007/s00227-020-03771-1

Choi S, Kim JH, Kang TH, An YH, Lee SJ, Hwang NS, Kim S-H (2024) Biomimetic marine-sponge-derived spicule-microparticle-mediated biomineralization and YAP/TAZ pathway for bone regeneration in vivo. Biomater Res 28:0056. https://doi.org/10.34133/bmr.0056

Chu JW, Leys SP (2010) High resolution mapping of community structure in three glass sponge reefs (Porifera, Hexactinellida). Mar Ecol Prog Ser 417:97–113. https://doi.org/10.3354/meps08794

Chu JWF, Maldonado M, Yahel G, Leys SP (2011) Glass sponge reefs as a silicon sink. Mar Ecol Prog Ser 441:1–14. https://doi.org/10.3354/meps09381

Chun C (1900) Aus den Tiefen des Weltmeeres. Gustav Fischer, Jena

Claus C (1868) Über Chondrocladia, eine neue Form von Schwämmen. Sitzungsber Akad Wiss Wien 58:663–668

Cohn SA, Nash J, Pickett-Heaps JD (1989) The effect of drugs on diatom valve morphogenesis. Protoplasma 149:130–143

Colares RV (2021) Biosilica from deep-sea marine sponge Geodia barretti for the development of 3D printed scaffolds for bone tissue regeneration applications. ProQuest dissertations & theses, Instituto Politécnico de Leiria (Portugal)

Colgren J, Nichols SA (2022) MRTF specifies a muscle-like contractile module in Porifera. Nat Commun 13(1):4178. https://doi.org/10.1038/s41467-022-31756-9

Conway KW, Barrie JV, Austin WC, Luternauer JL (1991) Holocene sponge bioherms on the western Canadian continental shelf. Cont Shelf Res 11(8):771–790. https://doi.org/10.1016/0278-4343(91)90079-L

Conway KW, Krautter M, Barrie JV, Neuweiler M (2001) Hexactinellid sponge reefs on the Canadian continental shelf: a unique “living fossil”. Geosci Can 28:65–72. https://doi.org/10.12789/geocanj.2001.28.065

Croce G, Frache A, Milanesio M, Viterbo D, Bavestrello G, Benatti U, Giovine M, Amenitsch H (2003) Fiber diffraction study of spicules from marine sponges. Microsc Res Tech 62(4):378–381. https://doi.org/10.1002/jemt.10403

Croce G, Frache A, Milanesio M, Marchese L, Causà M, Viterbo D, Barbaglia A, Bolis V, Bavestrello G, Cerrano C, Benatti U, Pozzolini M, Giovine M, Amenitsch H (2004) Structural characterization of siliceous spicules from marine sponges. Biophys J 86:526–534. https://doi.org/10.1016/S0006-3495(04)74131-4

Dayton PK, Kim S, Jarrell SC, Oliver JS, Hammerstrom K, Fisher JL, O'Connor K, Barber JS, Robilliard G, Barry J, Thurber AR, Conlan K (2013) Recruitment, growth and mortality of an Antarctic hexactinellid sponge, *Anoxycalyx joubini*. PLoS One 8(2):e56939. https://doi.org/10.1371/journal.pone.0056939

de Voogd NJ, Alvarez B, Boury-Esnault N, Cárdenas P, Díaz MC, Dohrmann M, Downey R, Goodwin C, Hajdu E, Hooper JNA, Kelly M, Klautau M, Lim SC, Manconi R, Morrow C, Pinheiro U, Pisera A, Ríos P, Rützler K, Schönberg CHL, Vacelet J, van Soest RWM (2025) World Porifera Database. Accessed at https://www.marinespecies.org/porifera on 2025-03-30. https://doi.org/10.14284/359

De Vos L, Rützler K, Boury-Esnault N, Donadey C, Vacelet J (1991) Atlas of sponge morphology. Smithsonian Institution Press, Washington, DC

Deflandre G (1950) Contribution à l'étude des Silicoflagellidés actuels et fossiles. Microscopie 2:72–78, 117–142, 191–210

Dendy A (1921) The tetraxonid sponge spicule: a study in evolution. Acta Zool 2:95–152

Dendy A (1925) The origin of sponge-spicules. Nature 115

Dendy A (1926) On the origin, growth, and arrangement of sponge-spicules: a study in symbiosis. J Cell Sci s2–70(277):1–74

Dendy A, Burton M (1926) Report on some deep-sea sponges from the Indian Museum collected by R.I.M.S. "Investigator". Rec Indian Mus 28:225–248

Desqueyroux-Faúndez R, Van Soest RWM (1996) A review of Iophonidae, Myxillidae and Tedaniidae occurring in the South East Pacific (Porifera: Poecilosclerida). Rev Suisse Zool 103(1):3–79

Dohrmann M, Reiswig HM, Kelly M, Mills S, Schätzle S, Reverter M, Niesse N, Rohde S, Schupp P, Wörheide G (2023) Expanded sampling of New Zealand glass sponges (Porifera: Hexactinellida) provides new insights into biodiversity, chemodiversity, and phylogeny of the class. PeerJ 11:e15017. https://doi.org/10.7717/peerj.15017

Domingos C, Moraes FC, Muricy G (2013) Four new species of Plakinidae (Porifera: Homoscleromorpha) from Brazil. Zootaxa 3718(6):501–523

Donati V (1753) Auszug seiner Natur-Geschichte des Adriatischen Meers. CP Franckens, Halle

Donati G (2016) Bright sponges. Nat Photonics 10:625. https://doi.org/10.1038/nphoton.2016.199

Dos Santos Jorge Sousa K, Parisi JR, de Souza A, de Almeida Cruz M, Erbereli R, de Araújo Silva J, do Espirito Santo G, do Amaral GO, Martignago CCS, Fortulan CA, Granito RN, Renno ACM (2023) 3D printed scaffolds manufactured with biosilica from marine sponges for bone healing in a cranial defect in rats. Mar Biotechnol (NY) 25(2):259–271. https://doi.org/10.1007/s10126-023-10202-x

Drozdov AL, Zemnukhova LA, Panasenko AE, Polyakova NV, Slobodyuk AB, Ustinov AY, Didenko NA, Tyurin SA (2021) Silicon compounds in sponges. Appl Sci 11(14):6587. https://doi.org/10.3390/app11146587

Drozdov AL, Zhuravlev AY, Karpenko AA, Kolesnikov KA, Ivantsov AY (2022) The morphology and chemical composition of siliceous spicules of the early Cambrian sponge *Lenica unica* Gorjansky, 1977. Russ J Mar Biol 48(1):33–38. https://doi.org/10.1134/S1063074022010059

Drum RW (1968) Electron microscopy of siliceous spicules from the freshwater sponge *Heteromyenia*. J Ultrastruct Res 22:12–21

Du et al (2015) Potential Ediacaran sponge gemmules from the Yangtze Gorges area in South China. Gondwana Res 28(3):1246–1254

Duncan PM (1881) On some remarkable enlargements of the axial canals of sponge spicules and their causes. J R Microsc Soc 2(1):557–572

Egedeuzu CS, Taylor PG, Wong LS (2025) Investigating silicatein selectivity and specificity in silicon–oxygen bond condensation and metathesis. Cat Sci Technol 15(4):1009–1015. https://doi.org/10.1039/D4CY00985A

Ehrlich H (2010) Chitin and collagen as universal and alternative templates in biomineralization. Int Geol Rev 52:661–699. https://doi.org/10.1080/00206811003679521

Ehrlich H (2019) Marine biological materials of invertebrate origin. Springer International Publishing, Cham, pp 1–329. https://doi.org/10.1007/978-3-319-92483-0

Ehrlich H (2022) Actin and the realization of unique biosilica-based architectures in sponges. In: van der Wind N, Paix B, de Voog NJ, Hrab J (eds) 11th World sponge conference. Book of abstracts. Naturalis Biodiversity Center, p 247

Ehrlich H, Krautter M, Hanke T, Simon P, Knieb C, Heinemann S, Worch H (2007) First evidence of the presence of chitin in skeletons of marine sponges. Part II. Glass sponges (Hexactinellida: Porifera). J Exp Zool B Mol Dev Evol 308B:473–483. https://doi.org/10.1002/jez.b.21174

Ehrlich H, Deutzmann R, Brunner E, Cappellini E, Koon H, Solazzo C, Yang Y, Ashford D, Thomas-Oates J, Lubeck M, Baessmann C, Langrock T, Hoffmann R, Wörheide G, Reitner J, Simon P, Tsurkan M, Ereskovsky AV, Kurek D, Bazhenov VV, Hunoldt S, Mertig M, Vyalikh DV, Molodtsov SL, Kummer K, Worch H, Smetacek V, Collins MJ (2010) Mineralization of the meter-long biosilica structures of glass sponges is templated on hydroxylated collagen. Nat Chem 2:1084–1088. https://doi.org/10.1038/nchem.899

Ehrlich H, Brunner E, Simon P et al (2011) Calcite reinforced silica-silica joints in the biocomposite skeleton of the deep-sea glass sponge. Adv Funct Mater 21:3473–3481

Ehrlich H, Maldonado M, Parker AR, Kulchin YN, Schilling J, Koehler B, Skrzypczak U, Simon P, Reiswig HM, Tsurkan MV, Brunner E, Voznesenskiy SS, Bezverbny AV, Golik SS, Nagorny IG, Vyalikh DV, Makarova AA, Molodtsov SL, Kummer K, Mertig M, Erler C, Kurek DV, Bazhenov VV, Natalio F, Kovalev AE, Gorb SN, Stelling AL, Heitmann J, Born R, Meyer DC, Tabachnick KR (2016) Supercontinuum generation in naturally occurring glass sponges spicules. Adv Opt Mater 4(10):1608–1613. https://doi.org/10.1002/adom.201600454

Ehrlich H, Luczak M, Ziganshin R, Mikšík I, Wysokowski M, Simon P, Baranowska-Bosiacka I, Kupnicka P, Ereskovsky A, Galli R, Dyshlovoy S, Fischer J, Tabachnick KR, Petrenko I, Jesionowski T, Lubkowska A, Figlerowicz M, Ivanenko VN, Summers AP (2022) Arrested in glass: actin within sophisticated architectures of biosilica in sponges. Adv Sci 9:2105059. https://doi.org/10.1002/advs.202105059

Ehrlich H, Voronkina A, Tabachnick K, Kubiak A, Ereskovsky A, Jesionowski T (2024) Silactins and structural diversity of biosilica in sponges. Biomimetics 9(7):393. https://doi.org/10.3390/biomimetics9070393

Evans R (1899) The structure and metamorphosis of the larva of Spongilla lacustris. Q J Microsc Sci 42:363–476

Fairhead M, Johnson KA, Kowatz T, McMahon SA, Carter LG, Oke M, Liu H, Naismith JH, van der Walle CF (2008) Crystal structure and silica condensing activities of silicatein alpha-cathepsin L chimeras. Chem Commun (Camb) 15:1765–1767. https://doi.org/10.1039/b718264c

Falcucci G, Amati G, Bella G et al (2024) Adapting to the abyss: passive ventilation in the deep-sea glass sponge Euplectella aspergillum. Phys Rev Lett 132:208402

Farfan GA, McKeown DA, Post JE (2022) Mineralogical characterization of biosilicas versus geological analogs. Geobiology 20(1):59–77. https://doi.org/10.1111/gbi.12553

Federwisch L, Janussen D, Richter C (2020) Macroscopic characteristics facilitate identification of common Antarctic glass sponges (Porifera, Hexactinellida, Rossellidae). Polar Biol 43(2):91–110. https://doi.org/10.1007/s00300-019-02612-2

Fell PE (1974) Diapause in the gemmules of the marine sponge, Haliclona loosanoffi, with a note on the gemmules of *Haliclona oculata*. Biol Bull 147(2):333–351

Fell PE (2005) Salinity tolerance and desiccation resistance of the gemmules of the brackish-water sponge, *Haliclona loosanoffi*. J Exp Zool 194(2):409–412

Ferguson AM, Davis AR (2008) Heart of glass: spicule armament and physical defense in temperate reef sponges. Mar Ecol Prog Ser 372:77–86

Fernandes MC, Saadat M, Cauchy-Dubois P, Inamura C, Sirota T, Milliron G, Haj-Hariri H, Bertoldi K, Weaver JC (2021) Mechanical and hydrodynamic analyses of helical strake-like ridges in a glass sponge. J R Soc Interface 18(182):20210559. https://doi.org/10.1098/rsif.2021.0559

Ferretti A, Messori F, Di Bella M, Sabatino G, Quartieri S, Cavalazzi B, Italiano F, Barbieri R (2019) Armoured sponge spicules from Panarea Island (Italy): implications for their fossil preservation. Palaeogeogr Palaeoclimatol Palaeoecol 536:109379. https://doi.org/10.1016/j.palaeo.2019.109379

Frigeri LG, Radabaugh TR, Haynes PA, Hildebrand M (2006) Identification of proteins from a cell wall fraction of the diatom *Thalassiosira pseudonana*: insights into silica structure formation. Mol Cell Proteomics 5(1):182–193. https://doi.org/10.1074/mcp.M500174-MCP200

Fu G, Zhang C, Hall JW, Butler D (2023) Are sponge cities the solution to China's growing urban flooding problems? WIREs Water 10(1):e1613. https://doi.org/10.1002/wat2.1613

Fuchs I, Aluma Y, Ilan M, Mastai Y (2014) Induced crystallization of amorphous biosilica to cristobalite by silicatein. J Phys Chem B 118(8):2104–2111. https://doi.org/10.1021/jp4114317

Fuchs I, Aluma Y, Ilan M, Kityk IV, Mastai Y (2015) Photoinduced electro-optics measurements of biosilica transformation to cristobalite. J Solid State Chem 226:231–236. https://doi.org/10.1016/j.jssc.2015.03.001

Gabbai-Armelin PR, Kido HW, Cruz MA, Prado JPS, Avanzi IR, Custódio MR, Renno ACM, Granito RN (2019) Characterization and cytotoxicity evaluation of a marine sponge biosilica. Mar Biotechnol (NY) 21(1):65–75. https://doi.org/10.1007/s10126-018-9858-9

Galloway JM, Senior L, Fletcher JM, Beesley JL, Hodgson LR, Harniman RL, Mantell JM, Coombs J, Rhys GG, Xue WF, Mosayebi M, Linden N, Liverpool TB, Curnow P, Verkade P, Woolfson DN (2018) Bioinspired silicification reveals structural detail in self-assembled peptide cages. ACS Nano 12(2):1420–1432. https://doi.org/10.1021/acsnano.7b07785

Gao J, Nakamura F (2022) Actin-associated proteins and small molecules targeting the actin cytoskeleton. Int J Mol Sci 23(4):2118. https://doi.org/10.3390/ijms23042118

Gao K, Suleiman MT, Brown DG, Sadeghnejad A, Lin H (2024) Bio-inspired soil improvement using silicatein-α enzyme. Preprint (Version 1). Res Sq. https://doi.org/10.21203/rs.3.rs-3832275/v1

Godigamuwa K, Nakashima K, Tsujitani S, Yokoi K, Igarashi Y, Ohshiro T, Ooshima H, Kawasaki S (2021) Fabrication of silica on chitin in ambient conditions using silicatein fused with a chitin-binding domain. Bioprocess Biosyst Eng 44:1883–1890. https://doi.org/10.1007/s00449-021-02568-w

Goldstein J, Riisgård HU, Kealy RA, Funch P (2024) Particle loads, contractile responses and cleaning in the demosponge *Halichondria panicea*. J Exp Mar Biol Ecol 577:152021. https://doi.org/10.1016/j.jembe.2024.152021

Gong W, Liu L, Luo L, Ji L (2023) Preparation and characterization of a self-crosslinking sodium alginate-bioactive glass sponge. J Biomed Mater Res B Appl Biomater 111:173–183. https://doi.org/10.1002/jbm.b.35143

Görlich S, Samuel AJ, Best RJ, Seidel R, Vacelet J, Leonarski FK, Tomizaki T, Rellinghaus B, Pohl D, Zlotnikov I (2020) Natural hybrid silica/protein superstructure at atomic resolution. Proc Natl Acad Sci USA 117(49):31088–31093. https://doi.org/10.1073/pnas.2018489117

Gost M, Pinya S, Sureda A, Tejada S, Ferriol P (2023) Effect of alkalinity and light intensity on the growth of the freshwater sponge *Ephydatia fluviatilis* (Porifera: Spongillidae). Aquat Ecol 57:353–367. https://doi.org/10.1007/s10452-023-10014-0

Granito RN, Custódio MR, Rennó ACM (2017) Natural marine sponges for bone tissue engineering: the state of art and future perspectives. J Biomed Mater Res B Appl Biomater 105(6):1717–1727. https://doi.org/10.1002/jbm.b.33706

Gray JE (1868) Observations on sponges and on their arrangement and nomenclature. Ann Mag Nat Hist 4:161–172

Grinyó J, Aguzzi J, Ortenzi L, Kenchington E, Violino S, Hanz U, Santín A, Nattkemper TW, Mienis F (2024) What do glass sponges do when no one is looking? *Vazella pourtalesii*: responses to sediment deposition, passive locomotion, and contracting behavior. Deep-Sea Res I Oceanogr Res Pap 213:104388. https://doi.org/10.1016/j.dsr.2024.104388

Guo J, Li C, Ling S, Huang W, Chen Y, Kaplan DL (2017) Multiscale design and synthesis of biomimetic gradient protein/biosilica composites for interfacial tissue engineering. Biomaterials 145:44–55. https://doi.org/10.1016/j.biomaterials.2017.08.025

Gutiérrez-Castorena MC, Effland WR (2010) Pedogenic and biogenic siliceous features. In: Stoops G, Marcelino V, Mees F (eds) Interpretation of micromorphological features of soils and regoliths. Elsevier, pp 471–496

Haeckel E (1887) Report on the Radiolaria collected by H.M.S. Challenger during the years 1873–1876. Report of the scientific results voyage of H.M.S. Challenger. Zoology 18:1–1803

Han G, Chamberlain J, Webber C, Hannah C (2022) A new method of determining glass sponge reef adaptive management zones for the Hecate Strait and Queen Charlotte Sound Marine Protected Area. J Mar Sci Eng 10(7):971. https://doi.org/10.3390/jmse10070971

Hartman WD (1981) Form and distribution of silica in sponges. In: Simpson TL, Volcani BE (eds) Silicon and siliceous structures in biological systems. Springer, New York, pp 453–493. https://doi.org/10.1007/978-1-4615-9259-3_15

He M, Li Y, Yin J, Sun Q, Xiong W, Li S, Yang L, Hao L (2022) Compressive performance and fracture mechanism of bio-inspired heterogeneous glass sponge lattice structures manufactured by selective laser melting. Mater Des 214:110396. https://doi.org/10.1016/j.matdes.2021.110396

Hentschel E (1929) Die Kiesel- und Hornschwämme des Nordens. In: Grimpe G, Wagler E (eds) Die Tierwelt der Nord- und Ostsee, vol 11a. Akademische Verlagsgesellschaft, Leipzig, pp 1–92

Hill MS, Hill AL (2002) Morphological plasticity in the tropical sponge *Anthosigmella varians*: responses to predators and wave energy. Biol Bull (Woods Hole) 202:86–95

Hill MS, Lopez NA, Young KA (2005) Antipredator defenses in western North Atlantic sponges with evidence of enhanced defense through interactions between spicules and chemicals. Mar Ecol Prog Ser 291:93–102

Hinde GJ (1887–1893) A monograph of the British fossil sponges. Palaeontogr Soc Monogr 40–47:1–92

Hirabayashi S, Koike Y (2008) Silicone rubber sponge composition. US Patent Application. US20080214688A1, 4 Sept 2008

Hirose Y, Tang SL, Yamashiro H (2024) Arrangement and development of spicules in the coral-killing sponge, *Terpios hoshinota*. Zool Stud 63:e26

Humann P, DeLoach N (2002) Reef creature identification: Florida, Caribbean, Bahamas. New World Publications, Jacksonville

Ijima I (1895) On two new Hexactinellida from Sagami Bay. Zool Mag 7:93–96

Ijima I (1901) Studies on the Hexactinellida. Contribution I. Mem Coll Sci Imp Univ Tokyo 15:1–55

Ijima I (1902) Studies on the Hexactinellida. Contribution II. Mem Coll Sci Imp Univ Tokyo 17:1–55

Ijima I (1903) Studies on the Hexactinellida. Contribution III. Mem Coll Sci Imp Univ Tokyo 19:1–55

Ijima I (1927) The Hexactinellida of the Siboga expedition. Siboga Expeditie 3:1–383

Ilan M, Gugel J, Galil BS, Janussen D (2003) Small bathyal sponge species from east Mediterranean revealed by non-regular soft bottom sampling technique. Ophelia 57(3):145–160

Imsiecke G, Steffen R, Custódio MR, Borojevic R, Müller WEG (1995) Formation of spicules by sclerocytes from the freshwater sponge Ephydatia muelleri in short-term cultures in vitro. In Vitro Cell Dev Biol Anim 31(6):528–535. https://doi.org/10.1007/BF02634030

Janussen D, Tabachnick KR, Tendal OS (2004) Deep-sea Hexactinellida (Porifera) of the Weddell Sea. Deep Sea Res II 51:1857–1882

Jetton TL, Fell PE, Harrison FW (1987) Cytological and cytochemical investigations of development from dormant gemmules of the marine sponge *Haliclona loosanoffi*. J Morphol 193(1):99–116

Jesionowski T et al (2018) Marine Spongin: naturally prefabricated 3D scaffold-based biomaterial. Mar Drugs 16(3):88

Jia C, Li L, Liu Y, Fang B, Ding H, Song J, Liu Y, Xiang K, Lin S, Li Z, Si W, Li B, Sheng X, Wang D, Wei X, Wu H (2020) Highly compressible and anisotropic lamellar ceramic sponges with superior thermal insulation and acoustic absorption performances. Nat Commun 11:3732. https://doi.org/10.1038/s41467-020-17533-6

Jiang H, Dong Z, Xia X, Li X (2023) Cathepsins in oral diseases: mechanisms and therapeutic implications. Front Immunol 14:1203071. https://doi.org/10.3389/fimmu.2023.1203071

Jiang M, Deng W, Lin H (2024) Sustainability through biomimicry: a comprehensive review of bionic design applications. Biomimetics 9:507. https://doi.org/10.3390/biomimetics9090507

Jones WC (1984) Spicule dimensions as taxonomic criteria in the identification of haplosclerid sponges from the shores of Anglesey. Zool J Linnean Soc 80:239–259. https://doi.org/10.1111/j.1096-3642.1984.tb01976.x

Jørgensen CB (1944) On the spicule-formation of *Spongilla lacustris* (L.). I. The dependence of the spicule-formation on the content of dissolved and solid silicic acid of the milieu. K Dan Vid Selsk Biol Medd 19(7):1–45

Kahn AS, Yahel G, Chu JWF, Tunnicliffe V, Leys SP (2018) Benthic grazing and carbon sequestration by deep-water glass sponge reefs. Limnol Oceanogr 63(1):260–273. https://doi.org/10.1002/lno.10002

Kaluzhnaya OV, Belikova AS, Podolskaya EP, Krasko AG, Müller WEG, Belikov SI (2007) Identification of silicateins in freshwater sponge *Lubomirskia baicalensis*. Mol Biol 41(4):554–558. https://doi.org/10.1134/S002689330704005X

Karcz J, Woźnica A, Binkowski M, Klonowska-Olejnik M, Bernas T, Karczewski J, Migula P (2015) SEM-EDS i mikrotomografia rentgenowska jako narzędzia do analizy struktury szkieletu i ciała słodkowodnej gąbki *Spongilla lacustris* z jeziora Goczałkowickiego. Folia Histochem Cytobiol 53(1):88–95. https://doi.org/10.5603/FHC.a2015.0002

Kent WS (1870) On the "Hexactinellidae" or Hexradiate Spiculed Silicious Sponges taken in the "Norna" expedition off the coast of Spain and Portugal. With description of new species, and revision of the order. Mon Microsc J 4:241–252, pls LXIII–LXV

Keren R, Mayzel B, Lavy A, Polishchuk I, Levy D, Fakra SC, Pokroy B, Ilan M (2017) Sponge-associated bacteria mineralize arsenic and barium on intracellular vesicles. Nat Commun 8:14393. https://doi.org/10.1038/ncomms14393

Kertmen A, Petrenko I, Schimpf C, Rafaja D, Petrova O, Sivkov V, Nekipelov S, Fursov A, Stelling AL, Heimler K, Rogoll A, Vogt C, Ehrlich H (2021) Calcite nanotuned chitinous skeletons of giant *Ianthella basta* marine demosponge. Int J Mol Sci 22:12588. https://doi.org/10.3390/ijms222212588

Kim B-H (2006) Silicone sponge and preparation method and usage thereof. US Patent Application. US20060063846A1, 23 Mar 2006

Kirkpatrick R (1907) Preliminary report on the Monaxonellida of the National Antarctic Expedition. Ann Mag Nat Hist 20(7):271–291. https://doi.org/10.1080/00222930709487333

Kozhemyako VB, Veremeichik GN, Shkryl YN, Kovalchuk SN, Krasokhin VB, Rasskazov VA, Zhuravlev YN, Bulgakov VP, Kulchin YN (2009) Silicatein genes in spicule-forming and nonspicule-forming Pacific demosponges. Mar Biotechnol 12:403–409. https://doi.org/10.1007/s10126-009-9225-y

Krautter M, Conway KW, Barrie JV (2006) Recent hexactinosidan sponge reefs (silicate mounds) off British Columbia, Canada: frame-building processes. J Paleontol 80:38–48

Kröger N, Deutzmann R, Bergsdorf C, Sumper M (2000) Species-specific polyamines from diatoms control silica morphology. Proc Natl Acad Sci USA 97(26):14133–14138. https://doi.org/10.1073/pnas.260496497

Kulchin YN et al (2007) Optical properties of natural biominerals—the spicules of the glass sponges. Opt Mem Neural Networks 16:189–197

Kulchin YN, Bezverbny AV, Bukin OA, Voznesensky SS, Galkina AN, Drozdov AL, Nagorny IG (2009) Optical and nonlinear optical properties of sea glass sponge spicules. Prog Mol Subcell Biol 47:315–340. https://doi.org/10.1007/978-3-540-88552-8_14

Lage A, Muricy G, Ruiz C, Pérez T (2018) New sciaphilic plakinids (Porifera, Homoscleromorpha) from the Central-Western Pacific. Zootaxa 4466:8–38. https://doi.org/10.11646/zootaxa.4466.1.4

Langenbruch PF (1983) Body structure of marine sponges. Mar Biol 75:319–325. https://doi.org/10.1007/BF00406018
Larsen PS, Riisgård HU (1994) The sponge pump. J Theor Biol 168(1):53–63. https://doi.org/10.1006/jtbi.1994.1087
Lavrov AI, Ereskovsky AV (2022) Studying Porifera WBR using the Calcerous sponges Leucosolenia. In: Blanchoud S, Galliot B (eds) Whole-body regeneration, Methods in molecular biology, vol 2450. Springer, New York, pp 69–93. https://doi.org/10.1007/978-1-0716-2172-1_4
Leys SP (2003) Comparative study of spiculogenesis in demosponge and hexactinellid larvae. Microsc Res Tech 62(4):300–311
Leys SP, Yahel G, Reidenbach MA, Tunnicliffe V, Shavit U, Reiswig HM (2011) The sponge pump: the role of current induced flow in the design of the sponge body plan. PLoS One 6(12):e27787. https://doi.org/10.1371/journal.pone.0027787
Li L, Li B, Wu L, Zhao X, Zhang J (2014) Magnetic, superhydrophobic and durable silicone sponges and their applications in removal of organic pollutants from water. Chem Commun 50(58):7831–7833. https://doi.org/10.1039/c4cc03045a
Li L, Hu T, Yang Y, Zhang J (2019) Strong, compressible, bendable and stretchable silicone sponges by solvent-controlled hydrolysis and polycondensation of silanes. J Colloid Interface Sci 540:554–562. https://doi.org/10.1016/j.jcis.2019.01.059
Li L, Guo C, Chen Y, Chen Y (2020) Optimization design of lightweight structure inspired by glass sponges (Porifera, Hexacinellida) and its mechanical properties. Bioinspir Biomim 15(3):036006. https://doi.org/10.1088/1748-3190/ab6ca9
Liang X, Zhang J, Ou H, Chen J, Mitragotri S, Chen M (2020) Skin delivery of siRNA using sponge spicules in combination with cationic flexible liposomes. Mol Ther Nucleic Acids 20:639–648. https://doi.org/10.1016/j.omtn.2020.04.003
Liang X, Zhang J, Zhang C, Zhai H, Yang P, Chen M (2024) Mesoporous silica coated spicules for photodynamic therapy of metastatic melanoma. J Nanobiotechnology 22:179. https://doi.org/10.1186/s12951-024-02471-y
Liu Y, Zou M, Qi Y, Chen L, Wang Z, Song J, He L (2024) Bionic design of thin-walled bilinear tubes with excellent crashworthiness inspired by glass sponge structures. Bioinspir Biomim 19(4):046002. https://doi.org/10.1088/1748-3190/ad580a
Loomis SH (2010) Diapause and estivation in sponges. In: Arturo Navas C, Carvalho J (eds) Aestivation, Progress in molecular and subcellular biology, vol 49. Springer, Berlin, Heidelberg. https://doi.org/10.1007/978-3-642-02421-4_11
Lopez-Acosta M, Leynaert A, Grall J, Maldonado M (2018) Silicon consumption kinetics by marine sponges: an assessment of their role at the global scale. Limnol Oceanogr 63(6):2653–2669. https://doi.org/10.1002/lno.10956
Łukowiak M (2020) Utilizing sponge spicules in taxonomic, ecological and environmental reconstructions: a review. PeerJ 8:e10601. https://doi.org/10.7717/peerj.10601
Łukowiak M, Pisera A, Stefanska T, Stefanskyi V (2022a) High diversity of siliceous sponges in Western Tethyan areas during the Eocene: palaeobiogeographical, ecological and taxonomic significance. Pap Palaeontol 8(1):e1416. https://doi.org/10.1002/spp2.1416
Łukowiak M, Van Soest R, Klautau M, Pérez T, Pisera A, Tabachnick K (2022b) The terminology of sponge spicules. J Morphol 283(12):1517–1545. https://doi.org/10.1002/jmor.21520
Łukowiak M et al (2023) Miocene sponge assemblages in the face of the Messinian Salinity Crisis—new data from the Atlanto-Mediterranean seaway. Peer J 11:e16277
Macchiagodena M, Fragai M, Gallo A, Pagliai M, Ravera E (2024) The role of lysozyme in the formation of bioinspired silicon dioxide. Chemistry 30(38):e202401249. https://doi.org/10.1002/chem.202401249
Maldonado M et al (2010) Selective feeding by sponges on pathogenic microbes: a reassessment of potential for abatement of microbial pollution. Mar Ecol Prog Ser 403:75–89
Maldonado M, Navarro L, Grasa A, González A, Vaquerizo I (2011) Silicon uptake by sponges: a twist to understanding nutrient cycling on continental margins. Sci Rep 1:30. https://doi.org/10.1038/srep00030

Maldonado M, Ribes M, van Duyl FC (2012) Nutrient fluxes through sponges: biology, budgets, and ecological implications. Adv Mar Biol 62:113–182. https://doi.org/10.1016/B978-0-12-394283-8.00003-5
Maldonado M, López-Acosta M, Abalde S, Martos I, Ehrlich H, Leynaert A (2022) On the dissolution of sponge silica: assessing variability and biogeochemical implications. Front Mar Sci 9:1005068. https://doi.org/10.3389/fmars.2022.1005068
Manconi R, Pronzato R (2007) Gemmules as a key structure for the adaptive radiation of freshwater sponges: a morpho-functional and biogeographical study. In: Custódio MR, Lôbo-Hajdu G, Hajdu E, Muricy M (eds) Porifera research: biodiversity, innovation, sustainability, Ser Livros, vol 28. Museu Nacional, Rio de Janeiro, pp 61–77
Manconi R, Cubeddu T, Pronzato R (2016) Australian freshwater sponges with a new species of Pectispongilla (Porifera: Demospongiae: Spongillida). Zootaxa 4196(1):143–159. https://doi.org/10.11646/zootaxa.4196.1.3
Manconi R, Copeland J, Kunigelis S, Pronzato R (2022) Biodiversity of Nearctic inland water: discovery of the genus *Heterorotula* (Porifera, Spongillida, Spongillidae) in the Appalachian Mountains, with biogeographical implications and description of new species. ZooKeys 1110:103–120
Marron AO, Ratcliffe S, Wheeler GL, Goldstein RE, King N, Not F, de Vargas C, Richter DJ (2016) The evolution of silicon transport in eukaryotes. Mol Biol Evol 33(12):3226–3248. https://doi.org/10.1093/molbev/msw209
Marshall W (1875) Untersuchungen über Hexactinelliden. Z Wiss Zool 25(Supplement 2):142–243, pls XI–XVII
Masse S, Pisera A, Laurent G, Coradin T (2016) A solid state NMR investigation of recent marine siliceous sponge spicules. Minerals 6(1):21. https://doi.org/10.3390/min6010021
Matsunaga S, Sakai R, Jimbo M, Kamiya H (2007) Long-chain polyamines (LCPAs) from marine sponge: possible implication in spicule formation. ChemBioChem 8(14):1729–1735. https://doi.org/10.1002/cbic.200700305
Mayzel B, Aizenberg J, Ilan M (2014) The elemental composition of Demospongiae from the Red Sea, Gulf of Aqaba. PLoS One 9(4):e95775. https://doi.org/10.1371/journal.pone.0095775
McCartney K, Witkowski J, Harwood DM (2014) New insights into skeletal morphology of the oldest known silicoflagellates: Variramus, Cornua and Gleserocha gen. nov. Rev Micropaléontol 57:75–91
McCutchin CA, Edgar KJ, Chen CL, Dove PM (2025) Silica–biomacromolecule interactions: toward a mechanistic understanding of silicification. Biomacromolecules 26:43–84. https://doi.org/10.1021/acs.biomac.4c00674
Mcheik A, Cassaignon S, Livage J, Gibaud A, Berthier S, Lopez PJ (2018) Optical properties of nanostructured silica structures from marine organisms. Front Mar Sci 5:123. https://doi.org/10.3389/fmars.2018.00123
McMurray SE, Blum JE, Pawlik JR (2008) Redwood of the reef: growth and age of the giant barrel sponge Xestospongia muta in the Florida Keys. Mar Biol 155:159–171
Meng S, Wu N, Fang J, Yu Y, Tang X, Wang Y, Deng X, Qi C, Kong T, Ding T, Liu Z (2025) Construction of a biomimetic tubular scaffold inspired by sea sponge structure: sponge-like framework and cell guidance. Adv Sci (Weinh) 12:e2416627. https://doi.org/10.1002/advs.202416627
Miao X, Sun D (2009) Graded/gradient porous biomaterials. Materials 2(2):613–650. https://doi.org/10.3390/ma3010026
Müller WEG (2006) Siliceous spicules in marine demosponges (example Suberites domuncula). Micron 37(2):107–120
Müller WEG, Wang X, Belikov SI et al (2007a) Silicateins, the major biosilica-forming enzymes present in demosponges: protein analysis and phylogenetic relationship. Gene 395:62–71
Müller WE, Schlossmacher U, Eckert C et al (2007b) Analysis of the axial filament in spicules of the demosponge Geodia cydonium: different silicatein composition in microscleres (asters) and megascleres (oxeas and triaenes). Eur J Cell Biol 86(8):473–487

Müller WE, Boreiko A, Schlossmacher U et al (2007c) Fractal-related assembly of the axial filament in the demosponge Suberites domuncula: relevance to biomineralization and the formation of biogenic silica. Biomaterials 28(30):4501–4511

Minchin EA (1900) Chapter III. Sponges. In: Lankester ER (ed) A treatise on zoology. Part II. The Porifera and Coelenterata. Adam and Charles Black, London, pp 1–178

Minchin EA (1909) Sponge-spicules. A summary of present knowledge. Ergebn Fortschr Zool 2:171–274

Minchin EA (1910) On some Parasites observed in the Rat-flea (Ceratophyllus fasciatus). Festschrift zum sechzigsten Geburtstag Richard Hertwigs 1:289–302, plate 23

Miserez A, Weaver JC, Thurner PJ, Aizenberg J, Dauphin Y, Fratzl P, Morse DE, Zok FW (2008) Effects of laminate architecture on fracture resistance of sponge biosilica: lessons from nature. Adv Funct Mater 18:1–8. https://doi.org/10.1002/adfm.200701135

Mohri K, Nakatsukasa M, Masuda Y, Agata K, Funayama N (2008) Toward understanding the morphogenesis of siliceous spicules in freshwater sponge: differential mRNA expression of spicule-type-specific silicatein genes in *Ephydatia fluviatilis*. Dev Dyn 237:3024–3039. https://doi.org/10.1002/dvdy.21708

Monn MA (2018) Insights into the mechanical functions of glass sponge spicules through a characterization of their strength and toughness properties. PhD thesis, Brown University. https://repository.library.brown.edu/studio/item/bdr:792721/

Monn M, Kesari H (2017) A new structure-property connection in the skeletal elements of the marine sponge *Tethya aurantia* that guards against buckling instability. Sci Rep 7:39547. https://doi.org/10.1038/srep39547

Monn MA, Weaver JC, Zhang T, Aizenberg J, Kesari H (2015) New functional insights into the internal architecture of the laminated anchor spicules of *Euplectella aspergillum*. Proc Natl Acad Sci USA 112(16):4976–4981. https://doi.org/10.1073/pnas.1415502112

Monn MA, Vijaykumar K, Kochiyama S, Kesari H (2020) Lamellar architectures in stiff biomaterials may not always be templates for enhancing toughness in composites. Nat Commun 11:373. https://doi.org/10.1038/s41467-019-14128-8

Morankar S, Sundaram Singaravelu AS, Niverty S, Mistry Y, Penick CA, Bhate D, Chawla N (2022) Tensile and fracture behavior of silica fibers from the Venus flower basket (*Euplectella aspergillum*). Int J Solids Struct 253:111622. https://doi.org/10.1016/j.ijsolstr.2022.111622

Morankar S, Kumar A, Luktuke A, Chawla N (2023) Influence of hydration on the micromechanical properties of silica spicules from the Venus flower basket (*Euplectella aspergillum*). JOM 75:3816–3826. https://doi.org/10.1007/s11837-023-05966-5

Morganti TM, Purser A, Rapp HT, German CR, Jakuba MV, Hehemann L, Blendl J, Slaby BM, Boetius A (2021) In situ observation of sponge trails suggests common sponge locomotion in the deep central Arctic. Curr Biol 31(8):R368–R370. https://doi.org/10.1016/j.cub.2021.03.014

Mugnaioli E, Natalio F, Schloßmacher U, Wang X, Müller WEG, Kolb U (2009) Crystalline nanorods as possible templates for the synthesis of amorphous biosilica during spicule formation in Demospongiae. ChemBioChem 10(4):683–689. https://doi.org/10.1002/cbic.200800623

Müller WEG, Rothenberger M, Boreiko A, Tremel W, Reiber A, Schröder HC (2005) Formation of siliceous spicules in the marine demosponge *Suberites domuncula*. Cell Tissue Res 321:285–297. https://doi.org/10.1007/s00441-005-1141-5

Müller WEG, Boreiko A, Schloßmacher U, Wang X, Eckert C, Kropf K, Li J, Schröder HC (2008a) Identification of a silicatein(-related) protease in the giant spicules of the deep-sea hexactinellid *Monorhaphis chuni*. J Exp Biol 211(3):300–305. https://doi.org/10.1242/jeb.008193

Müller WEG, Wang X, Kropf K, Boreiko A, Schloßmacher U, Brandt D, Schröder HC, Wiens M (2008b) Silicatein expression in the hexactinellid *Crateromorpha meyeri*: the lead marker gene restricted to siliceous sponges. Cell Tissue Res 333:339–351. https://doi.org/10.1007/s00441-008-0624-6

Müller WEG, Schröder HC, Muth S, Gietzen S, Korzhev M, Grebenjuk VA, Wiens M, Schloßmacher U, Wang X (2013) The silicatein propeptide acts as inhibitor/modulator

of self-organization during spicule axial filament formation. FEBS J 280(7):1693–1708. https://doi.org/10.1111/febs.12183
Müller WEG, Link T, Schröder HC, Korzhev M, Neufurth M, Brandt D, Wang X (2014) Dissection of the structure-forming activity from the structure-guiding activity of silicatein: a biomimetic molecular approach to print optical fibers. J Mater Chem B 2:5368–5377. https://doi.org/10.1039/C4TB00801D
Müller WEG, Neufurth M, Wang S, Schröder HC, Wang X (2024) The physiological inorganic polymers biosilica and polyphosphate as key drivers for biomedical materials in regenerative nanomedicine. Int J Nanomedicine 19:1303–1337. https://doi.org/10.2147/IJN.S446405
Muricy G (2011) Diversity of Indo-Australian Plakortis (Demospongiae: Plakinidae), with description of four new species. J Mar Biol Assoc UK 91:303–319
Muricy G, Boury-Esnault N, Bézac C, Vacelet J (1998) Taxonomic revision of the Mediterranean *Plakina* Schulze (Porifera, Demospongiae, Homoscleromorpha). Zool J Linnean Soc 124:169–203. https://doi.org/10.1111/j.1096-3642.1998.tb00572.x
Murphy MB, Blashki D, Buchanan RM, Fan D, De Rosa E, Shah RN, Stupp SI, Weiner BK, Simmons PJ, Ferrari MF, Tasciotti E (2011) Multi-composite bioactive osteogenic sponges featuring mesenchymal stem cells, platelet-rich plasma, nanoporous silicon enclosures, and peptide amphiphiles for rapid bone regeneration. J Funct Biomater 2(2):39–66. https://doi.org/10.3390/jfb2020039
Murr MM, Morse DE (2005) Fractal intermediates in the self-assembly of silicatein filaments. Proc Natl Acad Sci USA 102:11657–11662. https://doi.org/10.1073/pnas.050396810
Naik RR, Whitlock PW, Rodriguez F, Brott LL, Glawe DD, Clarson SJ, Stone MO (2003) Controlled formation of biosilica structures in vitro. Chem Commun 9:238–239. https://doi.org/10.1039/B210635C
Natalio F, Wiese S, Friedrich N, Werner P, Tahir MN (2014) Localization and characterization of ferritin in Demospongiae: a possible role on spiculogenesis. Mar Drugs 12(8):4659–4676. https://doi.org/10.3390/md12084659
Noll FC (1888) Beiträge zur Naturgeschichte der Kieselschwämme. Abh Senckenb Naturforsch Ges 15:58
Ogawa T, Koyama S, Omori T, Kikuchi K, de Maleprade H, Goldstein RE, Ishikawa T (2024) The architecture of sponge choanocyte chambers maximizes mechanical pumping efficiency. bioRxiv. https://doi.org/10.1101/2024.02.22.581376
Okada Y (1928) On the development of a hexactinellid sponge *Farrea sollasii*. J Fac Sci Imp Univ Tokyo Sect IV Zool 2(1):1–27
Orani AM, Barats A, Vassileva E, Thomas OP (2018) Marine sponges as a powerful tool for trace elements biomonitoring studies in coastal environment. Mar Pollut Bull 131:633–645. https://doi.org/10.1016/j.marpolbul.2018.04.073
Owen R (1841) Description of a new genus and species of sponge (*Euplectella aspergillum*). Proc Zool Soc Lond 9:3–5. https://www.biodiversitylibrary.org/part/232717
Palmer WG (1930) The action of aqueous hydrofluoric acid on silica. J Chem Soc (Resumed):1656–1664. https://doi.org/10.1039/JR9300001656
Pinheiro U, Silva C, Calheira L (2015) Sanidaster in freshwater sponges: an unexpected spicule for the birotuled genus *Corvoheteromeyenia* Ezcurra de Drago, 1979. Zoosystema 37(3):449–456. https://doi.org/10.5252/z2015n3a2
Pisera A (2006) Palaeontology of sponges—a review. Can J Zool 84(2):242–261. https://doi.org/10.1139/z05-169
Polini A, Pagliara S, Camposeo A, Cingolani R, Wang XH, Schröder HC, Müller WEG, Pisignano D (2012) Optical properties of in-vitro biomineralised silica. Sci Rep 2:607. https://doi.org/10.1038/srep00607
Polley C, Distler T, Scheufler C, Detsch R, Lund H, Springer A, Schneidereit D, Friedrich O, Boccaccini AR, Seitz H (2023) 3D printing of piezoelectric and bioactive barium titanate–bioactive glass scaffolds for bone tissue engineering. Mater Today Bio 21:100719. https://doi.org/10.1016/j.mtbio.2023.100719

Povarova NV, Baranov MS, Kovalchuk SN, Semiletova IV, Lukyanov KA, Kozhemyako VB (2015) A novel water-soluble substrate for silicateins. Russ J Bioorg Chem 41(3):338–339. https://doi.org/10.1134/S1068162015030073

Povarova NV, Barinov NA, Baranov MS, Markina NM, Varizhuk AM, Pozmogova GE, Klinov DV, Kozhemyako VB, Lukyanov KA (2018) Efficient silica synthesis from tetra(glycerol)orthosilicate with cathepsin. Sci Rep 8:16849. https://doi.org/10.1038/s41598-018-34965-9

Pozdnyakov IR, Sokolova AM, Ereskovsky AV, Karpov SA (2018) Kinetid structure in sponge choanocytes of Spongillida in the light of evolutionary relationships within Demospongiae. Zool J Linnean Soc 184(2):255–272. https://doi.org/10.1093/zoolinnean/zlx109

Pozzolini M, Miksik I, Ghignone S, Oliveri C, Tassara E, Giovine M (2022) The lysosome origin of biosilica machinery in the Demospongiae model *Petrosia ficiformis* (Poiret, 1789). Front Mar Sci 9:850653. https://doi.org/10.3389/fmars.2022.850653

Radons J, Biewusch U, Grässel S, Geuze HJ, Hasilik A (1994) Distinctive inhibition of the lysosomal targeting of lysozyme and cathepsin D by drugs affecting pH gradients and protein kinase C. Biochem J 302:581–586. https://doi.org/10.1042/bj3020581

Rahmanisa S, Prajatelistia E, Wibowo I, Barlian A (2022) 3D biosilica scaffolds from *Melophlus sarasinorum* and *Xestospongia testudinaria* Indonesian sponges are biocompatible for cell growth and differentiation of human Wharton's jelly mesenchymal stem cell in bone tissue engineering. Indones Biomed J 14(4):329–441. https://doi.org/10.18585/inabj.v14i4.1895

Rasbold GG, Calheira L, Domingos-Luz L, Pessenda LCR, Pinheiro U, McGlue MM (2023) A morphological guide of neotropical freshwater sponge spicules for paleolimnological studies. Front Ecol Evol 10:1067432. https://doi.org/10.3389/fevo.2022.1067432

Reiswig HM (1971) Particle feeding in natural populations of three marine demosponges. Biol Bull 141(3):568–591. https://doi.org/10.2307/1540270

Reiswig HM (1975) The aquiferous systems of three marine demospongiae. J Morphol 145(3):493–502. https://doi.org/10.1002/jmor.1051450407

Reiswig HM (2002) Family Tretodictyidae Schulze, 1886. In: Hooper JNA, Van Soest RWM (eds) Systema Porifera: a guide to the classification of sponges. Springer, New York, pp 1341–1354. https://doi.org/10.1007/978-1-4615-0747-5_137

Reiswig H, Kelly M (2018) The marine fauna of New Zealand: euplectellid glass sponges (Hexactinellida, Lyssacinosida, Euplectellidae). NIWA Biodivers Mem 130:1–170.

Ribes M, Yahel G, Romera-Castillo C, Mallenco R, Morganti TM, Coma R (2023) The removal of dissolved organic matter by marine sponges is a function of its composition and concentration: An in situ seasonal study of four Mediterranean species. Sci Total Environ 871:161991. https://doi.org/10.1016/j.scitotenv.2023.161991

Riesgo A, Maldonado M, Durfort M (2007) Dynamics of gametogenesis, embryogenesis, and larval release in a Mediterranean homosclerophorid demosponge. Mar Freshw Res 58(4):398–417. https://doi.org/10.1071/MF06052

Riesgo A, Maldonado M, López-Legentil S, Giribet G (2015) A proposal for the evolution of cathepsin and silicatein in sponges. J Mol Evol 80:278–291. https://doi.org/10.1007/s00239-015-9682-z

Rieuwpassa FJ, Tomasoa AM, Palawe JFP, Rieuwpassa F, Mege RA, Balansa W (2023) A new and practical method for measuring sponge spicules. J Ilm Platax 11(2):322–332. https://doi.org/10.35800/jip.v11i2.47882

Riisgård HU, Larsen PS (1995) Filter-feeding in marine macro-invertebrates: pump characteristics, modelling and energy cost. Biol Rev 70(1):67–106. https://doi.org/10.1111/j.1469-185X.1995.tb01440.x

Riisgård HU, Larsen PS (2022) Actual and model-predicted growth of sponges-With a bioenergetic comparison to other filter-feeders. J Mar Sci Eng 10:607

Rohde S, Schupp PJ (2012) Growth and regeneration of the elephant ear sponge *Ianthella basta*. Hydrobiologia 687:219–226. https://doi.org/10.1007/s10750-011-0774-5

Rützler K, Piantoni C, Van Soest RWM, Díaz C (2014) Diversity of sponges (Porifera) from cryptic habitats on the Belize barrier reef near Carrie Bow Cay. Zootaxa 3805:1–129. https://doi.org/10.11646/zootaxa.3805.1.1
Sabella C, Faszewski E, Himic L, Colpitts KM, Kaltenbach J, Burger MM, Fernàndez-Busquets X (2007) Cyclosporin A suspends transplantation reactions in the marine sponge *Microciona prolifera*. J Immunol 179:5927–5935. https://doi.org/10.4049/jimmunol.179.9.5927
Sánchez-Puig N, Cuéllar-Cruz M, Islas SR, Tapia-Vieyra JV, Arreguín-Espinosa RA, Moreno A (2021) The influence of silicateins on the shape and crystalline habit of silica carbonate biomorphs of alkaline earth metals (Ca, Ba, Sr). Crystals 11:438. https://doi.org/10.3390/cryst11040438
Sandford F (2003) Physical and chemical analysis of the siliceous skeletons in six sponges of two groups (Demospongiae and Hexactinellida). Microsc Res Tech 62(4):336–355
Santín A et al (2021) Unique spicules may confound species differentiation: taxonomy and biogeography of Melonanchora Carter, 1874 and two new related genera (Myxillidae: Poecilosclerida) from the Okhotsk Sea. Peer J 9:e12515
Sarikaya M, Fong H, Sunderland N, Flinn BD, Mayer G, Mescher A, Gaino E (2001) Biomimetic model of a sponge-spicular optical fiber—mechanical properties and structure. J Mater Res 16(5):1420–1428. https://doi.org/10.1557/JMR.2001.0198
Schill RO, Pfannkuchen M, Fritz G, Köhler HR, Brümmer F (2006) Quiescent gemmules of the freshwater sponge, *Spongilla lacustris* (Linnaeus, 1759), contain remarkably high levels of Hsp70 stress protein and hsp70 stress gene mRNA. J Exp Zool A Comp Exp Biol 305(5):449–457. https://doi.org/10.1002/jez.a.281
Schmid AMM (1980) Valve morphogenesis in diatoms: a pattern-related filamentous system in pennates and the effect of APM, colchicine, and osmotic pressure. Nova Hedwigia 33:811–847
Schmidt WJ (1926) Über das Wesen der Lamellierung und das gegenseitige Verhalten von organischer und anorganischer Substanz bei den Kieselschwammnadeln. Zool Jb Abt Anat Ontogenie Tiere 48:311–364
Schoeppler V, Reich E, Vacelet J, Rosenthal M, Pacureanu A, Rack A, Zaslansky P, Zolotoyabko E, Zlotnikov I (2017) Shaping highly regular glass architectures: a lesson from nature. Sci Adv 3(10):eaao2047. https://doi.org/10.1126/sciadv.aao2047
Schönberg CHL (2021) No taxonomy needed: sponge functional morphologies inform about environmental conditions. Ecol Indic 129:107806. https://doi.org/10.1016/j.ecolind.2021.107806
Schröder K (1936) Beiträge zur Kenntnis der Spiculabildung, der Larven-Spiculation und der Variationsbreite der Gerüstnadeln von Süsswasserschwämmen. Z Morph Ökol Tiere 31:245–267
Schulze FE (1880) Untersuchungen über den Bau und die Entwicklung der Spongien. Z Wiss Zool 34:407–451
Schulze FE (1885) Über den Bau und das System der Hexactinelliden. Abh Königl Preuss Akad Wiss:1–97
Schulze FE (1886) Über den Bau und das System der Hexactinelliden. II. Abh Königl Preuss Akad Wiss:1–74
Schulze FE (1887) Report on the Hexactinellida collected by H.M.S. "Challenger" during the years 1873–76. Report on the scientific results of the voyage of H.M.S. Challenger. Zoology 21(53):1–514
Schulze FE (1896) Hexactinelliden des Indischen Ozeans: II. Theil Die Hexasterophora. Abh Preuss Akad Wiss 3:1–92.
Schulze FE (1899) Amerikanische Hexactinelliden, nach dem Materiale der Albatross-Expedition. Fischer, Jena, pp 1–126, pls I–XIX
Schulze FE (1900) Hexactinelliden des Indischen Oceanes. III. Theil. Abh Preuss Akad Wiss Berlin 1900:1–46, pls I–VII
Schulze P (1923) Beiträge zur Kenntnis der Kieselnadelbildung, besonders bei den Spongilliden. Arch Zellforsch 17:105–130
Schulze P (1925) Zum morphologischen Feinbau der Kieselschwammnadeln. Z Morph Ökol Tiere 4:615–625

Schulze FE (1904) Hexactinellida. Wissenschaftliche Ergebnisse der Deutschen Tiefsee-Expedition auf dem Dampfer "Valdivia" 1898–1899. Separated Atlas 4:1–266

Shimizu K, Morse DE (2018) Silicatein: a unique silica-synthesizing catalytic triad hydrolase from marine sponge skeletons and its multiple applications. Methods Enzymol 605:429–455. https://doi.org/10.1016/bs.mie.2018.02.025

Shimizu K, Cha J, Stucky GD, Morse DE (1998) Silicatein α: cathepsin L-like protein in sponge biosilica. Proc Natl Acad Sci USA 95:6234–6238. https://doi.org/10.1073/pnas.95.11.6234

Shimizu K, Amano T, Bari MR, Weaver JC, Arima J, Mori N (2015) Glassin, a histidine-rich protein from the siliceous skeletal system of the marine sponge *Euplectella*, directs silica polycondensation. Proc Natl Acad Sci USA 112(37):11449–11454. https://doi.org/10.1073/pnas.1506968112

Shimizu K, Nishi M, Sakate Y, Kawanami H, Bito T, Arima J, Leria L, Maldonado M (2024) Silica-associated proteins from hexactinellid sponges support an alternative evolutionary scenario for biomineralization in Porifera. Nat Commun 15(1):181. https://doi.org/10.1038/s41467-023-44226-7

Shkryl YN, Bulgakov VP, Veremeichik GN, Kovalchuk SN, Kozhemyako VB, Kamenev DG, Semiletova IV, Timofeeva YO, Shchipunov YA, Kulchin YN (2016) Bioinspired enzymatic synthesis of silica nanocrystals provided by recombinant silicatein from the marine sponge *Latrunculia oparinae*. Bioprocess Biosyst Eng 39:53–58. https://doi.org/10.1007/s00449-015-1488-2

Shoham S, Keren R, Lavy A, Polishchuk I, Pokroy B, Ilan M (2024) Out of the blue: hyperaccumulation of molybdenum in the Indo-Pacific sponge *Theonella conica*. Sci Adv 10(29):eadn3923. https://doi.org/10.1126/sciadv.adn3923

Shore RE (1972) Axial filament of siliceous sponge spicules: its organic components and synthesis. Biol Bull 143:689–698. https://doi.org/10.2307/1540191

Sidor LM, Beaulieu MM, Rasskazov I, Acarturk BC, Ren J, Jenen E, Kamoen L, Vázquez Vitali M, Carney PS, Schmidt GR, Srubar WV III, Abbondanzieri EA, Meyer AS (2024) Engineered bacteria that self-assemble bioglass polysilicate coatings display enhanced light focusing. Proc Natl Acad Sci USA 121(51):e2409335121. https://doi.org/10.1073/pnas.2409335121

Simpson TL (1989) Silification processes in sponges: Geodia asters and the problem of morphogenesis of spicule shape. In: Crick RE (ed) Origin, evolution and modern aspects of biomineralization in plants and animals. Kluwer, New York, pp 125–136. https://doi.org/10.1007/978-1-4757-6114-6_9

Simpson TL, Gilbert JJ (1974) Gemmulation, gemmule hatching, and sexual reproduction in freshwater sponges. II. Life cycle events in young, larva-produced sponges. Trans Am Microsc Soc 93(4):437–451. https://doi.org/10.2307/3225217

Simpson TL, Langenbruch PF, Garrone R (1985) Sponge gemmule coats: germanium (Ge) modification of a collagenous structure. In: Bairati A, Garrone R (eds) Biology of invertebrate and lower vertebrate collagens, NATO ASI series, vol 93. Springer, Boston, pp 279–290. https://doi.org/10.1007/978-1-4684-7636-1_13

Sola-Rabada A, Michaelis M, Oliver DJ, Roe MJ, Colombi Ciacchi L, Heinz H, Perry CC (2018) Interactions at the silica-peptide interface: influence of the extent of functionalization on the conformational ensemble. Langmuir 34(28):8255–8263. https://doi.org/10.1021/acs.langmuir.8b00874

Sollas WJ (1888) Report on the Tetractinellida collected by H.M.S. Challenger during the years 1873–1876. Report on the scientific results of the voyage of H.M.S. Challenger during the years 1873–76. Zoology 25(63):1–458, pls 1–44

Stevenson A, Archer SK, Schultz JA, Dunham A, Marliave JB, Martone P, Harley CDG (2020) Warming and acidification threaten glass sponge *Aphrocallistes vastus* pumping and reef formation. Sci Rep 10:8176. https://doi.org/10.1038/s41598-020-65220-9

Stone RP, Conway KW, Csepp DJ et al (2013) The boundary reefs: glass sponge (Porifera: Hexactinellida) reefs on the international border between Canada and the United States. U.S. Department of Commerce, NOAA Technical Memorandum NMFS-AFSC 264, 31 p

Strunge K, Hoinkis N, Lutz H, Alamdari S, Roeters SJ, Lu H, Pfaendtner J, Weidner T (2022) Peptide mimic of the marine sponge protein silicatein fabricates ultrathin nanosheets of silicon dioxide and titanium dioxide. Langmuir 38(26):8087–8093. https://doi.org/10.1021/acs.langmuir.2c00918
Suarez PA, Leys SP (2022) The sponge pump as a morphological character in the fossil record. Paleobiology 48(3):446–461. https://doi.org/10.1017/pab.2021.43
Sundar VC, Grazul J, Aizenberg J (2003) Structural and initial optical characterization of basalia spicules in the glass sponge *Euplectella*. MRS Online Proc Libr 774:43
Tabachnick KR (2002a) Family Leucopsacidae Ijima, 1903. In: Hooper JNA, Van Soest RWM (eds) Systema Porifera: a guide to the classification of sponges. Springer, New York, pp 1435–1440
Tabachnick KR (2002b) Family Rossellidae Schulze, 1885. In: Hooper JNA, Van Soest RWM (eds) Systema Porifera: a guide to the classification of sponges. Springer, New York, pp 1441–1505
Tabachnick KR, Collins AG (2008) Glass sponges (Porifera, Hexactinellida) of the northern Mid-Atlantic Ridge. Mar Biol Res 4:25–47
Tabachnick KR, Reiswig HM (2002) Dictionary of Hexactinellida. In: Hooper JNA et al (eds) Systema Porifera: a guide to the classification of sponges. Springer, New York, pp 1224–1230
Tabachnick KR, Menshenina LL, Pisera A, Ehrlich H (2011) Revision of *Aspidoscopulia* Reiswig, 2002 (Porifera: Hexactinellida: Farreidae) with description of two new species. Zootaxa 2883(1):1–22. https://doi.org/10.11646/zootaxa.2883.1.1
Tabachnick K, Janussen D, Menshenina L (2017) Cold biosilicification in metazoan: psychrophilic glass sponges. In: Ehrlich H (ed) Extreme biomimetics. Springer, Cham, pp 53–80
Tahir MN, Théato P, Müller WEG, Schröder HC, Borejko A, Faiss S, Janshoff A, Huth J, Tremel W (2005) Formation of layered titania and zirconia catalysed by surface-bound silicatein. Chem Commun 44:5533–5535. https://doi.org/10.1039/B510113A
Tang Q, Wan B, Yuan X, Muscente AD, Xiao S (2019) Spiculogenesis and biomineralization in early sponge animals. Nat Commun 10:3348. https://doi.org/10.1038/s41467-019-11297-4
Tesson B, Hildebrand M (2010) Extensive and intimate association of the cytoskeleton with forming silica in diatoms: control over patterning on the meso- and micro-scale. PLoS One 5(12):e14300. https://doi.org/10.1371/journal.pone.0014300
Thoulet J (1884) Étude sur les spongiaires siliceux de l'Éocène du Cotentin. Bull Soc Linn Normandie 8:1–40
Tréguer P, Pondaven P (2000) Silica control of carbon dioxide. Nature 406(6794):358–359. https://doi.org/10.1038/35019236
Tsurkan D, Wysokowski M, Petrenko I, Voronkina A, Khrunyk Y, Fursov A, Ehrlich H (2020) Modern scaffolding strategies based on naturally pre-fabricated 3D biomaterials of poriferan origin. Appl Phys A Mater Sci Process 126:382. https://doi.org/10.1007/s00339-020-03564-9
Turner E (2021) Possible poriferan body fossils in early Neoproterozoic microbial reefs. Nature 596(7870):87–91. https://doi.org/10.1038/s41586-021-03773-z
Uriz MJ, Turon X, Becerro MA, Agell G (2003) Siliceous spicules and skeleton frameworks in sponges: origin, diversity, ultrastructural patterns, and biological functions. Microsc Res Tech 62(4):279–299. https://doi.org/10.1002/jemt.10395
Vacelet J, Boury-Esnault N (1995) Carnivorous sponges. Nature 373(6512):333–335. https://doi.org/10.1038/373333a0
Vacelet J, Duport E (2004) Prey capture and digestion in the carnivorous sponge *Asbestopluma hypogea* (Porifera: Demospongiae). Zoomorphology 123(4):179–190. https://doi.org/10.1007/s00435-004-0100-0
Vacelet J, Perez T, Hooper JNA (2002) Demospongiae incertae sedis: Myceliospongia Vacelet & Perez, 1998. In: Hooper JNA, Van Soest RWM, Willenz P (eds) Systema Porifera. Springer, Boston, MA
Van de Meene AML, Pickett-Heaps JD (2004) Valve morphogenesis in the centric diatom *Rhizosolenia setigera* (Bacillariophyceae, Centrales) and its taxonomic implications. Eur J Phycol 39(1):93–104. https://doi.org/10.1080/09670260310001646522

van den Heuvel DB, Stawski TM, Tobler DJ, Wirth R, Peacock CL, Benning LG (2018) Formation of silica-lysozyme composites through co-precipitation and adsorption. Front Mater 5:19. https://doi.org/10.3389/fmats.2018.00019
Van Soest RW, Beglinger EJ, De Voogd NJ (2010) Skeletons in confusion: a review of astrophorid sponges with (dicho-)calthrops as structural megascleres (Porifera, Demospongiae, Astrophorida). Zookeys 68:1–88
Van Soest R, Carballo JL, Hooper J (2012) Polyaxone monaxonids: revision of raspailiid sponges with polyactine megascleres (Cyamon and Trikentrion). ZooKeys 239:1–70. https://doi.org/10.3897/zookeys.239.3734
Varamogianni-Mamatsi D, Anastasiou TI, Vernadou E, Papandroulakis N, Kalogerakis N, Dailianis T, Mandalakis M (2021) A multi-species investigation of sponges' filtering activity towards marine microalgae. Mar Drugs 20(1):24. https://doi.org/10.3390/md20010024
Verbrugghe N, Rubinacci E, Khan AZ (2023) Biomimicry in architecture: a review of definitions, case studies, and design methods. Biomimetics 8:107. https://doi.org/10.3390/biomimetics8010107
Villani M (2008) Use of spongilla spicules as a dermabrasion device or resurfacing modality. WO2008043071A2. World Intellectual Property Organization
Vinogradov AP (1953) The elementary chemical composition of marine organisms. Mem Sears Found Mar Res 2:1–647
Vogel S (1977) Current-induced flow through living sponges in nature. Proc Natl Acad Sci USA 74(5):2069–2071. https://doi.org/10.1073/pnas.74.5.2069
Vogel S, Bretz WL (1972) Interfacial organisms: passive ventilation in the velocity gradients near surfaces. Science 175(4018):210–211. https://doi.org/10.1126/science.175.4018.210
von Lendenfeld R (1894) A monograph of the horny sponges. Trans Linnean Soc Lond Zool 6:1–936
von Lendenfeld R (1915) The sponges. E. Schweizerbart'sche Verlagsbuchhandlung, Stuttgart
Voronkina A, Romańczuk-Ruszuk E, Przekop RE, Lipowicz P, Gabriel E, Heimler K, Rogoll A, Vogt C, Frydrych M, Wienclaw P, Stelling AL, Tabachnick K, Tsurkan D, Ehrlich H (2023) Honeycomb biosilica in sponges: from understanding principles of unique hierarchical organization to assessing biomimetic potential. Biomimetics 8(2):234. https://doi.org/10.3390/biomimetics8020234
Voronkina A, Cárdenas P, Adam J, Meissner H, Nowacki K, Joseph Y, Tabachnick KR, Ehrlich H (2025) Biosilica 3D micromorphology of Geodiidae sponge spicules is patterned by F-actin. Microsc Res Tech 88(2):202–213. https://doi.org/10.1002/jemt.24798
Vosmaer GJ (1902) On the shape of some siliceous spicules of sponges. Proc Sect Sci K Ned Akad Wet 5:104–114
Vosmaer GCJ, Wijsman HP (1905) On the structure of some siliceous spicules of sponges. I. The styli of Tethya lyncurium. Kon Ned Akad Wetensch Proc 8:15–28
Wallace AF, DeYoreo JJ, Dove PM (2009) Kinetics of silica nucleation on carboxyl- and amine-terminated surfaces: insights for biomineralization. J Am Chem Soc 131(14):5244–5250. https://doi.org/10.1021/ja809486b
Wang X, Schröder HC, Müller WEG (2009) Giant siliceous spicules from the deep-sea glass sponge *Monorhaphis chuni*. Int Rev Cell Mol Biol 273:69–115. https://doi.org/10.1016/S1937-6448(08)01803-0
Wang XH, Schloßmacher U, Jochum KP, Gan L, Stoll B, Uriz MJ, Müller WEG (2010) Silica-protein composite layers of the giant basal spicules from *Monorhaphis*: basis for their mechanical stability. Pure Appl Chem 82(1):175–193. https://doi.org/10.1351/PAC-CON-09-01-07
Wang X, Schloßmacher U, Wiens M, Batel R, Schröder HC, Müller WEG (2012) Silicateins, silicatein interactors and cellular interplay in sponge skeletogenesis: formation of glass fiber-like spicules. FEBS J 279:1721–1736. https://doi.org/10.1111/j.1742-4658.2012.08533.x
Wang X, Tolba E, Schröder HC, Neufurth M, Feng Q, Diehl-Seifert B, Müller WEG (2014) Effect of bioglass on growth and biomineralization of SaOS-2 cells in hydrogel after 3D cell bioprinting. PLoS One 9(11):e112497. https://doi.org/10.1371/journal.pone.0112497

Wang S, Xia H, Liu Y (2022) Energy absorption characteristics of polygonal bio-inspired honeycomb column thin-walled structure under quasi-static uniaxial compression loading. Biomimetics 7:201. https://doi.org/10.3390/biomimetics7040201

Wang Y, Huang H, Zhao Y et al (2020) Self-assembly of ultralight and compressible inorganic sponges with hierarchical porosity by electrospinning. Ceram Int 46:768–774

Weaver JC, Morse DE (2003) Molecular biology of demosponge axial filaments and their role in biosilicification. Microsc Res Tech 62:256–367

Weaver JC, Pietrasanta LI, Hedin N, Chmelka BF, Hansma PK, Morse DE (2003) Nanostructural features of demosponge biosilica. J Struct Biol 144:271–281. https://doi.org/10.1016/j.jsb.2003.09.031

Weaver J et al (2007) Hierarchical assembly of the siliceous skeletal lattice of the hexactinellid sponge Euplectella aspergillum. J Struct Biol 158(1):93–106

Weaver JC, Milliron GW, Allen P, Miserez A, Rawal A, Garay J, Thurner PJ, Seto J, Mayzel B, Friesen LJ, Chmelka BF, Fratzl P, Aizenberg J, Dauphin Y, Kisailus D, Morse DE (2010) Unifying design strategies in demosponge and hexactinellid skeletal systems. J Adhes 86(1):72–95. https://doi.org/10.1080/00218460903417917

Werner P, Blumtritt H, Natalio F (2017) Organic crystal lattices in the axial filament of silica spicules of Demospongiae. J Struct Biol 198(3):186–195

Weissenfels N (1992) The filtration apparatus for food uptake in freshwater sponges (Porifera, Spongillidae). Zoomorphology 112:51–55. https://doi.org/10.1007/BF01632994

Wilkinson CR, Garrone R (1980) Ultrastructure of siliceous spicules and microsclerocytes in the marine sponge *Neofibularia irata* N. sp. J Morphol 166(1):51–63. https://doi.org/10.1002/jmor.1051660105

Winder SJ, Ayscough KR (2005) Actin-binding proteins. J Cell Sci 118:651–654. https://doi.org/10.1242/jcs.01667

Woesz A, Weaver JC, Kazanci M, Dauphin Y, Aizenberg J, Morse DE, Fratzl P (2006) Micromechanical properties of biological silica in skeletons of deep-sea sponges. J Mater Res 21:2068–2078. https://doi.org/10.1557/jmr.2006.0251

Woodland W (1907) Studies in spicule formation. VIII. Some observations on the scleroblastic development of Hexactinellid and other siliceous sponge spicules. Q J Microsc Sci s2-52(205):139–151. https://doi.org/10.1242/jcs.s2-52.205.139

Woźnica A, Karczewski J, Gwiazda R, Karcz J, Bernaś T, Libera M, Babczyńska A, Łozowski B (2023) Combined complementary imaging techniques in morphological analysis of *Spongilla lacustris* (Linnaeus 1759). Limnol Oceanogr Methods 21(3):141–151. https://doi.org/10.1002/lom3.10534

Wulff J (2016) Sponge contributions to the geology and biology of reefs: past, present, and future. In: Hubbard D, Rogers C, Lipps J, Stanley G Jr (eds) Coral reefs at the crossroads, Coral reefs of the world, vol 6. Springer, Dordrecht, pp 195–224. https://doi.org/10.1007/978-94-017-7567-0_5

Wysokowski M, Jesionowski T, Ehrlich H (2018) Biosilica as a source for inspiration in biological materials science. Am Mineral 103:665–691. https://doi.org/10.2138/am-2018-6259

Wysokowski M, Machałowski T, Petrenko I, Schimpf C, Rafaja D, Galli R, Ziętek J, Pantović S, Voronkina A, Kovalchuk V, Ivanenko VN, Hoeksema BW, Díaz C, Khrunyk Y, Stelling AL, Giovine M, Jesionowski T, Ehrlich H (2020) 3D chitin scaffolds of marine demosponge origin for biomimetic mollusk hemolymph-associated biomineralization ex-vivo. Mar Drugs 18(2):123. https://doi.org/10.3390/md18020123

Xu C, Wei Z, Gao H, Bai Y, Liu H, Yang H, Lai Y, Yang L (2017) Bioinspired mechano-sensitive macroporous ceramic sponge for logical drug and cell delivery. Adv Sci (Weinh) 4(6):1600410. https://doi.org/10.1002/advs.201600410

Yadati T, Houben T, Bitorina A, Shiri-Sverdlov R (2020) The ins and outs of cathepsins: physiological function and role in disease management. Cells 9(7):1679. https://doi.org/10.3390/cells9071679

Yahel G, Eerkes-Medrano DI, Leys SP (2006) Size-independent selective filtration of ultraplankton by hexactinellid glass sponges. Aquat Microb Ecol 45(2):181–194. https://doi.org/10.3354/ame045181

Zhang Y, Reed BW, Chung FR, Koski KJ (2016) Mesoscale elastic properties of marine sponge spicules. J Struct Biol 193(1):67–74. https://doi.org/10.1016/j.jsb.2015.11.009

Zhang K, Yu L, Li FR, Li X, Wang Z, Zou X, Zhang C, Lv K, Zhou B, Mitragotri S, Chen M (2020) Topical application of exosomes derived from human umbilical cord mesenchymal stem cells in combination with sponge spicules for treatment of photoaging. Int J Nanomedicine 15:2859–2872. https://doi.org/10.2147/IJN.S249751

Zhang C, Duan J, Huang Y, Chen M (2021) Enhanced skin delivery of therapeutic peptides using spicule-based topical delivery systems. Pharmaceutics 13(12):2119. https://doi.org/10.3390/pharmaceutics13122119

Zhang X et al (2024) A 3D-printed cuttlefish bone elastomeric sponge rapidly controlling noncompressible hemorrhage. Small 20(19):e2307041

Zimmermann P, Schulze P, Beck-Sickinger AG, Khrunyk Y (2024) Design and biofunctionalization of cloud sponge-inspired scaffolds for enhanced bone cell performance. ACS Appl Bio Mater 7(12):8281–8293. https://doi.org/10.1021/acsabm.4c01065

Zittel KA (1877) Studien über fossile Spongien. Abh Bayer Akad Wiss 13:1–63

Zlotnikov I, Masic A, Dauphin Y, Fratzl P, Zolotoyabko E (2014) Composition and mechanical properties of a protein/silica hybrid material forming the micron-thick axial filament in the spicules of marine sponges. Adv Eng Mater 16(9):1073–1077. https://doi.org/10.1002/adem.201300485

Chapter 3
From Biomechanics to Biomimetics: Lessons from the Molluscan Radula

Wencke Krings and Stanislav N. Gorb

Abstract Radular teeth of molluscs, with their extraordinary functional differentiation, represent the pinnacle of sophisticated integration of form and function across the scales, shaped by millions of years of meticulous evolutionary refinement. In this paper, we reviewed their astonishingly diverse morphologies, material compositions, and biomechanical properties that provide a treasure trove of knowledge that can be applied to biomimetic innovation. In the recent literature, researchers studied the radula's structural intelligence and its exceptional optimization for strength, flexibility, and efficiency, which could lead to the design of prototypes of advanced materials and tools with durability, precision, and adaptability—particularly at the micro- and nanoscale. This exploration into the form and function of radular teeth offers transformative directions for future advancements in biomimetic engineering, contributing to the development of revolutionary solutions across diverse industries.

Keywords Functional morphology · Mechanical properties · Teeth · Feeding · Chitin · Composition

W. Krings (✉)
Department of Cariology, Endodontology and Periodontology, Leipzig University, Leipzig, Germany

Department of Functional Morphology and Biomechanics, Zoological Institute, Kiel University, Kiel, Germany

Department of Electron Microscopy, University of Hamburg, Hamburg, Germany

Section of Mammalogy and Paleoanthropology, Leibniz Institute for the Analysis of Biodiversity Change, Hamburg, Germany
e-mail: wencke.krings@uni-hamburg.de

S. N. Gorb
Department of Functional Morphology and Biomechanics, Zoological Institute, Kiel University, Kiel, Germany

V. Bels et al. (eds.), *Biomimicry – A Functional Approach*, Biologically-Inspired Systems 20, https://doi.org/10.1007/978-3-032-11949-0_3

3.1 Introduction

Nature has long served as an inspiration for human innovation, providing countless examples of structures and systems optimized for function, efficiency, and resilience. Biological systems—ranging from the microscopic scale of cellular components to the macroscopic forms of organisms—offer sophisticated solutions to complex challenges. These natural designs, refined over millions of years through evolution, have become a cornerstone of biomimetics, the field dedicated to emulating biological principles for technological advancement [e.g. Bhushan 2009; Banken and Oeffner 2023; du Plessis et al. 2019; Yang and Kotov 2024; Hwang et al. 2015; Suresh Kumar et al. 2020; Bhushan 2018; Nosonovsky and Rohatgi 2012; Vincent 2009].

Biological structures exhibit unique characteristics, such as hierarchical organization, multifunctionality, adaptability, and energy efficiency. For instance, the lightweight yet robust architecture of bird bones has inspired the development of advanced aerospace materials and structures [e.g. Sullivan et al. 2017], while the self-cleaning properties of lotus leaves have led to the development of superhydrophobic technical surfaces [e.g. Collins and Safiuddin 2022]. Additionally, the study of spider silk, known for its extraordinary tensile strength and elasticity, has paved the way for innovations in high-performance polymeric materials (Chan et al. 2020). Studies on attachment devices of animals resulted in the new class of adhesive materials capable of adhesion without glue (Gorb et al. 2007; Heepe and Gorb 2014). In this context, biomimetic approaches leverage the insights gained from biological structures to address challenges in diverse fields, including engineering, medicine, robotics, and sustainable design. By understanding and applying these principles, researchers and engineers aim to develop solutions that are not only highly effective but also environmentally friendly, mirroring nature's inherent efficiency.

The radula, the feeding apparatus of molluscs, represents a remarkable example of biological adaptation and functionality, exemplifying the connection between form, material properties, and function in a biological system. Comprising rows of intricately arranged microscopic teeth embedded in a flexible, chitinous membrane, the radula combines durability, flexibility, and efficiency to facilitate a broad diversity of tasks, such as scraping, cutting, or grinding a vast range of food and substrates. It is important to note that different types of radulae are tailored to the needs of each particular species representing an Eldorado of biological solutions. This manuscript reviews recent literature on different aspects of the functional morphology of radula to show its potential to contribute to human technology.

3.2 Functional Morphology of Radulae

The phylum Mollusca has long evolutionary history spanning over 550 million years, allowing for a remarkable diversification of species. This is especially evident within the class Gastropoda, which includes approximately 80,000 extant species

(Bouchet et al. 2005; Parkhaev 2008). This class nicely exemplifies adaptability, thriving in extraordinarily diverse habitats, from marine and freshwater environments to extreme settings, such as deserts, hydrothermal vents, and urban areas. A key factor driving this incredible diversification is the radula—an organ vital for mechanical food acquisition and processing, enabling species to occupy a vast array of ecological niches.

The radula consists of a chitinous membrane with rows of embedded chitinous teeth, supported by structures called odontophoral cartilages or radular bolsters, which provide mechanical support and aid in the failure prevention of the radula (Guralnick and Smith 1999; Neustadter et al. 2002a; Montroni et al. 2019; Krings and Gorb 2021a; Krings et al. 2021a; Krings et al. 2021b). Together with the surrounding muscles, stabilizing cartilages, and sometimes the jaws, these components form the molluscan feeding apparatus or buccal mass (Fig. 3.1a–c). Each tooth is subdivided into three morphological components: (1) the basis, anchoring the tooth to the radular membrane; (2) the cusp, directly interacting with food; and (3) the stylus, an elongated structure connecting the basis and cusp. During foraging, the cusps engage directly with natural ingesta—food, substrate, minerals, and other materials—resulting in some degree of tooth wear. However, continuous secretion in the radular building zone by epithelial cells regenerates both teeth and membrane [e.g. Runham 1962; Runham and Isarankura 1966; Mackenstedt and Märkel 1987; Padilla et al. 1996; Shaw et al. 2008], ensuring that the radula remains exceptionally functional.

Initially considered a universal trait among molluscs, the radula was later recognized as a distinctively defining feature of the phylum and a pivotal tool for taxonomic classification. Over time, classification has evolved to emphasize highly distinct radular types. This system categorizes teeth by their form into "tooth types" in relation to their location (central, lateral, or marginal) and their quantity within each transversal tooth row (Gray 1853; Hyman 1967; Steneck and Watling 1982; Nixon 1995; Haszprunar and Götting 2007). Some distinct and highly specialized key types of radulae include the following:

(a) Stenoglossan radulae, found in Neogastropoda (Caenogastropoda), consist of one central tooth and two laterals, forming a precisely functional configuration (e.g. *Anentome helena* and *Buccinum undatum*; Fig. 3.1d, e).
(b) Homodont radulae, present in Cephalopoda, are characterized by a central tooth flanked by one lateral and two marginal teeth, all of similar shape, exemplifying functional uniformity (e.g. *Loligo vulgaris*; Fig. 3.1g).
(c) Taenioglossan radulae, common in most Caenogastropoda, feature one central tooth flanked by one lateral and two marginal teeth on each side (e.g. *Cleopatra johnstoni*, *Leloupiella minima*, and *Littorina littorea*; Fig. 3.1h–i, k).
(d) Isodont radulae, present in various Heterobranchia species, are distinguished by dozens to hundreds of similarly shaped teeth in each row (e.g. *Planorbella duryi* and *Felimare picta*; Fig. 3.1l, m).
(e) Rhipidoglossan radulae, found in Vetigastropoda and Neritimorpha, feature numerous intricately small marginal teeth, a larger central tooth, and usually

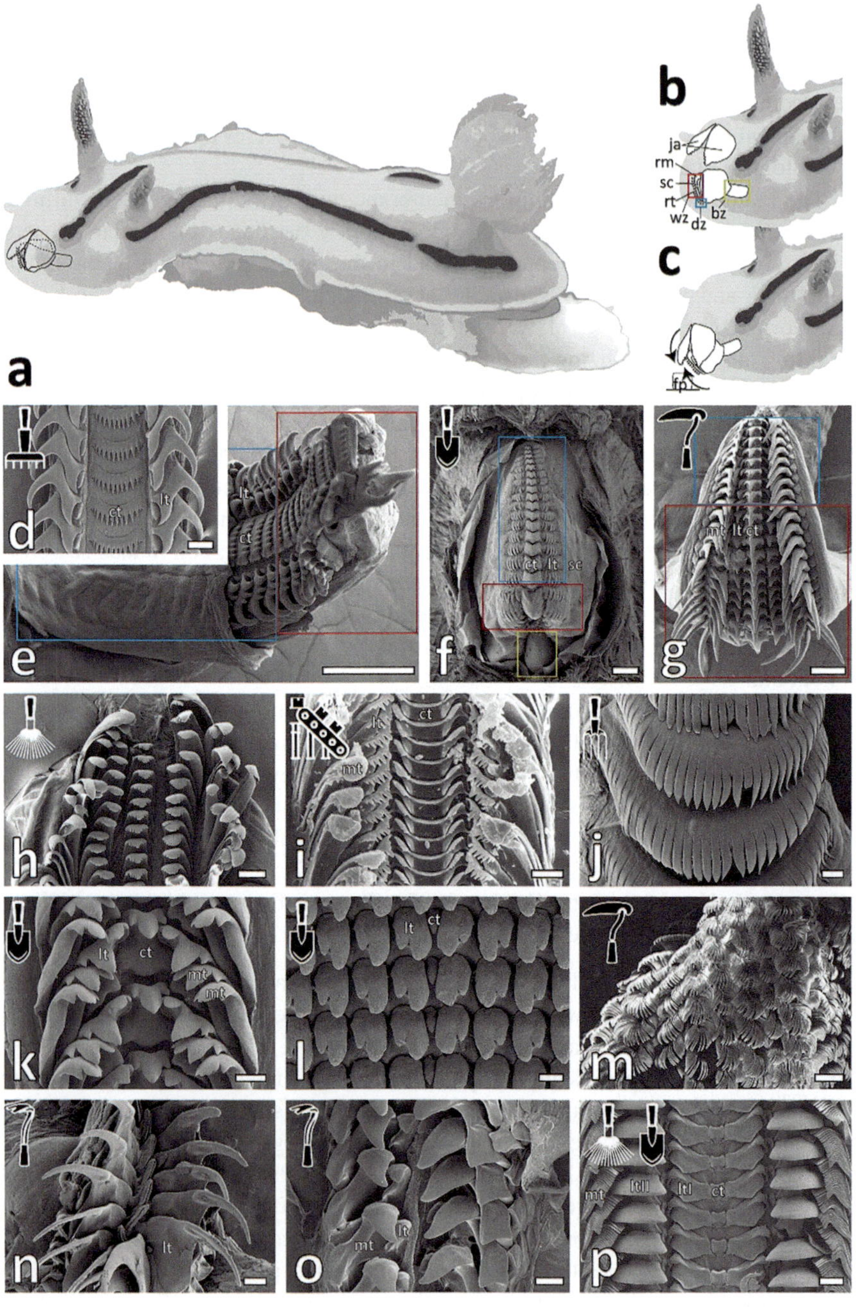

Fig. 3.1 Ontogeny and general structure of radular types. (**a–c**) Schematic illustration of the buccal mass at rest (**a**) and during feeding (**c**) in the gastropod *Chromodoris willani*. (**b**) The major buccal mass components are highlighted, including the radula and, present in some taxa, the jaws. The yellow box marks the building zone, where initial secretion occurs; the red box represents the

five lateral teeth, forming a meticulously optimized design (e.g. *Vittina turrita*; Fig. 3.1p).

(f) Docoglossan radulae, found in Polyplacophora and Patellogastropoda, typically have up to 13 teeth per row, including two large and prominent "dominant teeth or lateral teeth II" (e.g. *Lepidochitona cinerea*; Fig. 3.2a).

Many radulae, such as those found in certain gastropods like *Dendronotus lacteus* (Fig. 3.1f), *Aeolidia papillosa* (Fig. 3.1j), *Onchidoris bilamellata* (Fig. 3.1n), *Polycera quadrilineata* (Fig. 3.1o), display exquisitely unique morphologies that do not fit these existing categories.

Tooth morphology is not only influenced by phylogeny, but also by feeding ecology, exhibiting extraordinary variation in size, shape, and aspect ratios based on the diet and enabling optimal and highly specialized feeding performance [e.g. Steneck and Watling 1982; Solem 1974; Walsby 1975; Burch and Jeong 1984; Blinn et al. 1989; Hawkins et al. 1989; Duda et al. 2001; Padilla 2003; Ukmar-Godec et al. 2015; Ekimova et al. 2019; Krings et al. 2021c; Krings and Gorb 2024a, b].

For example, in highly specialized carnivorous species (Fig. 3.1d–g, j, m), such as the gastropods *Anentome helena*, *Buccinum undatum*, *Dendronotus lacteus*, *Aeolidia papillosa*, *Felimare picta*, or the cephalopod *Loligo vulgaris*, the teeth are

Fig. 3.1 (continued) working zone, the only active area used for foraging; and the blue box indicates the degenerative zone, where worn teeth are discarded. (**d**–**p**) Scanning electron microscopy (SEM) images showcase various radular types. Icons illustrate tools with functions analogous to specific radular teeth. (**d**) Gastropod *Anentome helena* feeds on molluscs and other small animals. Teeth likely act like rakes, piercing and tearing flesh. (**e**) Gastropod *Buccinum undatum*: carnivorous radula with visible cartilage supporting structures. (Reprinted with permission from Krings and Gorb (2024b). Copyright (2024), American Vacuum Society). (**f**) Gastropod *Dendronotus lacteus* feeds on Cnidaria. Central teeth resemble shovels for loosening polyps, while lateral teeth pull prey pieces inward. (Reprinted with permission from Krings and Gorb (2024b). Copyright (2024), American Vacuum Society). (**g**) Cephalopod *Loligo vulgaris* feeds on fish and crustaceans. The radula bends in a feeding position, enabling teeth to puncture and tear prey, akin to a sickle. (Reprinted with permission from Krings and Gorb (2024b). Copyright (2024), American Vacuum Society). (**h**) Gastropod *Cleopatra johnstoni* feeds on algae from soft sand surfaces. Teeth likely function like soft rakes or brushes. (**i**) Gastropod *Leloupiella minima*: forages algae from stones. Central teeth act as shovels and conveyor belts, while marginal teeth rake small particles. (**j**) Gastropod *Aeolidia papillosa* feeds on Cnidaria. Teeth function as rakes to pierce and transport prey parts. (**k**) Gastropod *Littorina littorea* feeds on algae from stones. Teeth operate like shovels. (**l**) Gastropod *Planorbella duryi* feeds on algae from hard surfaces. Teeth likely act as shovels. (**m**) Gastropod *Felimare picta* feeds on Cnidaria. Teeth resemble sickles for puncturing and tearing prey. (**n**) Gastropod *Onchidoris bilamellata* feeds on barnacles. Teeth act like nail claws to pry open prey carapaces. (**o**) Gastropod *Polycera quadrilineata* feeds on Bryozoans. Teeth function like nail claws to tear through Bryozoan exoskeletons. (**p**) Gastropod *Vittina turrita* feeds from stone surfaces. Central and lateral teeth I stabilize the structure during feeding, while lateral teeth II act as shovels to loosen food, and marginal teeth function as brushes to collect particles. Abbreviations: bz, building zone; ct, central tooth; dz, degenerative zone; fp, food particle; ja, jaw sheaths; lt, lateral tooth; ltI, lateral tooth I; ltII, lateral tooth II; mt, marginal tooth; rm, radular membrane; rt, radular teeth; sc, subradular cartilage; wz, working zone. Scale bars: (**d**, **f**, **h**, **m**, **o**, **p**) 100 μm; (**e**) 400 μm; g, 200 μm; (**i**) 10 μm; (**j**, **k**, **n**) 40 μm; (**l**) 20 μm

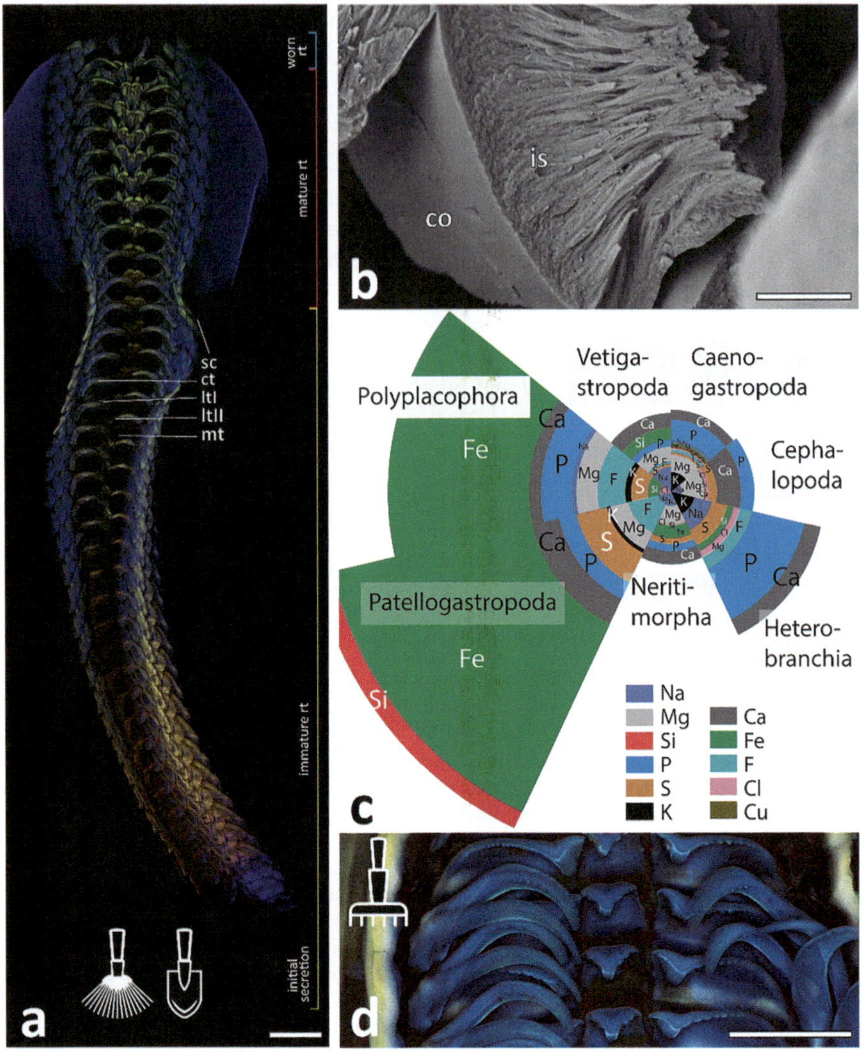

Fig. 3.2 Variability in radular material composition and tooth structure. (**a**) Confocal laser scanning microscopy (CLSM) image of the radula from the Polyplacophora *Lepidochitona cinerea*. The marginal teeth function as brushes to collect loosened particles, while the lateral teeth II act like shovels to dislodge algae from stones. The blue signal corresponds to regions with high protein content (in the building zone), unsclerotized chitin (in the alary processus), or calcium deposits (in the teeth and lateral plates). The green signal indicates the medium-sclerotized chitinous material (in the immature teeth) or presence of silica (in the mature teeth), while the black signal represents the iron content found in the lateral teeth II. (**b**) Microstructure of a broken tooth from the gastropod *Vittina turrita*. The teeth are composed of a matrix of chitin fibres and associated proteins, arranged in a complex architecture. Tooth surfaces exhibit a thin, homogeneous coating rich in mineral content. (**c**) Elemental composition of radular teeth across major molluscan groups. Proportions of elements (atomic %) were analysed in the radular teeth of Patellogastropoda, Polyplacophora, Vetigastropoda, Caenogastropoda, Cephalopoda, Heterobranchia, and Neritimorpha. Data is based

pointy, sharper and highly optimized for cutting, piercing, and tearing actions [e.g. Haszprunar and Götting 2007; Krings and Gorb 2024a, b; Hackethal et al. 2024; Krings et al. 2023a, 2024]. In the zoovorous nudibranch *Onchidoris bilamellata*, teeth are used to break the barnacle carapace (Fig. 3.1n). In *Polycera quadrilineata*, teeth are used to puncture and tear Bryozoan exoskeletons (Fig. 3.1o). Interestingly, teeth are separated by a thick membrane or other reduced teeth, which allow them to potentially clamp objects between the teeth. The extreme case represents the radula of cone snails possessing only a singly harpoon-like tooth per row. This tooth is spectacularly adapted for hunting: it is shot out of the mouth, to puncture prey and deliver venom (Moustafa et al. 2024).

In herbivorous molluscs, the remarkable diversity of radular morphology reflects adaptations not only to dietary preferences, but also to the substrate on which food is found. Species that feed from harder substrates, such as rocks or plant surfaces, tend to have broader, thicker teeth that allow for more forceful interactions with their environment, whereas species that feed on soft substrates typically possess long, slender teeth adorned with numerous delicate denticles, ideally suited for raking through sand and gathering food [e.g. Steneck and Watling 1982; Padilla 2003; Krings et al. 2021c; Krings 2020]. This is because thicker, broader teeth are more resistant to deformation under stress, providing superior stability and force transmission to the underlying radular membrane. This has been shown by Finite Element Analysis (FEA) (Krings et al. 2020a, 2021d), a computational mechanics technique used to simulate and predict mechanical behaviour under external forces. In contrast, longer, thinner teeth experience variations in mechanical behaviours such as stress and strain values in FEA, making them perfectly adapted for tasks requiring exceptional precision, like gathering particles from sand.

Herbivorous hard substrate feeders include notable examples of heterobranch gastropods, such as *Planorbella duryi* (Fig. 3.1l), where the whole radula is densely spiked with similarly shaped ("monofunctional radula"), broad teeth optimized for loosening algae and bacterial biofilm covers from rocky surfaces. In Polyplacophora (Fig. 3.2a) and Neritimorpha (Fig. 3.1p), the lateral teeth II are typically broad and thick, designed for scraping algae from solid surfaces [e.g. Krings et al. 2021a, 2022a; Ukmar-Godec et al. 2015]. These taxa also possess slender, thin teeth on the radula's sides, used to brush surfaces in a highly coordinated and complex motion, rendering this type of radula as "multifunctional", excellently capable of performing multiple tasks. In neritimorph taxa, the central and lateral teeth exhibit

Fig. 3.2 (continued) on the elemental analysis of 24 species, resulting in the depicted comparative diagram showing different elemental patterns in different lineages (Krings et al. 2022b). (**d**) CLSM image of the radula from the gastropod *Crepidula fornicata*, a filter-feeding species. It produces a mucous net to capture small particles. Subsequently, the mucus, along with trapped particles, is collected by teeth functioning as rakes (Filippov et al. 2025). The blue colour relates to calcium depositions in the outer tooth coating. Abbreviations: co, tooth coating; ct, central tooth; is, inner tooth structure; ltI, lateral tooth I; ltII, lateral tooth II; mt, marginal tooth; rt, radular teeth; sc, subradular cartilage. Scale bars: (**a**) 200 μm; (**b**) 2 μm; (**d**) 100 μm the remarkable additional function of spanning the radula open during feeding, showcasing the complex interplay of the entire system (Krings et al. 2021a). Species, which catch particles from the water columns as *Crepidula fornicata*, produce slime nets [e.g. Filippov et al. 2025] which are raked in after particle collection by pointy teeth (Fig. 3.2d).

The paludomid gastropods from the African Great Lake Tanganyika, which is shared among four the countries Tanzania, the Democratic Republic of the Congo, Burundi, and Zambia, and its surrounding river systems are extraordinarily unique. They display the above-described diversity in radular tooth morphologies corresponding to the different feeding substrates in a very close phylogenetic context (Krings et al. 2021c; Krings 2020; Gorb and Krings 2021). Here, these differences are particularly reflected in the morphology of the central and lateral teeth, which are broader and thicker in solid substrate feeders whereas all teeth in soft substrate feeders exhibit a slender, uniform design ("monofunctional radula"). These thin teeth with small thin denticles, as seen in species, such as *Cleopatra johnstoni* (Fig. 3.1h), facilitate highly effective particle collection from the substrate, as their fine denticles help in raking food. In contrast, species that feed on harder substrates, such as *Leloupiella minima* (Fig. 3.1i), exhibit lateral teeth with thicker, more prominent denticle, which may provide a larger interaction surface, transferring force directly to the ingesta. The central teeth in these species are often broad and spatulate, maximizing force distribution onto the substrate to break down harder material from large area. The marginal teeth are thin and slender, used to rake or brush food particles from surfaces after the centrals and laterals have performed scratching action ("multifunctional radula").

The attachment areas of the teeth to the membrane further enhance this extraordinary functionality (Krings et al. 2020b). The areas of the central and lateral teeth in solid substrate feeders have a broader cross-sectional area. This probably distributes the applied forces over a larger volume, reducing localized stress concentrations that could lead to fracture or deformation. Also, it could enhance the mechanical support at the interface with the radular membrane, allowing for more effective transmission of force from the tooth tip through to the underlying musculature. This might also improve the structural stability, especially in resisting bending or torsional loads that occur during scraping on tough surfaces. By contrast, the marginal teeth and those of soft substrate feeders have smaller attachment areas, fostering higher flexibility and exceptional adaptability in tooth movement. This finely tuned design enables each radular type to achieve optimal performance in its respective ecological niche.

This astonishing diversity in radular tooth morphology highlights the extraordinary versatility of these structures, offering a range of functional options for different feeding strategies. The strong variety of tooth shapes, from pointy teeth for cutting material, intricately denticled teeth for cleaning surfaces and collecting particles, to robust, force-transmitting teeth tailored to polish surfaces can inspire a range of advanced biomimetic designs for differentiated, multifunctional devices. This could be interesting for tool manufacturing (e.g. knifes or shovels) or home appliances (e.g. brooms or microfibre cloths) requiring varying levels of force transmission and flexibility. The radula's natural optimization for precise motion in contact with different substrate textures offers insights, for example, gripping devices in robotics. Its self-renewing capabilities, where teeth are continuously replaced throughout the mollusc's life, further offer insights into the design of self-healing or recyclable materials in material science.

3.3 Material Structure and Properties of the Radula and Their Significance

The material composition of radular teeth further contributes to their specific functions. It exemplifies a composite material, exhibiting structural organization that mitigates fracture formation and propagation. Composed of chitin fibres reinforced with associated proteins (Guralnick and Smith 1999; Lowenstam and Weiner 1985), its organic matrix varies in orientation, density, and arrangement (Evans et al. 1994; Wealthall et al. 2005; Grunenfelder et al. 2014; Wang et al. 2014; Krings et al. 2022b). This structural diversity at the micro- and nano-scale corresponds to functional demands: the outer material of the tooth forms a uniform coating, while underlying fibres are oriented perpendicularly to the surface (Lowenstam and Weiner 1985; Krings et al. 2022b; van der Wal et al. 1999; Shaw et al. 2010), enhancing wear resistance and fracture mitigation. The outer coating can contain high proportions of silicon or calcium on the interacting surfaces, which presumably contributes to the wear reduction (Krings and Gorb 2023a, 2024a, b; Hackethal et al. 2024; Krings et al. 2022b, 2023a, b, 2024).

The mechanical parameter hardness H refers to a material's resistance to surface deformation or indentation, and gives insight into how the material will respond to wear, contact stress, and mechanical damage. A material with higher H values has a higher fracture resistance than material with lower values. Stiffness or Young's modulus E describes its resistance to elastic deformation under an applied load. A material with higher E values is more effective at bearing weight, transmitting forces, and maintaining shape during movement or impact. Hardness can be tested by Vickers, Brinell, Knoop, or Rockwell tests. Materials of very small volumes (as radular teeth) or thin coatings (e.g. the radular tooth surface) can be tested by nanoindentation, which presses a sharp Berkovich diamond tip into the surface under a precisely controlled load and records a load vs. displacement curve. From the unloading curve of the load–displacement plot, specifically the initial unloading stiffness, E can be calculated using the Oliver–Pharr method (Oliver and Pharr 2004).

Previously it was found by nanoindentation, that variations in these mechanical property values within radular teeth and tooth rows align with ecological niches (e.g. feeding on rock, plant surfaces, or sand) and specific tooth functions (e.g. loosening food or collecting particles).

For example, among the paludomids of Lake Tanganyika, those foraging on harder substrates have harder and stiffer teeth to reduce wear and resist failure, whereas species feeding on softer substrates exhibit more flexible and softer teeth (Krings et al. 2021c; Krings 2020; Gorb and Krings 2021). Additionally, radular teeth can show heterogeneities in material properties: the material incorporates regions of differing hardness or stiffness, such as harder and stiffer cusps for cutting or piercing and more flexible and soft bases for maintenance of resilience under stress (Krings et al. 2019a, 2020a, 2021c, d, 2022c, 2023a, b, 2024; Krings and Gorb 2023a, 2024a, b; Hackethal et al. 2024; Gorb and Krings 2021; Weaver et al. 2010; Brooker and Shaw 2012; Lu and Barber 2012; Barber et al. 2015; Ukmar-Godec 2016; Ukmar-Godec et al. 2017; Pohl et al. 2020).

Within Mollusca, there are different mechanisms in radular teeth to increase E and H values. The first one is mineralization. Polyplacophora and Patellogastropoda, such as *Patella vulgata* and *Lepidochitona cinerea* (Fig. 3.2a), incorporate high levels of iron and silicon (e.g. silica, goethite, iron oxides, and apatite) into their prominent teeth (Krings and Gorb 2024a, b; Krings et al. 2022a; Lowenstam and Weiner 1985; van der Wal et al. 1999; Shaw et al. 2010; Brooker and Shaw 2012; Lu and Barber 2012; Barber et al. 2015; Ukmar-Godec 2016). These highly mineralized teeth are among the hardest and stiffest biological materials, with E values of 30–130 GPa and H values of 4–12 GPa in Polyplacophora (Krings et al. 2022a; Brooker and Shaw 2012). The dominant teeth in *P. vulgata* can even exceed these values (E: 52–150 GPa; H: 3–7 GPa) (Barber et al. 2015; Ukmar-Godec 2016). Most molluscan radular teeth are less mineralized, with limited studies on mineral presence and abundance. While Fe content is minimal in most gastropod radulae ($\leq$1 atomic %), it has been detected in small amounts in various snail taxa (Ukmar-Godec et al. 2015; Krings et al. 2022c, d, 2023a, b; Sollas 1907; Brütt et al. 2022). Similarly, Si, likely as silica, is found in low proportions ($\leq$2 atomic %) in non-patellogastropod taxa (Ukmar-Godec et al. 2015; Krings et al. 2022c, d, 2023a, b; Brütt et al. 2022). Elements like phosphorus, magnesium, and calcium (e.g. in forms, such as iron phosphate and apatitic calcium phosphate) enhance mechanical properties as well (Brooker and Shaw 2012; Ukmar-Godec 2016; Evans et al. 1992; Macey et al. 1996; Lee et al. 2000) (Fig. 3.2c).

The second mechanism to enhance material properties is the cross-linking of organic matrix. Certain ions of some elements (e.g. Ca, Cu, Fe, Mg, Si, Zn) act as cross-linkers in chitinous materials (Degtyar et al. 2014), enhancing mechanical properties (Ukmar-Godec et al. 2017; Pohl et al. 2020; Krings et al. 2022d; Gordon and Joester 2011). For instance, Mg and Ca ions in the internal tooth structure of *Megathura crenulata* lead to an increase of E to 16 GPa (Ukmar-Godec et al. 2017). These ions have also been detected in the radulae of other molluscan taxa and seem to be widely present [e.g. Pohl et al. 2020; Sollas 1907; Krings et al. 2022d; Brütt et al. 2022; Gordon and Joester 2011] (Fig. 3.2c).

The third mechanism to enhance material properties is the increase of tanning degree and protein abundance. The degree of tanning as well as the specific protein composition influence mechanical properties, akin to arthropod cuticle. Autofluorescence induced by confocal laser scanning microscopy (CLSM) (Fig. 3.2a, d) is used to relate composition with mechanical properties in radulae (Krings and Gorb 2024a, b; Hackethal et al. 2024; Krings et al. 2022c, 2023a, b, 2024): sclerotized, hard/stiff chitin-containing material correlates with red autofluorescence, as seen in unmineralized radulae. Weakly sclerotized material emits a green signal, indicating flexibility and toughness. This may also indicate Si incorporation in radulae with high inorganic content (Fig. 3.2a). Unsclerotized, soft/flexible material shows a blue signal. In mineralized radulae, similar signal may indicate high concentration of Ca (Fig. 3.2d). A brown or black signal may indicate Fe presence in mineralized radulae, as seen in polyplacophoran teeth (Fig. 3.2a).

As a result of these various factors, the radula's teeth achieve exceptional hardness and abrasion resistance while being embedded in a flexible, chitin-based organic membranous matrix, allowing them to withstand repeated use against tough surfaces. By mimicking these features, it is possible to design advanced cutting, cleaning, and grinding tools capable of maintaining sharpness under prolonged use, reducing material wear and operational downtime. This has previously also been proposed by (van der Wal et al. 1999), presenting the radular teeth of polyplacophoran and patellogastropod species as models for industrial cutting devices, due to the teeth's self-sharing characteristics. Applications of this biomimetic approach span industries such as mining, construction, and food processing, where efficient and sustainable mechanisms of material and surface processing are essential. Additionally, the structure–functional relationship of different radulae provides the possibility of the development of durable micro-tools using various technologies, such as 3d printing, micro-etching, and microlithography, specifically in biomedical engineering. Also, various types of coating technologies might be used to enhance performance of artificial radula-inspired systems. This includes protective coatings for heavy machinery or components for medical implants. In the field of structural engineering, the chitin architecture in connection with the tooth coating could inspire the design of light-weight houses with protective and durable surfaces. The arrangement of the chitin fibres and its cross-linking could be replicated in textiles, enhancing durability while maintaining the material's flexibility, or in the development of chitosan-based composites (see Stegbauer et al. 2021).

3.4 Mechanical Behaviour and Function of Radula

During feeding, numerous muscles manipulate the radular membrane, stretching and pulling it over the odontophore to bring the teeth into the right position for the interaction with the ingesta [e.g. Katsuno and Sasaki 2008]. Textbook descriptions often emphasize the licking-like feeding motion of Heterobranchia, involving radular protrusion to loosen ingesta [e.g. Neustadter et al. 2002a,b; Montroni et al. 2019; Mackenstedt and Märkel 2001; Kehl et al. 2019; Krings et al. 2019b]. However, studies on buccal musculature [e.g. Graham 1973; Fretter and Graham 1976; Crampton 1977; Golding et al. 2011; Simone 2011] and feeding tracks [e.g. Krings et al. 2021a; Mackenstedt and Märkel 2001; Ankel 1936, 1938; Eigenbrodt 1941; Hickman and Morris 1985; Janssen and Triebskorn 1987] reveal highly diverse and complex motion systems. The radula, traditionally likened to a flat ribbon or a "belt-and-pulley" system (Solem 1974; Ankel 1938; Eigenbrodt 1941; Hickman 1980), is now recognized as a far more intricate structure. Radular movements occur in all three dimensions, with bending and stretching enabling complex tooth–tooth and radula–jaw interactions.

During radular motion, different structures were found that function as counter bearings, enabling the molluscs to grab and tear larger pieces of ingesta (Scheel et al. 2020) (Fig. 3.3). These findings indicate that counter bearing mechanisms have evolved convergently across Mollusca. These structures include the following:

(a) The prominent lateral teeth II (e.g. in polyplacophoran species, such as *Lepidochitona cinerea*, Fig. 3.3a). Initially, the radula extends toward the mouth opening, while the teeth spread outward in a fan-like arrangement. This curved flexion reverses as the radula retracts. The rotation of the teeth, particularly the lateral teeth II, facilitates the gripping of larger food items, with the lateral teeth II themselves functioning as counter bearings (Scheel et al. 2020).
(b) The radula and the jaws, which are hard chitinous structures located at the ventral and anterior part of the mouth opening [e.g. Montroni et al. 2019; Märkel 1957]. This motion pattern can be seen, for example, in heterobranch species as *Planorbella duryi* (Fig. 3.3b). Following the radula's protrusion, its lateral edges are drawn caudally, forming a concave, spoon-like shape. As the radula moves anteriorly, it loosens and collects ingesta until it reaches its counter bearing, the jaw. Together, these structures facilitate the ability to pull and tear large food items (Neustadter et al. 2002a,b; Montroni et al. 2019; Mackenstedt and Märkel 2001; Kehl et al. 2019; Krings et al. 2019b, 2021e; Eigenbrodt 1941; Scheel et al. 2020; Wägele 1983).
(c) The two sides, that is, left and right side, of the radula (e.g. in the Heterobranch species *Stenophysa marmorata*, Fig. 3.3c). In this pattern, the radula bends about the anterior–posterior radular axis situated along the central teeth, with the lateral portions of the radula being pulled towards the substrate forming a U-shape. The lateral wings are then folded towards the radular centre, gripping and holding the ingesta, until the radula, along with the captured food, is retracted into the oral cavity (Eigenbrodt 1941; Scheel et al. 2020).
(d) The tooth rows from anterior and posterior radular regions (e.g. in caenogastropod species as *Littorina littorea*, Fig. 3.3d). First, the anterior portion of the radula (purple region in Fig. 3.3d) is extended and moved in anterior–ventral direction, while the posterior region of the radula (pink region in Fig. 3.3d) is stretched into posterior–ventral direction. The intermediate radular region is pulled dorsally. Afterwards, the radula interacts with the food, while the anterior portion of the radula is pulled towards posterior direction and the posterior portion of the radula is pulled towards anterior direction. This allows the teeth from two different radular regions to act together, grabbing food pieces and transporting them into the oral tube (Eigenbrodt 1941; Scheel et al. 2020; Wägele 1983).

These distinct motion patterns likely arise from differences in the structural organization of radular and buccal mass muscles [e.g. Graham 1973; Fretter and

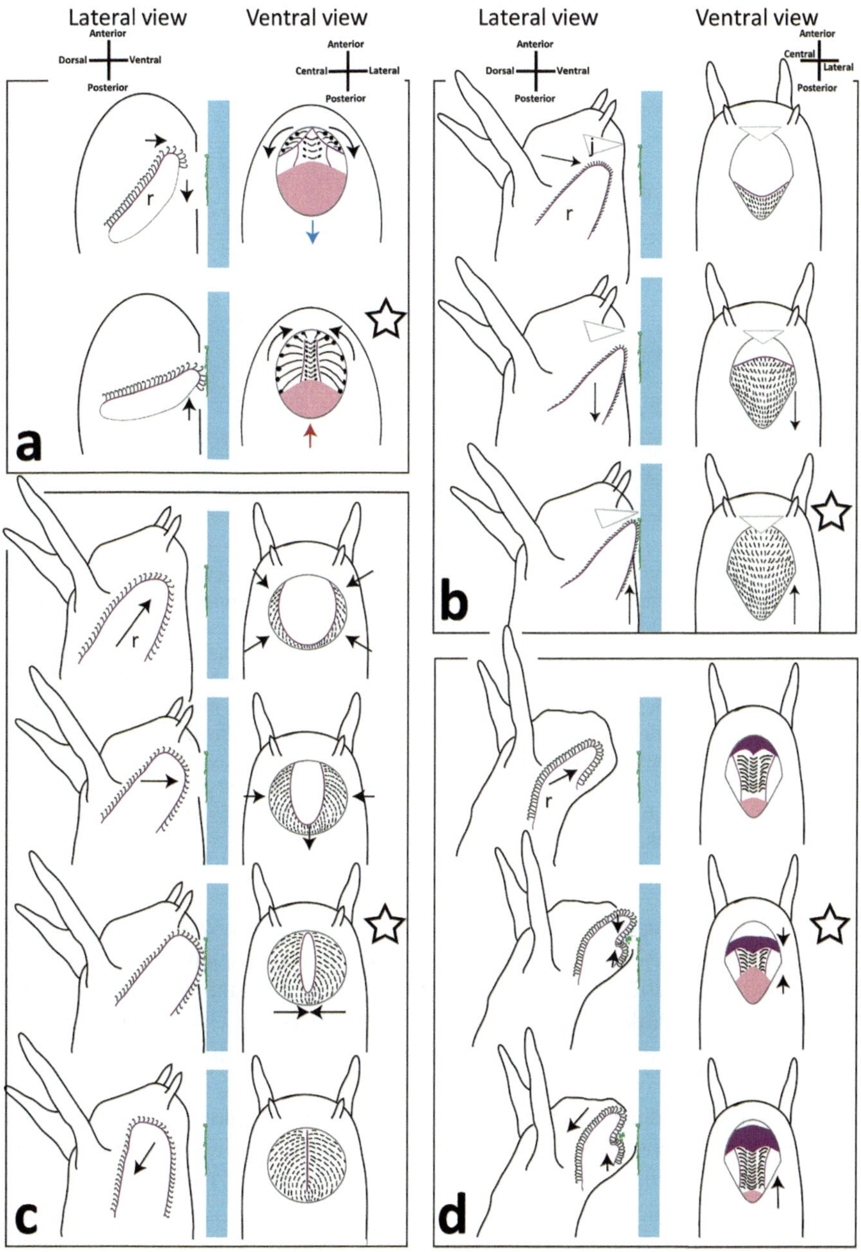

Fig. 3.3 Radulae are moved in different manners to grab and tear food (see Scheel et al. 2020). Sub images should be read from above to below. (**a**) The prominent lateral teeth II in the polyplacophoran species *Lepidochitona cinerea* rotate which enables the gripping of larger food (marked with a star). (**b**) The radula and the jaws, for example, in the gastropod *Planorbella duryi*, pull and tear large food items (marked with a star). (**c**) The two sides of the radula, for example, in the gastropod *Stenophysa marmorata*, fold towards the radular centre, gripping and holding the ingesta (marked with a star). (**d**) The tooth rows from anterior and posterior radular regions, for example, in the gastropod *Littorina littorea*, act together and grab food pieces (marked with a star)

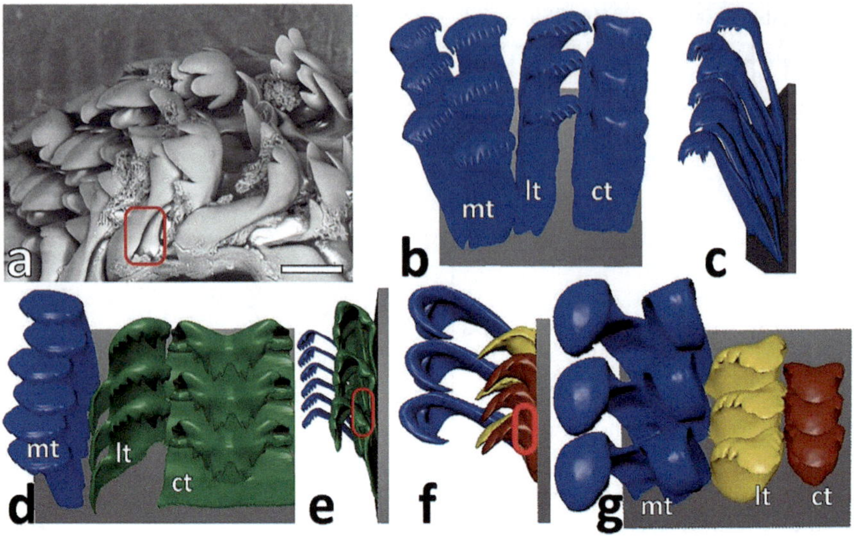

Fig. 3.4 Radular teeth can interlock to different degrees ("collective effect"), transferring stresses and reducing structural failure. (**a**) The teeth of the gastropod *Littorina littorea*, some interlocking teeth are highlighted with red boxes. (**b–g**) 3d models of paludomid gastropod teeth. The teeth were computer-modelled by hand always comparing SEM images with the digital model. (Images in **d** and **e** are reprinted with permission from Krings and Gorb (2024b). Copyright (2024), American Vacuum Society). (**b**, **c**) The teeth of the gastropod *Cleopatra johnstoni*, foraging from soft surfaces. The teeth do not interlock. (**d**, **e**) The teeth of the gastropod *Bridouxia grandidieriana*. The central and lateral teeth (green colour) interlock when loaded (red box). (**f**, **g**) The teeth of the gastropod *Lavigeria grandis*. The central and lateral teeth (yellow and red colour) interlock tightly and very tightly, respectively when loaded (red box). Abbreviations: ct, central tooth; lt, lateral tooth; mt, marginal tooth. Scale bar: (**a**) 50 μm

Graham 1976; Crampton 1977; Golding et al. 2011; Simone 2011] as well as the supporting structures of the radula, which vary in shape, quantity, volume [e.g. Krings and Gorb 2021a; Krings et al. 2021a, b, e; Katsuno and Sasaki 2008; Mackenstedt and Märkel 2001; Golding et al. 2011], and, likely, bending capacity. The intricate interaction of these components is critical to feeding functionality and performance.

During these radular motions, the individual teeth do not act independently on ingesta, but rather transmit and distribute forces and stresses collectively by interlocking in certain directions [e.g. Montroni et al. 2019; Padilla 2003; Solem 1972; Morris and Hickman 1981; Hickman 1984; Herrera et al. 2015; Krings and Gorb 2021b; Krings et al. 2021f, g]. The interaction between the teeth on the radula (Fig. 3.4a) is another key aspect of their functionality. Adjacent teeth support each other, with the specifically shaped bases of the lateral teeth stabilizing each other while interacting with the substrate. This interconnected system ensures that the forces exerted during feeding are more homogenously distributed, preventing tooth damage and maximizing feeding performance.

The ability of teeth to interlock leads to the reliance of the teeth from one tooth row onto adjacent tooth rows. This kind of mechanical behaviour ("collective effect") depends on tooth morphology and arrangement (Krings et al. 2021d, f, g; Hickman 1984; Krings and Gorb 2021b). For example, in solid substrate feeders (Fig. 3.4f, g) and species, which forage occasionally from solid surfaces (mixed substrate feeders; Fig. 3.4d, e), the morphology of central and lateral teeth allows the posterior part of one front tooth to fit tightly with the anterior part of the hind tooth. This interlocking mechanism resists high shear forces and distributes stress uniformly and structural failure is only observed in areas not supported by adjacent teeth (Krings et al. 2021d, f, g). In contrast, the marginal teeth of solid substrate feeders and all teeth of soft substrate feeders (Fig. 3.4b, c), typically do not interlock with other teeth. Instead, their long and slender styli allow for significant motion of the cusps, including deformation and twisting under shear force.

In addition to morphology, the mechanical properties of teeth and the radular membrane were found to contribute to the degree of the collective effect (Gorb and Krings 2021). Soft substrate feeder teeth are relatively homogeneous, with similar hardness H and Young's modulus E values across all tooth parts, resulting in soft, flexible teeth capable of bending and twisting over larger deflection degrees. This likely also aids in the interaction with larger obstacles without structural failure due to larger deformability. This, however, on the other hand, does not allow for a collective effect. In contrast, solid substrate feeder central and lateral teeth exhibit pronounced gradients in H and E, with cusps being the hardest and stiffest, followed by the stylus, and the basis as the softest and most flexible region. This gradient of mechanical properties enables specific functional adaptations: teeth puncture ingesta with their hard and stiff cusps, while the regional flexibility and softness of the tooth bases act as shock absorption against mechanical impacts (Gorb and Krings 2021; Pohl et al. 2020; Herrera et al. 2015). This regional flexibility and softness also enable the interlocking of the teeth. Mixed substrate feeders exhibit not as pronounced H and E values, allowing a medium degree of interlocking (Gorb and Krings 2021; Krings et al. 2021g). This design seems to balance the energy-demanding construction of hard, stiff materials with functional efficiency, which as result enables them to occasionally forage on challenging surfaces, but with a higher risk of potential structural failure.

All these biomechanical principles and structural adaptations of the radula could enhance the design of robotic grippers with advanced anchoring mechanisms. By mimicking its flexible base and robust, patterned surface of the teeth, gripping effectively, it is possible to develop multifunctional gripping systems in robotics capable of handling delicate or irregularly shaped objects while applying controlled force without damage of the target surface structure. The radula's efficient and sustainable anchoring principles highlight its relevance to modern technological challenges, particularly in environments, where traditional adhesives or fasteners are impractical. The capability of teeth to interlock, when loaded, leads to the redistribution of stress, which reduces the failure of the whole system and could be potentially implemented in biomimetic tools. Radula's self-adjusting properties will allow biomimetic tools to be more energy-efficient. Applications include surgical robotics, underwater exploration, agricultural harvesting, and industrial automation, where precision, adaptability, and robustness are critical.

3.5 Radular Systems and Function

To deeply understand the radular systems with all their subunits, different kinds of numerical simulations could be applied (Krings et al. 2020a, 2021d; van der Wal et al. 1999; Pohl et al. 2020; Filippov et al. 2023). As one strong method, FEA enables the examination of radular tooth performance by visualizing stress and strain distributions within 3d models of teeth (Krings et al. 2020a, 2021d; van der Wal et al. 1999; Pohl et al. 2020) (Fig. 3.5). These analyses may incorporate gradients in mechanical properties, attachment to the membrane, and interactions between teeth. Additionally, simulations utilizing a cellular automata approach with MATLAP provide insights into radular performance (Filippov et al. 2023). These models simulate the complex interactions among food particles, substrate, and radular teeth, factoring in gradients of mechanical properties and numerous other factors, such as adhesion to the tooth surface, gravity, and particle interactions. These simulations yield data on particle collection efficiency, particle motion patterns, and the system's work requirements for gathering food. These metrics depend on surface pressure (indicating substrate contact) and the partial work exerted by the teeth (which may be correlated with the degree of wear or structural failure).

The adaptability of numerical simulations makes them valuable as a first step towards biomimetic applications. By varying parameters, such as tooth angle, spacing, substrate properties, and feeding behaviours, these controlled virtual experiments enable researchers to identify optimal configurations for future synthetic designs. This knowledge can guide the development of technologies that mimic the radula's efficiency in particle manipulation and material collection.

Physical models, which are the second step of the biomimetic developments, also play an important role. 3d models of teeth (Fig. 3.5) can be simplified to various degrees (Fig. 3.6a, b), their degree of interlocking can be manipulated to fit the problem (Fig. 3.6c–f), and the structures can be variously assembled (Fig. 3.6g) to perform multifunctional tasks. The 3d-printed radulae (Fig. 3.6h) can be used as physical dynamic radular models (Fig. 3.6j) offering a platform for studying the interaction between radular components and ingesta (Krings et al. 2021b; Mangan et al. 2005). As these models evolve, they can more closely replicate the geometry, mechanical property gradients, and structural arrangements of natural radulae. Experiments using 3d-printed radular teeth (Fig. 3.6h) have already demonstrated how slight morphological modifications impact mechanical behaviour and performance outcomes. For example, pull tests using 3d-printed teeth on substrates covered with agar gels of varying densities allow researchers to quantify material removal efficiency depending on force and work applied (Krings et al. 2025). Additionally, printed upscale models can be used as educational tools (Fig. 3.6i).

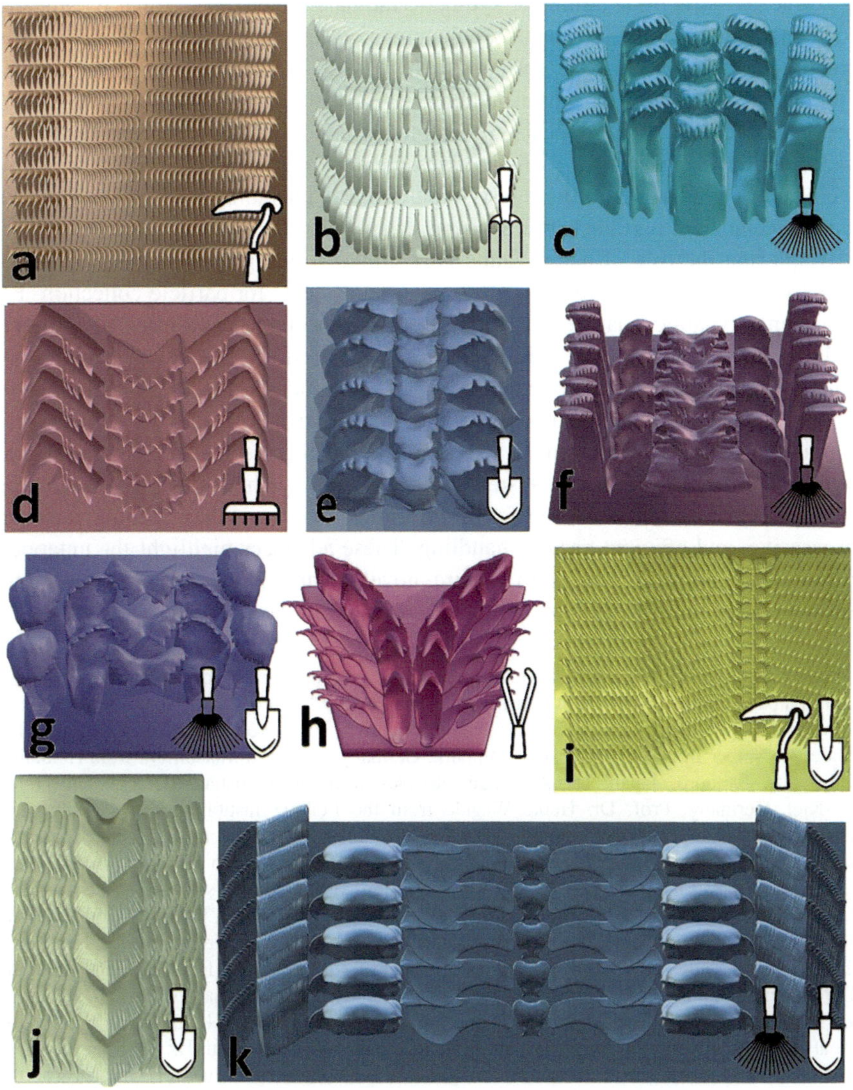

Fig. 3.5 3d models of different radular types. Each was computer-modelled by hand always comparing the model to the SEM images. The teeth can be compared to different tools in their functionality. (**a**) Radula of the gastropod *Felimare picta*, which teeth resemble sickles for puncturing and tearing prey. (**b**) Radula of the gastropod *Aeolidia papillosa*, which teeth function as rakes to pierce and transport prey parts. (**c**) Radula of the gastropod *Cleopatra johnstoni*, which teeth function like soft rakes or brushes. (**d**) Radula of the gastropod *Anentome helena*, which teeth act like rakes, piercing and tearing flesh. (**e**) Lateral and central teeth of the paludomid gastropod *Lavigeria grandis*, which teeth function like shovels on solid substrate. (**f**) Radula of the paludomid gastropod *Bridouxia grandidieriana*, which marginal teeth function like brushes. (**g**) Radula of the paludomid gastropod *Spekia zonata*, which central and lateral teeth function like shovels and the marginal teeth like brushes on solid substrate. (**h**) Radula of the gastropod

Wear and structural failure of synthetic radular models are critical parameters for considerations of biomimetic applications. Wear experiments, where models are dragged across abrasive surfaces, can provide further insights into material efficiency and durability, while breaking-stress tests aid in measuring structural resilience. These methods can also be adapted for studying real radulae under native wet conditions, ensuring a realistic assessment of mechanical performance. The inclusion of snail saliva in these tests is particularly intriguing, as its high adhesive capacity enhances particle capture during feeding (Krings and Gorb 2023b).

Exploring the role of natural adhesives similar to the properties of snail saliva could inspire new technologies for bio-inspired systems for particle collection. By emulating dynamic adaptability and surface-conforming capabilities of artificial radulae in connection with fluid adhesives, it is possible to develop mechanisms suited for effective particle capture in 2d- and 3d-space. The integration of computational and physical modelling with insights from biological systems offers a unique approach to biomimetics. By translating the mechanical principles and structural dynamics of radular systems (Fig. 3.7) into synthetic designs, researchers can develop innovative solutions for future applications in surface cleaning, material processing, and efficient particle handling. These advances highlight the untapped potential of natural systems to inspire novel technologies and underscore the importance of interdisciplinary approaches in biomimetic research.

Acknowledgements We would like to express our gratitude to our collaborators who have contributed to our radular research over the past years: Dr. Jordi Marcé-Nogué from the Universitat Rovira i Virgili, Tarragona, Spain; Dr. Frank Friedrich, Elke Woelken, Peter Stutz, and Dr. Marco T. Neiber from the Universität Hamburg, Germany; Dr. Yoko Matsumura from Hokkaido University, Sapporo, Japan; Dr. Alexander Kovalev from the Christian-Albrechts-Universität zu Kiel, Germany; Prof. Dr. Heike Wägele from the Leibniz Institute for the Analysis of Biodiversity Change, Bonn, Germany; Prof. Dr. Alexander Filippov from the Donetsk Institute for Physics and Engineering, Ukraine; and Dr. Ellen Schulz-Kornas from the Universität Leipzig, Germany. We also appreciate the valuable contributions of our former students and co-authors: Leon Gladziejewski, Jan-Ole Brütt, Taissa Faust, Charlotte Neumann, Christine Hempel, Hasan Karabacak, Carolin Scheel, Lisa Siemers, Henrik Schreiber, Holger Krohn, Svenja Hackethal, Tamina Riesel, and Nicole Wolff. We are deeply thankful for the support from the members of the Department of Functional Morphology and Biomechanics at the Christian-Albrechts-Universität zu Kiel and the Section of Mammalogy and Paleoanthropology at the Leibniz Institute for the Analysis of Biodiversity Change. We are grateful for the discussion with Dr. Mohsen Jafarpour and Dr. Shahab Eshghi, both from the Christian-Albrechts-Universität zu Kiel, Germany, on the results of this manuscript. We would like to thank the editor of this book, Prof. Vincent Bels, for the coordination of the book.

Fig. 3.5 (continued) *Gastropteron rubrum* (feeding on hard Foraminifera), which teeth function like a clamp. (**i**) Radula of the gastropod *Armina tigrina* (feeding on coral polyps), which central teeth function like shovels to loosen ingesta parts whereas the outer teeth function as sickles to puncture and tear the prey. (**j**) Radula of the gastropod *Dendronotus lacteus*, which central teeth resemble shovels, while lateral teeth pull pieces inward. (**k**) Radula of the gastropod *Vittina turrita*, which central and lateral teeth I unfold and stabilize the radula during feeding; its lateral teeth II act as shovels and marginal teeth as brushes

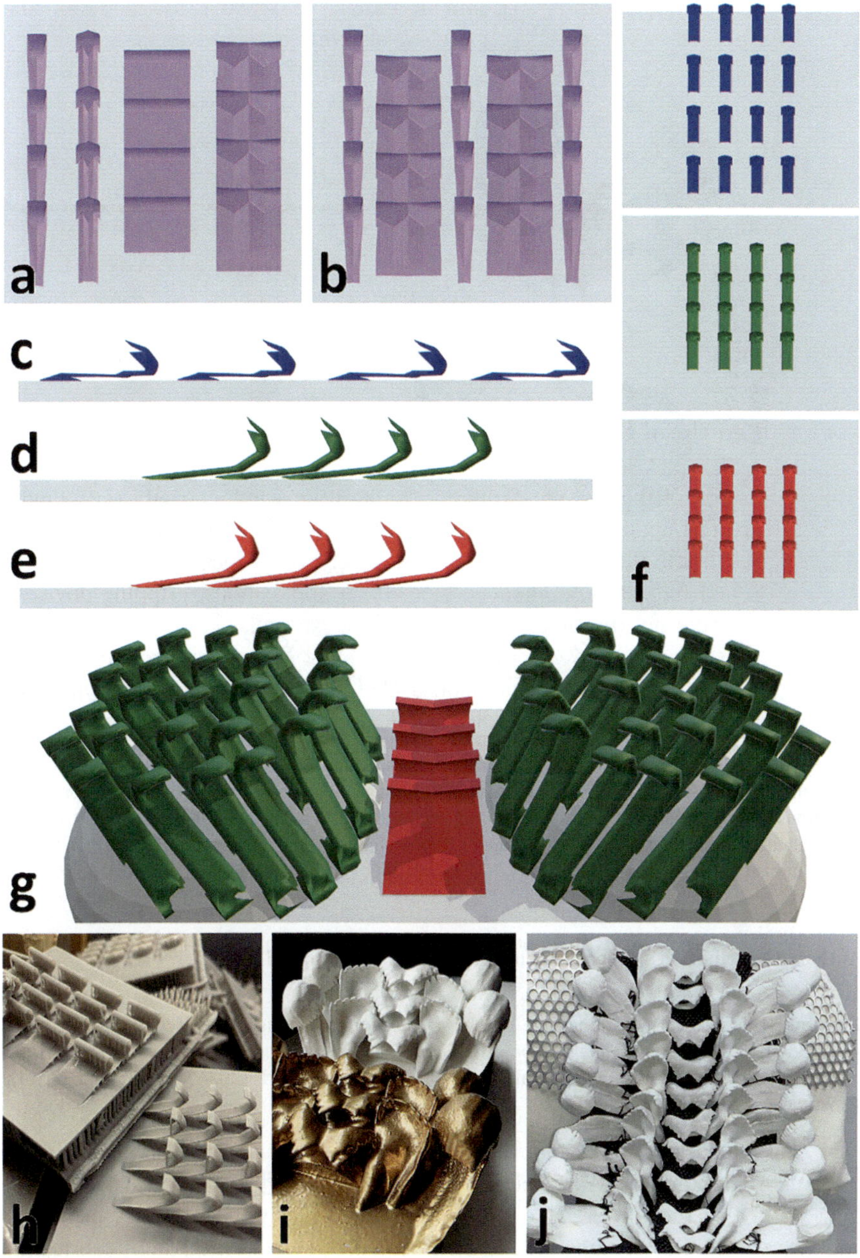

Fig. 3.6 (**a–g**) Abstracted radular models, which can have different functions depending on the morphology and the degree of their interlocking (blue, no interlocking; green, intermediate degree of interlocking; red, high degree of interlocking). (**h–j**) 3d-printed radular models. (**h**) 3d-printed with a stereolithography 3d printer to test the influence of morphology on the performance of the feeding system (see Krings et al. 2025). (**i**) Powder-printed radulae for educational purposes. (**j**) A physical radular model designed to mimic the interlocking of teeth in rows and between the neighbours under different motion conditions and to study the complete 3d mechanical behaviour of the system (see Krings et al. 2021b)

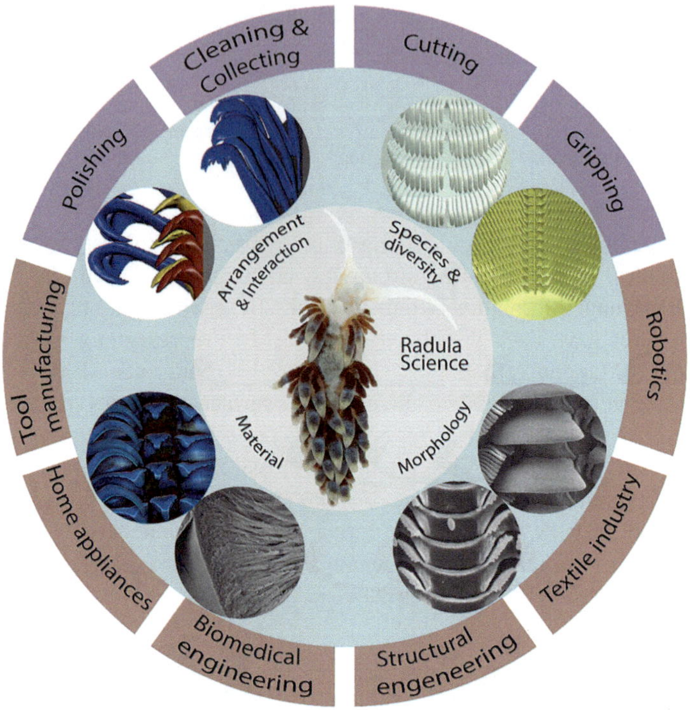

Fig. 3.7 Tooth morphologies, their chemical composition, microstructure, arrangement and degree of interlocking; can differ between species showing strong diversity in their functions (polishing, cleaning and collecting, cutting, or gripping). All these aspects offer transformative directions for future advancements in biomimetic engineering, specifically in the fields of tool manufacturing, the design of home appliances, biomechanical engineering, structural engineering, textile industry, or robotics

Authors Contribution Both authors initiated, wrote, and designed this paper.

Funding This project was financed by the Deutsche Forschungsgemeinschaft (DFG) grant 470833544 to WK. *Permission to reproduce material from other sources*. Images Figs. 3.1e–g and 3.4d, e were reprinted with permission from Krings and Gorb (2024b). Copyright (2024), American Vacuum Society

Data Availability The data that support the findings of this study are available from the corresponding author upon reasonable request

Competing Interests The are no competing interests

References

Ankel WE (1936) Die Fraßspuren von *Helcion* und *Littorina* und die Funktion der Radula. Verh Dtsch Zool Ges Freib 38:174

Ankel WE (1938) Erwerb und Aufnahme der Nahrung bei den Gastropoden. Verh Dtsch Zool Ges, Zoologischer Anzeiger 11:223–295

Banken E, Oeffner J (2023) Biomimetics for innovative and future-oriented space applications – a review. Front Space Technol 3:1000788

Barber AH, Lu D, Pugno NM (2015) Extreme strength observed in limpet teeth. J R Soc Interface 12:20141326

Bhushan B (2009) Biomimetics: lessons from nature–an overview. Phil Trans R Soc A 367:1445–1486

Bhushan B (2018) Biomimetics. Bioinspired hierarchical-structured surfaces for green science and technology. Springer, Cham

Blinn W, Truitt RE, Pickart A (1989) Feeding ecology and radular morphology of the freshwater limpet *Ferrissia fragilis*. J N Am Benthol Soc 8(3):237–242

Bouchet P, Rocroi J-P, Frýda J, Hausdorf B, Ponder W, Valdes A, Warén A (2005) Classification and nomenclator of gastropod families. Malacologia 47(1–2):1–368

Brooker LR, Shaw JA (2012) The chiton radula: a unique model for biomineralization studies. In: Seto J (ed) Advanced topics in biomineralization. Intech Open, Rijeka, pp 65–84

Brütt J-O, Gorb SN, Krings W (2022) The ontogeny of elements: distinct ontogenetic patterns in the radular tooth mineralization of gastropods. Sci Nat 109:58

Burch JB, Jeong KH (1984) The radula teeth of selected Planorbidae. Malacol Rev 17(1–2):67–84

Chan NJA, Gu D, Tan S, Fu Q, Pattison TG, O'Connor AJ, Qiao GG (2020) Spider-silk inspired polymeric networks by harnessing the mechanical potential of β-sheets through network guided assembly. Nat Commun 11:1630

Collins CM, Safiuddin M (2022) Lotus-leaf-inspired biomimetic coatings: different types, key properties, and applications in infrastructures. Infrastructures 7(4):46

Crampton D (1977) Functional anatomy of the buccal apparatus of *Onchidoris bilamellata* (Mollusca: Opisthobranchia). Trans Zool Soc Lond 34(1):45–86

Degtyar E, Harrington MJ, Politi Y, Fratzl P (2014) The mechanical role of metal ions in biogenic protein-based materials. Angew Chem Int Ed 53(45):12026–12044

du Plessis A, Broeckhoven C, Yadroitsava I, Yadroitsev I, Hands CH, Kunju R, Bhate D (2019) Beautiful and functional: a review of biomimetic design in additive manufacturing. Addit Manuf 27:408–427

Duda TF, Kohn AJ, Palumbi SR (2001) Origins of diverse feeding ecologies within *Conus*, a genus of venomous marine gastropods. Biol J Linn Soc 73(4):391–409

Eigenbrodt H (1941) Untersuchungen über die Funktion der Radula einiger Schnecken. Z Morphol Oekol Tiere 37:735–791

Ekimova I, Valdés Á, Chichvarkhin A, Antokhina T, Lindsay T, Schepetov D (2019) Diet-driven ecological radiation and allopatric speciation result in high species diversity in a temperate-cold water marine genus *Dendronotus* (Gastropoda: Nudibranchia). Mol Phylogenet Evol 141:106609

Evans LA, Macey DJ, Webb J (1992) Calcium biomineralization in the radula teeth of the chiton, *Acanthopleura hirtosa*. Calcif Tissue Int 51:78–82

Evans LA, Macey DJ, Webb J (1994) Matrix heterogeneity in the radular teeth of the chiton *Acanthopleura hirtosa*. Acta Zool 75(1):75–79

Filippov AE, Gorb SN, Kovalev AE, Krings W (2023) Dynamical numerical model of the molluscan radula in its interaction with the substrate and food particles. Adv Theory Simul 6:2300055

Filippov AE, Gorb SN, Krings W (2025) Dynamic numerical model of mucus feeding in Gastropoda. Adv Theory Simul 8(4):2401405

Fretter V, Graham A (1976) A functional anatomy of Invertebrates. Academic Press, London

Golding RE, Ponder WF, Byrne M (2011) Three-dimensional reconstruction of the odontophoral cartilages of Caenogastropoda (Mollusca: Gastropoda) using micro-CT: morphology and phylogenetic significance. J Morphol 270(5):558–587

Gorb SN, Krings W (2021) Mechanical property gradients of taenioglossan radular teeth are associated with specific function and ecological niche in Paludomidae (Gastropoda: Mollusca). Acta Biomater 134:513–530

Gorb SN, Varenberg M, Peressadko A, Tuma J (2007) Biomimetic mushroom-shaped fibrillar adhesive microstructure. J R Soc Interface 4(13):271–275

Gordon L, Joester D (2011) Nanoscale chemical tomography of buried organic–inorganic interfaces in the chiton tooth. Nature 469:194–198

Graham A (1973) The anatomical basis of function in the buccal mass of prosobranch and amphineuran molluscs. J Zool Lond 169(3):317–348

Gray JE (1853) On the division of ctenobranchous gasteropodous Mollusca into larger groups and families. Ann Mag Nat Hist 11(2):124–133

Grunenfelder LK, de Obaldia EE, Wang Q, Li D, Weden B, Salinas C, Wuhrer R, Zavattieri P, Kisailus D (2014) Biomineralization: stress and damage mitigation from oriented nanostructures within the radular teeth of *Cryptochiton stelleri*. Adv Funct Mater 24(39):6093–6104

Guralnick R, Smith K (1999) Historical and biomechanical analysis of integration and dissociation in molluscan feeding, with special emphasis on the true limpets (Patellogastropoda: Gastropoda). J Morphol 241(2):175–195

Hackethal S, Schulz-Kornas E, Gorb SN, Krings W (2024) Wear patterns of radular teeth in *Loligo vulgaris* (Cephalopoda; Mollusca) are related to their structure and mechanical properties. Interface Focus 14:20230082

Haszprunar G, Götting KJ (2007) Mollusca, Weichtiere. In: Westheide W, Rieger G (eds) Spezielle Zoologie. Teil 1. Einzeller und wirbellose Tiere. Springer, Berlin

Hawkins SJ, Watson DC, Hill AS, Harding SP, Kyriakides MA, Hutchinson S, Norton TA (1989) A comparison of feeding mechanisms in microphagous, herbivorous, intertidal, prosobranchs in relation to resource partitioning. J Molluscan Stud 55(2):151–165

Heepe L, Gorb SN (2014) Biologically inspired mushroom-shaped adhesive microstructures. Annu Rev Mater Res 44:173–203

Herrera SA, Grunenfelder L, Escobar E, Wang Q, Salinas C, Yaraghi N, Geiger J, Wuhrer R, Zavattieri P, Kisailus D (2015) Stylus support structure and function of radular teeth in *Cryptochiton stelleri*. In: Proc 20th international conference on composite materials, Copenhagen

Hickman CS (1980) Gastropod radulae and the assessment of form in evolutionary paleontology. Paleobiology 6(3):276–294

Hickman CS (1984) Implications of radular tooth-row functional-integration for archaeogastropod systematics. Malacologia 25:143–160

Hickman CS, Morris TE (1985) Gastropod feeding tracks as a source of data in analysis of the functional morphology of radulae. Veliger 27(4):357–365

Hwang J, Jeong Y, Park JM, Lee KH, Hong JW, Choi J (2015) Biomimetics: forecasting the future of science, engineering, and medicine. Int J Nanomedicine 10:5701–5713

Hyman LH (1967) Mollusca I. Aplacophora, Polyplacophora, Monoplacophora. Gastropoda, the coelomate Bilateria. The invertebrates, vol 6. McGraw-Hill Book Company, New York

Janssen HH, Triebskorn R (1987) Comparative morphology of the radulae in *Pomatia elegans* and in *Littorina littorea* (Gastropoda: Taenioglossa). Zool Anz 219(1/2):73–82

Katsuno S, Sasaki T (2008) Comparative histology of radula-supporting structures in Gastropoda. Malacologia 50(1–2):13–56

Kehl CE, Wu J, Lu S, Neustadter DM, Drushel RF, Smoldt RK, Chiel HJ (2019) Soft-surface grasping: radular opening in *Alysia californica*. J Exp Biol 222(16):jeb191254

Krings W (2020) Trophic specialization of paludomid gastropods from 'ancient' Lake Tanganyika reflected by radular tooth morphologies and material properties. Dissertation, University of Hamburg

Krings W, Gorb SN (2021a) Substrate roughness induced wear pattern in gastropod radulae. Biotribology 26:100164
Krings W, Gorb SN (2021b) Radula packing and storage facilitated by tooth morphology in selected taenioglossan Gastropoda. J Molluscan Stud 87:eyab007
Krings W, Gorb SN (2023a) Foraging on sand particles: coatings of soft radular teeth (Paludomidae, Gastropoda, Mollusca). Invertebr Biol 142(3):e12410
Krings W, Gorb SN (2023b) Particle binding capacity of snail saliva. J Chem Phys 159:185101
Krings W, Gorb SN (2024a) Diversity of mineral systems, material compositions and mechanical properties in molluscan radulae. In: Lehnert MS (ed) Metals and their functional role in the structures of invertebrates. Biologically-inspired systems, vol 18. Springer, Cham, pp 173–214
Krings W, Gorb SN (2024b) Performance of biological food processing interfaces: perspectives on the science of mollusc radula. Biointerphases 19(3):030801
Krings W, Kovalev A, Glaubrecht M, Gorb SN (2019a) Differences in the Young modulus and hardness reflect different functions of teeth within the taenioglossan radula of gastropods. Zoology 137:125713
Krings W, Faust T, Kovalev A, Neiber MT, Glaubrecht M, Gorb SN (2019b) In slow motion: radula motion pattern and forces exerted to the substrate in the land snail *Cornu aspersum* (Mollusca, Gastropoda) during feeding. R Soc Open Sci 6:190222
Krings W, Marcé-Nogué J, Karabacak H, Glaubrecht M, Gorb SN (2020a) Finite element analysis of individual taenioglossan radular teeth (Mollusca). Acta Biomater 115:317–332
Krings W, Brütt J-O, Gorb SN, Glaubrecht M (2020b) Tightening it up: diversity of the chitin anchorage of radular-teeth in paludomid freshwater-gastropods. Malacologia 63(1):77–94
Krings W, Hempel C, Siemers L, Neiber MT, Gorb SN (2021a) Feeding experiments on *Vittina turrita* (Mollusca, Gastropoda, Neritidae) reveal tooth contact areas and bent radular shape during foraging. Sci Rep 11:9556
Krings W, Karabacak H, Gorb SN (2021b) From the knitting shop: the first physical and dynamic model of the taenioglossan radula (Mollusca: Gastropoda) aids in unravelling functional principles of the radular morphology. J R Soc Interface 18(182):20210377
Krings W, Neiber MT, Kovalev A, Gorb SN, Glaubrecht M (2021c) Trophic specialisation reflected by radular tooth material properties in an 'ancient' Lake Tanganyikan gastropod species flock. BMC Ecol Evol 21:35
Krings W, Marcé-Nogué J, Gorb SN (2021d) Finite element analysis relating shape, material properties, and dimensions of taenioglossan radular teeth with trophic specialisations in Paludomidae (Gastropoda). Sci Rep 11:22775
Krings W, Neumann C, Neiber MT, Kovalev A, Gorb SN (2021e) Radular force performance of stylommatophoran gastropods (Mollusca) with distinct body masses. Sci Rep 11:10560
Krings W, Kovalev A, Gorb SN (2021f) Influence of water content on mechanical behaviour of gastropod taenioglossan radulae. Proc R Soc B 288:20203173
Krings W, Kovalev A, Gorb SN (2021g) Collective effect of damage prevention in taenioglossan radular teeth is related to the ecological niche in Paludomidae (Gastropoda: Cerithioidea). Acta Biomater 135:458–472
Krings W, Brütt J-O, Gorb SN (2022a) Ontogeny of the elemental composition and the biomechanics of radular teeth in the chiton *Lepidochitona cinerea*. Front Zool 19:19
Krings W, Brütt J-O, Gorb SN (2022b) Micro-cracks and micro-fractures reveal radular tooth architecture and its functional significance in the paludomid gastropod *Lavigeria grandis*. Philos Trans R Soc A 380:20210335
Krings W, Matsumura Y, Brütt J-O, Gorb SN (2022c) Material gradients in gastropod radulae and their biomechanical significance: a combined approach on the paludomid *Lavigeria grandis*. Sci Nat 109:52
Krings W, Brütt J-O, Gorb SN (2022d) Elemental analyses reveal distinct mineralization patterns in radular teeth of various molluscan taxa. Sci Rep 12:7499
Krings W, Wägele H, Neumann C, Gorb SN (2023a) Coping with abrasive food: diverging composition of radular teeth in two Porifera-consuming nudibranch species (Mollusca, Gastropoda). J R Soc Interface 20:20220927

Krings W, Neumann C, Gorb SN, Koehnsen A, Wägele H (2023b) Elemental composition and material properties of radular teeth in the heterobranch snail *Gastropteron rubrum* (Mollusca, Gastropoda, Cephalaspidea) foraging on hard organisms. Ecol Evol 13(8):e10332

Krings W, Gorb SN, Neumann C, Wägele H (2024) Radular tooth coating in members of Dendronotidae and Flabellinidae (Nudibranchia, Gastropoda, Mollusca). J Morphol 285:e21773

Krings W, Riesel T, Kaiser TM, Daasch A, Schulz-Kornas E, Gorb SN (2025) The influence of the collective effect of tooth–tooth interaction on the feeding efficiency of gastropods: a biomimetic approach. J Bionic Eng 22:1364–1380

Lee AP, Brooker LR, Macey DJ, van Bronswijk W, Webb J (2000) Apatite mineralization in teeth of the chiton *Acanthopleura echinata*. Calcif Tissue Int 67:408–415

Lowenstam HA, Weiner S (1985) Transformation of amorphous calcium phosphate to crystalline dahllite in the radula teeth of chitons. Science 227:51–52

Lu D, Barber AH (2012) Optimized nanoscale composite behaviour in limpet teeth. J R Soc Interface 9(71):1318–1324

Macey DJ, Brooker LR, Webb J, St Pierre TG (1996) Structural organisation of the cusps of the radular teeth of the chiton *Plaxiphora albida*. Acta Zool 77(4):287–294

Mackenstedt U, Märkel K (1987) Experimental and comparative morphology of radula renewal in pulmonates (Mollusca, Gastropoda). Zoomorphology 107:209–239

Mackenstedt U, Märkel K (2001) Radular structure and function. In: Barker GM (ed) The biology of terrestrial mollusc. CABI Publishing, Oxon, pp 213–236

Mangan EV, Kingsley DA, Quinn RD, Sutton GP, Mansour JM, Chiel HJ (2005) A biologically inspired gripping device. Ind Robot 32(1):49–54

Märkel K (1957) Bau und Funktion der Pulmonaten-Radula. Z Wiss Zool 160:213–289

Märkel K (1964) Modell-Untersuchungen zur Klärung der Arbeitsweise der Gastropodenradula. Verh Dtsch Zool Ges 18:232–243

Montroni D, Zhang X, Leonard J, Kaya M, Amemiya C, Falini G, Rolandi M (2019) Structural characterization of the buccal mass of *Ariolimax californicus* (Gastropoda; Stylommatophora). PLoS One 14(8):e0212249

Morris TE, Hickman CS (1981) A method for artificially protruding gastropod radulae and a new model of radula function. Veliger 24:85–89

Moustafa AY, Abdel-Wahab M, El-Masry SM (2024) Radular teeth of *Conus* (Mollusca: Neogastropoda: Conidae) from the Red Sea: comparative morphology and character analysis. Molluscan Res 44:118–134

Neustadter DM, Drushel RF, Crago PE, Adams BW, Chiel HJ (2002a) A kinematic model of swallowing in *Aplysia californica* based on radula/odontophore kinematics and in vivo magnetic resonance images. J Exp Biol 205(20):3177–3206

Neustadter DM, Drushel RF, Chiel HJ (2002b) Kinematics of the buccal mass during swallowing based on magnetic resonance imaging in intact, behaving *Aplysia californica*. J Exp Biol 205(7):939–958

Nixon M (1995) A nomenclature for the radula of the Cephalopoda (Mollusca) – living and fossil. J Zool 236(1):73–81

Nosonovsky M, Rohatgi PK (2012) Biomimetics in materials science. Self-healing, self-lubricating, and self-cleaning materials. Springer, Cham

Oliver WC, Pharr GM (2004) Measurement of hardness and elastic modulus by instrumented indentation: advances in understanding and refinements to methodology. J Mater Res 19:3–20

Padilla DK (2003) Form and function of radular teeth of herbivorous molluscs: focus on the future. Am Malacol Bull 18(1–2):163–168

Padilla DK, Dittman DE, Franz J, Sladek R (1996) Radular production rates in two species of *Lacuna* Turton (Gastropoda: Littorinidae). J Molluscan Stud 62(3):275–280

Parkhaev PY (2008) The early Cambrian radiation of Mollusca. In: Ponder WF, Lindberg DR (eds) Phylogeny and evolution of the Mollusca. University of California Press, Berkeley, pp 33–69

Pohl A, Herrera SA, Restrepo D, Negishi R, Jung J-Y, Salinas C, Wuhrer R, Yoshino T, McKittrick J, Arakaki A, Nemoto M, Zavattieri P, Kisailus D (2020) Radular stylus of *Cryptochiton stelleri*:

a multifunctional lightweight and flexible fiber-reinforced composite. J Mech Behav Biomed Mater 111:103991
Runham NW (1962) Rate of replacement of the molluscan radula. Nature 194:992–993
Runham NW, Isarankura K (1966) Studies on radula replacement. Malacologia 5:73
Scheel C, Gorb SN, Glaubrecht M, Krings W (2020) Not just scratching the surface: distinct radular motion patterns in Mollusca. Biol Open 9(10):bio055699
Shaw JA, Macey DJ, Brooker LR (2008) Radula synthesis by three species of iron mineralizing molluscs: production rate and elemental demand. J Mar Biol Assoc UK 88(3):597–601
Shaw JA, Macey DJ, Brooker LR, Clode PL (2010) Tooth use and wear in three iron-biomineralizing mollusc species. Biol Bull 218(2):132–144
Simone LRL (2011) Phylogeny of the Caenogastropoda (Mollusca), based on comparative morphology. Arq Zool 42(4):161–323
Solem A (1972) Malacological applications of scanning electron microscopy II. Radular structure and functioning. Veliger 14:327–336
Solem A (1974) The shell makers: introducing mollusks. John Wiley & Sons, New York
Sollas IB (1907) The molluscan radula: its chemical composition, and some points in its development. Q J Microsc Sci 51:115–136
Stegbauer L, Smeets PJM, Free R, Wallace SG, Hersam MC, Alp EE, Joester D (2021) Persistent polyamorphism in the chiton tooth: from a new biomineral to inks for additive manufacturing. Proc Natl Acad Sci U S A 118(23):e2020160118
Steneck RS, Watling L (1982) Feeding capabilities and limitation of herbivorous molluscs: a functional group approach. Mar Biol 68:299–319
Sullivan TN, Wang B, Espinosa HD, Meyers MA (2017) Extreme lightweight structures: avian feathers and bones. Mater Today 20(7):377–391
Suresh Kumar N, Padma Suvarna R, Chandra Babu Naidu K, Banerjee P, Ratnamala A, Manjunatha H (2020) A review on biological and biomimetic materials and their applications. Appl Phys A Mater Sci Process 126(6):445
Ukmar-Godec T (2016) Mineralization of goethite in limpet radular teeth. In: Faivre D (ed) Iron oxides: from nature to applications. Wiley-VCH, Weinheim, pp 207–224
Ukmar-Godec T, Kapun G, Zaslansky P, Faivre D (2015) The giant keyhole limpet radular teeth: a naturally-grown harvest machine. J Struct Biol 192(3):392–402
Ukmar-Godec T, Bertinetti L, Dunlop JWC, Godec A, Grabiger MA, Masic A, Nguyen H, Zlotnikov I, Zaslansky P, Faivre D (2017) Materials nanoarchitecturing via cation-mediated protein assembly: making limpet teeth without mineral. Adv Mater 29(27):1701171
van der Wal P, Giesen HJ, Videler JJ (1999) Radular teeth as models for the improvement of industrial cutting devices. Mater Sci Eng C 7(2):129–142
Vincent JF (2009) P Biomimetics — a review. Proc Inst Mech Eng H 223(8):919–939
Wägele H (1983) Rasterelektronenmikroskopische Untersuchungen an Radulae einiger Nordseeschnecken (Gastropoda: Prosobranchia) mit Anmerkungen zur Funktionsmorphologie. Drosera 83(1):68–78
Walsby JR (1975) Feeding and the radula in the marine pulmonate limpet *Trimusculus reticulatus*. Veliger 18:139–145
Wang C, Li QY, Wang SN, Qu SX, Wang XX (2014) Microstructure and self-sharpening of the magnetite cap in chiton tooth. Mater Sci Eng C 37:1–8
Wealthall RJ, Brooker LR, Macey DJ, Griffin BJ (2005) Fine structure of the mineralized teeth of the chiton *Acanthopleura echinata* (Mollusca: Polyplacophora). J Morphol 265(2):165–175
Weaver JC, Wang Q, Miserez A, Tantuccio A, Stromberg R, Bozhilov KN, Maxwell P, Nay R, Heier ST, DiMasi E (2010) Analysis of an ultra hard magnetic biomineral in chiton radular teeth. Mater Today 13(1–2):42–52
Yang M, Kotov NA (2024) Quantitative biomimetics of high-performance materials. Nat Rev Mater 10:382–395

Chapter 4
Getting Beyond the Sticking Point: A Biodiversity-Aware Approach to Gecko-Inspired Adhesion Offers Potential Pathways for the Improvement of Future Biomimetic Designs

Austin M. Garner and Anthony P. Russell

Abstract It has been 25 years since Autumn and colleagues published their seminal research examining the adhesive forces generated by isolated gecko adhesive setae and identifying van der Waals forces as the major mechanism of adhesion in geckos. These findings served as the intellectual stimulus for thousands of studies aimed at synthetically replicating the form, function, and properties of gecko attachment, thereby establishing geckos as one of the most popular sources for biomimetic models to date. Despite these attempts to develop gecko-based synthetic adhesives (GSAs), several authors have noted that GSAs still fail to capture the performance and multifunctional properties exhibited by the natural system. In this chapter, we posit that the field of gecko-inspired adhesion represents an "inverted pyramid of observation and application," whereby the explosive interest in replicating this organismal attachment system is founded upon a rather small number of empirical studies of limited taxonomic representation, species examined, and direct observations of setal attachment mechanics.

We follow this with an overview of the immense diversity in the form and function of squamate fibrillar adhesive systems to provide researchers with the impetus to broaden the foundations upon which gecko-inspired adhesion is built. We then systematically review the 100 most-cited attempts to fabricate GSAs and assess their success in achieving the performance and properties of their natural counterpart. Subsequently, we employ this assessment to identify major successes and shortcomings of approaches to gecko-inspired adhesion and suggest avenues

A. M. Garner (✉)
Department of Biology and BioInspired Institute, Syracuse University, Syracuse, NY, USA

Present address: Department of Biological Sciences, Florida Atlantic University, Boca Raton, FL, USA
e-mail: garnera@fau.edu

A. P. Russell
Department of Biological Sciences, University of Calgary, Calgary, AB, Canada

V. Bels et al. (eds.), *Biomimicry – A Functional Approach*, Biologically-Inspired Systems 20, https://doi.org/10.1007/978-3-032-11949-0_4

for future "biodiversity-aware" research that we propose will improve biomimetic designs and their functional outcomes.

Keywords Gekkota · Anolidae · Scincidae · Fibrillar adhesion · Bio-inspired adhesion

4.1 Introduction

In this chapter, we set forth the idea that a more biodiversity-aware approach to gecko-inspired adhesion might be advantageous for permitting further progress in the technological application of the remarkable attachment properties of these animals (and other lizards with similar attachment abilities). Not only geckos, but also anoles (of the genus *Anolis*) and some lineages of skinks exhibit adhesive capabilities based upon similar modifications of the integument of their digits (and, in some geckos and skinks, their tails also). Our motivation for suggesting this more biodiversity-conscious approach is that the adhesive systems exhibited by these three, rather distantly related, lineages of squamates exhibit (both within and between them) not only diversity but also disparity. Focusing narrowly on a few favored exemplars and generalizing, based upon a relatively small number of detailed assessments of the adhesive properties and mechanics of their adhesively competent integumentary modifications, is likely to have constrained our abilities to tease out general principles. There seemingly has been too much reliance placed upon a generalized model assumed to be representative of "geckos" as a whole (notwithstanding that anoles and skinks have barely registered in these considerations).

Recent breakthroughs have resulted from approaches applying nanotechnological techniques to explore squamate-based phenomena that were previously largely intractable. Impressive as this progress has been, we suggest that endeavors to duplicate the adhesive capabilities of geckos (Russell and Johnson 2009), an evolutionary "innovation" that arose at least 100 million years ago (Bauer et al. 2005; Arnold and Poinar 2008; Daza et al. 2016; Fontanarrosa et al. 2018), requires a more complete understanding of the enormous diversity that squamates display in their adhesive systems and the ways in which they develop, grow, and deploy their adhesive structures. Over the past 20 years, a modest degree of reciprocity has occurred between pure and applied research as they relate to this issue, but increased interdisciplinarity is needed to attack crucial issues and maintain progress. We contend that reliance upon data accumulated almost entirely from a single species (the Tokay gecko, *Gekko gecko*—see below) as the foundation for biomimetic simulacra may have constrained approaches endeavoring to mimic the properties of such structures and fabricate synthetic representations. It is our contention that (pun intended) a sticking point, "a situation that is likely to cause an impasse" (Editors of the American Heritage Dictionaries 2016), has likely been reached in biomimetic approaches to the employment of gecko, and other squamate, adhesive capabilities because of an over-reliance on the aforementioned single "model" that has been over-generalized as being genuinely representative of filamentous squamate adhesives and their fundamental properties. Furthermore, the "model" that has been adopted is seemingly more structurally complex than is desirable for teasing out basic principles and properties.

In a review of progress toward understanding the integrative aspects of gecko adhesion, Russell et al. (2019) laid out five grand challenges that will be of vital importance for harnessing "the underlying principles for human applications." These are as follows: (1) employ the diversity of geckos to understand how the adhesive system is tuned to environmental challenges; (2) explore developmental pathways that determine setal structure associated with diversity exhibited both taxonomically and across the extent of the setal fields of individual species; (3) enhance our understanding of how geckos use their adhesive system in nature to improve our conceptualization of how gecko setae function; (4) improve our understanding of how gecko-inspired synthetic adhesives perform under combinations of variable environmental conditions; and (5) compare, contrast, and synthesize the diversity of fibrillar adhesives in the natural world. In this contribution, we concentrate on the first, and to a lesser extent the fifth, of the above-mentioned grand challenges, with the hope that a broader appreciation of the diversity and disparity exhibited by the adhesive structures of geckos, anoles and skinks, from the macroscopic to the nanoscopic scale, will shed light on how to further explore them and how to approach the remaining three grand challenges.

We begin by reviewing the discoveries made about gecko adhesion in the first decade of this millennium and summarizing the enormous influence that these observations had on the conceptualization of biomimetic designs and applications. We conclude this part of our contribution with what we perceive to be the current state of affairs in the field of gecko-inspired adhesion and why a greater appreciation of the diversity and disparity of the adhesive apparatus of squamates is likely essential for continued progress in this endeavor.

Following this, we take a step back to present a historical overview of scientifically based fascination with the adhesive capabilities of geckos, indicating the early importance that the diversity of digit form displayed by this group had on the appreciation of its collective adhesive prowess, prior to focus becoming narrowed to a single exemplar for biomimetic inspiration. We then further expound upon how a more broad-scale exploration of the biodiversity and structural disparity of the squamate adhesive apparatus might be profitable for identifying its fundamental design principles that can be more readily adapted for various biomimetic applications.

The core of this contribution is devoted to the documentation of the form and function of the adhesive systems of the three squamate lineages that possess them (geckos, anoles, and skinks). For each we identify the major phylogenetic lineages expressing these features, the diversity of toe pad morphology displayed by each, the diversity of integumentary derivatives that play a role in adhesion, the intraindividual variation in these derivatives (where this is known), and comment upon what is known about interspecific adhesive performance. The amount of coverage afforded each of these squamate lineages is, of necessity, disproportionate.

Much more is known about this for geckos than anoles and skinks, but by surveying what is known for each group, we can begin to tease out the common and fundamental principles that underpin all three. We compare the structure of these adhesion-promoting devices in these three clades, noting that they have originated independently in each, and in geckos, and possibly skinks, on multiple occasions within the evolutionary history of these lineages.

We conclude by synthesizing our findings about the commonalities between the adhesive mechanisms of geckos, anoles, and skinks and use that information to identify major findings, successes, and shortcomings of gecko-inspired adhesion to date, and to suggest future research directions that might help improve biomimetic designs and applications.

4.2 The Impetus for Interest in Gecko-Inspired Adhesion

Through measurement of attachment forces generated by isolated individual setae, applied to a surface at a specific orientation and under tensile loading, Autumn et al. (2000) found that these were consistent with intermolecular attachment via van der Waals forces, but not wholly consistent with suction, friction, and/or micro-interlocking (see below). Because van der Waals forces are universal and do not require the presence of a permanent dipole, an adhesive system configured to exploit them would be capable of adhering to almost any polarizable surface. Autumn et al.'s (2002) subsequent observations refuted the potential differential ability of setae to attach to hydrophilic versus hydrophobic surfaces, with Hiller's (1968, 1969) findings being attributed to the experimental conditions employed rather than preferential capability. The elucidation of van der Waals forces as being responsible for gecko attachment triggered an explosion of interest in this topic, not only of biologists but also physicists, chemists, computer scientists, engineers, materials scientists, and biomimeticists.

4.2.1 Limitations Imposed by the Employment of a Single Exemplar Species on the Conceptualization of Biomimetic Designs

Xu et al. (2018) state that "the hierarchical structures of the adhesion mechanisms [of geckos] have been studied extensively" and that "many research groups have studied the gecko adhesive system." They then note that, "many gecko-inspired materials and devices have been fabricated for tailor designed purposes." In reality, however, very few studies have been conducted on the setae of geckos with regard to their functional properties. Although over 2300 species of gecko have been described (Uetz et al. 2025), approximately 60% of which possess adhesive toe pads, essentially only one species, *Gekko gecko*, has been explored in association with the adhesive properties of its individual setae. *Hemidactylus frenatus* was ostensibly investigated by Sun et al. (2005), but its adhesion forces per spatula were obtained from whole digit preparations, rather than from isolated setae, and the setae of this species were not documented with regard to their dimensions, density, or branching patterns. Furthermore, all of the adhesively related physical attributes of the setae of *Hemidactylus frenatus* were considered to be identical to those of *Gekko gecko* (Sun et al. 2005).

Estimates of the combined attributes and properties of the setal fields of *Gekko gecko* have, therefore, been based upon extrapolations from an "idealized" seta (with regard to length, stalk diameter, setal tip dimensions, and branches per seta) that has been assumed to repeat in an identical fashion throughout the entirety of the toe pad (Autumn et al. 2000). The initial assessments of the high adhesive force able to be generated by a single seta (Autumn et al. 2000) were likely attributable to the difficulty of effectively calibrating the miniscule cantilevers used with microelectromechanical systems (MEMS) force sensors (Hill et al. 2011), and more recent assessments have yielded much lower values. The smaller range of variation for the adhesion and pull-off forces (Huber et al. 2005a, b, 2007; Xu et al. 2018) may represent actual variation or may be due, to some extent, to a combination of their contact with different surfaces, the potential effects of focused ion beam micromachining of the setae to trim away all but a few spatulae (Huber et al. 2005a, b, 2007; Xu et al. 2015) and differences in spatula tip diameter (Autumn et al. 2000, 2002; Huber et al. 2005b, 2007; Sun et al. 2005). Overall, a body of data has been assembled for the dimensions and forces generated by the setae of *Gekko gecko* (Table 4.1), but it is questionable that these measurements can be considered typical of geckos as a whole, or even all of the setae of *Gekko gecko* (see below).

4.2.2 The Current State of Affairs of Gecko-Inspired Adhesion

The increase in investigative interest sparked by the findings of Autumn et al. (2000) led to reassessment of the role of certain types of frictional forces in gecko attachment (Autumn et al. 2006a; Tian et al. 2006). At the molecular level, the strength of a frictional force is proportional to the true area of contact between two surfaces (Bowden and Tabor 1966; Bhushan et al. 1995), where surfaces are not smooth but instead bear many asperities, limiting true contact between surfaces to discrete contact points. These investigations found that the same conditions that promote attachment via van der Waals forces also favor friction at the nanoscale (Huber et al. 2005b; Tian et al. 2006; Puthoff et al. 2010). Capillary forces, resulting from the surface tension of a layer of liquid between two contacting surfaces (Petrucci and Harwood 1993), have also been proposed to contribute to gecko adhesion, wherein they can act as an adhesive if surface wettability is high (Huber et al. 2005b; Sun et al. 2005; Kim and Bhushan 2008). The capillary bridges formed between two surfaces are resistant to perpendicular forces, but will slide along the

Table 4.1 Dimensions of toe pads and setal dimensions as well as measured adhesive forces recorded for individual isolated setae of geckos. All data reported, save for one value, are attributable to *Gekko gecko*. Cells with gray fill indicate that no pertinent data for that category of information were provided by the study represented by each column

	Autumn et al. (2000)	Autumn et al. (2002)	Huber et al. (2005a)	Huber et al. (2005b)	Sun et al. (2005)	Sun et al. (2005)	Huber et al. (2007)	Hill et al. (2011)	Xu et al. (2015)	Xu et al. (2018)
Species	*Gekko gecko*	*Gekko gecko*	*Gekko gecko*	*Gekko gecko*	*Gekko gecko*	*Hemidactylus frenatus*	*Gekko gecko*	*Gekko gecko*	*Gekko gecko*	*Gekko gecko*
Total toe pad area for 20 digits	400 mm^2				227 mm^2					
Number of setae per digit										Millions
Seta length	30–130 µm		100 µm							
Setal density	5000 setae/mm^2	14400 setae/mm^2			5300 setae/mm^2					
Seta diameter			2–10 µm				10 µm			
Number of spatulae per seta	100 - 1000	100s	100-1000	100-1000	100s		100s–1000s			100s
Estimated total number of spatulae for 20 digits	1 billion		1 billion	1 billion						
Spatula tip radius	0.2–0.5 µm	0.23–0.44 µm	0.2 µm		0.1 µm		0.2 µm			
Estimated total seta force for 20 digits	40 N			10 N	20 N					
Adhesive force per seta	194 ± 25 µN	40.4– 1.3 µN			20 µN			14 µN		
Estimated force range per seta	40–400 µN									
Adhesion force per spatula			14.0 ± 1.7 nN	~ 10.0 nN		11.8 ± 2.2 nN				7.0 – 10.5 nN
Pull-off force per spatula			18.4 ± 3.5 nN	10.0 nN			15.6 nN			

surface, and thus they function best on rough surfaces, with asperities that can act as anchor points for the fluid. More recently, it has been advocated that the relatively poorly understood phenomenon of contact electrification (Izadi et al. 2014) may also contribute, and perhaps dominate.

The observation that geckos are unable to stick to hydrophobic, weakly polarizable surfaces (such as polytetrafluoroethylene, commonly known as Teflon™) could be due to either reduced capillary adhesion or reduced van der Waals forces (Autumn et al. 2002). To resolve this, polarizability was separated from hydrophilicity/hydrophobicity, revealing that single setae can attach equally well to both hydrophilic and hydrophobic polarizable surfaces, suggesting that van der Waals forces, rather than capillary forces, are the dominant forces of attachment in geckos. Additional support for attachment via van der Waals forces resulted from the discovery that the setal fields of geckos are superhydrophobic (Autumn and Hansen 2006; Liu et al. 2012), which likely reduces the effectiveness of capillary adhesion. Spatular pull-off forces increase with increasing relative humidity, however, indicating that a water layer between spatulae and the substrate may play a role in enhancing adhesion (Huber et al. 2005b; Sun et al. 2005; Pesika et al. 2009), although a large drop in adhesive strength occurs when the adhesive system is submerged in water and attached to a hydrophilic substrate (Stark et al. 2012, 2013).

Overall, a consistent picture of gecko frictional adhesion, dominated by van der Waals forces, is the currently favored explanation (Bhushan and Sayer 2007b; Autumn and Gravish 2008; Autumn et al. 2014; Niewiarowski et al. 2016). Most biomimetic endeavors are based upon this interpretation. Central to most biomimetic approaches is the seemingly canonical statement that a single seta of *Gekko gecko* can produce up to 200μN of force, this being consistent with estimated values of setal force of 40–400μN obtained from a mathematical model based on the Johnson–Kendall–Roberts theory of van der Waals forces (Johnson et al. 1971; Israelachvili 1992). For such a seta, a shear force of at least double the adhesive force is required for the adhesive force to be maintained (Autumn et al. 2006a), and measurements from isolated setae of 200μN of shear force and 40μN of adhesive force (Autumn et al. 2000, 2002) suggest that that this factor may often be exceeded. Setae are unidirectional structures with a non-adhesive default state, preventing inappropriate adhesion to surfaces (Hansen and Autumn 2005; Autumn and Hansen 2006; Hu et al. 2012).

The frictional adhesion model also helps to explain how detachment of setae occurs. The relationship between shear and adhesion forces is consistent with the existence of a critical detachment angle, which for *Gekko gecko* is approximately 30° (Autumn et al. 2000). At shear forces less than twice the adhesive force, and at an angle of greater than the critical detachment angle, adhesive forces can no longer be maintained, and setae will release from the surface with no measurable detachment forces (Autumn et al. 2006a).

4.3 Revisiting the History of Fascination with Gecko Adhesion: Insights into the Importance of Considering Biodiversity

The remarkable locomotor abilities of geckos have intrigued observers for millennia. Aristotle commented on them (Autumn and Peattie 2002) and during the very late eighteenth and early nineteenth centuries, when biology was formally established as a field of scholarly endeavor (Russell 1916), interest in their remarkable capabilities burgeoned. By 1830, the generic names *Cyrtodactylus*, *Gymnodactylus*, *Hemidactylus*, *Phyllodactylus*, *Platydactylus*, *Ptyodactylus*, *Sphaerodactylus*, *Stenodactylus*, and *Thecadactylus* were part of the gekkotan taxonomic lexicon (Kluge

1993; Russell and Gamble 2019), the diversity of digit form exhibited by geckos clearly influencing how they were perceived (Merrem 1820; Wagler 1830; Duméril and Bibron 1836; Fitzinger 1843; Russell and Bauer 2002).

Prior to anatomical exploration of gecko digits, various speculative interpretations of the relationship between their peculiar morphological features and their functional roles were put forth. For example, Lacépède (1788, 1802) reported, in reference to *Gekko gecko* (his account of "the gecko" is actually an amalgam of observations on *G. gecko* and additional, unnamed species) that their toes were the source of noxious fluids that could be injurious or fatal to people. However, he also noted, in his account of *Tarentola mauritanica* (under the name "Geckoette"), that it was able to climb nimbly on walls, the underside of roofing beams of houses, and the vaulted ceilings of churches using its claws and the scales on the ventral surfaces of its digits. Schneider (1812) referred to geckos as "*Wandkletterer*" (wall climbers) (Bauer 2019), but it was not until Home (1816a), pondering the ability of house flies to walk inverted upon the ceiling, turned to recently made behavioral observations of *Gekko gecko* by Joseph Banks on Java, combining them with his own morphological explorations. Banks reported (Home 1816a) that *Gekko gecko*, "comes out of an evening, from the roofs of the houses, and walks down the smooth hard polished chinam walls, in search of the flies that settle upon them, which are its natural food, and then runs up again to the roof of the house." Home (1816a) concluded that the ability to adhere to, and swiftly navigate, such smooth surfaces was due to the digits generating and breaking vaccua, these being sealed by "the rows of a beautiful fringe" (later to be identified as setae) (Home 1816b).

Following Home's (1816a, b) pioneering anatomical investigations, histological observations of the digits of geckos (Cartier 1872; Braun 1878; Haase 1900; Schmidt 1912, 1913) revealed the presence, general structure (including branching), and distribution of subdigital setae and also established that the toe pads lack glands, thereby refuting the involvement of secretions (see below). These observations led to various hypotheses about how attachment to surfaces was brought about. Suction, initially advanced by Home (1816a), was further elaborated upon (Wagler 1830; Simmermacher 1884; Tornier 1899; Gadow 1901; Kunitzky 1903). Competing hypotheses based upon sticky secretions (Cartier 1872), electrostatic attraction (Haase 1900; Schmidt 1905), friction (Hora 1923), and the interlocking of setae with surface irregularities (a more nuanced version of frictional interactions) (Mahendra 1941; Altevogt 1954) were advanced.

The suggestion that setae have a very high coefficient of friction (Hora 1923) led to the hypothesis that geckos attach by way of frictional forces, and this idea remained dominant for many years. Mahendra (1941) visualized setae as structures capable of microinterlocking with substrate asperities, but this was rejected (Maderson 1964) on structural grounds and later by the demonstration that gecko adhesion can occur on molecularly smooth surfaces (Autumn et al. 2000).

The contemporaneous ability to test many of the aforementioned hypotheses was hampered by the lack of applicable technology and/or appropriately developed physical theory. Dellit (1934, 1949), employing a combination of detailed morphological investigation and carefully controlled experiments on living geckos, was able to convincingly refute suction as a contributor, confirming Weitlaner's (1902) rejection of this agency. Dellit (1934, 1949) also cast doubt on theories involving friction and electrostatic attraction, the latter having already been criticized by Hora (1923). Later, however, Maderson (1964) and Hiller (1968) advocated that electrostatic attraction might be supplementary to other agents of attachment (see below).

Inspired by the work of Haase (1900), who had first suggested that molecularly based adhesion forces were responsible for clinging in geckos, Hiller (1968, 1976) predicated his investigations and interpretations on the recently elucidated concept of free surface energy and the molecular polarity of surfaces. He developed ways of measuring and manipulating the surface energy of specific substrates (polytetrafluoroethylene films) and recording the clinging ability of geckos under varying free surface energy conditions (Russell et al. 2019). Hiller's (1968, 1969) experiments suggested that a gecko's ability to adhere was directly related to the surface energy of the substrate. He discovered that setae could stick to hydrophilic surfaces but were less capable of adhering to hydrophobic ones. This suggested that the adhesive force generated by setae increases with increasing surface energy. Based on this, Hiller (1968, 1969, 2000) concluded that attachment occurs via intermolecular forces, but was unable to conclusively determine which type of intermolecular forces dominated. Hiller's (1968, 1969, 2000) explanations remained the most compelling in the understanding of these phenomena for the next 30 years (Russell et al. 2014).

The application of electron microscopy (Altevogt 1954; Ruibal and Ernst 1965; Hiller 1968, 1976; Peterson 1983a), to the investigation of gecko (and anole) setae revealed more detail about their structure and geometry and furnished information about interspecific variation in setal form. The structure of the tips of gecko setae was resolved by Ruibal and Ernst (1965) and led to their suggestion that the large total surface area they contributed was compatible with their operation via friction. Hiller's (1968) investigations resulted in the re-examination of intermolecular forces (true "adhesive" forces) being responsible for gecko attachment. The highly branched nature of setae and their very numerous, very small spatulate tips were hypothesized to provide increased contact area (Niewiarowski et al. 2016) and thereby promote adhesion as well as friction between them and the substratum.

The β-keratin of which setae are composed is a very stiff material with a fairly high elastic modulus (Autumn et al. 2006b). Fine subdivision of setal tips and the high aspect ratio of the setal shafts, however, result in a considerable decrease in the effective elastic modulus of the β-keratin, allowing setae to act as flexible attachment structures and to conform more easily to irregular surfaces (Autumn et al. 2006b, 2014).

Setae have dimensions (Russell et al. 2007) that are conducive to their behavior as pressure-sensitive adhesive structures, enhancing their ability to conform and adhere to microscopically rough surfaces and to comply with surface irregularities (Jagota and Bennison 2002; Campolo et al. 2003; Persson 2003; Persson and Gorb 2003; Persson et al. 2004; Spolenak et al. 2005; Autumn et al. 2006b). Such properties exploit both the high tensile strength of β-keratin (Bonser and Purslow 1995; Bonser

2000; Dalla Valle et al. 2007), and the reduced elastic modulus that results from the elaboration of setae into cantilevered beams (Zhao et al. 2009; Hu and Greaney 2014) that enhance adhesion via van der Waals interactions.

Maderson (1964, 1966) and Russell (1975, 1981, 1986) used *Gekko gecko* as foundation for their establishment of the general morphology and anatomy of the gekkotan adhesive system, this species being selected because of its large size and its relative ease of availability. Subsequently, Russell (1976, 1979) expanded this coverage to reveal the range of variation of the adhesive apparatus in the various gekkotan clades. Despite this wide range of form, *Gekko gecko* remained the main species of focal interest for those conducting experimental studies on gecko adhesion, from the macroscopic to the nanoscopic (e.g., Autumn et al. 2000, 2002; Zaaf et al. 2001a, b; Autumn and Peattie 2002; Huber et al. 2005b; Alibardi 2009, 2018; Stark et al. 2012, 2013, 2015). Thus, most of our understanding of the mechanism of adhesion in geckos in general is based on just a few species and dominated by data pertaining to *Gekko gecko*. There is, of course no reason to suppose that other geckos do not employ the same basic principles (Bauer 2019), but such a narrow focus may have constrained our understanding of the fundamentals through unwarranted generalization.

4.4 Exploring the Biodiversity of the Squamate Adhesive "Apparatus" to Identify Fundamental Design Principles

Given the enormous variation in the basic dimensions of gekkotan setae (Table 4.2) and the variable results emanating from studies of the dimensions (Russell and Garner 2021, Table 4.1) and adhesive properties of isolated setae of *Gekko gecko* (Table 4.1), it is likely that the variation in setal form of *Gekko gecko* (Russell et al. 2007; Russell and Garner 2021; Fig. 4.3), the subject of all investigations into the adhesive properties of isolated gekkotan setae, has contributed to the variation of the adhesively related data reported for this species. None of the studies cited in Table 4.1 indicate where the examined setae were isolated from on the toe pad. Reliance on such data has led to highly reductionistic statements about the adhesive properties of what have been referred to in a very generalized sense as "gecko lizards" (Xu et al. 2018). These conceptualizations contribute to the great oversimplification of the likely considerable variation in adhesive attributes of the setae both within the toe pads of individual gecko species and throughout this highly speciose and geographically widespread radiation (Table 4.2). This level of generalization, however, has served as the foundation for hundreds of studies devoted to the fabrication of synthetic adhesives based upon the fibrillar attachment structures of geckos. This constitutes an inverted pyramid in which a small amount of foundational evidence about the "idealized" setae of a single species is used to infer generalities about geckos, as a whole, in large numbers of applied studies. We contend that for biomimetic progress on this fascinating system to get beyond

Table 4.2 Setal dimensions (stalk length, setal basal diameter, stalk density) recorded for gekkotans

Toepad pattern	Species	Family	Stalk length µm	Stalk basal diameter µm	Stalk density per mm²	Source
c	*Diplodactylus vittatus*	Diplodactylidae	26	No data	No data	Russell (1972)
	Strophurus intermedius	Diplodactylidae	42	No data	No data	Russell (1972)
	Heteronotia binoei	Gekkonidae	8.14–34.07	0.39–1.23	90000–1960000[a]	Riedel et al (2021)
d	*Amalosia lesueurii*	Diplodactylidae	17	1.5–3.0	150000	Schleich and Kästle (1986)
	Oedura tryoni	Diplodactylidae	46	No data	No data	Russell (1972)
e	*Sphaerodactylus cinereus*	Sphaerodactylidae	10–85	No data	No data	Ruibal and Ernst (1965) and Röll (1995)
f	*Ptyodactylus hasselquistii*	Phyllodactylidae	78.5	No data	No data	Russell (1972)
g	*Naultinus elegans*	Diplodactylidae	15	0.8	621000	Bauer (1998)
	Naultinus rudis	Diplodactylidae	21	0.8	490000	Bauer (1998)
h	*Eurydactylodes vieillardi*	Diplodactylidae	19	No data	363000	Bauer (1998)
	Correlophus ciliatus	Diplodactylidae	32.7	1.1	300000	Griffing et al (2021)
	Correlophus sarasinorum	Diplodactylidae	36	1.3	134000	Bauer (1998)
	Rhacodactylus auriculatus	Diplodactylidae	38	0.6	172000	Bauer (1998); Griffing et al (2021)
	Gekko gecko	Gekkonidae	10–130	1.0–10.0	5300–14400	Alexander (2006); Alibardi and Toni (2005); Pugno and Lepore (2008); Hansen and Autumn (2005); Huber et al (2005b); Niewiarowski et al (2016); Sun et al (2005); Russell et al (2007); Russell and Garner (2021).

(continued)

Table 4.2 (continued)

	Gekko kuhli	Gekkonidae	24–91	No data	No data	Schleich and Kästle (1986)
	Gekko vittatus	Gekkonidae	37–78	3.5	25600	Schleich and Kästle (1986)
	Chondrodactylus bibronii	Gekkonidae	20–124	1.39–6.26	9000–64000	Schleich and Kästle (1986), Webster et al (2009), and Johnson and Russell (2009)
	Tarentola annularis	Phyllodatylidae	114	No data	No data	Schleich and Kästle (1996)
	Tarentola caboverdiana	Phyllodactylidae	27–68	4.0–4.5	26000	Schleich and Kästle (1986)
i	*Bavayia cyclura*	Diplodactylidae	28–32	1.3	429000	Russell (1972); Bauer (1998)
	Bavayia sauvagii	Diplodactylidae	31	1.5	335000	Bauer (1998)
j	*Pseudothecadactylus lindneri*	Diplodactylidae	37	1.2	142000	Bauer (1998)
	Thecadactylus rapicauda	Phyllodactylidae	50–60	1.8	6000	Schleich and Kästle (1986)
m	*Rhoptropus afer*	Gekkonidae	8–105	0.82–4.27	9000–36000	Johnson and Russell (2009)
	Rhoptropus barnardi	Gekkonidae	33.5–96.5	1.2–6.7	9000–36000	Johnson and Russell (2009)
	Rhoptropus biporosus	Gekkonidae	30–114	1.2–6.54	9000–36000	Johnson and Russell (2009)
	Rhoptropus boultoni	Gekkonidae	16–102	1.34–7.02	9000–25000	Johnson and Russell (2009)
	Rhoptropus bradfieldi	Gekkonidae	10-119	0.85–7.09	9000–49000	Johnson and Russell (2009)
	Rhoptropus diporus	Gekkondae	9–107	0.83–6.01	9000–49000	Johnson and Russell (2009)
o	*Gonatodes ocellatus*	Sphaerodactylidae	3.23–5.26	0.52–0.61	530468–557749	Russell et al (2015)
p	*Cyrtodactylus novaeguineae*	Gekkonidae	11	No data	No data	Russell (1972); (1976)
	Cyrtodactylus louisiadensis	Gekkonidae	17.2	No data	No data	Russell 1(972); 1976
	Gonatodes humeralis	Sphaerodactylidae	4.78–15.12	0.51–0.67	299991–505140	Russell et al (2015)
q	*Dactylocnemis pacificus*	Diplodactylidae	17–20	1.5	280000	Schleich and Kästle (1986); Bauer (1998)

(continued)

Table 4.2 (continued)

	Toropuku stephensi	Diplodactylidae	17	0.8	458000	Bauer (1998)
	Woodworthia maculata	Diplodactylidae	13	0.6	342000	Bauer (1998)
	Aristelliger praesignis	Sphaerodactylidae	46-57	1.3	No data	Ruibal and Ernst (1965)
r	*Hemidactylus bouvieri*	Gekkonidae	10–50	1.5–2.5	110000	Schleich and Kästle (1986)
	Hemidactylus frenatus	Gekkonidae	84	No data	No data	Dellit (1934)
	Hemidactylus mabouia	Gekkonidae	117	No data	No data	Dellit (1934)
	Hemidactylus persicus	Gekkonidae	107	No data	No data	Russell (1972)
	Hemidactylus turcicus	Gekkonidae	95	No data	No data	Dellit (1934)

[a]The family (Fig. 4.1) to which each species belongs is indicated, and the digital morphological pattern exhibited by each species is indicated by a lower-case letter that corresponds to the pattens illustrated in Fig. 4.2. Subjectively assigned relative stalk length (short, medium, long), basal diameter (narrow, moderate, broad) and density (low, moderate, high) is indicated by the color-coding convention of blue, green, red in ascending order for each. Stalk length: <45 μm = short; 46–90 μm = medium; >91 μm = long. Stalk basal diameter: <2 μm = narrow; 3–5 μm = moderate; >6 μm = broad. Stalk density: 0–100000 setae/mm^2 = low; 101000 – 300000 setae/mm^2 = moderate; >300000 setae/mm^2 = high adensity estimated through values of setal spacing

its current sticking point, taxa exhibiting structurally less elaborate setae than those taken to be typical of *Gekko gecko* (Autumn et al. 2000, 2002; Huber et al. 2005a, b, 2007; Sun et al. 2005; Hill et al. 2011; Xu et al. 2015, 2018) become the focus of extensive investigation in an attempt to reassess the fundamentals of dimensions, mechanics, and adhesive attributes of squamate filamentous adhesives. It is noteworthy that Griffing et al. (2021) report that the diplodactylid gecko *Correlophus ciliatus*, with relatively short and moderately densely packed setae (Table 4.2) that are much less profusely branched than those of *Gekko gecko,* can generate whole animal adhesive forces on par with those of the latter. Similarly, anoles (see below) can generate considerable adhesive forces from their fields of densely packed, short, unbranched setae (Garner et al. 2021).

4.4.1 Diversity in Form and Function of Gecko Adhesive Systems

4.4.1.1 Major Groups

The extant members of the Gekkota are currently assigned to one of the seven family-level clades: Diplodactylidae; Carphodactylidae; Pygopodidae; Eublepharidae; Sphaerodactylidae; Phyllodactylidae; and Gekkonidae. These families are interpreted as having the pattern of cladistic relationships depicted in Fig. 4.1. Gamble et al. (2012) posited that adhesive toe pads have been gained and lost, with approximately equal frequency, on multiple independent occasions during the phylogenetic history of the Gekkota, with at least one origin in each of

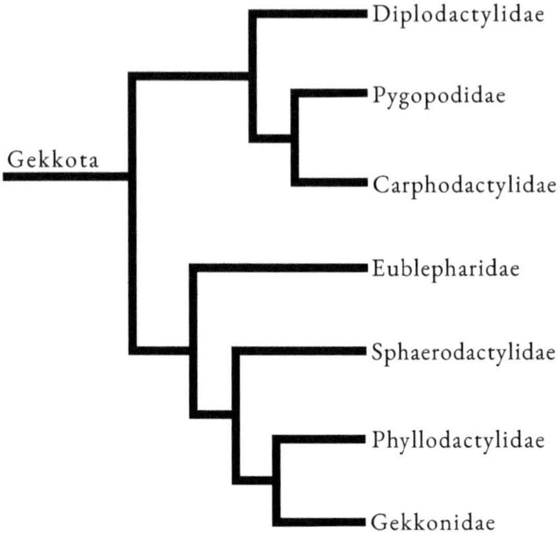

Fig. 4.1 Phylogenetic relationships of the gekkotan families. Note that this tree is not time calibrated

the Diplodactyldae, Sphaerodactylidae, Phyllodactylidae, and Gekkonidae. The Carphodactylidae and Eublepharidae were interpreted to have never possessed adhesive toe pads (Gamble et al. 2012). The Pygopodidae includes exclusively functionally limbless species that lack digits. A more comprehensive analysis of the anatomy of the digits of all six limb-bearing families of the extant Gekkota (Russell and Gamble 2019) revealed a great deal of variation in the structure of the adhesive toe pads, both within and between families, strengthening the likelihood of multiple evolutionary origins of this functional complex. Russell and Gamble (2019, Fig. 4.1) indicated that the ancestral state for the crown group Gekkota was the absence of adhesive toe pads and that minimally there had been one origin of adhesively competent digits in the Diplodactylidae, four in the Sphaerodactylidae, five in the Phyllodactylidae, and four in the Gekkonidae.

It is evident, also, that some gekkotans of the Cretaceous and Lower Eocene (Bauer et al. 2005; Arnold and Poinar 2008; Daza et al. 2016) possessed toe pads with similar morphology to those of some extant lineages, and that digital modifications associated with scansoriality were evident in gekkotans as far back as the Jurassic (Simões et al. 2017). Currently, there is no evidence that the toe pads of stem gekkotans (Daza et al. 2016; Gamble et al. 2017) were inherited by the most recent common ancestor of the crown group Gekkota, thus indicating an extended history of the evolutionary transition to toe pads in this clade. Furthermore, recent evidence of the transition to adhesively competent incipient toe pads within the extant gekkotan radiation (Russell et al. 2015; Higham et al. 2017a; Riedel et al. 2021, 2024) provides further evidence of multiple origins of toe pad-based adhesion in geckos.

All known instances of adhesive toe pads in geckos employ filamentous modifications of the Oberhäutchen layer (Maderson 1970) of the subdigital (and in some instances the subcaudal—Tornier 1899; Vitt and Ballinger 1982; Bauer 1998; Alibardi and Bonfitto 2019; Griffing et al. 2021) epidermis, known as setae (Russell and Garner 2021). Anatomical and behavioral specializations combine to permit controlled setal application to and release from the locomotor substratum (Russell 2002), although different expressions of these basic attributes occur in different lineages (Russell and Gamble 2019). Given the multiple transitions to adhesive competence of the digits of geckos and their associated anatomical and behavioral differences, variation in the form of the setae is to be expected (e.g., Riedel et al. 2021). Comparison of toe pad morphology across the Gekkota and exploration of setal variation should, therefore, be informative in determining the basic attributes of these features and help to provide insights that might improve the design and success of biomimetic facsimiles.

4.4.1.2 Diversity of Gekkotan Toe Pad Morphology

The multiple origins of the digital (and caudal) adhesive system of geckos (Russell 2002; Russell and Gamble 2019) is seemingly reflected by the wide array of configuration of the external morphology of their digits. Digital features have figured extensively in the naming of genera, indicative of this diversity across the Gekkota. It is evident that internal digital anatomy adds even more variation to this mix and that similar patterns of external morphology are expressed on multiple occasions within and between families (Russell and Gamble 2019). Here we document the major morphological expressions of digit form, all genera of geckos being able to be assigned to one (sometimes two, where there is considerable intrageneric variation) of these patterns, although we caution that there is considerable variation in the general appearance between genera within each category. The illustrations that accompany this account (Fig. 4.2) have been chosen, wherever possible, because there is accompanying data (Table 4.2) about the form and distribution of their subdigital micro-ornamentation (spinules, setae and the like, considered more fully in the next section).

In their most basic form, gecko digits resemble those of other lizards in being relatively straight and slender, and expressing no regional enlargement of subdigital scales or any evidence of the expression of adhesion-promoting modifications (Fig. 4.2a). This condition is shared with other lizard lineages and represents the ancestrally padless condition for the Gekkota as a whole (see above). Some species have straight, slender, ancestrally padless digits that bear lateral fringes associated with their psammophilous habits (Fig. 4.2b). Several genera possess a pair of terminal leaf-like, undivided seta-bearing adhesive scansors that symmetrically embrace the claw at the distal end of the digit (Fig. 4.2c). This condition is hypothesized to have preceded that in which scansors encroached more proximally along the digit (Russell and Gamble 2019) by recruiting additional adhesive plates (Fig. 4.2d). In other lineages; the terminal, leaf-like pads are asymmetrically disposed (Fig. 4.2e). Some lineages with terminally situated toe pads have them divided into a fan-like arrangement of scansors without further proximal encroachment (Fig. 4.2f).

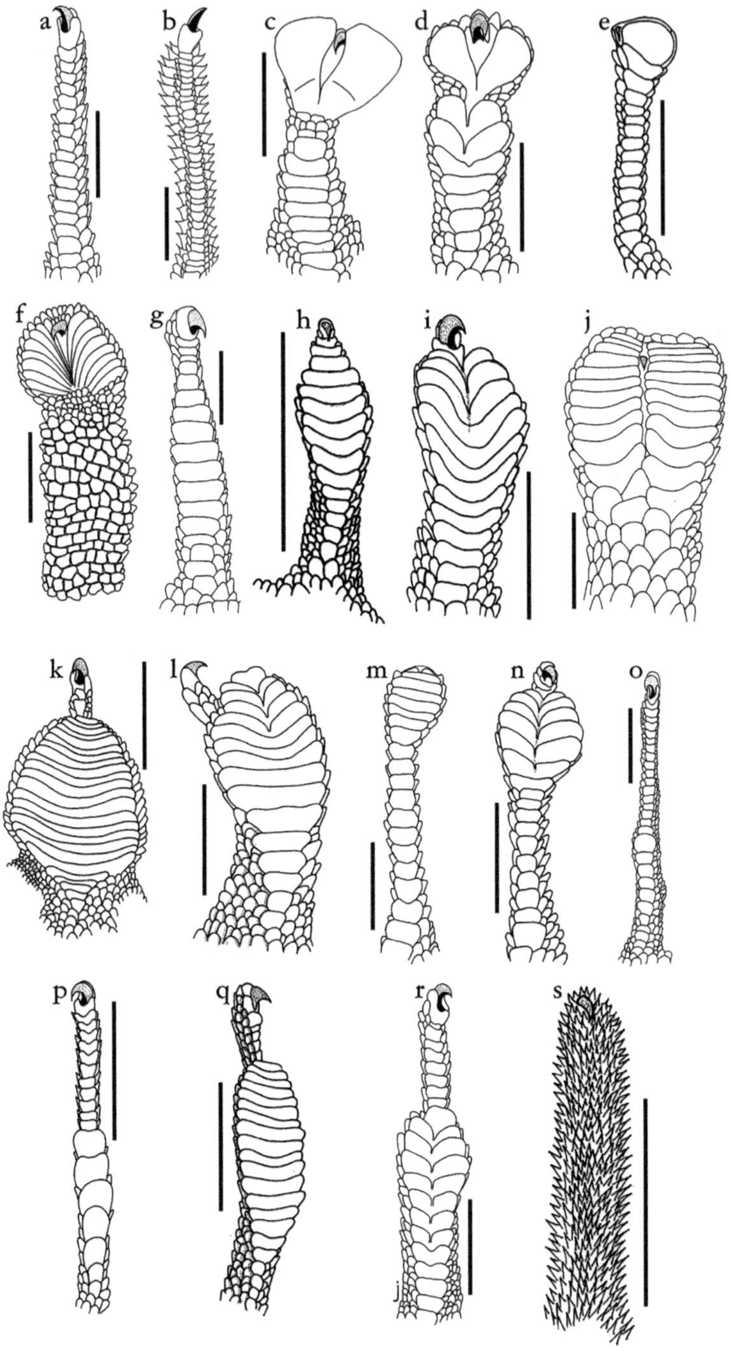

Fig. 4.2 Diversity of digit form in the Gekkota. (**a**) *Homonota horrida* (Phyllodactylidae). 98.7.7.1. Ancestrally padless condition with no noticeably enlarged subdigital scales.

Proximal encroachment of scales modified to become seta-bearing structures yields digits with a long series of scansors that approach the base of the digit, and a claw/ungual phalanx (in cases where the claw is reduced) situated at the distal end of the toe pad and the penultimate phalanx being encapsulated within the confines of the toe pad. An incipient form of such a pad (Fig. 4.2g) is exhibited by some genera, in which some digits, especially the first, retain small terminal leaves together with the incipient basal pads investing the majority of the ventral surfaces of the digits. Further elaboration of the extensive subdigital toe pad (the situation familiar in the oft illustrated *Gekko gecko*) yields pads that may have undivided (Fig. 4.2h) or distally cleft plates (Fig. 4.2i). In some instances, this pattern is further modified by the penultimate phalanx and the claw being free of the toe pad (Fig. 4.2k, l), in a cantilever-like fashion. In some, the cleft may be further excavated into a deep sulcus into which the claw can be retracted (Fig. 4.2j). Some taxa exhibit a distal

Fig. 4.2 (continued) (**b**) *Ptenopus garrulus* (Gekkonidae). 1946.8.23.54. Ancestrally padless condition with no noticeably enlarged subdigital scales but with enlarged, laterally fringing scales. (**c**) *Urocotyledon palmatus* (Gekkonidae). 1957.1.13.93. Pair of terminal leaf-like, undivided seta-bearing adhesive scansors symmetrically embracing the claw. (**d**) *Amalosia lesueurii* (Diplodactylidae). XXII.3c. Leaf-like scansors encroaching more proximally along the digit. (**e**) *Sphaerodactylus copei* (Sphaerodactyidae). 1961.1024. Terminal, leaf-like pads asymmetrically disposed. (**f**) *Uroplatus ebenaui* (Gekkonidae). 1946.6.1.5. Terminally situated toe pads divided into a fan-like arrangement but without further proximal encroachment. (**g**) *Naultinus rudis* (Diplodactylidae). 1946.8.22.37. Incipient toepad of expanded plates occupying the middle portion of the digit. (**h**) *Gekko japonicus* (Gekkonidae). 88.1.30.8. Extensive subdigital toepad with undivided scansors and with the penultimate phalanx and claw bound to the toepad. (**i**) *Lepidodactylus lugubris* (Gekkonidae). 75.12.31.3. Extensive subdigital toepad with scansors cleft distally. (**j**) *Thecadactylus rapicauda* (Phyllodactylidae). 1940.3.11.33. Cleft between scansors excavated into a deep sulcus into which the claw can be retracted; penultimate phalanx and claw subsumed within the confines of the toepad. (**k**) *Gehyra marginata* (Gekkonidae). 1946.8.12.72. Extensive subdigital toepad with undivided scansors and with the penultimate phalanx and claw cantilevered free of the toepad. (**l**) *Gehyra australis* (Gekkonidae). XXII.55d. Extensive subdigital toepad with divided scansors distally and with the penultimate phalanx and claw cantilevered free of the toepad. (**m**) *Rhoptropus afer* (Gekkonidae). 1937.12.3.55. Distal displacement of extensive toepad through disproportionate elongation of the proximal most phalanges, scansors undivided. (**n**) *Lygodactylus bonsi* (Gekkonidae). 1963.130. Distal displacement of extensive toepad through disproportionate elongation of the proximal most phalanges, scansors divided. (**o**) *Gonatodes ocellatus* (Sphaerodactylidae). 1945.11.1.3. Ancestrally padless condition with enlarged friction plate beneath the mid-digit inflection. (**p**) *Gonatodes humeralis* (Sphaerodactylidae). 1947.1047. Incipient toepad expressed beneath a mid-digit inflection, plates undivided. (**q**) *Aristelliger lar* (Sphaerodactylidae) Specimen number unknown. Basal toepad extensive with undivided scansors and a long, free distal portion of the digit not encompassed within the confines of the toepad. (**r**) *Hemidactylus forbesii* (Gekkonidae). 1946.8.25.47. Basal toepad extensive with divided scansors and a long, free distal portion of the digit not encompassed within the confines of the toepad. (**s**) *Lucasium damaeum* (Diplodactylidae). 1954.1.1.70. Secondary absence of toepads. Panels **a**, **d**, **e**, **f**, **g**, **h**, **I**, **k**, **l**, **q** and **s** depict digit IV, left pes; panels **b**, **c**, **m**, **o**, **p** and **r** depict digit IV, right pes; panel **j** depicts digit V, right pes. All registered numbers (e.g., 98.7.7.1) refer to specimens in the collections of The Natural History Museum, London, UK. All scale bars represent 2 mm except for panels **h**, **k** and **q** where they represent 5 mm displacement of this type of toe pad through disproportionate elongation of the proximalmost phalanges—again the scansors may be entire (Fig. 4.2m) or divided in the midline (Fig. 4.2n).

Many genera exhibit digits that have enlarged, traction-enhancing (Arnold 1998; Russell et al. 2015) friction plates (Padian and Olsen 1984; Russell and Bauer 1990) beneath mid-digital inflection points (Fig. 4.2o). This form of ancestrally padless digit is an evolutionary precursor in several genera of incipient toe pads that are expressed beneath a mid-digit inflection (Fig. 4.2p). At least some of these have been demonstrated to bear setae resembling those of more demonstratively pad-bearing taxa (Russell et al. 2015) and to enable adhesively assisted locomotion on smooth, vertically oriented substrates (Higham et al. 2017a). This type of incipient basal pad is, in turn, hypothesized to be the evolutionary precursor of toe pads that are situated relatively more proximally on the digits (Russell and Gamble 2019), with a long, free distal portion of the digit again cantilevered beyond the toe pad, a configuration similar to the situation encountered in *Anolis* and pad-bearing skinks (see below). Such toe pads may carry either entire (Fig. 4.2q) or divided (Fig. 4.2r) scansors. Lastly, there are some instances in which the toe pads have been secondarily lost (Fig. 4.2s) (Russell and Gamble 2019) in association with a readoption of a terrestrial existence (Russell 1976; Lamb and Bauer 2006).

4.4.1.3 Interspecific and Intraindividual Diversity of Gekkotan Setal Morphology and Configuration

Gekkotan setae exhibit a wide array of structural variation (Peattie 2007; Russell and Garner 2021) in features such as stalk length, basal diameter, and density (Table 4.2). No information is available about the location on the toe pad from which setae documented in most of the studies cited in Table 4.2 were sourced, therefore, comparison between them is somewhat hampered. There may be species-specific differences in the way in which these parameters interrelate, prompting the question of what this enormous variation might mean.

By admittedly subjectively partitioning the data in Table 4.2 into categories (short, moderate, and long for setal stalk length; narrow, moderate, and broad for setal basal diameter; and low, moderate, and high for setal density—Table 4.2), a pattern of clustering emerges. Taxa possessing relatively short setae generally exhibit narrow or moderately wide setal stalks and high or moderate stalk density. Species with long setal stalks exhibit setae with relatively wide bases that are present in low densities. Only species exhibiting setae with the latter combination of attributes have had their setae surveyed and documented for the entirety of the toe pad (Russell et al. 2007; Johnson and Russell 2009; Webster et al. 2009; Russell and Garner 2021). For the taxa for which such studies have been undertaken, it is evident that even in a single individual of a single species setae may be of short, moderate, or long length and have narrow, moderate, or wide bases, although their setae always exhibit low density (Table 4.2) (see below for more about intraindividual variation of setal morphology and configuration). For the only two species (*Gonatodes humeralis* and *Heteronotia binoei*) surveyed to date that exhibit incipient, adhesively competent toe pads (Russell et al. 2015; Higham et al. 2017a; Riedel et al. 2021), there is variation in setal length (although always in the short category), setal basal diameter (although again always in the narrow category), and setal density (ranging from high to low) (Table 4.2). These data suggest that variation in setal dimensions is common to geckos with adhesive toe pads. There is, therefore, much variation in the expression of setal form and density between species (Table 4.2). This is also likely the case for spatula diameter at the free ends of the setae (Russell and Garner 2021, Table 4.1). Although the data available for setal tips are sparse, generalizations about the numbers of terminal branches (hundreds to thousands) and their tip diameters abound.

Transects of setal form and dimensions along the entire length of toe pads have revealed extensive gradational variation proximodistally, not only along the toe pad as a whole, but also along each scansor (Russell et al. 2007; Johnson and Russell 2009; Johnson et al. 2009; Webster et al. 2009; Russell and Johnson 2014; Russell and Garner 2021). Thus, within the toe pad of a given individual, setae of different form, dimensions, and mechanical properties (Garner and Russell 2021) contribute to the global clinging force measured for a given species (Irschick et al. 1996).

An example of this variation is provided for *Gekko gecko* (Fig. 4.3), the species investigated by Autumn et al. (2000). How the output of a single, isolated seta of *Gekko gecko* (Autumn et al. 2000) relates to the adhesive performance of the intact entire organism (Irschick et al. 1996), with its extensive and regimented array of setal form (Fig. 4.3) remains unknown.

The setal form characterized by Autumn et al. (2000) is the most complex expression for *Gekko gecko*, in terms of branching pattern and setal tip density (Russell et al. 2007), and it is this that has become the default model for biomimetic inspiration. It is evident, however, that there is no "typical" seta for any species for which the entire toe pad has been surveyed (Garner and Russell 2021; Russell and Garner 2021). The comparable adhesive and mechanical properties of simpler forms of setal morphology of *Gekko gecko* (or any other species) remains unknown, although total adhesive performance of toe pads is predicated upon extrapolations based upon toe pad area and the assumption that the entire toe pad is populated with setae identical to that described by Autumn et al. (2000). There is a large discrepancy between the maximum potential adhesive performance of the toe pads of *Gekko gecko* based upon such extrapolations (Autumn et al. 2000; Autumn 2006a) and the actual realized output of intact animals (Irschick et al. 1996). How much of this discrepancy is related to the potential patchiness of contact between setae and the substrate (Russell and Johnson 2007) as opposed to differential contribution of setae of different form, dimensions (Fig. 4.3), and mechanical properties along the length of the toe pads (Russell et al. 2007) remains unexplored. The length and diameter of the setae, for example, influences their effective elastic modulus and bending stiffness, with long, thin setae, typical of the distal region of distal scansors, having a very low effective elastic modulus and effective bending stiffness whereas setae at the proximal end of the same scansor have a much higher values for these properties (Russell and Johnson 2014). For all such variables of setal form along the length

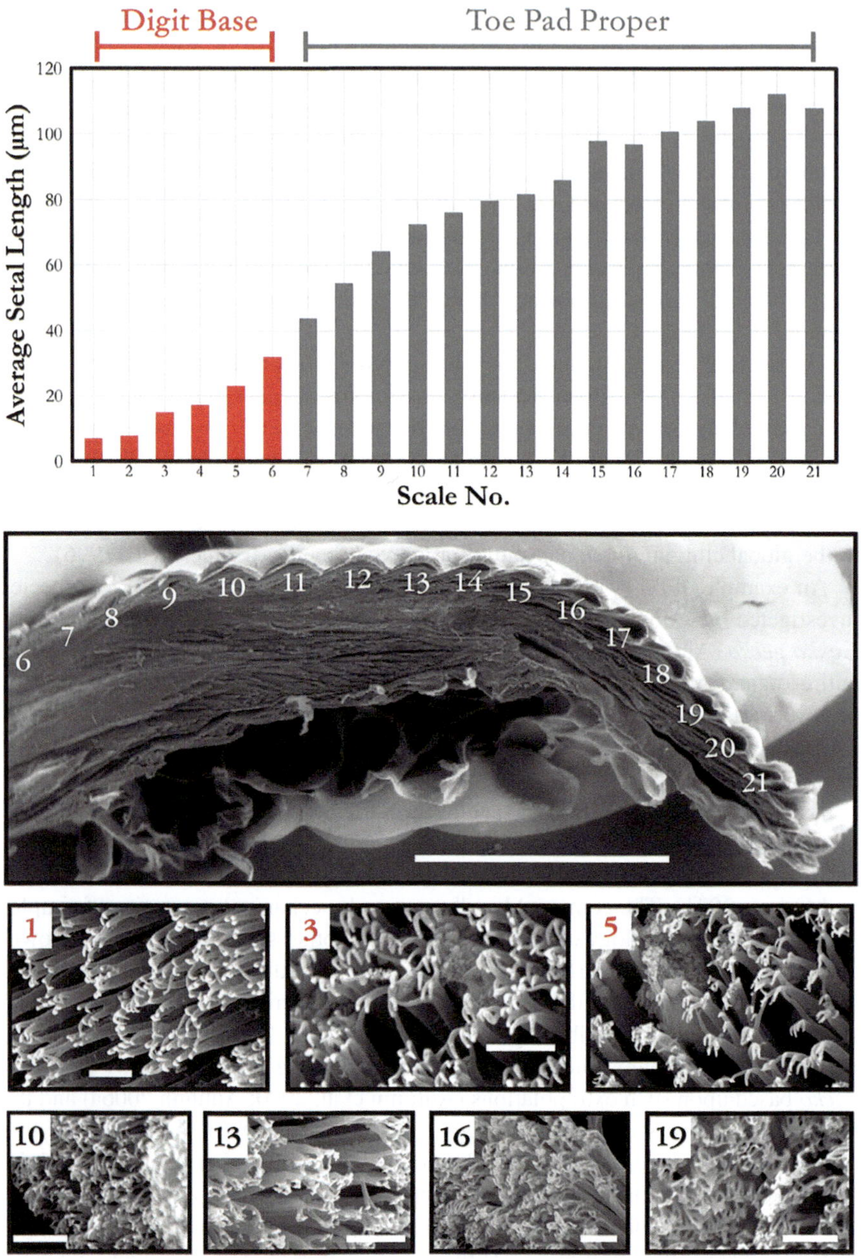

Fig. 4.3 Variation in epidermal microornamentation along the length of digit IV, left pes of *Gekko gecko*. The central panel depicts a scanning electron micrograph of a longitudinal section through the digit beneath a histogram indicating the average length of setae (in μm) along the length of the digit. The proximal end of the digit is to the left. On the histogram, plates situated on the digit base are indicated in red and plates (scansors) constituting the toepad proper (from plate 7

of the toe pads, and the individual scansors, there is no single set of values typical of a particular species (Russell and Johnson 2014) that can be applied for the entire adhesive pad.

To understand the most fundamental adhesive properties of the setae of this (and other) species, it will be instructive to explore these attributes for setae of simpler form (Garner and Russell 2021). Such exploration may well provide profitable insights for biomimetic simulacra that rely upon more simple representations of these structures and the subsequent manufacture of fields of them for particular applications.

4.4.1.4 Interspecific Diversity of Gecko Adhesive Performance

Authors investigating the mechanisms involved in effecting adhesion in geckos have employed a variety of smooth artificial surfaces (Hiller 1968, 1969; Irschick et al. 1996; Vanhooydonck et al. 2005; Autumn 2006a, b), enabling the strength of the bond established by intact animals to be assessed. Irschick et al. (1996) conducted the first comparative study of digital clinging performance in lizards with adhesively competent modifications of the subdigital integument (geckos, anoles, and skinks). They found that for the six species of gecko they investigated, clinging ability, as measured using a near-vertically oriented force platform, with the lizards attaching to an acetate sheet by their two forefeet, was tightly correlated with toe pad area and that all species examined were functionally similar, when normalized for body mass (Newtons of force per gram body mass), in their clinging abilities (as previously suggested by Ruibal and Ernst 1965 for a much smaller sample of species). Examination of the scaling of pad area to body mass (Bauer and Good 1986; Irschick et al. 1996), has yielded seemingly conflicting outcomes. Under isometry, pad area should scale to mass$^{0.66}$, but under functional similarity, the scaling exponent should be 1. Among species of *Gekko*, pad area was found to scale with body mass$^{0.59}$ (Bauer and Good 1986). Irschick et al. (1996) examined a wider array of pad-bearing lizard species, and calculated a scaling exponent of mass$^{0.75–0.78}$, which represents positive allometry but a lack of functional similarity, indicative of a decrease in adhesive capacity with increasing body size. Webster et al. (2009),

Fig. 4.3 (continued) onward) are indicated in black. Surrounding panels, numbered 1, 3, 5, 10, 13, 16, 19 show details of the epidermal filaments on selected plates along the digit, from proximal to distal. Panels 1, 3, and 5 show seta-like filaments on scales of the digit base and panels 10, 13, 16, 19 depict setae at progressively more distal locations along the length of the toepad. Data for outgrowths in the panels are as follows (nd indicates no data for that category): Stalk diameter/Tip width/Stalk density. Panel 1. 0.9 μm/0.06 μm/48000 per mm^2; Panel 3. nd/nd/nd; Panel 5. 1.9 μm/0.07 μm/24000–60,000 per mm^2; Panel 10. 3.2 μm/0.17 μm/24000 per mm^2; Panel 13. 3.45 μm/0.17 μm/nd; Panel 16. 3.7 μm/0.19 μm/nd; Panel 19. 3.9 μm/0.24 μm/16000 per mm^2. Scale bar for longitudinally sectioned digit is 2 mm. Scale bars for panels 1, 3, 5, 10, 13, 16, 19 represent 2 μm examining an ontogenetic growth series of the gekkonid *Chondrodactylus bibronii*, found that toe pad area scaled with body mass with a slope of 0.62 (not significantly different from isometry), but considerably below the 1.0 indicative of functional similarity, a result in accord with the congeneric cross-species findings of Bauer and Good (1986). Conversely, Higham et al. (2017a) examined the scaling of frictional adhesive force with body mass among anoles and geckos and revealed a scaling exponent of 1.08 (0.95 when phylogenetic relationships were incorporated), indicative of the maintenance of functional similarity as growth occurs. The apparent lack of scaling of toe pad area to keep pace with what would be needed for functional similarity (Bauer and Good 1986; Irschick et al. 1996; Webster et al. 2009) may be compensated for by some other means if clinging ability is maintained at a functionally similar level (Higham et al. 2017a). All six species of gecko that Irschick et al. (1996) examined belong to the Family Gekkonidae, and their toe pads represent five of the patterns depicted in Fig. 4.2 (*Hemidactylus frenatus* and *H. turcicus*—Fig. 4.2 pattern r; *Gehyra oceanica*—Fig. 4.2 pattern k; *Gehyra mutilata*—Fig. 4.2 pattern l; *Gekko gecko*—Fig. 4.2 pattern h; and *Lepidodactylus lugubris*—Fig. 4.2 pattern i). More detailed ontogenetic studies of the scaling of toe pad area and clinging force within single species are required to provide greater resolution of these relationships (particularly because the actual boundaries of what constitutes the toe pad can be more effectively established; Russell 1986). Cross-species comparisons may mask differences in these relationships between different lineages of geckos that exhibit quite disparate expressions of toe pad configuration (Fig. 4.2).

Geckos are able to hold station for long periods (Russell and Oetelaar 2016) on vertical surfaces, indicating that frictional loading can be maintained passively (Hiller 1968; Bergmann and Irschick 2005). This applies even after death (Hiller 1968; Vinson and Vinson 1969; Stewart and Higham 2014). Irschick et al. (1996) noted that all geckos included in their study were able to readily climb vertical and overhanging surfaces, indicating that they not only exhibit strong static attachment forces but also that their dynamic employment of the adhesive system permits agile locomotion in such situations. This contrasts with anoles and skinks that are either able to scale overhanging surfaces with difficulty, or not at all (Irschick et al. 1996). Thus, although anole static clinging ability is not functionally dissimilar from that of geckos (see below, skinks performing less well), the dynamic employment of their adhesive system is less versatile. Kinematic analysis of a wider range of adhesively competent lizards negotiating low friction surfaces is needed to explore the relationship between static and dynamic effectiveness of the adhesive system.

Although the clinging performance of whole animals (Irschick et al. 1996) has ostensibly been correlated with toe pad area (see above), it is evident from transect surveys of toe pads (see above) that setae at different locations differ in stalk length, stalk diameter, stalk density, aspect ratio (and therefore bending stiffness), and extent of branching. This variation likely contributes to variation in clinging ability. Whole animal observations of clinging force per digit result in performance values lower by an order of magnitude or more (Irschick et al. 1996) than those predicted from the extrapolation based upon a combination of the adhesive potential of an exemplar seta and an assessment of the total number of such setae carried on a single digit (Autumn et al. 2000, 2002; Autumn and Peattie 2002). The discrepancy between these two approaches may be due, to some extent, to variability in the number of spatulate tips per seta along the length of the toe pad (Russell and Garner 2021), differences in setal density at different stations on the toe pad (Fig. 4.3; Table 4.2), and the number of setae actually making contact with the surface (which may be reduced based on the surface roughness of the substrate—Russell and Johnson 2007). Regardless of the measured or estimated clinging ability of geckos, there is a large discrepancy between the total adhesive force output and the force needed to support the body mass of a given individual. Such a discrepancy has led to the observation that the gekkotan adhesive system is apparently "overbuilt" in relation to the body mass to be supported (by anywhere from several thousand-fold (Autumn 2006b), to several hundred-fold (Irschick et al. 1996), to many tens of times (Huber et al. 2005b)). At the upper estimated limit of force output, a 50 g tokay gecko would need only 0.04% of its setae to support its own mass (Autumn 2006b). Even for the maximum clinging performance measured for *G. gecko* on a smooth surface (Irschick et al. 1996), the adhesive force generated is several hundred-fold the force required to support its 50 g mass.

Recent ventures into the investigation of toe pad structure and performance in geckos that are much less renowned for their adhesive prowess have been revealing. *Gonatodes humeralis* (Russell et al. 2015; Higham et al. 2017a) possesses incipient toe pads (Figs. 4.2p and 4.4 pattern p) and was previously regarded as lacking adhesive capabilities. The size-corrected frictional adhesive force able to be generated by this species was found to be significantly smaller than that of other pad-bearing geckos, significantly greater than that of skinks, but not significantly different from that of anoles (Higham et al. 2017a). In comparison, *Gonatodes vittatus* (Figs. 4.2o and 4.4 pattern o) generates no measurable adhesive attachment forces (Higham et al. 2017a). Table 4.2 indicates that the setal stalks of *Gonatodes humeralis* are very short, their diameter is very narrow, and their density is moderate to very high, and that these parameters do not differ greatly from those of its non-adhesively competent (Garner and Russell 2021) congener, *G. ocellatus*. *Gonatodes humeralis* bears spatulate tips on its setae, but the branching of these filaments is sparse when compared to that of the "typical" setae of *Gekko gecko* (Autumn et al. 2000). Thus, adhesively assisted locomotion (Higham et al. 2017a; Garner and Russell 2021) can occur in taxa in which the setae are much less "elaborate" than those advocated to be the "norm" for geckos (Autumn et al. 2000; Fig. 4.3, scansor 19; Table 4.2).

Tellingly, Griffing et al. (2021) have demonstrated that the diplodactylid *Correlophus ciliatus*, with short, slender, moderately densely packed (Table 4.2), and not profusely branched setae is able to generate absolute digital adhesive forces that rank among the highest recorded for geckos, with a maximum output for both forefeet being recorded as 8.12 N. Adhesive performance of the manus scaled with mass$^{1.04}$. Such findings indicate that substantial adhesive performance among geckos is attainable in the absence of long, diametrically broad, relatively widely spaced, and profusely branched setae.

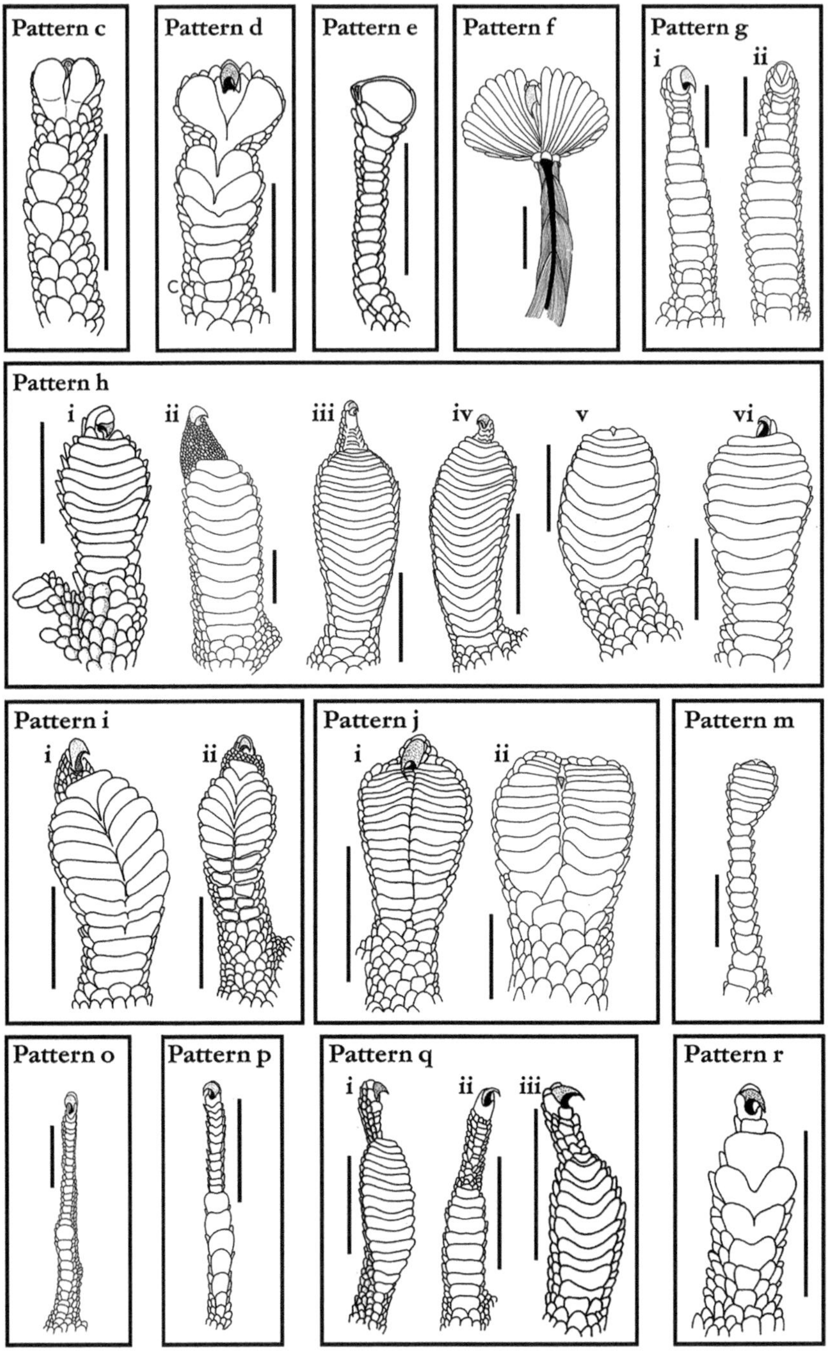

Fig. 4.4 Gekkotan digital configurations (see Fig. 4.2) for which setal dimensions have been reported in the literature (see Table 4.2 for measurement data and pattern

4.4.2 Diversity of Form and Function of the Anolid Adhesive System

4.4.2.1 Phylogenetics and Ecomorphology

Unlike geckos, all anoles bearing adhesive toe pads (and one documented case of secondary loss) belong to a single genus, *Anolis*, the only currently recognized genus in the Family Anolidae (formerly Dactyloidae) of the Superfamily Iguania. The Family Anolidae and genus *Anolis* currently contains 434 recognized species (Uetz et al. 2025). Despite its straightforward appearance, anolid taxonomy and evolutionary history has only recently been simplified; the literature on these issues is replete with debates regarding the origin and validity of family and genus names. These discussions, although important, are largely inconsequential for our review and analysis. Thus, we point readers to other literature should they be interested in the taxonomic and phylogenetic history of the genus *Anolis* (Etheridge 1960; Williams 1976a, b; Cannatella and de Queiroz 1989; Guyer and Savage 1992; Poe 2004, 2013; Nicholson et al. 2012, 2014; Poe et al. 2017) and the Family Anolidae (de Queiroz 2022, 2023).

Anoles are popular models in the field of evolutionary ecology (Losos 1994, 2011; Irschick et al. 2006) largely because of their centrality in relation to the concept of ecological morphology (or ecomorphology), the relationships between an organism's morphological traits and its environment (Wainwright and Reilly 1994). Caribbean anoles are textbook examples of ecomorphology (Fig. 4.5)

Fig. 4.4 (continued) assignment). Pattern c—*Diplodactylus vittatus* (Diplodactylidae). 1909.7.2.1. Pattern d—*Amalosia lesueurii* (Diplodactylidae). XXII.3c. Pattern e—*Sphaerodactylus copei* (Sphaerodactylidae) [for *S. cinereus*]. 1961.1024. Pattern f—*Ptyodactylus hasselquistii* (Phyllodactylidae). 1954.1.5.40. Pattern g—(i) *Naultinus elegans* (Diplodactylidae). 1946.8.22.38. (ii) *Naultinus rudis* (Diplodactylidae). 1946.8.27.37. Pattern h—(i) *Eurydactylodes viellardii* (Diplodactylidae). 1926.9.17.7. (ii) *Rhacodactylus auriculatus* (Diplodactylidae). 86.3.11.6. (iii) *Gekko smithii* (Gekkonidae) [for *G. gecko*]. 91.8.29.23. (iv) *Gekko vittatus* (Gekkonidae). 1910.4.26.15. (v) *Chondrodactylus bibronii* (Gekkonidae). Specimen number unknown. (vi) *Tarentola ephippiata* (Phyllodactylidae) [for *T. caboverdiana*]. 65.5.3.67. Pattern i—(i) *Bavayia cyclura* (Diplodactylidae). 1926.9.17.9. (ii) *Bavayia sauvagii* (Diplodactylidae). 1926.9.17.25. Pattern j—(i) *Pseudothecadactylus lindneri* (Diplodactylidae) [for *P. australis*]. 77.3.3.12. (ii) *Thecadactylus rapicauda* (Phyllodactylidae). 1940.3.11.33. Pattern m—*Rhoptropus afer* (Gekkonidae). 1937.12.3.55. Pattern o—*Gonatodes ocellatus* (Sphaerodactylidae). 1945.11.1.3. Pattern p—*Gonatodes humeralis* (Sphaerodactylidae). 1971.1049. Pattern q—(i) *Dactylocnemis pacificus* (Diplodactylidae). 1946.9.8.14. (ii) *Woodworthia maculata* (Diplodactylidae). 1946.8.22.71. (iii) *Aristelliger lar* (Sphaerodactylidae) [for. *A. praesignis*]. Specimen number unknown. Pattern r—*Hemidactylus bouvieri* (Gekkonidae). 66.4.12.3. All illustrations depict ventral views of digit IV, right pes except for patterns d, e, g(ii), h(vi), i(i), i(ii), j(i), p, q(i), q(iii), and r, which represent digit IV, left pes and patterns f, and h(i) which depict digit III, right pes. All registered numbers (e.g., 1909.7.2.1) refer to specimens in the collections of The Natural History Museum, London, UK. All scale bars represent 2 mm except for patterns h(iii), h(iv), j(i), and q(iii) where they represent 5 mm

Fig. 4.5 Ecomorphology of Caribbean *Anolis*. (**a**) *Anolis cuvieri*, a crown-giant ecomorph. (Photo by Matt McElroy). (**b**) *Anolis stratulus*, a trunk-crown ecomorph. (Photo by Alberto López). (**c**) *Anolis occultus*, a twig ecomorph. (Photo by D. Luke Mahler). (**d**) *Anolis distichus*, a trunk ecomorph. (Photo by Pierson Hill). (**e**) *Anolis lineatopus*, a trunk-ground ecomorph. (Photo by D. Luke Mahler). (**f**) *Anolis pulchellus*, a grass-bush ecomorph. (Photo by Kristiina Ovaska). (All photos obtained from Huie et al. (2021) by public domain)

because several morphological and behavioral traits covary with microhabitat use, irrespective of geographic location or evolutionary history (Williams 1972; Losos 1990a, b; Mahler et al. 2013). These repeatedly and convergently evolved behaviors and morphologies between anoles that share microhabitats allow them to be categorized into so-called *ecomorph* categories. Currently, six such categories of Caribbean anoles are recognized and are named based on the occupancy of different microhabitats: (1) crown-giant; (2) trunk-crown; (3) twig; (4) trunk; (5) trunk-ground; and (6) grass-bush (Losos 1990a, 2011). Crown-giant anoles, for example, are characterized by their large body size, relatively large adhesive toe pads, long tails, and short limbs—these morphological features are considered adaptations to their highly arboreal lifestyle in the crowns of trees. Trunk-ground anoles, on the other hand, are intermediately sized, have relatively smaller toe pads, long limbs, and long tails—adaptations that allow success on the broad substrates that they typically exploit on the bases of tree trunks and the ground (Cartmill 1985; Losos and Irschick 1996; Beuttell and Losos 1999; Irschick and Losos 1999; Losos 2011).

4.4.2.2 Diversity of Anolid Toe Pad Morphology

All species but one in the genus *Anolis* possess adhesive toe pads, with the sand dune specialist *Anolis onca* being the only species to have secondarily lost adhesive toe pads (Peterson and Williams 1981). Anole toe pads are composed of laterally expanded subdigital scales called lamellae that are carried on the basal portions of their digits (Russell and Eslinger 2017). The claws of *Anolis* are typically separated by a considerable gap from the toe pads because they are supported by a penultimate phalanx that arches dorsally from the toe pad's distal extent (Russell and Garner 2023) and extends far beyond it. Lamellae carry the batteries of adhesive setae, many of which are borne upon the epidermal free margins of the lamellae (Ruibal and Ernst 1965; Peterson and Williams 1981; Williams and Peterson 1982; Peterson 1983a, b; Russell and Eslinger 2017; Garner et al. 2021). The epidermal free margin is devoid of subepidermal tissue and able to bend and conform to the substrate (Russell and Eslinger 2017).

Toe pad morphology is much less diverse in *Anolis* than in gekkotans. The most obvious differences in toe pad morphology are typically expressed in terms of gross morphometric parameters, such as toe pad width/length, area, and number of lamellae carried on the toe pad (e.g., Pounds 1988; Glossip and Losos 1997; Macrini et al. 2003; Elstrott and Irschick 2004). Most of these characterizations have been made in the context of ecomorphology. Toe pad area and number of lamellae typically scale positively with perch height, the distance above the ground at which anoles typically live. Comparative data reveal that anole species occupying sectors of the habitat higher from the ground tend to have larger toe pads with a greater number of lamellae, in both an absolute and size-corrected sense, because anole body size also tends to increase with increases in perch height (Pounds 1988; Glossip and Losos 1997; Macrini et al. 2003; Elstrott and Irschick 2004). Other than dimensions, area, and/or lamella number, anole toe pads are of quite stereotypical form between taxa (Fig. 4.6). As described above, the diversity and disparity of toe pad shapes and configurations in geckos is likely the result of multiple independent origins within the Gekkota (Gamble et al. 2012; Russell and Gamble 2019), reflective of their evolutionary age and circumglobal tropical and temperate distribution. Anoles have purportedly evolved toe pads once only, diverged evolutionarily much more recently than geckos (~70–78 million years ago; Zheng and Wiens 2016; Poe et al. 2017; Griffing et al. 2022; Miller and Stroud 2022), and are geographically much more restricted in their distribution.

4.4.2.3 Interspecific and Intraindividual Diversity of Anolid Setal Morphology and Configuration

Anolid setae are much simpler in their basic form than those generally reported for gekkotans—they are composed of a single stalk that terminates in a single, wide spatulate tip (Ruibal and Ernst 1965; Peterson and Williams 1981; Williams and Peterson 1982; Peterson 1983a, b; Garner et al. 2019, 2021, 2022). Relative to geckos with complex manifestations of setal form (e.g., *Gekko gecko*, Fig. 4.3),

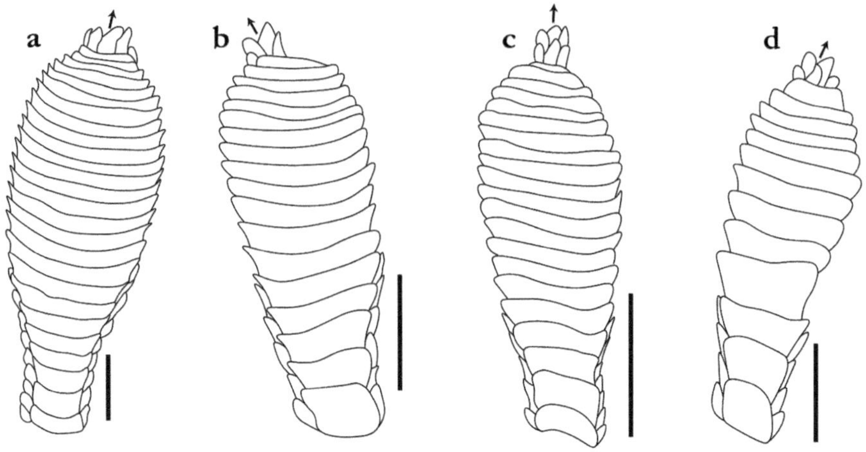

Fig. 4.6 Comparative morphology of the toe pads of four Jamaican *Anolis* species. (**a**) *Anolis garmani*, a crown-giant ecomorph; R-73381. (**b**) *Anolis grahami*, a trunk-crown ecomorph; R-192656. (**c**) *Anolis valencienni*, a twig ecomorph; R-192635. (**d**) *Anolis lineatopus*, a trunk-ground ecomorph; R-192710. *Anolis garmani* (**a**) is the most arboreal of the species shown here and possesses, both relatively and absolutely, the largest toe pad (in terms of area, length, and width) carrying the greatest number of lamellae. *Anolis lineatopus* (**d**), on the other hand, has similar toe pad area to *A. grahami* (**b**) and *A. valencienni* (**c**) but has substantially fewer lamellae on its relatively long and slender toe pad. Illustrations terminate distally just beyond the distal extent of the toe pad—arrows indicate the location at which the phalanges not supporting the toe pad continue, ultimately terminating in a claw. Numbers proceeding each species above (e.g., R-73381) correspond to specimen accession numbers in the herpetology collection of Harvard University's Museum of Comparative Zoology (MCZ). All scale bars represent 1 mm

exemplar anolid setae are thinner and shorter, with tapered stalks, and exist at much higher densities (Garner et al. 2021). Garner et al. (2019) tabulated all known average setal morphometrics of anolids at the time, and their findings are reproduced in Table 4.3. Garner et al. (2021, 2022) added additional data for three of the species listed in Table 4.3 (*A. valencienni, A. equestris*, and *A. lineatopus*) and provided newly collected data for three other species (*A. garmani, A. grahami*, and *A. opalinus*) during their exploration of setal morphometric diversity along longitudinal transects of entire toe pads (see below). Ranges of setal morphometrics observed in these studies are reported in Table 4.4. As is the case for gekkotans, the setal morphometrics of relatively few species of anolids have been documented, representing only about 4% of anolid species (Garner et al. 2019). For most species assayed, the location along the toe pads from which average setal morphometrics were obtained was not specified (Table 4.3). Given the previous discussion of intraspecific morphological variability in gekkotan setae (and the forthcoming section regarding such variability in anolids), these data are not conducive to robust interspecific comparisons. Nevertheless, anolid setae appear to be less morphometrically diverse than those of geckos. Setal densities are generally reported to be around 1 seta per μm^2, setal lengths are generally 15–20 μm, setal

Table 4.3 Comparative setal and spatular dimensions of *Anolis* as of 2019. Note that only 15 species are represented and that the complete set of characters has only been reported for two species (denoted by **). Cells with gray fill indicate that no pertinent data for that parameter were provided by the study or studies

Species	Density (setae/μm²)	Length (μm)	Base diameter (μm)	Tip width (μm)	Tip area (μm²)	Sources
Anolis chameleonides	1	18.4-18.6	0.56–0.58		0.206–0.377	(1)
Anolis porcus	1	22.3	0.53		0.264	(1)
Anolis barbouri	0.5–0.6	5.0–8.4	0.47–0.56		0.6–1.472	(1)
Anolis heterodermus	1.2	13.7–14.8	0.41–0.46		0.214–0.308	(1)
Anolis cuvieri	0.9–1.4	22.4–27.2	0.51–0.65		0.184–0.253	(1)
*Anolis cuvieri***	1	22.4	0.51	0.729	0.229	(2)
Anolis sheplani	1.1–1.2	11.8–12.6	0.39–0.41		0.220–0.279	(1)
Anolis occultus	1.4	11	0.49		0.237	(1)
Anolis sp. n. near *eulaemus*	1	20.4	0.57		0.593	(1)
Anolis valencienni	1.1-1.4	15.3-17.2	0.40–0.47		0.171–0.209	(1)
Anolis carolinensis	0.83	21	0.5	0.87		(3,4)
Anolis equestris	0.7	30				(3,4)
Anolis lineatopus	0.51			1		(3,4)
Anolis sagrei	1.7	20		0.75		(3,4)
Anolis homolechis		20				(3)
*Anolis annectens***	1.0-2.0	20	0.48	0.73	0.211	(5)

Sources: (1) Peterson (1983b); (2) Williams and Peterson (1982); (3) Ruibal and Ernst (1965); (4) Peattie and Full (2007); and (5) Peterson and Williams (1981). Table reproduced from Garner et al. (2019) by permission of the Society for Integrative and Comparative Biology

diameters are reported around 500 nm, tip widths are around 750 nm–1 μm, and tip areas are typically around 0.25 μm^2.

Garner et al. (2022) recently revealed potential interspecific differences in anolid setal morphometrics in their examination of the setal morphology of five species of the Jamaican radiation of anoles. The species examined represent a group of anoles that are all closely related and from the same geographic location but exploit different microhabitats (representing 4 of the 6 "ecomorph" categories; Fig. 4.5). By quantifying setal morphometric diversity along longitudinal sections of entire toe pads within each species, Garner et al. (2022) were able to make appropriate

Table 4.4 Setal and spatular morphometrics of *Anolis* species obtained from studies surveying longitudinal sections of entire toe pads. Cells with gray fill indicate that data for that parameter were not collected by the study

Species	Density (setae/μm²)	Length (μm)	Base diameter (μm)	Apex diameter (μm)	Resting angle (°)	Tip width (μm)	Tip area (μm²)	Source
Anolis equestris	1.0–1.9	11.6–27.2	0.5–0.9	0.2–0.5	26–93			(1)
Anolis garmani	0.9–2.1	9.6–28.3	0.4-0.8	0.1–0.5	43–90	0.485–0.502	0.091–0.093	(2)
Anolis grahami	0.6–2.2	12.6–27.1	0.2–0.7	0.2–0.4	39–90	0.585–0.635	0.110–0.132	(2)
Anolis valencienni	0.8-1.8	10.5–25.4	0.4–0.7	0.2–0.5	34–92	0.491–0.737	0.102–0.186	(2)
Anolis opalinus	0.6-1.6	12.7–24.3	0.5–0.7	0.2–0.3	47–80	0.583–0.707	0.120–0.145	(2)
Anolis lineatopus	0.4-1.6	11.1–23.5	0.2–0.9	0.2–0.5	23–92	0.569–0.780	0.129–0.187	(2)

Sources: (1) Garner et al. (2021); and (2) Garner et al. (2022)

comparisons between species. Interestingly, they found that Jamaican *Anolis* that exploit habitats farther from the ground carry setae that are significantly more densely packed, exhibit significantly smaller spatula areas, but, even so, have similar total area available for contact relative to species that exploit habitats closer to the ground. Whether or not this trend is representative of the other Caribbean *Anolis* communities requires additional work. Jamaican anoles also differ in the patterns of setal length variation along the proximodistal axis of their toe pads (more on this below), with more terrestrial species exhibiting less variation in setal length along the toe pad compared to more arboreal species which exhibit a pattern described below (Garner et al. 2022).

Anolid setal morphometrics vary markedly and predictably along the proximodistal axis of the toe pad. Garner et al. (2021) explored setal morphometrics along longitudinal sections of toe pads of a highly arboreal anole, *Anolis equestris*, of similar size to the Tokay gecko. Along the toe pad, setae are significantly shorter on proximal lamellae, while they are longer on intermediate and distal lamellae. Setae generally decrease in diameter and resting angle proximodistally. Setal density remains consistent along the length of the toe pad. Within a lamella, setae are longest at intermediate locations relative to distal and proximal regions of these plates and decrease in diameter and resting angle proximodistally. During contact with the substrate, setae carried on the proximal regions of lamellae are likely covered by the epidermal free margin of the next most proximal lamella, effectively rendering the cline of setal length opposite to that observed in geckos (Russell et al. 2007;

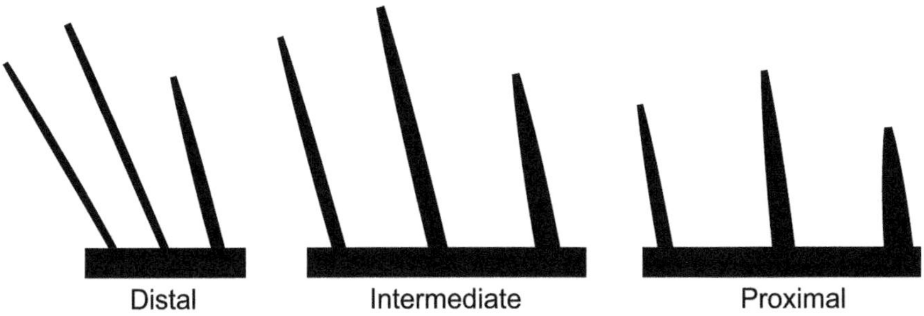

Fig. 4.7 Schematic representation of the variation in setal length, setal base and apex diameter, setal resting angle, and lamella length in *Anolis equestris* along the proximodistal axis of toe pads and/or lamellae. (Reused with permission from Garner et al. (2021))

Johnson and Russell 2009; Garner et al. 2021). In other words, setal length increases proximodistally in most geckos examined thus far and decreases proximodistally in *Anolis equestris*. This may reflect differences in peeling mechanics between the two groups of lizards. Most geckos actively peel their toe pads distoproximally, while anolids passively peel their toe pads proximodistally during routine locomotion. Gradational variation in setal length is predicted to allow all setae of a given scansor/lamella to reach their critical detachment angle simultaneously as opposed to sequentially, thereby potentially reducing detachment time and force. Modeling of setal bending stiffness revealed that effective bending stiffness of *Anolis equestris* setae decreases proximodistally, meaning that anolid setae become progressively and effectively softer further distally along the entire toe pad and along each individual lamella (Garner et al. 2021) (Fig. 4.7).

4.4.2.4 Interspecific Diversity of Anolid Adhesive Performance

The first comparative and quantitative study of anolid adhesive performance (often referred to as clinging ability in the anole-focused literature) is attributed to Losos (1990b) in his seminal study exploring connections between morphology, performance, and behavior in Caribbean *Anolis*. Despite considerable variation in maximum clinging force among 15 species of *Anolis*, this variation was largely explained by differences in body size and did not vary significantly with any aspects of morphology or ecology assayed in his study. Lamella number, however, was the only toe pad morphological character he measured. Subsequent studies have since demonstrated significant correlations between clinging ability, toe pad area, and habitat use (Macrini et al. 2003; Elstrott and Irschick 2004; Crandell et al. 2014). Elstrott and Irschick (2004), for example, explored correlations between morphology, clinging ability, and habitat use in 12 species of Caribbean *Anolis* and found that species that perched higher from the ground had significantly greater toe pad area and clinging abilities than those that perched closer to the ground, even when all parameters were corrected for variation in body size. Additionally, they found that both clinging ability and toe pad area scale isometrically with body mass (i.e., a scaling exponent of mass$^{0.67}$) indicating that small and large anoles have functionally and morphologically dissimilar toe pads, whereby large anoles possess relatively smaller toe pads with lesser clinging ability for their mass compared to those of their smaller counterparts.

Although toe pad area is often used to explain variation in clinging ability, toe pads of identical size could theoretically exhibit differences in attachment performance by varying the morphology and performance of the actual agents of attachment, the setae and their spatulae. Indeed, Elstrott and Irschick (2004) found that only 45% of the variation in size-corrected clinging ability was explainable by size-corrected toe pad area. Garner et al. (2022) found that Jamaican *Anolis* differ significantly in the size and configuration of their setal arrays—anoles exploiting habitats higher from the ground had setal arrays with greater densities and smaller spatulae than those that perched closer to the ground. These findings are consistent with the concept of contact splitting, whereby subdivision of an adhesive area into finer, more numerous contacts can result in an adhesive performance advantage even if total contact area is the same. Furthermore, Jamaican *Anolis* differ in the variation in setal length along toe pads and lamellae, with arboreal species exhibiting clines in setal length similar to those observed in the arboreal crown-giant anole *Anolis equestris* and more terrestrial species exhibiting no significant variation in setal length. These differences may reflect differences in the peeling mechanics between these species. Thus, dynamic use of toe pads during scaling of smooth, vertical substrates may differ considerably. Collectively, these data suggest that it is not only toe pad area that contributes to variation in clinging ability. Future studies should also examine setal morphology and configuration when attempting to explain this functional variation (Garner et al. 2022).

4.4.3 Diversity of Form and Function of the Scincid Adhesive System

A few scincid genera have subdigital toe pads somewhat reminiscent in appearance (Fig. 4.8) to those of anoles and some gekkotan lineages. Smith (1937) was the first to examine such with regard to their potential adhesive functionality. He noted that they were characterized by transverse expansion of proximal subdigital lamellae (Fig. 4.8) and posited that they represented the earliest stage in the expression of more fully differentiated toe pads of anoles and geckos. Of the species that he examined (*Dasia vittata, Emoia sandfordi*, and *Prasinohaema virens*) exhibiting such subdigital scale modification, only *Prasinohaema virens* was found to possess setae.

Maderson (1970) added further observations to those of Smith (1937), and employed histology to explore epidermal structure, scale surface microarchitecture,

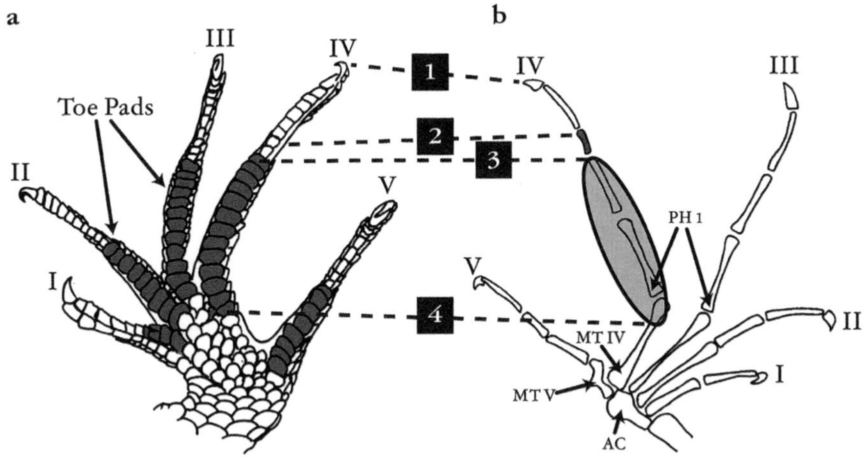

Fig. 4.8 Morphology of the scincid adhesive system. (**a**) Ventral aspect of the left pes of *Prasinohaema virens*, redrawn and modified from Smith (1937, Fig. 1e). Scales on the ventral surface of the digits and distal part of the plantar face of the foot bearing setae shaded in gray, thereby indicating the extent of the toepads. (**b**) Ventral aspect of the skeleton of the right pes of *Lipinia longipes*, redrawn and modified from Shea and Greer (2002, Fig. 4.8b), indicating the reduced relative length of the antepenultimate phalanx of digit IV, shaded in gray. Dashed lines between panels **a** and **b** link 1, the distal tips of digit IV in each; 2 and 3, the distal and proximal ends, respectively, of the antepenultimate phalanx of digit IV of *L. longipes* with their approximately equivalent locations in *P. virens*; and 4 The proximal extent of the toepad of digit IV of *P. virens* with its approximately equivalent location (indicated by a gray-filled oval) superimposed upon the pedal skeleton of *L. longipes*. Abbreviations: AC astragalocalcaneum, MT IV, MT V metatarsals IV and V, PH 1 phalanx 1 of digits III and IV, I–V digits I to V

and the form of epidermal outgrowths exhibited by those arboreal skinks appearing to possess incipiently expressed toe pads. He noted that, as for other skinks, they all exhibit the lamellate/dentate pattern of scale surface microarchitecture (Peterson 1984), as opposed to the spinulate pattern displayed by geckos and anoles. Maderson (1970) measured the lengths of the subdigital epidermal outgrowths of eight species of skink—*Lipinia infralineolata* 1.5 μm, *L. leptosoma* 15.0 μm, *L. longiceps* 20.0 μm, *Ornithuroscincus noctua* 1.5 μm, *Prasinohaema flavipes* 25.0 μm, *P. prehensicauda* 20.0 μm, *P. virens* 50.0 μm, and "*Lipinia miotus*" 1.5 μm (the latter a synonym of *Ornithuroscincus noctua*). These observations indicated that the species of *Prasinohaema* examined, as well as *Lipinia leptosoma*, possessed setae as long as those of some geckos (Table 4.2) and anoles (Table 4.3). Subsequent investigations (see below), however, have reported setae only in *Prasinohaema virens*.

The next advance in characterizing the form of the incipient toe pads in skinks was made by Williams and Peterson (1982), investigating these structures using scanning electron microscopy. They examined *Lipinia leptosoma*, *L. longiceps*, *Ornithuroscincus noctua*, *Prasinohaema flavipes*, *P. prehensicauda*, and *P. virens*, noting that all exhibit subdigital pads but that only the latter possesses setae. All other species examined have the toe pads clad in folds oriented across the long axis of the digit, with fringed, ruffle-like borders (Williams and Peterson 1982). The combined area of the folds and the spaces between them were found to be greater than the area bounded by the lamellate/dentate margins of the underlying epidermal patterning, suggestive that the superimposed reticulating pattern of the epidermis was the likely source of the folds and ruffles characterizing the toe pad surface (Williams and Peterson 1982). They showed that the setae of *P. virens* are very similar in form to those of *Anolis*, being unbranched, of similar length, and terminating in a single spatulate tip but suggested that the setae were less densely packed and the stalks much wider than those of anoles (Williams and Peterson 1982). Although unbranched, the total tip area per seta was stated to be more similar to that of geckos and to be much greater than that of anoles, suggesting a combination of features constituting a third pattern of setal form among squamates (Williams and Peterson 1982; Peterson 1983b).

The most recent morphological exploration of the structure of scincid toe pads was conducted by Austin (1995). He examined *Lipinia rouxi*, *Lobulia elegans*, *L. brongersmai, Papuascincus stanleyanus*, *Prasinohaema flavipes, P. prehensicauda*, *P. semoni*, *P. virens*, *Sphenomorphus jobiensis*, and "*Sanguviridis tudaki*" (the latter not able to be referred to the synonymy of any currently recognized scincid species—Uetz et al. 2025). Austin (1995) confirmed the observations of Williams and Peterson (1982) that setae are only encountered in *Prasinohaema virens* among skinks (Pianka and Sweet 2005) and that other scincid taxa with ostensibly adhesive toe pads exhibit different integumentary morphologies. Although all scincid species with toe pads were described as lacking the extensively expanded structures of geckos and anoles (Fig. 4.8), they were regarded as furnishing a relatively large surface area for bearing their adhesive microstructures (Austin 1995). Like Williams and Peterson (1982), Austin (1995) compared the setal dimensions of *P. virens* with those deemed to be "typical" of geckos and anoles (rather than the more extensive comparative data currently available—Table 4.2), noting the same differences alluded to by Williams and Peterson (1995). For those scincid taxa with demonstrably differentiated toe pads but lacking setae, Austin (1995) recognized three distinctive morphological conditions: (a) folded (present in "*Sanguviridis tudaki*" and all species of *Prasinohaema* except *P. virens*)—folds 12–18 μm in height, separated by gaps of 10–18 μm, oriented orthogonally to the long axis of the digit, such folds also being present on the plantar surfaces of the feet; (b) ridged (present in *Lipinia leptosoma*)—ridges about 8 μm in height arranged in 8–10 roughly concentric semicircles separated by gaps of 11–16 μm; and (c) shaved (present in *Lipinia longiceps*)—integumentary flakes raised in an irregular pattern perpendicular to the long axis of the digit.

Based on the wide phylogenetic separation between geckos, anoles, and skinks, the occurrence of setae (or seta-like) integumentary attachment structures in each is regarded convergent (Maderson 1970; Williams and Peterson 1982; Austin 1995; Irschick et al. 1996; Larson and Losos 1996; Hagey et al. 2017; Griffing et al. 2022; Miller and Stroud 2022). Although anoles are currently regarded as having experienced only one evolutionary origin of toe pads (Hagey et al. 2017; Miller and Stroud 2022), the differences in morphological expression of the integumentary modifications in skinks, and their phylogenetic distribution, are consistent with repeated origins within this lineage (Williams and Peterson 1982; Austin 1995), reminiscent of the situation advocated for geckos (Gamble et al. 2012).

The presence of a lamellate/dentate pattern of the outer epidermal generation of skinks (Peterson 1984), rather than the spinulate pattern of the Oberhäutchen layer (Maderson 1970) of geckos and anoles, suggests that setae have arisen via a different integumentary developmental pathway than is evident for other seta-bearing lizards (Bauer 1998; Arnold 2002). Williams and Peterson (1982) posited that elaborations of epidermal cell boundaries may underlie setal development in scincids. Austin (1995) further elaborated upon this by intimating that setae in *Prasinohaema virens* may have been elaborated from an ancestral condition via the splitting, division, and subsequent elongation of the "folded" pattern of subdigital microarchitecture (see above).

Little is known of the mechanics of operation of the scincid adhesive system. Peterson (1983b) noted that the lamellae of *Prasinohaema virens* have a free margin, similar to that found in anoles (Russell and Eslinger 2017). Shea and Greer (2002) documented the relative shortening of the antepenultimate phalanx in the fourth digit of genera of skinks bearing toe pads (Fig. 4.8) and likened this to a similar condition encountered in several lineages of gecko (Russell 1977; Russell et al. 2015) that is associated with the mechanical operation of toe pads. Irschick et al. (1996) recorded low adhesive output for *Lipinia laptosoma*, *Prasinohaema flavipes*, *P. prehensicauda*, and *P. virens*. The greatest absolute force was recorded for *P. flavipes* but *P. virens*, the seta-bearing species, produced the greatest force in relation to body mass. All skinks examined, however, generated adhesive forces well below those of geckos and anoles of equivalent body mass (Irschick et al. 1996). Smith (1937) indicated that none of the scincid species he examined were able to climb "perfectly smooth surfaces" (in contrast to the adhesive abilities of geckos and anoles), but conversely McCoy (2006) noted that *Prasinohaema virens* can scale the "wall of a glass aquarium" and Brown and Fehlmann (1958) stated that *Lipinia leptosoma* can climb on vertical and overhanging glass surfaces. The development of adhesive abilities in skinks has been associated with arboreality (Slavenko et al. 2021), permitting climbing and leaping between the smooth surfaces of leaves (Williams and Peterson 1982; Coddington 1988), reminiscent of the exploitation of adhesive capacities by some geckos (Higham et al. 2017b).

4.5 Assessment of the Attempts to Fabricate Gecko-Inspired Synthetic Adhesives

Similar to our approach of summarizing the diversity of the form and function of squamate fibrillar adhesive systems, we explore the diversity in morphology, replication techniques, and multifunctional performance of the top 100 cited examples (see Table 4.5 for the list of studies included) of gecko-inspired synthetic adhesives (GSAs). We begin this section by describing our approach to reviewing this literature and follow this with an overview of our major findings. At the end of this section, we use our results to identify the major successes and shortcomings of gecko-inspired adhesion. Thereafter, we synthesize the findings of our biological and technological reviews to outline major lines of research that are ripe for future inquiry.

4.5.1 Overview of Our Approach

We used Clarivate's Web of Science^{TM} to seek articles matching the following search terms: ("gecko" OR "geckos") AND ("adhesion" OR "adhesive") OR ("gecko-inspired" OR "gecko inspired"). We selected these search terms to maximize the potential for identifying articles actually focused on gecko-inspired adhesion, while minimizing the return of unrelated articles. Simpler searches either overlooked substantive contributions to the field of gecko-inspired adhesion or returned far too many articles having minimal or no relevance. Our final search produced a list of 1594 published articles which were then sorted by number of citations, starting with the highest. We then manually examined each of the articles to identify whether a GSA was designed and fabricated in the report. If so, the article was included in the list of publications that we would later review. We continued this approach until a list of the top 100 cited articles exploring gecko-inspired adhesion was compiled (Table 4.5). A total of 296 articles were examined before this list was complete—most of the excess 196 articles not selected for further review focused on the following: (1) exploration of gecko adhesion with living animals or isolated gecko setal arrays; (2) robotic or mathematical models that did not incorporate a GSA into their design; (3) robotic models that incorporated a GSA into their design but provided minimal-to-no details of their design, fabrication, or performance for use in our review; (4) synthetic adhesive designs that were not actually inspired by geckos but mentioned the terms *gecko*, *geckos*, *adhesion*, or *adhesive* somewhere in the article or references cited; and (5) were otherwise unrelated to the field of gecko-inspired adhesion but included one or more of the search terms in the article or references cited.

Once our list of the top 100 cited articles (Table 4.5) of gecko-inspired adhesion was compiled, we reviewed each contribution and tabulated the following data: (1) article metadata; (2) GSA structure and morphometrics; (3) GSA replication method and materials; (4) images of the GSAs; (5) GSA property and performance assessment; and (6) GSA end use. Article metadata included the first author's last name, year of publication, journal, issue, and volume. The shape of the fiber's stalk and tip, as well as the fiber diameter, tip diameter, fiber length, fiber density (or center-to-center spacing), and fiber angle were documented if applicable and reported to enable description of the basic form and dimensions of the GSAs. Many articles deliberately varied fiber dimensions or left them uncontrolled and

Table 4.5 Top 100 cited studies exploring biomimetic gecko-inspired adhesion in order of decreasing number of citations in Clarivate's Web of Science^{TM} core database (as of 4 December 2024)

Rank	Authors	Year	Journal	Title	No. of citations
1	Lee et al.	2007	Nature	A reversible wet/dry adhesive inspired by mussels and geckos	1704
2	Geim et al.	2003	Nature Materials	Microfabricated adhesive mimicking gecko foot-hair	1068
3	Jin et al.	2005	Advanced Materials	Superhydrophobic aligned polystyrene nanotube films with high adhesive force	678
4	Qu et al.	2008	Science	Carbon nanotube arrays with strong shear binding-on and easy normal lifting-off	622
5	Kim et al.	2008	IEEE Transactions on Robotics	Smooth vertical surface climbing with directional adhesion	504
6	Jeong et al.	2009	Proceedings of the National Academy of Sciences	A nontransferring dry adhesive with hierarchical polymer nanohairs	473
7	Mahdavi et al.	2008	Proceedings of the National Academy of Sciences	A biodegradable and biocompatible gecko-inspired tissue adhesive	442
8	Sitti and Fearing	2003	Journal of Adhesion Science and Technology	Synthetic gecko foot-hair micro/nano-structures as dry adhesives	415
9	Murphy et al.	2009a, b	Small	Gecko-inspired directional and controllable adhesion	385
10	Greiner et al.	2007	Langmuir	Adhesion of bioinspired micropatterned surfaces: Effects of pillar radius, aspect ratio, and preload	371
11	del Campo et al.	2007	Langmuir	Contact shape controls adhesion of bioinspired fibrillar surfaces	370

(continued)

Table 4.5 (continued)

12	Ge et al.	2007	Proceedings of the National Academy of Sciences	Carbon nanotube-based synthetic gecko tapes	363
13	Kim et al.	2016	ACS Nano	Bioinspired, highly stretchable, and conductive dry adhesives based on 1D-2D hybrid carbon nanocomposites for all-in-one ECG electrodes	334
14	Liu et al.	2012	Nanoscale	Superhydrophobic gecko feet with high adhesive forces towards water and their bio-inspired materials	319
15	Murphy et al.	2009a, b	ACS Applied Materials & Interfaces	Enhanced adhesion by gecko-inspired hierarchical Fibrillar adhesives	308
16	Reddy et al.	2007	Advanced Materials	Bioinspired surfaces with switchable adhesion	283
17	Kim and Sitti	2006	Applied Physics Letters	Biologically inspired polymer microfibers with spatulate tips as repeatable fibrillar adhesives	270
18	Yurdumakan et al.	2005	Chemical Communications	Synthetic gecko foot-hairs from multiwalled carbon nanotubes	251
19	Bartlett et al.	2012	Advanced Materials	Looking beyond Fibrillar features to scale gecko-like adhesion	244
20	Drotlef et al.	2017	Advanced Materials	Bioinspired composite microfibers for skin adhesion and signal amplification of wearable sensors	234
21	Jiang et al.	2017	Science Robotics	A robotic device using gecko-inspired adhesives can grasp and manipulate large objects in microgravity	223

(continued)

Table 4.5 (continued)

Rank	Authors	Year	Journal	Title	No. of citations
22	Parness et al.	2009	Journal of the Royal Society Interface	A microfabricated wedge-shaped adhesive array displaying gecko-like dynamic adhesion, directionality and long lifetime	210
23	Sethi et al.	2008	Nano Letters	Gecko-inspired carbon nanotube-based self-cleaning adhesives	203
24	Cho and Choi	2008	Advanced Functional Materials	Fabrication of hairy polymeric films inspired by geckos: Wetting and high adhesion properties	199
25	Drotlef et al.	2014	Advanced Materials	Magnetically actuated patterns for bioinspired reversible adhesion (dry and wet)	196
26	Qu and Dai	2007	Advanced Materials	Gecko-foot-mimetic aligned single-walled carbon nanotube dry adhesives with unique electrical and thermal properties	190
27	Greiner et al.	2009	Advanced Materials	Hierarchical gecko-like adhesives	189
28	Zhao et al.	2006	Journal of Vacuum Science & Technology B	Interfacial energy and strength of multiwalled-carbon-nanotube-based dry adhesive	181
29	Northen et al.	2008	Advanced Materials	A gecko-inspired reversible adhesive	177
30	Song et al.	2017	Proceedings of the National Academy of Sciences	Controllable load sharing for soft adhesive interfaces on three-dimensional surfaces	171
31	Murphy et al.	2007	Journal of Adhesion Science and Technology	Adhesion and anisotropic friction enhancements of angled heterogeneous micro-fiber arrays with spherical and spatula tips	164

(continued)

Table 4.5 (continued)

32	Ma et al.	2018	Advanced Materials	Remote control over underwater dynamic attachment/detachment and locomotion	161
33	Song and Sitti	2014	Advanced Materials	Soft grippers using micro-fibrillar adhesives for transfer printing	156
34	Shahsavan et al.	2017	Advanced Materials	Thermally active liquid crystal network gripper mimicking the self-peeling of gecko toe pads	153
35	Cheng et al.	2008	Advanced Functional Materials	Tunable adhesive Superhydrophobic surfaces for superparamagnetic microdroplets	153
36	Bae et al.	2013	Advanced Healthcare Materials	Enhanced skin adhesive patch with modulus-tunable composite micropillars	151
37	Mengüc et al.	2012	Advanced Functional Materials	Gecko-inspired controllable adhesive structures applied to micromanipulation	150
38	Kim et al.	2009a, b, c	Advanced Materials	Stooped Nanohairs: Geometry-controllable, unidirectional, reversible, and robust gecko-like dry adhesive	150
39	Lee et al.	2008a, b	Applied Physics Letters	Directional adhesion of gecko-inspired angled microfiber arrays	139
40	Glass et al.	2008	IEEE Transactions on Biomedical Engineering	A legged anchoring mechanism for capsule endoscopes using micropatterned adhesives	138
41	Davies et al.	2009	International Journal of Adhesion and Adhesives	A practical approach to the development of a synthetic gecko tape	136
42	Xie et al.	2008	Chemistry of Materials	Self-peeling reversible dry adhesive system	134
43	Kizilkan et al.	2017	Science Robotics	Bioinspired photocontrollable microstructured transport device	133

(continued)

Table 4.5 (continued)

Rank	Authors	Year	Journal	Title	No. of citations
44	Menon et al.	2004	Proceedings of the IEEE International Conference on Robotics and Biomimetics	Gecko inspired surface climbing robots	128
45	Yu et al.	2011	Advanced Functional Materials	Gecko-inspired dry adhesive for robotic applications	124
46	Aksak et al.	2008	2008 IEEE International Conference on Robotics and Automation	Gecko inspired micro-fibrillar adhesives for wall climbing robots on micro/nanoscale rough surfaces	121
47	Xu et al.	2015	Nature Communications	Robust self-cleaning and micromanipulation capabilities of gecko spatulae and their bio-mimics	118
48	King et al.	2014	Advanced Materials	Creating gecko-like adhesives for real world surfaces	113
49	Ruffatto et al.	2014	Journal of the Royal Society Interface	Improving controllable adhesion on both rough and smooth surfaces with a hybrid electrostatic/gecko-like adhesive	113
50	Cui et al.	2012	Advanced Materials	Bioinspired actuated adhesive patterns of liquid crystalline elastomers	112
51	Kim et al.	2009a, b, c	Soft Matter	Reversible dry micro-fibrillar adhesives with thermally controllable adhesion	112
52	Tao et al.	2017	Advanced Functional Materials	Controllable anisotropic dry adhesion in vacuum: Gecko inspired wedged surface fabricated with Ultraprecision diamond cutting	110

(continued)

Table 4.5 (continued)

53	Lee et al.	2008a, b	Journal of the Royal Society Interface	Sliding-induced adhesion of stiff polymer microfibre arrays. I. Macroscale behaviour	110
54	Röhrig et al.	2012	Small	3D direct laser writing of Nano- and Microstructured hierarchical gecko-mimicking surfaces	109
55	Tan et al.	2019	Small	Switchable adhesion of micropillar adhesive on rough surfaces	105
56	Kim et al.	2009a, b, c	Langmuir	Wet self-cleaning of biologically inspired elastomer mushroom shaped Microfibrillar adhesives	104
57	Jeong et al.	2010	Applied Physics Letters	Effect of leaning angle of gecko-inspired slanted polymer nanohairs on dry adhesion	103
58	Lee and Fearing	2008	Langmuir	Contact self-cleaning of synthetic gecko adhesive from polymer microfibers	103
59	Wang et al.	2014	ACS Applied Materials & Interfaces	Fabrication of well-defined mushroom-shaped structures for biomimetic dry adhesive by conventional photolithography and molding	102
60	Cheng et al.	2010	Langmuir	Tip geometry controls adhesive states of Superhydrophobic surfaces	101
61	Shahsavan et al.	2015	Advanced Materials	Smart muscle-driven self-cleaning of biomimetic microstructures from liquid crystal elastomers	92
62	Jin et al.	2012	Langmuir	Design and fabrication of gecko-inspired adhesives	91

(continued)

Table 4.5 (continued)

Rank	Authors	Year	Journal	Title	No. of citations
63	Jin et al.	2014	Advanced Functional Materials	Biomimetic bidirectional switchable adhesive inspired by the gecko	90
64	Purtov et al.	2015	ACS Applied Materials & Interfaces	Switchable adhesion in vacuum using bio-inspired dry adhesives	89
65	Wang	2018	ACS Nano	Slanted functional gradient micropillars for optimal bioinspired dry adhesion	88
66	Zhou et al.	2013	ACS Applied Materials & Interfaces	Controllable interfacial adhesion applied to transfer light and fragile objects by using gecko inspired mushroom-shaped pillar surface	88
67	Gravish et al.	2010	Journal of the Royal Society Interface	Rate-dependent frictional adhesion in natural and synthetic gecko setae	88
68	Gillies et al.	2013	Advanced Functional Materials	Controllable particle adhesion with a magnetically actuated synthetic gecko adhesive	86
69	Kustandi et al.	2007a, b	Advanced Functional Materials	Self-assembled nanoparticles-based fabrication of gecko foot-hair-inspired polymer nanofibers	86
70	Barreau et al.	2016	Advanced Functional Materials	Fibrillar elastomeric micropatterns create tunable adhesion even to rough surfaces	84
71	Wang et al.	2020	Materials Today	Bioinspired footed soft robot with unidirectional all-terrain mobility	83
72	Chen et al.	2013	Advanced Functional Materials	Buckling-based strong dry adhesives via interlocking	80
73	Fischer et al.	2017	ACS Applied Materials & Interfaces	Composite pillars with a tunable Interface for adhesion to rough substrates	78

(continued)

Table 4.5 (continued)

74	Ho et al.	2011	ACS Nano	Fabrication and analysis of gecko-inspired hierarchical polymer Nanosetae	78
75	Kwak et al.	2011	Small	Anisotropic adhesion properties of triangular-tip-shaped micropillars	76
76	Lee et al.	2009	Langmuir	Gecko-inspired combined lamellar and Nanofibrillar Array for adhesion on nonplanar surface	76
77	Du et al.	2017	Small	Bio-inspired design and fabrication of micro/Nano-brush dual structural surfaces for switchable oil adhesion and antifouling	74
78	Raut et al.	2018	ACS Applied Materials & Interfaces	Gecko-inspired dry adhesive based on micro-nanoscale hierarchical arrays for application in climbing devices	73
79	Mengüc et al.	2014	Journal of the Royal Society Interface	Staying sticky: Contact self-cleaning of gecko-inspired adhesives	73
80	Kim et al.	2007a, b	Applied Physics Letters	Enhanced friction of elastomer microfiber adhesives with spatulate tips	73
81	Liimatainen et al.	2020	Advanced Materials	Liquid-Superrepellent bioinspired Fibrillar adhesives	72
82	Schubert et al.	2008	Journal of the Royal Society Interface	Sliding-induced adhesion of stiff polymer microfibre arrays. II. Microscale behaviour	71
83	Li et al.	2020	ACS Applied Materials & Interfaces	Switchable adhesion for nonflat surfaces mimicking geckos' adhesive structures and toe muscles	69
84	Liu et al.	2019	Journal of Cleaner Production	Bioinspired design by gecko structure and mussel chemistry for bio-based adhesive system through incorporating natural fibers	69

(continued)

Table 4.5 (continued)

Rank	Authors	Year	Journal	Title	No. of citations
85	Lee et al.	2012	Soft Matter	Hierarchical gecko-inspired nanohairs with a high aspect ratio induced by nanoyielding	68
86	Yu et al.	2012	Langmuir	Friction and adhesion of gecko-inspired PDMS flaps on rough surfaces	67
87	Glass et al.	2010	Langmuir	Enhanced wet adhesion and shear of elastomeric micro-fiber arrays with mushroom tip geometry and a Photopolymerized p(DMA-co-MEA) tip coating	65
88	Wang et al.	2019a, b	Nanoscale	Gecko toe pads inspired in situ switchable superhydrophobic shape memory adhesive film	63
89	Kustandi et al.	2007a, b	Journal of Micromechanics and Microengineering	Fabrication of a gecko-like hierarchical fibril array using a bonded porous alumina template	63
90	Liu et al.	2021	Small	Adhesion enhancement of micropillar Array by combining the adhesive design from gecko and tree frog	62
91	Cañas et al.	2012	Acta Biomaterialia	Effect of nano- and micro-roughness on adhesion of bioinspired micropatterned surfaces	62
92	Kim et al.	2007a, b	Microsystem Technologies	Replication of high-aspect-ratio nanopillar array for biomimetic gecko foot-hair prototype by UV nano embossing with anodic aluminum oxide mold	62

(continued)

Table 4.5 (continued)

93	Glassmaker et al.	2005	Acta Biomaterialia	Adhesion enhancement in a biomimetic fibrillar interface	62
94	Zhang et al.	2021	Chemistry of Materials	Gecko's feet-inspired self-peeling switchable dry/wet adhesive	61
95	Li et al.	2011	Colloids and Surfaces A	Gecko-inspired synthesis of superhydrophobic ZnO surfaces with high water adhesion	61
96	Dayan et al.	2021	Advanced Materials	3D printing of elastomeric bioinspired complex adhesive microstructures	60
97	Rong et al.	2014	Advanced Materials	Bio-inspired hierarchical polymer fiber-carbon nanotube adhesives	59
98	Bhushan and Sayer	2007a, b	Microsystem Technologies	Surface characterization and friction of a bio-inspired reversible adhesive tape	59
99	Wang et al.	2019a, b	ACS Applied Materials & Interfaces	Strong wet and dry adhesion by cupped microstructures	57
100	Guo et al.	2015	ACS Applied Materials & Interfaces	Reverse adhesion of a gecko-inspired synthetic adhesive switched by an ion-exchange polymer-metal composite actuator	57

reported a range of dimensions; for these cases, minimum and maximum values of the aforementioned metrics were recorded.

Autumn et al. (2006b) summarized the major properties and performance capabilities of the natural gecko adhesive system and we used these to benchmark GSA properties and performance relative to their source of natural inspiration. Gecko adhesive toe pads are directional, exhibit strong attachment yet detach easily, attach to a variety of substrates, are self-cleaning, and are anti-matting. Gecko toe pads are also renowned for their ability to attach in non-ideal conditions (rough, hot, wet, humid environments), attach and detach repeatedly from a substrate, and for their superhydrophobic properties. We tabulated whether each article examined/observed these properties or performance metrics in their GSA, except for strong attachment performance. "Strong" is highly subjective, thus we recorded adhesive stress (force per unit area; either in shear or normal) if the article reported these values as a proxy for this performance metric. We also tabulated whether each article was considered

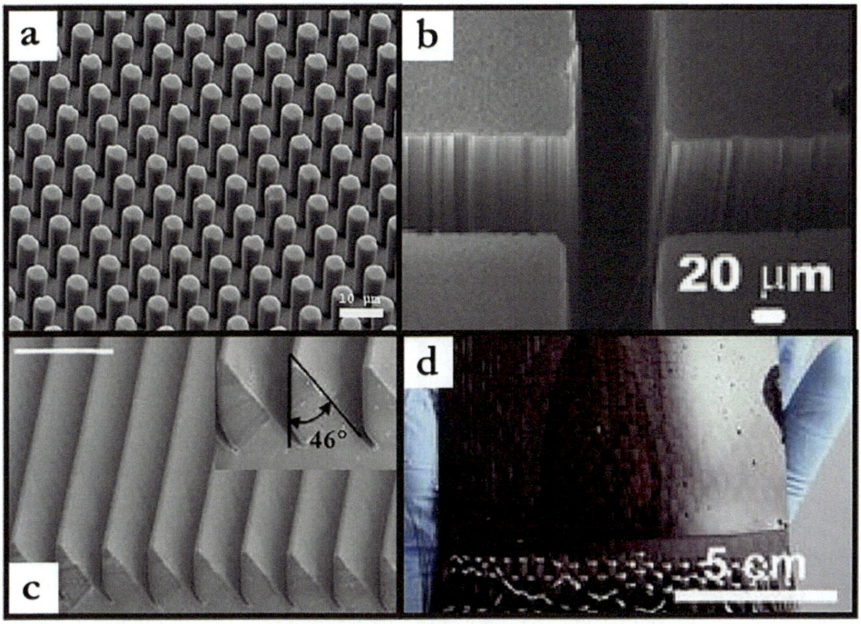

Fig. 4.9 Fiber shape variation in surveyed gecko-inspired synthetic adhesives (GSAs). (**a**) Cylindrical fibers. Scale bar represents 10 μm. (Adapted with permission from Greiner et al. (2007). Copyright 2007 American Chemical Society). (**b**) Rectangular prismatic fibers. (Adapted with permission from Sethi et al. (2008). Copyright 2008 American Chemical Society). (**c**) Wedge-shaped fibers. Scale bar represents 200 μm. (Adapted with permission from Tao et al. (2017)). (**d**) Non-fibrillar GSA. (Adapted with permission from Bartlett et al. (2012))

to fall within the field of basic or applied research, the latter assignment being based on whether the adhesive was integrated or tested in a specific applied context (e.g., biomedical adhesive, robotic model, gripping device). We also recorded potential areas of end use if the authors described such details in their article.

4.5.2 Assessment Findings

4.5.2.1 GSA Structure and Morphometrics

The top 100 articles concerning fabrication of GSAs reveal great diversity in form and dimensions of the simulacra. The stalks of most GSA fibers (80 of them; Fig. 4.9a) were of cylindrical form, although other shapes also appeared several times in the literature. Seven GSA fibers were made of rectangular prisms (Fig. 4.9b), and five GSA fibers were wedge-shaped (Fig. 4.9c). Interestingly, four GSAs lacked fibrillar structures entirely and instead incorporated principles inspired by other features of the gecko adhesive system (e.g., load-bearing tendons,

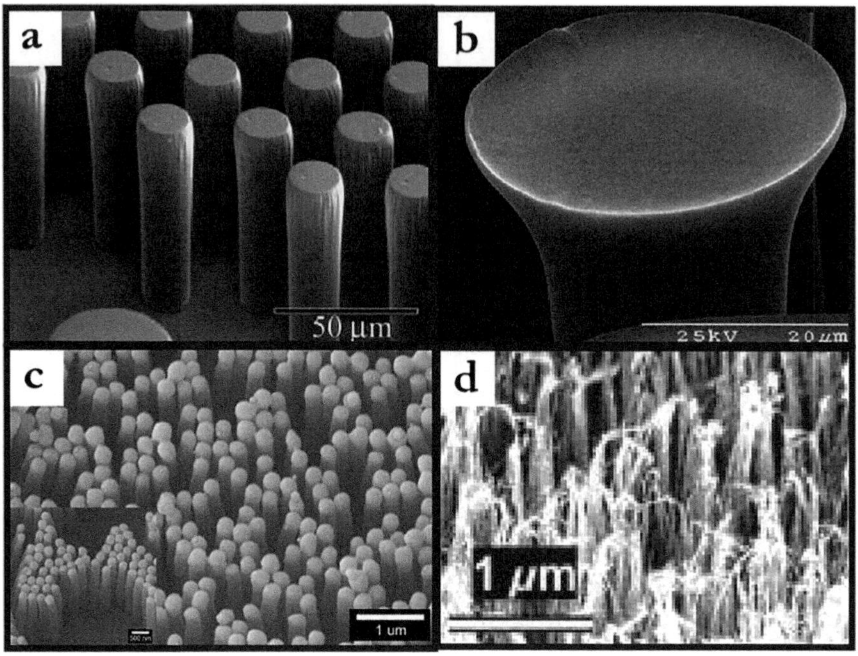

Fig. 4.10 Tip shape variation in surveyed gecko-inspired synthetic adhesives (GSAs). (**a**) Flat. (Adapted with permission from Reddy et al. (2007)). (**b**) Mushroom-shaped. (Reprinted with permission by publisher (Taylor and Francis Ltd) from Murphy et al. (2007)). (**c**) Rounded. Scale bar of inset represents 500 nm. (Adapted with permission from Kustandi et al. (2007b)). (**d**) Cylindrical, in the form of carbon nanotubes. (Reprinted with permission from Zhao et al. (2006), American Vacuum Society)

cushioning provided by blood-filled sinuses or adipose tissue; Fig. 4.9d). Cone-shaped, triangular prismatic, and hexagonal fibers each appeared twice in the literature, while paddle-shaped and ridged features, each appeared only once in our search. Totals exceed 100 because multiple articles produced and explored multiple fiber stalk shapes. The tips of GSA fibers were more diverse in shape than the fiber stalks (Fig. 4.10). GSA fiber tips were most commonly either flat (30 of them; Fig. 4.10a) or mushroom-shaped (24 of them; Fig. 4.10b), with a reasonable number of GSA tip shapes left uncontrolled or unspecified (15 of them). Round (Fig. 4.10c) and cylindrical (generally in the form of carbon nanotubes; Fig. 4.10d) fiber tips were represented on nine and eight occasions, respectively. Some GSA fibers terminated in a flat, thin film (2 of them) or did not possess fibrillar structures at all (4 of them). The remaining tip shapes were generally variations of one of the above, often incorporating angled or cupped tips or tips with multiple suction cavities.

Although all gecko setae observed to date branch into multiple nanoscopic spatulate tips (Garner and Russell 2021), only 12 of the GSAs surveyed incorporated some form of structural hierarchy (the remaining 91 employed unbranced fibers with

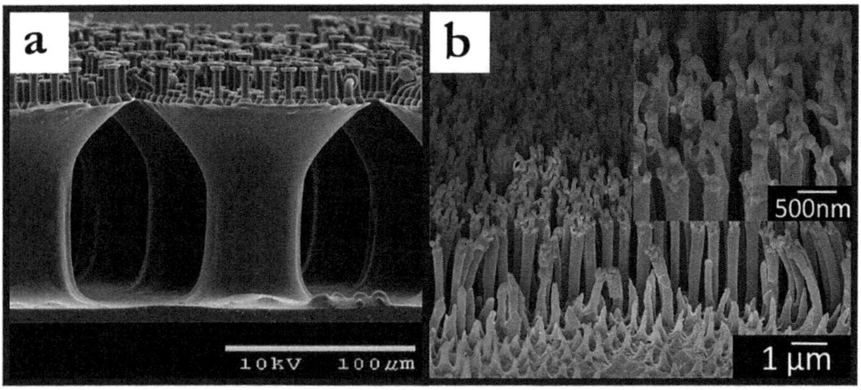

Fig. 4.11 Attempts to generate hierarchical and branched gecko-inspired synthetic adhesives. (**a**) Mushroom-shaped "pillars-on-pillars". (Adapted with permission from Murphy et al. (2009b). Copyright 2009 American Chemical Society). (**b**) Branched nanopillars fabricated via alumina membrane molding. (Adapted with permission from Ho et al. (2011). Copyright 2011 American Chemical Society)

a single tip). Typically, the hierarchical structures were not actually branched—either large fibers topped with smaller fibers were generated (Fig. 4.11a) or carbon nanotubes were bunched into larger fibers, with the nanotubes extending as the tips of the fibers. Some studies were able to generate branched nanostructures with remarkable similarity to the branched epidermal outgrowths located more proximally on Tokay gecko toe pads (Fig. 4.3; see also Garner and Russell 2021) by employing hierarchical alumina or polycarbonate molds (Fig. 4.11b). Regardless, the majority of GSAs tend to consist of single fibers terminating in a single tip.

Many of the selected studies designed and fabricated fibers with variable stalk diameters (27 of them), tip diameters (26 of them), and/or fiber length (29 of them). In these contributions, stalk and tip diameters ranged broadly from a minimum of 10 nm carbon nanotubes to 2 mm flat-tipped polymer fibers, while fiber lengths varied from 150 nm polyimide fibers to 500 μm carbon nanotubes.

To more closely explore the variability in GSA fiber and tip dimensions, we filtered our data set to consider only cylindrical fibers (by far the most common fiber shape) with single dimensional values reported (Fig. 4.12). Of the 52 GSAs that met these criteria for stalk diameter, 1 GSA (~2%) had stalk diameters in the millimeter range, 37 GSAs (~71%) had stalk diameters in the micrometer range, and 14 GSAs (~27%) were reported to have stalk diameters in the nanometer range (<1 μm). Of the 50 GSAs that met the above criteria for tip diameter, 33 GSAs (66%) had tip diameters in the micrometer range; 24 GSAs possessed tip diameters between 5 and 100 μm, 4 GSAs possessed tip diameters between 1 and 5 μm, and 5 GSAs possessed tip diameters greater than 100 μm. A total of 17 (34%) GSAs had tip diameters in the nanometer range (<1 μm). Of the 53 GSAs that met the above

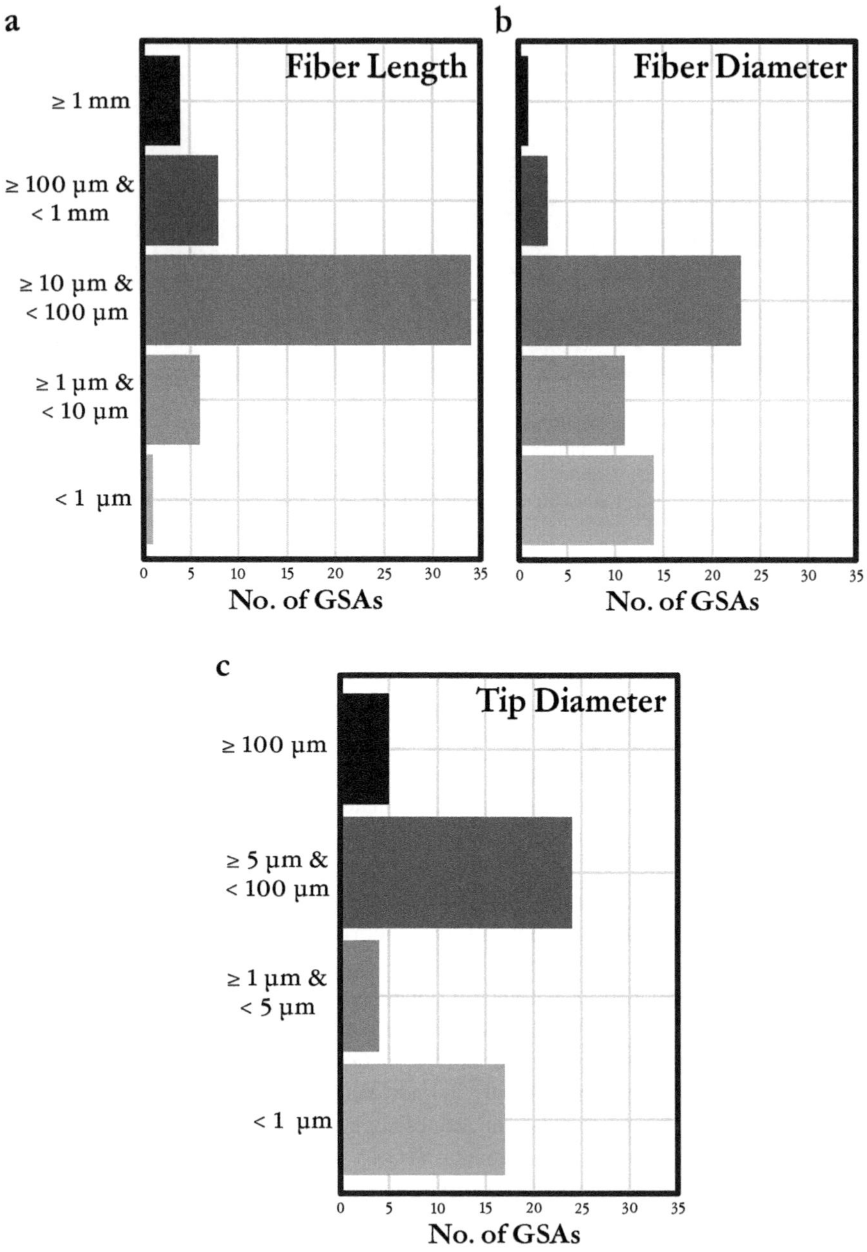

Fig. 4.12 Morphometric variation in the (**a**) fiber length, (**b**) fiber diameter, and (**c**) tip diameter of surveyed gecko-inspired synthetic adhesives (GSAs)

criteria for fiber length, 4 GSAs (~8%) had fiber lengths in the millimeter range, 48 GSAs (~90%) had fiber lengths in the micrometer range, and only 1 GSA (~2%) had fiber lengths in the nanometer range. Several articles did not report stalk diameters (6 of them), tip diameters (9 of them), or fiber lengths (4 of them).

Fewer of the selected articles designed and fabricated fibers with variable densities (5 of them). In these studies, fiber density varied from 676,000 fibers per mm^2 to 10^9 fibers per mm^2. Similarly to our approach outlined above, we further explored variability in GSA fiber density by filtering our data set to consider only cylindrical fibers with single density values (Fig. 4.13). Of the 9 GSAs that met these criteria, 2 GSAs (~22%) had fiber densities less than 10,000 fibers per mm^2, 3 GSAs (~33%) had fiber densities between 100,000 and 10^6 fibers per mm^2, and the remaining 4 GSAs (~45%) had fiber densities between 10^6 and 10^7 fibers per mm^2. The majority of articles (84 of them) did not report fiber density, but 37 of these articles instead reported some measure of fiber spacing. Some of these studies (8 of them) actively varied fiber spacing with spacings as low as 400 nm and as high as 75 μm. After filtering our data set to consider only cylindrical fibers with single fiber spacing values reported, 8 GSAs had fiber spacing greater than 100 μm, 20 GSAs had fiber spacing between 1 and 100 μm, and only 1 GSA had fiber spacing below 1 μm. The remaining 47 articles that did not provide fiber spacing details, reported no figures related to fiber density.

Some of our selected studies (15 of them) occasionally varied fiber angle, reported here as the angle relative to horizontal. In these studies, fiber angles varied from 0° in the case of magnetically actuated microridges to the more typical 90°. Similarly to our approach outlined above, we more closely explored fiber angle variation by filtering our data set to include only cylindrical fibers with single fiber angle values reported. Of the 75 GSAs that met these criteria, 68 GSAs exhibited fiber angles of 90°, 4 GSAs exhibited fiber angles between 70° and 80°, and 3 GSAs exhibited fiber angles of 45° or less. There were 6 articles that did not report fiber angle.

4.5.2.2 Methods of GSA Replication and Materials Employed

The majority of GSAs surveyed were produced using some form of lithographic techniques (57 of them), often with additional steps such as various etching techniques (12 GSAs), inking or dipping in another polymer (8 GSAs), machining (2 GSAs), or chemical vapor deposition (1 GSA). These additional steps were typically used to create more detailed or unique fiber morphologies such as mushroom-shaped, hierarchical, or angled fibers. Most commonly photolithography was the lithographic technique of choice (44 GSAs), but other lithographic techniques include multi-photon lithography, soft lithography, electron beam lithography, and nanoimprint lithography. Lithographic techniques were generally employed to generate negative molds into which a material of choice could then be deposited to create a positive cast of the fibrillar array (or single fiber in some cases). Alumina or polycarbonate membrane molding was also a common technique to

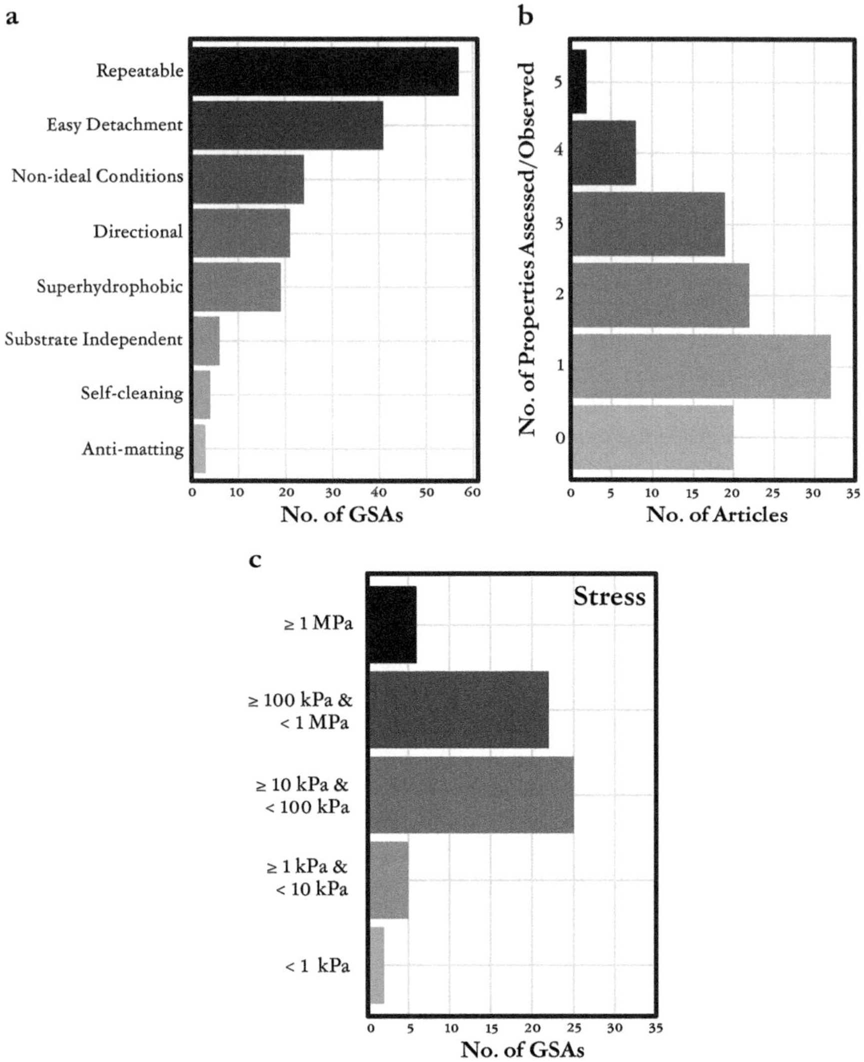

Fig. 4.13 Properties and performance of surveyed gecko-inspired synthetic adhesives (GSAs). (**a**) Number of GSAs observing/assessing the various multifunctional properties found in the gecko adhesive system. (**b**) Number of articles and the number of properties observed/assessed in a single study. (**c**) Distribution of reported adhesive stress (in normal or shear loading orientations) by surveyed GSAs

create single fibered, or even hierarchical, nanostructures (13 GSAs). Other less common replication techniques included machining, plasma etching, atomic force microscope nanoindentation, and track etching. Many articles were quite vague in their descriptions of replication techniques, often citing previous work, although clear alterations were made to the cited protocol.

Most GSAs surveyed were composed of polydimethylsiloxane or PDMS (38 of them), although polyurethane (23 GSAs), carbon nanotubes (9 GSAs), polypropylene (4 GSAs), polyimide (4 GSAs), polystyrene (3 GSAs), and polycarbonate (2 GSAs) were also common material choices. Several articles incorporated multiple materials into their fibrillar arrays. Some authors incorporated magnetic nanoparticles into their fibers to vary material properties along the length of fibers or to generate actuatable smart adhesives that respond to magnetic fields. Other articles coated or topped their GSAs with another adhesive substance, such as 3,4-dihydroxy-L-phenylalanine (DOPA)-containing materials, similar to those found in mussel adhesive secretions, or soft, viscoelastic materials like p(DMA-co-MEA).

4.5.2.3 Properties and Performance of GSAs

Of the major properties of the gecko adhesive system we indicated above, the most common property assessed and observed for GSAs, and shared with the natural system, was their ability to sustain repeated attachment and detachment cycles (57 GSAs; Fig. 4.13a). The next most common properties assessed and observed were easy detachment (41 GSAs), attachment in non-ideal conditions (24 GSAs), directional performance (21 GSAs), and superhydrophobicity (19 GSAs). The other properties were assessed and observed much less frequently with 6 GSAs attaching to a variety of substrates with variable properties (substrate independence), 4 GSAs possessing self-cleaning qualities, and 3 GSAs claiming anti-matting properties. Attachment in non-ideal conditions was generally assessed on rough or wet substrates and substrates under vacuum. Although many studies begin the introductions to their articles by describing the remarkable multifunctional properties of the gecko adhesive system, many (32 of them) only explore and report on one of these properties in a single article (Fig. 4.13b). A reasonable number of articles explore two (22 of them) or three (19 of them) properties in a single study, but far fewer explore four or five properties at once (8 and 2 of them, respectively). No studies examined more than five properties in a single study. Interestingly, no studies assessing a GSA for a particular property reported that they were unable to observe that property in their engineered material.

A total of 60 articles reported adhesive stress (in shear or normal loading orientations) of their GSAs (Fig. 4.13c). Adhesive stress varied markedly across these articles, with stresses as low as 0.028 kPa and as high as 4750 kPa being reported. Most GSAs surveyed generated adhesive stresses an order of magnitude lower than the value of approximately 100 kPa typically reported for Tokay geckos (*Gekko gecko*; Irschick et al. 1996), while a similar number were able to achieve performance of the same magnitude. Several GSAs (6 of them) were able to attach a whole order of magnitude better than Tokay geckos while others achieved performance two or three orders of magnitude lower (5 and 2 of them, respectively).

4.5.2.4 End Use of GSAs

Most GSAs (85 of them) were explored in a basic research capacity (i.e., fundamental questions about gecko-inspired adhesion were being investigated with no direct application), although a reasonable number (18 of them) were used in an applied context (e.g., robotic models, biomedical applications, gripping devices). Most GSAs were not discussed in the context of particular cases of end use (55 of them), but several were described in the context of potential robotics (20 of them), biomedical (12 of them), and industrial (11 of them) applications. Other end use cases that were less frequently discussed were pick-and-place transport, water transfer/transport, and space/microgravity.

4.5.3 Successes and Shortcomings in Gecko-Inspired Adhesion

Our review of the top 100 most-cited articles exploring gecko-inspired adhesion have revealed several clear successes in attempts to design and fabricate synthetic adhesives based on the gecko adhesive system:

1. Most GSAs we surveyed were able to attach with similar adhesive stresses under controlled laboratory conditions as live Tokay geckos, and several of them were able to achieve stresses far beyond what geckos can produce.
2. Many articles employed innovative techniques to generate hierarchical fibrillar structures and fibers with dimensions of similar magnitudes to those of geckos.
3. Several researchers were able to achieve robust adhesive performance using stiff materials similar in magnitude to the elastic modulus of β-keratin by fabricating high aspect ratio fibers.
4. Many GSAs were capable of repeated attachment–detachment cycles and directionality of attachment, permitting strong attachment but easy release.
5. Many researchers took advantage of the lack of evolutionary, developmental, physiological, or morphological constraints in engineered systems by incorporating multiple bio-inspired principles into their GSAs or manipulating parameters unable to, or not observed to, vary in the gecko adhesive system.

Our analysis also resulted in the identification of several shortcomings in gecko-inspired adhesion:

1. Clearly, GSAs are not capable of replicating all or even most of the multifunctional properties exhibited by the natural system. Of course, in most cases, researchers were not systematically testing the multifunctional properties of their GSAs and were instead focused on a subset of properties or did not aspire to achieve multifunctional performance. Some authors use the same fundamental fibrillar adhesive design and report separate findings for different properties in separate articles.

2. None of the articles we reviewed assessed a GSA for a particular property or performance metric only to report they were unable to achieve their desired outcomes. In other words, none of the gecko-inspired adhesion literature we reviewed reported "negative results."
3. The GSAs that perform on par with or beyond the natural system in terms of adhesive stress often incorporate other properties not found in the gecko adhesive system to achieve this. In several cases, authors opted to abandon fibers entirely and chose other structures or principles in the gecko adhesive system to replicate.
4. Despite encouraging progress, fabricating hierarchical, branched nanostructures is still technically difficult. However, alumina or polycarbonate nanomolding techniques have been used on several occasions to produce nanofibers with impressive branching patterns reminiscent of the condition observed in the gecko adhesive system.
5. Many of the articles reported exceptionally vague descriptions of their replication approaches and morphometrics of their GSAs, which makes it challenging to benchmark their success in replicating the natural system or even in simply replicating their experimental procedures.

4.6 Synthesis, Outlook, and Conclusions

Twenty-five years have passed since Autumn and colleagues isolated single setae of *Gekko gecko*, measured their adhesive performance (Autumn et al. 2000), and identified van der Waals forces as the major mechanism of gecko adhesion (Autumn et al. 2002). Their seminal work undoubtedly catalyzed the enormous interest in exploring both gecko adhesion and biomimetic gecko-inspired adhesion. Most of the work that followed, however, adopted a limited view of the form and function of squamate adhesive systems, largely based on that of a single species possessing a relatively complex adhesive apparatus (i.e., the Tokay gecko, *Gekko gecko*). Our review of the intraindividual and interspecific diversity of form and function of squamate adhesive systems revealed several factors of relevance for future attempts at biomimetic designs. First, there is no such thing as a "typical" adhesive toe pad or seta (Russell and Garner 2021). The toe pads of gekkotans, anolids, and scincids differ quite markedly in form, with geckos possessing the greatest morphological diversity compared to the other two groups (Russell and Gamble 2019). The functional consequences of this morphological variation are not well-resolved, although some work has identified performance benefits of different toe pad configurations under certain environmental conditions (Cole et al. 2005; Delaugerre et al. 2015; Higham et al. 2017a; Russell and Delaugerre 2017). With what little is known about variation in the form and dimensions of squamate setae/spatulae, it appears that geckos possess much greater interspecific diversity in the form and dimensions of their setae and spatulae than do anolids and scincids. Furthermore, studies of setal morphological and dimensional variation along longitudinal transects of toe pads in both geckos and anoles have identified that both setal form and dimensions vary considerably along the proximodistal axis of toe pads and adhesive scales (scansors or lamellae; Delannoy 2005; Russell et al. 2007; Johnson and Russell 2009; Russell and Johnson 2014; Garner et al. 2021, 2022; Russell and Garner 2021). Several functional hypotheses relating to this have been suggested but currently remain unexplored (Johnson and Russell 2009; Garner et al. 2021, 2022). Biomimetics is a useful approach to further probe form–function connections in ways that are not easy or feasible in organismal systems. Thus, one area of future work could involve biomimetic physical and/or theoretical representations of naturally observed morphological variation to explore how, or if, this variation impacts static and dynamic attachment processes under different environmental conditions.

Second, investigations of the intraindividual and interspecific variation in the form and dimensions of gekkotan and anolid setae have identified several instances of alternative configurations of setal arrays that produce similar functional outcomes (Irschick et al. 1996; Garner et al. 2021; Griffing et al. 2021). *Gekko gecko* setae, for example, are relatively long, wide, profusely branched, and less densely packed than those of *Correlophus ciliatus,* which are relatively short, thin, less branched, and more densely packed (Griffing et al. 2021; Table 4.1). Compared with *G. gecko*, the setae of *Anolis equestris* are very similar to the condition observed in *C. ciliatus*, although each seta of *A. equestris* has single spatula that is larger than the numerous, smaller spatulae of the multi-spatulae setae of *G. gecko* (Garner et al. 2021). Despite these morphological differences, these three species have functionally similar attachment performance on smooth substrates (Irschick et al. 1996; Griffing et al. 2021). Whether one configuration is superior in certain locomotor and/or environmental conditions than others, remains to be explored. As described above, anoles and geckos may have functionally similar toe pads when static clinging ability is measured on smooth substrates, but a few minutes of observing geckos like *G. gecko* or *C. ciliatus* actively deploying their toe pads during locomotion on similar substrates reveals that they are much more accomplished climbers than anoles like *A. equestris* or *A. carolinensis* (Irschick et al. 1996). Comparative studies of the mechanics and properties of isolated setae and setal arrays between different species and groups are desperately needed. Biomimetic devices with variable configurations would offer additional ways to verify and model results obtained from such comparative work.

Third, in some cases, morphological variation may indeed result in reduced performance. As described above, Irschick et al. (1996) explored comparative clinging ability across gekkotans, anolids, and scincids. Relative to the gekkotans included in the study, anoles were functionally similar when attaching statically to a smooth sheet of acetate transparency (anoles ~65.5 $\pm$ 7.6 kPa vs. geckos ~58.5 $\pm$ 9.4 kPa), but when compared to scincids (16.9 $\pm$ 3.4 kPa), anolids and gekkotans both achieved greater adhesive stress. These findings, in addition to our assertion regarding the differential dynamic capabilities of some gecko and anole toe pads, indicate that some adhesive system configurations are equivalent in some circumstances while others are better or worse than others (depending on the property being considered). As such, we believe that this morphological and functional diversity has enormous potential for biomimetic design, as practitioners can select the configurations that best meet their specific objectives. The adhesive toe pads of anoles, for example, could serve as an additional model for both biological and biomimetic fibrillar adhesion studies, particularly those focused on static adhesion, because it is functionally similar to those of geckos but is much simpler in morphology and bears a much closer resemblance to our current physical and theoretical models of fibrillar adhesion (Garner et al. 2019).

Our analysis of the top 100 most-cited gecko-inspired adhesives (GSAs) also suggests directions for potential implementation in future biomimetic studies of gecko adhesion. Notably, we believe that GSAs need to be tested more rigorously and systematically to better benchmark their multifunctional performance and properties relative to the natural systems they are attempting to mimic. In some ways, it is not particularly surprising that we have yet to manufacture a multifunctional GSA that performs similarly to the natural system (Niewiarowski et al. 2016; Russell et al. 2019), as most studies do not assess multifunctional performance and properties beyond high adhesive stress, directionality, and/or ease of release. Of course, this may be the result of some studies not aspiring to achieve multifunctionality instead being more narrowly focused on a specific application. In any case, additional studies focused on the self-cleaning, anti-matting, and substrate independent properties of GSAs are needed, in addition to studies exploring attachment under non-ideal conditions like hot, humid environments. We also found that some research groups used the same fibrillar adhesive design for assessing its multifunctionality over multiple publications, rendering it challenging to ascribe successes and shortcomings in multifunctionality and congruency with the natural system. One potential solution to this is that groups perform systematic testing and report it in a single publication. This, of course, is challenging to accomplish considering that both experimental and theoretical approaches are often combined in such work, leading to lengthy and unwieldy contributions. Alternatively, research groups focused on biomimetic fibrillar adhesives could periodically publish small reviews of their various attempts to capture the multifunctionality of the gecko adhesive system on a per design basis. Additionally, we found an enormous lack of biomimetic studies disclosing so-called *negative results*, when a GSA did not successfully mimic a desired function or property of the natural system. Thus, where feasible, researchers should also consider including such results in their publications or finding additional outlets that permit the dissemination of such results.

In summary, our review of the diversity of the form and function of squamate fibrillar adhesive systems and our analysis of the most-cited attempts to fabricate biomimetic models identified a number of successes in gecko-inspired adhesion, but also found several shortcomings that should be addressed in future work. Many GSAs have achieved strong, robust yet reversible attachment on par with or exceeding that observed in live geckos and have also managed to achieve attachment to rough and/or wet substrates. Despite this, GSAs fail to capture all the multifunctional properties and performance exhibited by the natural system.

Figure 4.14 displays the number of articles focused on gecko adhesion (based on our search terms used above) published per year since 2000. Only 5 years

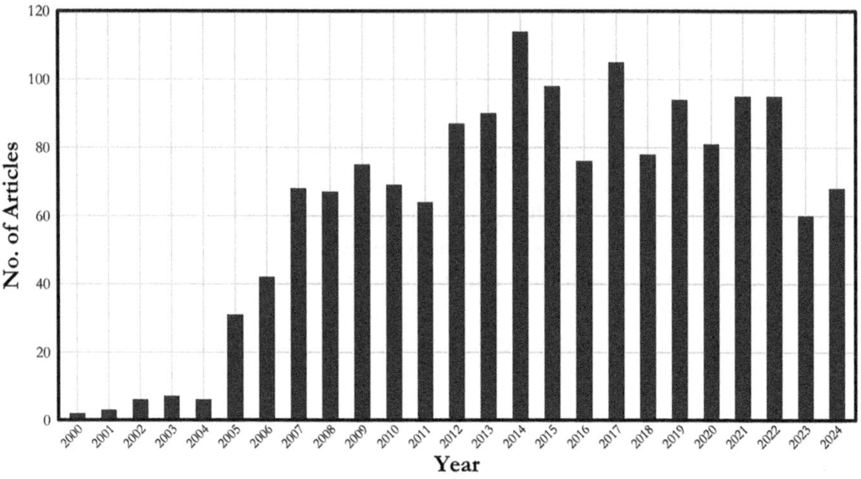

Fig. 4.14 Number of articles focused on gecko adhesion published per year since 2000. Articles included were returned following a Clarivate's Web of Science^{TM} search with the following search terms: ("gecko" OR "geckos") AND ("adhesion" OR "adhesive") OR ("gecko-inspired" OR "gecko inspired")

after the seminal work by Autumn et al. (2000), research on gecko adhesion exceeded 30 articles per year and by 2007 (just 2 years later), more than doubled. Between 2007 and 2011, 64–75 articles were consistently published on gecko adhesion per year. By 2012, research on gecko adhesion exceeded 80 articles per year and remained that way for another decade (except for 2 years when just under 80 articles were published). In 2023, there was a sudden ~37% reduction in the publication of gecko adhesion research (only 60 articles), representing the lowest publication rate in 16 years. This trend continued in 2024 with similarly fewer numbers of publications (only 68 articles). We believe this recent decline in gecko adhesion research to be evidence of the "sticking point" alluded to at the beginning of this contribution, where further advancements of GSAs have stalled. We posit that part of this results from the observation that the field of gecko-inspired adhesion is built upon an "inverted pyramid of observation and application," based on relatively few, somewhat limited studies focused on a single species with arguably the most complex manifestation of the squamate adhesive system. To rectify this, we draw attention to the enormous amount of intraindividual and interspecific variation in the form and configuration of both the actual agents of adhesion (i.e., setae/spatulae) and the anatomical features upon which they are carried (i.e., toe pads). Reassessment of the fundamental properties and mechanics driving multifunctional fibrillar adhesion, as well as what is minimally necessary to achieve desired performance, using simpler manifestations of squamate adhesive systems (e.g., geckos with less complex setal branching, anoles, and/or skinks) will likely be informative and instructive. In so doing, this will provide biomimeticists with a broader empirical foundation from which to select the critical features (e.g., form, dimensions, material properties) most appropriate for implementation into their synthetic simulacra. Future inter- and transdisciplinary investigations considering the biodiversity of squamate fibrillar adhesive systems are likely to produce biomimetic replicas that more closely mirror their natural counterparts and expand the morphological and functional diversity from which to gain inspiration.

References

Aksak B, Murphy M, Sitti M (2008) Gecko inspired micro-fibrillar adhesives for wall climbing robots on micro/nanoscale rough surfaces. In: 2008 IEEE international conference on robotics and automation. IEEE, pp 3058–3063

Alibardi L (2009) Cell biology of adhesive setae in gecko lizards. Zoology 112:403–424

Alibardi L (2018) Mapping proteins localized in adhesive setae of the tokay gecko and their possible influence on the mechanism of adhesion. Protoplasma 255:1785–1797

Alibardi L, Bonfitto A (2019) Morphology of setae in regenerating caudal adhesive pads of the gecko *Lygodactylus capensis* (Smith, 1849). Zoology 133:1–9

Altevogt R (1954) Probleme eines Fusses. Kosmos 50:428–430

Arnold EN (1998) Structural niche, limb morphology and locomotion in lacertid lizards (Squamata, Lacertidae); a preliminary survey. Bull Nat Hist Mus Zool 64:63–90

Arnold EN (2002) History and function of scale microornamentation in lacertid lizards. J Morphol 252:145–169

Arnold EN, Poinar G (2008) A 100 million year old gecko with sophisticated adhesive toe pads, preserved in amber from Myanmar. Zootaxa 1847:62–68

Austin CC (1995) Molecular, morphological, and physiological evolution in South Pacific Scincid lizards. Ph.D. Dissertation

Autumn K (2006a) How gecko toes stick. Am Sci 94:124–132

Autumn K (2006b) Properties, principles, and parameters of the gecko adhesive system. In: Smith AM, Calow JA (eds) Biological adhesives. Springer-Verlag, Berlin, pp 225–256

Autumn K, Gravish N (2008) Gecko adhesion: evolutionary nanotechnology. Philos Trans R Soc A Math Phys Eng Sci 366:1575–1590

Autumn K, Hansen W (2006) Ultrahydrophobicity indicates a non-adhesive default state in gecko setae. J Comp Physiol A 192:1205–1212

Autumn K, Peattie AM (2002) Mechanisms of adhesion in geckos. Integr Comp Biol 42:1081–1090

Autumn K, Liang YA, Hsieh ST, Zesch W, Chan WP, Kenny TW, Fearing R, Full RJ (2000) Adhesive force of a single gecko foot-hair. Nature 405:681–685

Autumn K, Sitti M, Liang YA, Peattie AM, Hansen WR, Sponberg S, Kenny TW, Fearing R, Israelachvili JN, Full RJ (2002) Evidence for van der Waals adhesion in gecko setae. Proc Natl Acad Sci USA 99:12252–12256

Autumn K, Dittmore A, Santos D, Spenko M, Cutkosky M (2006a) Frictional adhesion: a new angle on gecko attachment. J Exp Biol 209:3569–3579

Autumn K, Majidi C, Groff RE, Dittmore A, Fearing R (2006b) Effective elastic modulus of isolated gecko setal arrays. J Exp Biol 209:3558–3568

Autumn K, Niewiarowski PH, Puthoff JB (2014) Gecko adhesion as a model system for integrative biology, interdisciplinary science, and bioinspired engineering. Annu Rev Ecol Evol Syst 45:445–470

Bae W, Kim D, Kwak M, Ha L, Kang S, Suh K (2013) Enhanced skin adhesive patch with modulus-tunable composite micropillars. Adv Healthc Mater 2:109–113

Barreau V, Hensel R, Guimard N, Ghatak A, McMeeking R, Arzt E (2016) Fibrillar elastomeric micropatterns create tunable adhesion even to rough surfaces. Adv Funct Mater 26:4687–4694

Bartlett M, Croll A, King D, Paret B, Irschick D, Crosby A (2012) Looking beyond fibrillar features to scale gecko-like adhesion. Adv Mater 24:1078–1083
Bauer AM (1998) Morphology of the adhesive tail tips of carphodactyline geckos (Reptilia: Diplodactylidae). J Morphol 235:41–58
Bauer AM (2019) Gecko adhesion in space and time: a phylogenetic perspective on the scansorial success story. Integr Comp Biol 59:117–130
Bauer AM, Good DA (1986) Scaling of scansorial surface area in the genus *Gekko*. In: Studies in herpetology. Charles University, Prague, pp 363–366
Bauer AM, Böhme W, Weitschat W (2005) An early Eocene gecko from Baltic amber and its implications for the evolution of gecko adhesion. J Zool 265:327–332
Bergmann PJ, Irschick DJ (2005) Effects of temperature on maximum clinging ability in a diurnal gecko: evidence for a passive clinging mechanism? J Exp Zool Part A 303:785–791
Beuttell K, Losos JB (1999) Ecological morphology of Caribbean anoles. Herpetol Monogr 13:1–28
Bhushan B, Sayer R (2007a) Surface characterization and friction of a bio-inspired reversible adhesive tape. Microsyst Technol 13:71–78
Bhushan B, Sayer RA (2007b) Gecko feet: natural attachment systems for smart adhesion. In: Bhushan B, Fuchs H (eds) Applied scanning probe methods. biomimetics and industrial applications. Springer-Verlag, Heidelberg, pp 41–76
Bhushan B, Israelachvili JN, Landman U (1995) Nanotribology: friction, wear and lubrication at the atomic scale. Nature 374:607–616
Bonser RHC (2000) The Young's modulus of ostrich claw keratin. J Mater Sci Lett 19:1039–1040
Bonser RH, Purslow PP (1995) The Young's modulus of feather keratin. J Exp Biol 198:1029–1033
Bowden FP, Tabor D (1966) Friction, lubrication and wear: a survey of work during the last decade. Br J Appl Phys 17:1521
Braun M (1878) Zur Bedeutung der Cuticularborsten auf den Haftlappen der Geckotiden. Arb Zool Zoot Inst Würzburg 4:231–237
Brown WC, Fehlmann A (1958) A new genus and species of arboreal scincid lizards from the Palau Islands. Occas Pap Nat Hist Mus Stanf Univ 6:1–7
Campolo D, Jones S, Fearing RS (2003) Fabrication of gecko foot-hair like nano structures and adhesion to random rough surfaces. In: IEEE conference on nanotechnology. IEEE, pp 856–859
Cañas N, Kamperman M, Völker B, Kroner E, McMeeking R, Arzt E (2012) Effect of nano- and micro-roughness on adhesion of bioinspired micropatterned surfaces. Acta Biomater 8:282–288
Cannatella DC, de Queiroz K (1989) Phylogenetic systematics of the anoles: is a new taxonomy warranted? Syst Zool 38:57–69
Cartier O (1872) Studien über den feineren Bau der Haut bei den Reptilien. I Abteilung Epidermis der Geckotiden. Verh Phys Med Ges Würzburg 1:83–96
Cartmill M (1985) Climbing. In: Hildebrand M (ed) Functional vertebrate morphology. Harvard University Press, Cambridge, pp 73–88
Chen C, Chiang C, Lai C, Xie T, Yang S (2013) Buckling-based strong dry adhesives via interlocking. Adv Funct Mater 23:3813–3823
Cheng Z, Feng L, Jiang L (2008) Tunable adhesive superhydrophobic surfaces for superparamagnetic microdroplets. Adv Funct Mater 18:3219–3225
Cheng Z, Gao J, Jiang L (2010) Tip geometry controls adhesive states of superhydrophobic surfaces. Langmuir 26:8233–8238
Cho W, Choi I (2008) Fabrication of hairy polymeric films inspired by geckos: wetting and high adhesion properties. Adv Funct Mater 18:1089–1096
Coddington JA (1988) Cladistic tests of adaptational hypotheses. Cladistics 4:3–22
Cole NC, Jones CG, Harris S (2005) The need for enemy-free space: the impact of an invasive gecko on island endemics. Biol Conserv 125:467–474
Crandell KE, Herrel A, Sasa M, Losos JB, Autumn K (2014) Stick or grip? Co-evolution of adhesive toepads and claws in *Anolis* lizards. Zoology 117:363–369

Cui J, Drotlef D, Larraza I, Fernández-Blázquez J, Boesel L, Ohm C, Mezger M, Zentel R, del Campo A (2012) Bioinspired actuated adhesive patterns of liquid crystalline elastomers. Adv Mater 24:4601–4604
Dalla Valle L, Nardi A, Toffolo V, Niero C, Toni M, Alibardi L (2007) Cloning and characterization of scale beta-keratins in the differentiating epidermis of geckoes show they are glycine-proline-serine-rich proteins with a central motif homologous to avian beta-keratins. Dev Dyn 236:374–388
Davies J, Haq S, Hawke T, Sargent J (2009) A practical approach to the development of a synthetic gecko tape. Int J Adhes Adhes 29:380–390
Dayan C, Chun S, Krishna-Subbaiah N, Drotlef D, Akolpoglu M, Sitti M (2021) 3D printing of elastomeric bioinspired complex adhesive microstructures. Adv Mater 33:2103826
Daza JD, Stanley EL, Wagner P, Bauer AM, Grimaldi DA (2016) Mid-Cretaceous amber fossils illuminate the past diversity of tropical lizards. Sci Adv 2:e1501080
de Queiroz K (2022) The correct name for the taxon ranked as a family containing the genus *Anolis* under rank-based nomenclature and the author of the name *Anolis loysiana*. Herpetol Rev 53:418–420
de Queiroz K (2023) Authorship of the Name Anolidae. Herpetol Rev 54:211–212
del Campo A, Greiner C, Arzt E (2007) Contact shape controls adhesion of bioinspired fibrillar surfaces. Langmuir 23:10235–10243
Delannoy SM (2005) Subdigital setae of the Tokay gecko (*Gekko gecko*): variation in form and implications for adhesion. MSc Thesis
Delaugerre M, Gauthier A, Leoncini A (2015) One island, two geckos and some powder. Why and how a colonization process can fail. In: Presented at the X Congresso Nazionale della Societas Herpetologica Italica. Societas Herpetologica Italica, Pavia, pp 117–121
Dellit W (1934) Zur Anatomie und Physiologie der geckozehe. Jena Z Naturwiss 68:613–656
Dellit WD (1949) Zum Haftproblem der Geckoniden. Deutsche Aquarien- und Terrarien-Zeitschrift 2:56–58
Drotlef D, Blümler P, del Campo A (2014) Magnetically actuated patterns for bioinspired reversible adhesion (dry and wet). Adv Mater 26:775–779
Drotlef D, Amjadi M, Yunusa M, Sitti M (2017) Bioinspired composite microfibers for skin adhesion and signal amplification of wearable sensors. Adv Mater 29:1701353
Du T, Ma S, Pei X, Wang S, Zhou F (2017) Bio-inspired design and fabrication of micro/nano-brush dual structural surfaces for switchable oil adhesion and antifouling. Small 13:1602020
Duméril AMC, Bibron G (1836) Erpétologie générale ou histoire naturelle complète des reptiles, vol Tome 3. Librairie Encyclopédique de Roret, Paris
Editors of the American Heritage Dictionaries (2016) The American Heritage dictionary of the English language, 5th edn. Houghton Mifflin Harcourt Publishing Company, Boston
Elstrott J, Irschick DJ (2004) Evolutionary correlations among morphology, habitat use and clinging performance in Caribbean *Anolis* lizards. Biol J Linn Soc 83:389–398
Etheridge RE (1960) The relationships of the anoles (Reptilia: Sauria: Iguanidae): an interpretation based on skeletal morphology. Ph.D. Dissertation, University of Michigan
Fischer S, Arzt E, Hensel R (2017) Composite pillars with a tunable interface for adhesion to rough substrates. ACS Appl Mater Interfaces 9:1036–1044
Fitzinger LJ (1843) Systema reptilium. Fasciculus primus. Amblygossae. In: Adler K (ed) (1973) Facsimile reprints in herpetology. Society for the Study of Amphibians and Reptiles, Athens
Fontanarrosa G, Daza JD, Abdala V (2018) Cretaceous fossil gecko hand reveals a strikingly modern scansorial morphology: qualitative and biometric analysis of an amber-preserved lizard hand. Cretac Res 84:120–133
Gadow H (1901) Amphibia and reptiles. In: Harmer SF, Shipley AE (eds) Cambridge natural history, vol VIII. Macmillan and Co. Ltd, London
Gamble T, Greenbaum E, Jackman TR, Russell AP, Bauer AM (2012) Repeated origin and loss of adhesive toepads in geckos. PLoS One 7:e39429
Gamble T, Greenbaum E, Jackman TR, Russell AP, Bauer AM (2017) Repeated evolution of digital adhesion in geckos, a reply to Harrington and Reeder. J Evol Biol 30:1429–1436

Garner AM, Russell AP (2021) Revisiting the classification of squamate adhesive setae: historical, morphological and functional perspectives. R Soc Open Sci 8:202039
Garner AM, Wilson MC, Russell AP, Dhinojwala A, Niewiarowski PH (2019) Going out on a limb: how investigation of the anoline adhesive system can enhance our understanding of fibrillar adhesion. Integr Comp Biol 59:61–69
Garner AM, Wilson MC, Wright C, Russell AP, Niewiarowski PH, Dhinojwala A (2021) The same but different: the setal arrays of anoles and geckos indicate alternative approaches to achieving similar adhesive effectiveness. J Anat 238:1143–1155
Garner AM, Wilson MC, Wright C, Russell AP, Niewiarowski PH, Dhinojwala A (2022) Parameters of the adhesive setae and setal fields of the Jamaican radiation of anoles (Dactyloidae: *Anolis*): potential for ecomorphology at the microscopic scale. Biol J Linn Soc 137:85–99
Ge L, Sethi S, Ci L, Ajayan P, Dhinojwala A (2007) Carbon nanotube-based synthetic gecko tapes. Proc Natl Acad Sci USA 104:10792–10795
Geim A, Dubonos S, Grigorieva I, Novoselov K, Zhukov A, Shapoval S (2003) Microfabricated adhesive mimicking gecko foot-hair. Nat Mater 2:461–463
Gillies A, Kwak J, Fearing R (2013) Controllable particle adhesion with a magnetically actuated synthetic gecko adhesive. Adv Funct Mater 23:3256–3261
Glass P, Cheung E, Sitti M (2008) A legged anchoring mechanism for capsule endoscopes using micropatterned adhesives. IEEE Trans Biomed Eng 55:2759–2767
Glass P, Chung H, Washburn N, Sitti M (2010) Enhanced wet adhesion and shear of elastomeric micro-fiber arrays with mushroom tip geometry and a photopolymerized p(DMA-co- MEA) tip coating. Langmuir 26:17357–17362
Glassmaker N, Jagota A, Hui C (2005) Adhesion enhancement in a biomimetic fibrillar interface. Acta Biomater 1:367–375
Glossip D, Losos JB (1997) Ecological correlates of number of subdigital lamellae in anoles. Herpetologica 53:192–199
Gravish N, Wilkinson M, Sponberg S, Parness A, Esparza N, Soto D, Yamaguchi T, Broide M, Cutkosky M, Creton C, Autumn K (2010) Rate-dependent frictional adhesion in natural and synthetic gecko setae. J R Soc Interface 7:259–269
Greiner C, del Campo A, Arzt E (2007) Adhesion of bioinspired micropatterned surfaces: effects of pillar radius, aspect ratio, and preload. Langmuir 23:3495–3502
Greiner C, Arzt E, del Campo A (2009) Hierarchical gecko-like adhesives. Adv Mater 21:479
Griffing AH, Sanger TJ, Epperlein L, Bauer AM, Cobos A, Higham TE, Naylor E, Gamble T (2021) And thereby hangs a tail: morphology, developmental patterns and biomechanics of the adhesive tails of crested geckos (*Correlophus ciliatus*). Proc R Soc Lond B Biol Sci 288:20210650
Griffing AH, Gamble T, Cohn MJ, Sanger TJ (2022) Convergent developmental patterns underlie the repeated evolution of adhesive toe pads among lizards. Biol J Linn Soc 135:518–532
Guo D, Liu R, Cheng Y, Zhang H, Zhou L, Fang S, Elliott W, Tan W (2015) Reverse adhesion of a gecko-inspired synthetic adhesive switched by an ion-exchange polymer-metal composite actuator. ACS Appl Mater Interfaces 7:5480–5487
Guyer C, Savage JM (1992) Anole systematics revisited. Syst Biol 41:89–110
Haase A (1900) Untersuchungen über den Bau und die Entwicklung der Haftlappen bei den Geckotiden. Arch Naturgesch 66:321–345
Hagey TJ, Harte S, Vickers M, Harmon LJ, Schwarzkopf L (2017) There's more than one way to climb a tree: limb length and microhabitat use in lizards with toe pads. PLoS One 12:e0184641
Hansen WR, Autumn K (2005) Evidence for self-cleaning in gecko setae. Proc Natl Acad Sci 102:385–389
Higham TE, Gamble T, Russell AP (2017a) On the origin of frictional adhesion in geckos: small morphological changes lead to a major biomechanical transition in the genus *Gonatodes*. Biol J Linn Soc 120:503–517
Higham TE, Russell AP, Niklas KJ (2017b) Leaping lizards landing on leaves: escape-induced jumps in the rainforest canopy challenge the adhesive limits of geckos. J R Soc Interface 14:20170156

Hill GC, Soto DR, Peattie AM, Full RJ, Kenny TW (2011) Orientation angle and the adhesion of single gecko setae. J R Soc Interface 8:926–933
Hiller U (1968) Untersuchungen zum Feinbau und zur Funktion der Haftborsten von Reptilien. Z Morph Tiere 62:307–362
Hiller U (1969) Zusammenhang zwischen vorbehandelten Polyäthylen-Folien durch Korona-Entladung und dem Haftvermögen von *Tarentola m. muritanica* (Rept.). Forma et Functio 1:350–352
Hiller U (1976) Comparative studies on the functional morphology of two gekkonid lizards. J Bombay Nat Hist Soc 73:278–282
Hiller U (2000) Walking upside down: the mystery of climbing ability in gekkonid lizards. Gekko 1:36–40
Ho A, Yeo L, Lam Y, Rodriguez I (2011) Fabrication and analysis of gecko-inspired hierarchical polymer nanosetae. ACS Nano 5:1897–1906
Home E (1816a) IX. Some account of the feet of those animals whose progressive motion can be carried on in opposition to gravity. Philos Trans R Soc Lond 106:149–155
Home E (1816b) XXIII. Farther observations on the feet of animals whose progressive motion can be carried on against gravity. Philos Trans R Soc Lond 106:322–330
Hora SL (1923) The adhesive apparatus on the toes of certain geckos and tree frogs. J Asiat Soc Bengal 9:137–145
Hu C, Greaney PA (2014) Role of seta angle and flexibility in the gecko adhesion mechanism. J Appl Phys 116(7):074302
Hu S, Lopez S, Niewiarowski PH, Xia Z (2012) Dynamic self-cleaning in gecko setae via digital hyperextension. J R Soc Interface 9:2781–2790
Huber G, Gorb SN, Spolenak R, Arzt E (2005a) Resolving the nanoscale adhesion of individual gecko spatulae by atomic force microscopy. Biol Lett 1:2–4
Huber G, Mantz H, Spolenak R, Mecke K, Jacobs K, Gorb SN, Arzt E (2005b) Evidence for capillarity contributions to gecko adhesion from single spatula nanomechanical measurements. Proc Natl Acad Sci USA 102:16293–16296
Huber G, Gorb SN, Hosoda N, Spolenak R, Arzt E (2007) Influence of surface roughness on gecko adhesion. Acta Biomater 3:607–610
Huie JM, Prates I, Bell RC, de Queiroz K (2021) Convergent patterns of adaptive radiation between island and mainland *Anolis* lizards. Biol J Linn Soc 134:85–110
Irschick DJ, Losos JB (1999) Do lizards avoid habitats in which performance is submaximal? The relationship between sprinting capabilities and structural habitat use in Caribbean anoles. Am Nat 154:293–305
Irschick DJ, Austin CC, Petren K, Fisher RN, Losos JB, Ellers O (1996) A comparative analysis of clinging ability among pad-bearing lizards. Biol J Linn Soc 59:21–35
Irschick DJ, Herrel A, Vanhooydonck B (2006) Whole-organism studies of adhesion in pad-bearing lizards: creative evolutionary solutions to functional problems. J Comp Physiol A Neuroethol Sens Neural Behav Physiol 192:1169–1177
Israelachvili JN (1992) Intermolecular and surface forces. Academic Press, New York
Izadi H, Stewart KM, Penlidis A (2014) Role of contact electrification and electrostatic interactions in gecko adhesion. J R Soc Interface 11:1–4
Jagota A, Bennison SJ (2002) Mechanics of adhesion through a fibrillar microstructure. Integr Comp Biol 42:1140–1145
Jeong H, Lee J, Kim H, Moon S, Suh K (2009) A nontransferring dry adhesive with hierarchical polymer nanohairs. Proc Natl Acad Sci USA 106:5639–5644
Jeong H, Lee J, Kwak M, Moon S, Suh K (2010) Effect of leaning angle of gecko-inspired slanted polymer nanohairs on dry adhesion. Appl Phys Lett 96:043704
Jiang H, Hawkes E, Fuller C, Estrada M, Suresh S, Abcouwer N, Han A, Wang S, Ploch C, Parness A, Cutkosky M (2017) A robotic device using gecko-inspired adhesives can grasp and manipulate large objects in microgravity. Sci Robot 2:eaan4545
Jin M, Feng X, Feng L, Sun T, Zhai J, Li T, Jiang L (2005) Superhydrophobic aligned polystyrene nanotube films with high adhesive force. Adv Mater 17:1977

Jin K, Tian Y, Erickson J, Puthoff J, Autumn K, Pesika N (2012) Design and fabrication of gecko-inspired adhesives. Langmuir 28:5737–5742
Jin K, Cremaldi J, Erickson J, Tian Y, Israelachvili J, Pesika N (2014) Biomimetic bidirectional switchable adhesive inspired by the gecko. Adv Funct Mater 24:574–579
Johnson MK, Russell AP (2009) Configuration of the setal fields of *Rhoptropus* (Gekkota: Gekkonidae): functional, evolutionary, ecological and phylogenetic implications of observed pattern. J Anat 214:937–955
Johnson KL, Kendall K, Roberts AD (1971) Surface energy and the contact of elastic solids. Proc R Soc Lond A 324:301–313
Johnson M, Russell A, Delannoy S (2009) Surface characteristics of locomotor substrata and their relationship to gekkonid adhesion: a case study of *Rhoptropus* cf. *biporosus*. In: Gorb SN (ed) Functional surfaces in biology. Springer Netherlands, Dordrecht, pp 123–154
Kim TW, Bhushan B (2008) The adhesion model considering capillarity for gecko attachment system. J R Soc Interface 5:319–327
Kim S, Sitti M (2006) Biologically inspired polymer microfibers with spatulate tips as repeatable fibrillar adhesives. Appl Phys Lett 89:261911
Kim D, Lee H, Lee J, Kim S, Lee K, Moon W, Kwon T (2007a) Replication of high-aspect-ratio nanopillar array for biomimetic gecko foot-hair prototype by UV nano embossing with anodic aluminum oxide mold. Microsyst Technol 13:601–606
Kim S, Aksak B, Sitti M (2007b) Enhanced friction of elastomer microfiber adhesives with spatulate tips. Appl Phys Lett 91:221913
Kim S, Spenko M, Trujillo S, Heyneman B, Santos D, Cutkosky M (2008) Smooth vertical surface climbing with directional adhesion. IEEE Trans Robot 24:65–74
Kim S, Cheung E, Sitti M (2009a) Wet self-cleaning of biologically inspired elastomer mushroom shaped microfibrillar adhesives. Langmuir 25:7196–7199
Kim S, Sitti M, Xie T, Xia X (2009b) Reversible dry micro-fibrillar adhesives with thermally controllable adhesion. Soft Matter 5:3689–3693
Kim T, Jeong H, Suh K, Lee H (2009c) Stooped nanohairs: geometry-controllable, unidirectional, reversible, and robust gecko-like dry adhesive. Adv Mater 21:2276
Kim T, Park J, Sohn J, Cho D, Jeon S (2016) Bioinspired, highly stretchable, and conductive dry adhesives based on 1D-2D hybrid carbon nanocomposites for all-in-one ECG electrodes. ACS Nano 10:4770–4778
King D, Bartlett M, Gilman C, Irschick D, Crosby A (2014) Creating gecko-like adhesives for "real world" surfaces. Adv Mater 26:4345–4351
Kizilkan E, Strueben J, Staubitz A, Gorb S (2017) Bioinspired photocontrollable microstructured transport device. Sci Robot 2:eaak9454
Kluge AG (1993) Gekkonoid lizard taxonomy. International Gecko Society, San Diego
Kunitzky I (1903) Die Zeit der Entstehung der Borsten und Mechanismus der Bewegung bei den Geckotiden (*Ptychozoon homalocephalum* Creveldt). Bull Acad Imp Sci St Pétersb 18:21–24
Kustandi T, Samper V, Ng W, Chong A, Gao H (2007a) Fabrication of a gecko-like hierarchical fibril array using a bonded porous alumina template. J Micromech Microeng 17:N75–N81
Kustandi T, Samper V, Yi D, Ng W, Neuzil P, Sun W (2007b) Self-assembled nanoparticles-based fabrication of gecko foot-hair-inspired polymer nanofibers. Adv Funct Mater 17:2211–2218
Kwak M, Jeong H, Bae W, Jung H, Suh K (2011) Anisotropic adhesion properties of triangular-tip-shaped micropillars. Small 7:2296–2300
Lacépède BGED (1788) Histoire naturelle des quadrupèdes ovipares et des serpens, vol Tom. 1. Hôtel de Thou, Paris
Lacépède BGED (1802) The natural history of oviparous quadrupeds and serpents, Arranged and published from the Collection of Count de Buffon, by the Count de la Cepede. Alex Smellie, London, Edinburgh
Lamb T, Bauer AM (2006) Footprints in the sand: independent reduction of subdigital lamellae in the Namib–Kalahari burrowing geckos. Proc R Soc Lond B Biol Sci 273:855–864
Larson A, Losos JB (1996) Phylogenetic systematics of adaptation. In: Rose MR, Lauder GV (eds) Adaptation. Academic Press, San Diego, pp 187–220

Lee J, Fearing R (2008) Contact self-cleaning of synthetic gecko adhesive from polymer microfibers. Langmuir 24:10587–10591
Lee H, Lee B, Messersmith P (2007) A reversible wet/dry adhesive inspired by mussels and geckos. Nature 448:338–3U4
Lee J, Fearing R, Komvopoulos K (2008a) Directional adhesion of gecko-inspired angled microfiber arrays. Appl Phys Lett 93:191910
Lee J, Majidi C, Schubert B, Fearing R (2008b) Sliding-induced adhesion of stiff polymer microfibre arrays. I. Macroscale behaviour. J R Soc Interface 5:835–844
Lee J, Bush B, Maboudian R, Fearing R (2009) Gecko-inspired combined lamellar and nanofibrillar array for adhesion on nonplanar surface. Langmuir 25:12449–12453
Lee D, Lee D, Lee S, Cho K (2012) Hierarchical gecko-inspired nanohairs with a high aspect ratio induced by nanoyielding. Soft Matter 8:4905–4910
Li J, Liu X, Ye Y, Zhou H, Chen J (2011) Gecko-inspired synthesis of superhydrophobic ZnO surfaces with high water adhesion. Colloids Surf A Physicochem Eng Asp 384:109–114
Li S, Tian H, Shao J, Liu H, Wang D, Zhang W (2020) Switchable adhesion for nonflat surfaces mimicking geckos' adhesive structures and toe muscles. ACS Appl Mater Interfaces 12:39745–39755
Liimatainen V, Drotlef D, Son D, Sitti M (2020) Liquid-superrepellent bioinspired fibrillar adhesives. Adv Mater 32:2000497
Liu K, Du J, Wu J, Jiang L (2012) Superhydrophobic gecko feet with high adhesive forces towards water and their bio-inspired materials. Nanoscale 4:768–772
Liu X, Wang K, Gao Q, Zhang W, Zhou W, Shi S, Xia C, Li J (2019) Bioinspired design by gecko structure and mussel chemistry for bio-based adhesive system through incorporating natural fibers. J Clean Prod 236:117591
Liu Q, Tan D, Meng F, Yang B, Shi Z, Wang X, Li Q, Nie C, Liu S, Xue L (2021) Adhesion enhancement of micropillar array by combining the adhesive design from gecko and tree frog. Small 17:2005493
Losos JB (1990a) The evolution of form and function: morphology and locomotor performance in West Indian *Anolis* lizards. Evolution 44:1189–1203
Losos JB (1990b) Ecomorphology, performance capability, and scaling of West Indian *Anolis* lizards: an evolutionary analysis. Ecol Monogr 60:369–388
Losos JB (1994) Integrative approaches to evolutionary ecology: *Anolis* lizards as model systems. Annu Rev Ecol Syst 25:467–493
Losos JB (2011) Lizards in an evolutionary tree: ecology and adaptive radiation of anoles. University of California Press, Berkeley
Losos JB, Irschick DJ (1996) The effect of perch diameter on escape behaviour of *Anolis* lizards: laboratory predictions and field tests. Anim Behav 51:593–602
Ma Y, Ma S, Wu Y, Pei X, Gorb S, Wang Z, Liu W, Zhou F (2018) Remote control over underwater dynamic attachment/detachment and locomotion. Adv Mater 30:1801595
Macrini TE, Irschick DJ, Losos JB (2003) Ecomorphological differences in toepad characteristics between mainland and island anoles. J Herpetol 37:52–58
Maderson PFA (1964) Keratinized epidermal derivatives as an aid to climbing in gekkonid lizards. Nature 203:780–781
Maderson PFA (1966) Some macroscopic and microscopic observations on the foot-pads of the tokay (*Gekko gecko*). Mem Hong Kong Nat Hist Soc 7:10
Maderson PFA (1970) Lizard glands and lizard hands: models for evolutionary study. Forma et Functio 3:179–204
Mahdavi A, Ferreira L, Sundback C, Nichol J, Chan E, Carter D, Bettinger C, Patanavanich S, Chignozha L, Ben-Joseph E, Galakatos A, Pryor H, Pomerantseva I, Masiakos P, Faquin W, Zumbuehl A, Hong S, Borenstein J, Vacanti J, Langer R, Karp J (2008) A biodegradable and biocompatible gecko-inspired tissue adhesive. Proc Natl Acad Sci USA 105:2307–2312
Mahendra BC (1941) Contributions to the bionomics, anatomy, reproduction and development of the Indian house-gecko, *Hemidactylus flaviviridis* Rüppel. Part II The problem of locomotion. Proc Indian Acad Sci 13:288–306

Mahler DL, Ingram T, Revell LJ, Losos JB (2013) Exceptional convergence on the macroevolutionary landscape in island lizard radiations. Science 341:292–295
McCoy M (2006) Reptiles of the Solomon Islands. Pensoft Publisher, Sofia
Mengüç Y, Yang S, Kim S, Rogers J, Sitti M (2012) Gecko-inspired controllable adhesive structures applied to micromanipulation. Adv Funct Mater 22:1246–1254
Mengüç Y, Röhrig M, Abusomwan U, Hölscher H, Sitti M (2014) Staying sticky: contact self-cleaning of gecko-inspired adhesives. J R Soc Interface 11:20131205
Menon C, Murphy M, Sitti M (2004) Gecko inspired surface climbing robots. In: Proceedings of the IEEE international conference on robotics and biomimetics. IEEE, pp 431–436
Merrem B (1820) Versuch eines Systems der Amphibien: Tentamen systematis amphibiorum. Johann Christian Krieger, Marburg
Miller AH, Stroud JT (2022) Novel tests of the key innovation hypothesis: adhesive toepads in arboreal lizards. Syst Biol 71:139–152
Murphy M, Aksak B, Sitti M (2007) Adhesion and anisotropic friction enhancements of angled heterogeneous micro-fiber arrays with spherical and spatula tips. J Adhes Sci Technol 21:1281–1296
Murphy M, Aksak B, Sitti M (2009a) Gecko-inspired directional and controllable adhesion. Small 5:170–175
Murphy M, Kim S, Sitti M (2009b) Enhanced adhesion by gecko-inspired hierarchical fibrillar adhesives. ACS Appl Mater Interfaces 1:849–855
Nicholson KE, Crother BI, Guyer C, Savage JM (2012) It is time for a new classification of anoles (Squamata: Dactyloidae). Zootaxa 3477:1–108
Nicholson KE, Crother BI, Guyer C, Savage JM (2014) Anole classification: a response to Poe. Zootaxa 3814:109–120
Niewiarowski PH, Stark AY, Dhinojwala A (2016) Sticking to the story: outstanding challenges in gecko-inspired adhesives. J Exp Biol 219:912–919
Northen M, Greiner C, Arzt E, Turner K (2008) A gecko-inspired reversible adhesive. Adv Mater 20:3905
Padian K, Olsen PE (1984) Footprints of the Komodo monitor and the trackways of fossil reptiles. Copeia 1984:662–671
Parness A, Soto D, Esparza N, Gravish N, Wilkinson M, Autumn K, Cutkosky M (2009) A microfabricated wedge-shaped adhesive array displaying gecko-like dynamic adhesion, directionality and long lifetime. J R Soc Interface 6:1223–1232
Peattie AM (2007) The function and evolution of gekkotan adhesive feet. Ph.D. Dissertation, University of California, Berkeley
Persson BNJ (2003) On the mechanism of adhesion in biological systems. J Chem Phys 118:7614–7621
Persson BNJ, Gorb S (2003) The effect of surface roughness on the adhesion of elastic plates with application to biological systems. J Chem Phys 119:11437–11444
Persson BNJ, Albohr O, Tartaglino U, Volokitin AI, Tosatti E (2004) On the nature of surface roughness with application to contact mechanics, sealing, rubber friction and adhesion. J Phys Condens Matter 17:R1
Pesika NS, Zeng H, Kristiansen K, Zhao B, Tian Y, Autumn K, Israelachvili J (2009) Gecko adhesion pad: a smart surface? J Phys Condens Matter 21:464132
Peterson JA (1983a) The evolution of the subdigital pad in *Anolis*. I. Comparisons among the anoline genera. In: Advances in herpetology and evolutionary biology: essays in honor of Ernest E. Williams. Museum of Comparative Zoology, Harvard University, Cambridge
Peterson JA (1983b) The evolution of the subdigital pad of *Anolis* 2. Comparisons among the iguanid genera related to the anolines and a view from outside the radiation. J Herpetol 17:371–397
Peterson JA (1984) The scale microarchitecture of *Sphenodon punctatus*. J Herpetol 18:40–47
Peterson JA, Williams EE (1981) A case history in retrograde evolution: the *onca* lineage in anoline lizards: II. Subdigital fine structure. Bull Mus Comp Zool 149:215–268
Petrucci RH, Harwood WS (1993) General chemistry, 6th edn. MacMillan Publishing Co, New York

Pianka ER, Sweet SS (2005) Integrative biology of sticky feet in geckos. BioEssays 27:647–652
Poe S (2004) Phylogeny of anoles. Herpetol Monogr 18:37–89
Poe S (2013) 1986 Redux: new genera of anoles (Squamata: Dactyloidae) are unwarranted. Zootaxa 3626:295–299
Poe S, Nieto-Montes de Oca A, Torres-Carvajal O, De Queiroz K, Velasco JA, Truett B, Gray LN, Ryan MJ, Köhler G, Ayala-Varela F (2017) A phylogenetic, biogeographic, and taxonomic study of all extant species of *Anolis* (Squamata; Iguanidae). Syst Biol 66:663–697
Pounds JA (1988) Ecomorphology, locomotion, and microhabitat structure: patterns in a tropical mainland *Anolis* community. Ecol Monogr 58:299–320
Purtov J, Frensemeier M, Kroner E (2015) Switchable adhesion in vacuum using bio-inspired dry adhesives. ACS Appl Mater Interfaces 7:24127–24135
Puthoff JB, Prowse MS, Wilkinson M, Autumn K (2010) Changes in materials properties explain the effects of humidity on gecko adhesion. J Exp Biol 213:3699–3704
Qu L, Dai L (2007) Gecko-foot-mimetic aligned single-walled carbon nanotube dry adhesives with unique electrical and thermal properties. Adv Mater 19:3844
Qu L, Dai L, Stone M, Xia Z, Wang Z (2008) Carbon nanotube arrays with strong shear binding-on and easy normal lifting-off. Science 322:238–242
Raut H, Baji A, Hariri H, Parveen H, Soh G, Low H, Wood K (2018) Gecko-inspired dry adhesive based on micro-nanoscale hierarchical arrays for application in climbing devices. ACS Appl Mater Interfaces 10:1288–1296
Reddy S, Arzt E, del Campo A (2007) Bioinspired surfaces with switchable adhesion. Adv Mater 19:3833
Riedel J, Zozaya SM, Hoskin CJ, Schwarzkopf L (2021) Parallel evolution of toepads in rock-dwelling lineages of a terrestrial gecko (Gekkota: Gekkonidae: *Heteronotia binoei*). Zool J Linnean Soc 193:636–654
Riedel J, Eisele K, Gabelaia M, Higham TE, Wu J, Do QH, Nguyen TQ, Meneses CG, Brown RM, Ziegler T, Grismer LL, Russell AP, Rödder D (2024) Ecologically-related variation of digit morphology in *Cyrtodactylus* (Gekkota, Squamata) reveals repeated origins of incipient adhesive toepads. Funct Ecol 38:1630–1648
Röhrig M, Thiel M, Worgull M, Hölscher H (2012) 3D direct laser writing of nano- and microstructured hierarchical gecko-mimicking surfaces. Small 8:3009–3015
Rong Z, Zhou Y, Chen B, Robertson J, Federle W, Hofmann S, Steiner U, Goldberg-Oppenheimer P (2014) Bio-inspired hierarchical polymer fiber-carbon nanotube adhesives. Adv Mater 26:1456–1461
Ruffatto D, Parness A, Spenko M (2014) Improving controllable adhesion on both rough and smooth surfaces with a hybrid electrostatic/gecko-like adhesive. J R Soc Interface 11:20131089
Ruibal R, Ernst V (1965) The structure of the digital setae of lizards. J Morphol 117:271–293
Russell ES (1916) Form and function: a contribution to the history of animal morphology. John Murray, London
Russell AP (1975) A contribution to the functional analysis of the foot of the Tokay, *Gekko gecko* (Reptilia: Gekkonidae). J Zool 176:437–476
Russell AP (1976) Some comments concerning interrelationships amongst gekkonine geckos. In: d'Bellairs A, Cox C (eds) Morphology and biology of reptiles, Linnean symposium series. Academic Press, London, pp 217–244
Russell AP (1977) The phalangeal formula of *Hemidactylus* Oken, 1817 (Reptilia, Gekkonidae): a correction and a functional explanation. Anat Histol Embryol 6:332–338
Russell AP (1979) Parallelism and integrated design in the foot structure of gekkonine and diplodactyline geckos. Copeia 1979:1–21
Russell AP (1981) Descriptive and functional anatomy of the digital vascular system of the tokay, *Gekko gecko*. J Morphol 169:293–323
Russell AP (1986) The morphological basis of weight-bearing in the scansors of the tokay gecko (Reptilia: Sauria). Can J Zool 64:948–955
Russell AP (2002) Integrative functional morphology of the gekkotan adhesive system (Reptilia: Gekkota). Integr Comp Biol 42:1154–1163

Russell AP, Bauer AM (1990) Digit I in pad-bearing gekkonine geckos: alternate designs and the potential constraints of phalangeal number. Mem Queensl Mus 29:453–472
Russell AP, Bauer AM (2002) Underwood's classification of the geckos: a twenty-first century appreciation. Bull Nat Hist Mus Zool 68:113–121
Russell AP, Delaugerre M-J (2017) Left in the dust: differential effectiveness of the two alternative adhesive pad configurations in geckos (Reptilia: Gekkota). J Zool 301:61–68
Russell AP, Eslinger A (2017) A whole lamella perspective on the origin of the epidermal free margin of *Anolis* (Reptilia: Dactyloidae) toe pads. J Morphol 278:360–368
Russell AP, Gamble T (2019) Evolution of the gekkotan adhesive system: does digit anatomy point to one or more origins? Integr Comp Biol 59:131–147
Russell AP, Garner AM (2021) Setal field transects, evolutionary transitions and gecko–anole convergence provide insights into the fundamentals of form and function of the digital adhesive system of lizards. Front Mech Eng 6:1–17
Russell AP, Garner AM (2023) Solutions to a sticky problem: convergence of the adhesive systems of geckos and anoles (Reptilia: Squamata). In: Bels V, Russell AP (eds) Convergent evolution. Springer, Cham
Russell AP, Johnson MK (2007) Real-world challenges to, and capabilities of, the gekkotan adhesive system: contrasting the rough and the smooth. Can J Zool 85:1228–1238
Russell AP, Johnson MK (2009) The gecko effect: design principles of the gekkotan adhesive system across scales of organization. In: Favret EA, Fuentes NO (eds) Functional properties of bio-inspired surfaces. World Scientific Publishing Co, New Jersey, pp 103–132
Russell AP, Johnson MK (2014) Between a rock and a soft place: microtopography of the locomotor substrate and the morphology of the setal fields of Namibian day geckos (Gekkota: Gekkonidae: *Rhoptropus*). Acta Zool 95:299–318
Russell AP, Oetelaar GS (2016) Limb and digit orientation during vertical clinging in Bibron's gecko, *Chondrodactylus bibronii* (A. Smith, 1846) and its bearing on the adhesive capabilities of geckos. Acta Zool 97:345–360
Russell AP, Johnson MK, Delannoy SM (2007) Insights from studies of gecko-inspired adhesion and their impact on our understanding of the evolution of the gekkotan adhesive system. J Adhes Sci Technol 21:1119–1143
Russell AP, Tinius A, Werner YL (2014) Dr. rer. nat. Uwe Hiller (1942–2008): pioneer of experimental observation of the molecular basis of adhesion in geckos. Herpetol Rev 45:387–390
Russell AP, Baskerville J, Gamble T, Higham TE (2015) The evolution of digit form in *Gonatodes* (Gekkota: Sphaerodactylidae) and its bearing on the transition from frictional to adhesive contact in gekkotans. J Morphol 276:1311–1332
Russell AP, Stark AY, Higham TE (2019) The integrative biology of gecko adhesion: historical review, current understanding, and grand challenges. Integr Comp Biol 59:101–116
Schmidt HR (1905) Zur Anatomie und Physiologie der Geckopfote. Jenaische Zeitschrift für Naturwissenschaft 39:551–580
Schmidt WJ (1912) Studien am Integumentder Reptilien. I. Die Haut der Geckoniden. Z Wiss Zool 51:139–258
Schmidt WJ (1913) Studien am Integument der Reptilien: IV. *Uroplatus fimbriatus* (Schneid.) und die Geckoniden. Zool Jahrb Abt Anat Ontog Tiere 36:377–464
Schneider JG (1812) Kritische Uebersicht der einzelnen Arten aus der Gattung von Eidechsen, welche ich Wandkletterer nenne, Linné aber und andere, Geckonen. In: Denkschriften der Koniglichen Akademie der Wissenschaften zu Munchen 1811/1812, pp 31–71
Schubert B, Lee J, Majidi C, Fearing R (2008) Sliding-induced adhesion of stiff polymer microfibre arrays. II. Microscale behaviour. J R Soc Interface 5:845–853
Sethi S, Ge L, Ci L, Ajayan P, Dhinojwala A (2008) Gecko-inspired carbon nanotube-based self-cleaning adhesives. Nano Lett 8:822–825
Shahsavan H, Salili S, Jakli A, Zhao B (2015) Smart muscle-driven self-cleaning of biomimetic microstructures from liquid crystal elastomers. Adv Mater 27:6828
Shahsavan H, Salili S, Jakli A, Zhao B (2017) Thermally active liquid crystal network gripper mimicking the self-peeling of gecko toe pads. Adv Mater 29:1604021

Shea GM, Greer AE (2002) From *Sphenomorphus* to *Lipinia*: generic reassignment of two poorly known New Guinea skinks. J Herpetol 36:148–156
Simmermacher G (1884) Haftapparate bei Wirbeltieren. Zool Gart 25:289–301
Simões TR, Caldwell MW, Nydam RL, Jiménez-Huidobro P (2017) Osteology, phylogeny, and functional morphology of two Jurassic lizard species and the early evolution of scansoriality in geckoes. Zool J Linnean Soc 180:216–241
Sitti M, Fearing R (2003) Synthetic gecko foot-hair micro/nano-structures as dry adhesives. J Adhes Sci Technol 17:1055–1073
Slavenko A, Allison A, Meiri S (2021) Elevation is a stronger predictor of morphological trait divergence than competition in a radiation of tropical lizards. J Anim Ecol 90:917–930
Smith MA (1937) A review of the genus *Lygosoma* (Scincidae: Reptilia) and its allies. Rec Zool Surv India 39:213–234
Song S, Sitti M (2014) Soft grippers using micro-fibrillar adhesives for transfer printing. Adv Mater 26:4901
Song S, Drotlef D, Majidi C, Sitti M (2017) Controllable load sharing for soft adhesive interfaces on three-dimensional surfaces. Proc Natl Acad Sci USA 114:E4344–E4353
Spolenak R, Gorb S, Arzt E (2005) Adhesion design maps for bio-inspired attachment systems. Acta Biomater 1:5–13
Stark AY, Sullivan TW, Niewiarowski PH (2012) The effect of surface water and wetting on gecko adhesion. J Exp Biol 215:3080–3086
Stark AY, Badge I, Wucinich NA, Sullivan TW, Niewiarowski PH, Dhinojwala A (2013) Surface wettability plays a significant role in gecko adhesion underwater. Proc Natl Acad Sci USA 110:6340–6345
Stark AY, Palecek AM, Argenbright CW, Bernard C, Brennan AB, Niewiarowski PH, Dhinojwala A (2015) Gecko adhesion on wet and dry patterned substrates. PLoS One 10:e0145756
Stewart WJ, Higham TE (2014) Passively stuck: death does not affect gecko adhesion strength. Biol Lett 10:20140701
Sun W, Neuzil P, Kustandi TS, Oh S, Samper VD (2005) The nature of the gecko lizard adhesive force. Biophys J 89:L14–L17
Tan D, Wang X, Liu Q, Shi K, Yang B, Liu S, Wu Z, Xue L (2019) Switchable adhesion of micropillar adhesive on rough surfaces. Small 15:1904248
Tao D, Gao X, Lu H, Liu Z, Li Y, Tong H, Pesika N, Meng Y, Tian Y (2017) Controllable anisotropic dry adhesion in vacuum: gecko inspired wedged surface fabricated with ultraprecision diamond cutting. Adv Funct Mater 27:1606576
Tian Y, Pesika N, Zeng H, Rosenberg K, Zhao B, McGuiggan P, Autumn K, Israelachvili J (2006) Adhesion and friction in gecko toe attachment and detachment. Proc Natl Acad Sci USA 103:19320–19325
Tornier G (1899) Ein Eidechsenschwanz mit Saugscheibe. Biol Zentralbl 19:549–552
Uetz P, Freed P, Aguilar R, Reyes F, Kudera J, Hošek J (2025) The reptile database. https://www.reptiledatabase.org
Vanhooydonck B, Andronescu A, Herrel A, Irschick DJ (2005) Effects of substrate structure on speed and acceleration capacity in climbing geckos. Biol J Linn Soc 85:385–393
Vinson J, Vinson J-M (1969) The saurian fauna of the Mascarene Islands. Bull Mauritius Inst 6:203–320
Vitt LJ, Ballinger RE (1982) The adaptive significance of a complex caudal adaptation in the tropical gekkonid lizard *Lygodactylus klugei*. Can J Zool 60:2582–2587
Wagler JG (1830) Natürliches System der Amphibien: mit vorangehender Classification der Säugethiere und Vögel: ein Beitrag zur vergleichenden Zoologie Mūnchen. J.G. Cotta'schen Buchhandlung
Wainwright PC, Reilly SM (1994) Ecological morphology: integrative organismal biology. University of Chicago Press
Wang Z (2018) Slanted functional gradient micropillars for optimal bioinspired dry adhesion. ACS Nano 12:1273–1284

Wang Y, Hu H, Shao J, Ding Y (2014) Fabrication of well-defined mushroom-shaped structures for biomimetic dry adhesive by conventional photolithography and molding. ACS Appl Mater Interfaces 6:2213–2218
Wang Y, Lai H, Cheng Z, Zhang H, Zhang E, Lv T, Liu Y, Jiang L (2019a) Gecko toe pads inspired in situ switchable superhydrophobic shape memory adhesive film. Nanoscale 11:8984–8993
Wang Y, Rang V, Arzt E, Federle W, Hensel R (2019b) Strong wet and dry adhesion by cupped microstructures. ACS Appl Mater Interfaces 11:26483–26490
Wang X, Yang B, Tan D, Li Q, Song B, Wu Z, del Campo A, Kappl M, Wang Z, Gorb S, Liu S, Xue L (2020) Bioinspired footed soft robot with unidirectional all-terrain mobility. Mater Today 35:42–49
Webster NB, Johnson MK, Russell AP (2009) Ontogenetic scaling of scansorial surface area and setal dimensions of *Chondrodactylus bibronii* (Gekkota: Gekkonidae): testing predictions derived from cross-species comparisons of gekkotans. Acta Zool 90:18–29
Weitlaner F (1902) Eine Untersuchung über den Haftfuss des Gecko. Verh Zool Bot Ges Wien 52:328–332
Williams EE (1972) The origin of faunas. Evolution of lizard congeners in a complex island fauna: a trial analysis. In: Dobzhansky T, Hecht MK, Steere WC (eds) Evolutionary biology, vol 6. Springer, New York, pp 47–89
Williams EE (1976a) South American anoles: the species groups. Pap Avulsos Zool 29:259–268
Williams EE (1976b) West Indian anoles: a taxonomic and evolutionary summary. I. Introduction and a species list. Breviora 440:1–21
Williams EE, Peterson JA (1982) Convergent and alternative designs in the digital adhesive pads of scincid lizards. Science 215:1509–1511
Xie T, Xiao X (2008) Self-peeling reversible dry adhesive system. Chem Mater 20:2866–2868
Xu Q, Wan Y, Hu T, Liu T, Tao D, Niewiarowski P, Tian Y, Liu Y, Dai L, Yang Y, Xia Z (2015) Robust self-cleaning and micromanipulation capabilities of gecko spatulae and their bio-mimics. Nat Commun 6:8949
Xu Q, Wu X, Wang Z, Hu TS, Street J, Luo Y, Xia Z (2018) Temperature-induced tunable adhesion of gecko setae/spatulae and their biomimics. Mater Today Proc 5:25879–25893
Yu J, Chary S, Das S, Tamelier J, Pesika N, Turner K, Israelachvili J (2011) Gecko-inspired dry adhesive for robotic applications. Adv Funct Mater 21:3010–3018
Yu J, Chary S, Das S, Tamelier J, Turner K, Israelachvili J (2012) Friction and adhesion of gecko-inspired PDMS flaps on rough surfaces. Langmuir 28:11527–11534
Yurdumakan B, Raravikar N, Ajayan P, Dhinojwala A (2005) Synthetic gecko foot-hairs from multiwalled carbon nanotubes. Chem Commun 30:3799–3801
Zaaf A, Van Damme R, Herrel A, Aerts P (2001a) Limb joint kinematics during vertical climbing and level running in a specialist climber: *Gekko gecko* Linneus, 1758 (Lacertilia: Gekkonidae). Belg J Zool 131:173–182
Zaaf A, Van Damme R, Herrel A, Aerts P (2001b) Spatio-temporal gait characteristics of level and vertical locomotion in a ground-dwelling and a climbing gecko. J Exp Biol 204:1233–1246
Zhang Y, Ma S, Li B, Yu B, Lee H, Cai M, Gorb S, Zhou F, Liu W (2021) Gecko's feet-inspired self-peeling switchable dry/wet adhesive. Chem Mater 33:2785–2795
Zhao Y, Tong T, Delzeit L, Kashani A, Meyyappan M, Majumdar A (2006) Interfacial energy and strength of multiwalled-carbon-nanotube-based dry adhesive. J Vac Sci Technol B 24:331–335
Zhao B, Pesika N, Zeng H, Wei Z, Chen Y, Autumn K, Turner K, Israelachvili J (2009) Role of tilted adhesion fibrils (setae) in the adhesion and locomotion of gecko-like systems. J Phys Chem B 113:3615–3621
Zheng Y, Wiens JJ (2016) Combining phylogenomic and supermatrix approaches, and a time-calibrated phylogeny for squamate reptiles (lizards and snakes) based on 52 genes and 4162 species. Mol Phylogenet Evol 94:537–547
Zhou M, Tian Y, Sameoto D, Zhang X, Meng Y, Wen S (2013) Controllable interfacial adhesion applied to transfer light and fragile objects by using gecko inspired mushroom-shaped pillar surface. ACS Appl Mater Interfaces 5:10137–10144

Chapter 5
Taking Hints from Whales: The Mysticete Oral Baleen Filter and Biomimetic Applications to Microplastics Removal Systems

Alexander Werth, Jean Potvin, Amanda Cox, Jeddrick Gamilla, Welkin Huelsman, and David Wisbey

Abstract Filtration is widespread in the biological world at many levels, from cells and organs to wetland ecosystems. As a basic yet crucial process of separating solid material from a fluid medium, filtration is also ubiquitous in the engineered human world, with collected solids retained or eliminated. Although the principle is simple, filtration can be complex in natural and artificial systems to improve efficiency and prevent filter failure or clogging. Suspension feeding is a classic example of filtration in living and extinct aquatic animals, including many of the largest species of each group. Aquatic animals use modified gill rakers, teeth, or other oral or pharyngeal structures to separate organic detritus or whole organisms from water. Baleen, a novel oral tissue that evolved in whales 25–30 million years ago, is a remarkably effective filtering material that allowed mysticetes to become the largest animals that ever lived. Although exposed baleen is dead keratin, like hair and nails, it is an ever-growing material that replaces lost parts due to generative cells in the gingiva. Baleen combines rigid strength with pliant flexibility, enabling

A. Werth (✉)
Department of Biology, Hampden-Sydney College, Hampden Sydney, VA, USA
e-mail: awerth@hsc.edu

J. Potvin
Department of Physics, Saint Louis University, St. Louis, MO, USA

WATER Institute, Saint Louis University, St. Louis, MO, USA

A. Cox
WATER Institute, Saint Louis University, St. Louis, MO, USA

Department of Civil Engineering, Saint Louis University, St. Louis, MO, USA

J. Gamilla · W. Huelsman
Department of Aerospace and Mechanical Engineering, Saint Louis University, St. Louis, MO, USA

D. Wisbey
Department of Physics, Saint Louis University, St. Louis, MO, USA

V. Bels et al. (eds.), *Biomimicry – A Functional Approach*, Biologically-Inspired Systems 20, https://doi.org/10.1007/978-3-032-11949-0_5

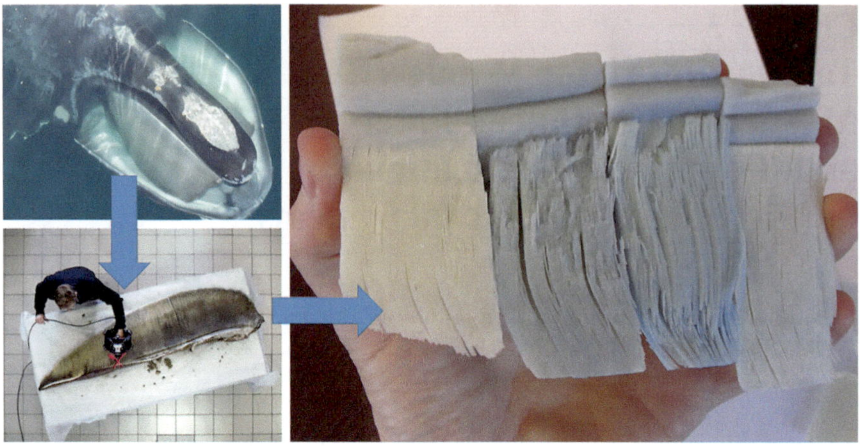

Taking Hints from Whales: The Mysticete Oral Baleen Filter and Biomimetic Applications to Microplastics Removal Systems

it to withstand pressures and forces of flow that would break most materials. Its unique construction as a fiber-reinforced composite keratinous matrix, with hollow horn tubes surrounded by intertubular keratin all sandwiched between two flat sheets of cortical keratin, provides an ideal biomimetic focus. Baleen withstands cracking except along its longitudinal axis, producing long, hair-like fringes as it slowly erodes. The combination of comb-like plates and their medial fibrous fringes together facilitate filtration and avoid breakage. Dynamic variable porosity based on prevailing flow, in throughput or tangential filtration, helps baleen filters to resist high forces without clogging and persist over a whale's entire life spanning 100–200 years. In addition to baleen's utility in traditional biomimetics—spurring creation of improved artificial filters—there are equally exciting opportunities to "complete the circle" via the opposite process: by reverse, retro, or reciprocal biomimetics. Controlled laboratory experiments and computational simulations enable testing of materials/conditions that do not exist in nature, permitting better understanding of how mysticete filtration truly operates in vivo. Traditional and reciprocal biomimetics complement and inform each other, and, in a tightening Archimedean spiral, collaborate to enhance biological understanding and product design.

Keywords Filter · Filtration · Separation · Baleen · Cetacea · Keratin

5.1 Introduction

Biomimicry, also known as biomimetics (Magna et al. 2013; Cohen et al. 2014), involves emulating natural systems or organisms to solve human problems at various scales ranging from consumer products to industrial processes. Janine Benyus, who coined the term *biomimicry* (Benyus 1997), originally envisioned innovative, nature-inspired solutions to major challenges facing society. Such endeavors include acquiring and storing energy, building strong shelters and bridges, feeding ever-increasing human populations, finding potable water, seeking cures for diseases, and storing vast troves of data. Along with such "hard problems," biomimicry has come to include smaller, consumer-oriented products such as Velcro, a remarkably effective yet simple hook-and-loop fastener inspired by burrs that transport seeds, and thus propagate plants, by affixing dried fruits to animal fur or human clothing.

Although biomimicry is now a fashionable strategy that has gained prominence in recent years (Cohen et al. 2014), the concept is nothing new. For hundreds of years, humans looked to bird wings as inspiration for designing contraptions capable of gliding or powered flight, such as those of Leonardo da Vinci, Otto Lilienthal, John Montgomery, and the Wright brothers (Goldstone 2015; Harwood and Fogel 2012; Burdick and Burdick 2017). As the ancient Greek Daedalus and Icarus myth shows, this strategy has persisted for thousands of years and presumably originated much earlier. Early humans likely based simple cutting tools on carnivore teeth and claws, shelters on animal nests, and clothing on animal pelage. Despite these ancient origins, modern biomimicry, which looks to nature for model systems, nonetheless now depends on cutting-edge engineering, such as 3D printing or other additive manufacturing, for solutions to technological challenges such as improving efficiency of locomotion or power generation.

Nature is an obvious place to seek inspiration for filtration, a ubiquitous physical process. At its simplest level, filtration involves separation of solid physical matter from a circulating liquid or gaseous fluid by use of a semipermeable porous structure: a filter (Sutherland 2005). For various reasons—most notably, (1) the sheer physical simplicity of filtration as a process, (2) the vast range of basic filtration strategies found in living organisms and natural systems, and (3) the profound importance of filtration in so many aspects of modern human life—filtration is especially amenable to biomimetic solutions, and therefore attracts considerable research scrutiny (Adelmann et al. 2022; Bianciardi and Cascini 2022; Clark and San-Miguel 2021; Dou et al. 2017; Goel et al. 2021; Hamann and Blanke 2022; Liao et al. 2024; Masselter et al. 2023; Schroeder et al. 2019; Sun et al. 2024; Xu et al. 2023; Zhang et al. 2022).

A common biological use of hydrosol filtration is suspension feeding (Jorgensen 1966; Rubenstein and Koehl 1977; LaBarbera 1984; Shimeta and Jumars 1991; Espinosa-Gayosso et al. 2012, 2021; Hamann and Blanke 2022). This occurs in a wide range of animals, from tiny plankton to gigantic whales and other vertebrates (Sanderson and Wassersug 1993). As organs or other components of animal bodies, filters can operate and be structured within the body, as in the oral or pharyngeal gill rakers of large filterers such as basking and whale sharks and manta rays, or external to the body, as in the case of filter-feeding appendages of copepods and other aquatic invertebrates (Sanderson and Wassersug 1990). The type of filter profiled in this chapter involves the internal oral keratinous filter, called baleen, in toothless mysticete whales.

Baleen is a novel tissue, analogous to other feeding filters such as teeth and gill rakers but not homologous to any other tissues (Gatesy and O'Leary 2001; Deméré et al. 2008; Fudge et al. 2009; Gatesy et al. 2013; Thewissen et al. 2017). There is considerable speculation surrounding baleen's origins in archaeocete whale ancestors with fully functional dentition. It remains unclear how early proto-baleen functioned in early toothed mysticetes (Thewissen 1998; Thewissen and Bajpai 2001; Deméré et al. 2008; Thewissen et al. 2009, 2017; Fitzgerald 2010; Uhen 2010; Gatesy et al. 2013; Marx and Fordyce 2015; Tsai and Fordyce 2015; Berta et al. 2015, 2016; Boessenecker and Fordyce 2015; Marx et al. 2016a, b, 2017; Geisler et al. 2017; Hocking et al. 2017a, b; Lambert et al. 2017; Peredo et al. 2017; Ekdale and Deméré 2021; Gatesy et al. 2022). Nonetheless, it is clear that baleen originated as a neomorphism 25–30 million years ago (Gioncada et al. 2016; Bosio et al. 2021; Bisconti et al. 2023) and that, based on the enlarging body size and taxonomic diversity of mysticetes, it was a major factor in the evolutionary success of crown Mysticeti, which includes the largest animals to have ever lived (Pyenson and Vermeij 2016; Slater et al. 2017; Fordyce and Marx 2018; Goldbogen and Madsen 2018). Not only has baleen filtration been documented as the source of whales' success, but it is also widely known to underlie crucial global nutrient cycling phenomena on which all oceanic life—and by extension, all life on Earth—depends (Fordyce 1980; Fordyce and Barnes 1994; Pearson et al. 2023). In short, this remarkable keratinous tissue lies at the heart of some of the biggest (literally) biological processes known to exist. Could baleen filtration not be equally important, at least in schematic form, to human-inspired filtration?

Like swimsuits inspired by the placoid denticles of sharkskin or turbine blades inspired by the tubercles on humpback whale flippers (Fish et al. 2011; Fish 2020; Johari et al. 2007), whale baleen offers fertile ground for biomimicry (Trakumas et al. 2001). Baleen filtration has been called the most dynamic biomechanical event in the world (Goldbogen 2010) because it involves staggering masses and volumes of prey-laden water easily approaching or exceeding 100 m^3 (100,000 L) in fully grown 25 m-long blue whales, which is then filtered through a filter with surface area of approximately 4 m^2 in roughly 30 s, with peak pressures potentially reaching 5–10 kPa, the equivalent of 10^4 N/m^2 (Werth 2013). Further, this filter must retain structural integrity and remain functional for the whale's entire lifespan (up to a century or more; George et al. 2021). Baleen must filter incredibly capacious volumes of water and hold massive volumes of water-borne schooling prey, at astonishing pressures, without breaking, clogging, or otherwise failing. This is an astonishing feat, and engineers have much to learn and gain from studying and incorporating key components of baleen filters and filtration.

Perhaps baleen's most notable feature is its robust, durable resilience. In essence, baleen is unbreakable. It is virtually guaranteed to work regardless of congenital or acquired structural defects, just with compromised efficiency. This presumably facilitated baleen's evolutionary progression to the tissue and forms it possesses today. Ironically, what makes baleen so ingenious—if the concept of genius can be applied to a feature that evolved via an algorithmic process of natural selection—is that its resilient "unbreakability" results from a key design feature: breakability, in the sense that baleen naturally cracks and erodes, due to its composite construction, solely along its vertical axis. As will be shown in detail, this axial construction and "intentional" failure helps baleen to resist forces and prevents breakage along any axes apart from the dorsoventral axis. Further, baleen's eventual, inevitable erosion creates bristly fringes, adding an additional integral layer of filter form and function. Finally, the baleen filter does not rely on sticky mucus for particle capture, as in many biological filtration systems, which offers a clear advantage in biomimetic filter design.

Following a basic review of filtration, this chapter presents five fundamental elements of baleen that make it an excellent candidate for biomimetics:

1. Material: specifically, baleen's biomechanical properties as an ever-growing anisotropic composite with durable resilience
2. Structure: baleen's metafilter arrangement incorporates comb-like racks comprising hundreds of individual major and minor plates eroded into hair-like, bristly fringes
3. Function: baleen's dynamic porosity based on differential flow speeds and directions
4. Robustness: its cross-flow filtration mode works, albeit with varying particle capture efficiency, regardless of incoming flow speed and filtration surface porosity
5. Natural practicality: as a low-power organ using energy of incurrent flow, and which can be cleaned and maintained simply, without replacement

Although the earlier sections primarily review previously published material, the section on baleen function also reports entirely new data, including results from computer simulations and experiments with multiple flow tanks studying flow in models or actual baleen tissue to replicate functional aspects of baleen filtration.

This chapter concludes with a discussion of ways that baleen can prove useful both in traditional biomimetics (i.e., inspiring creation and use of improved artificial filters) as well as in an opposite process of reciprocal biomimetics. The latter involves controlled laboratory experiments and computational simulations enabling testing of 3D-printed or other artificial materials/conditions that do *not* exist in nature, permitting better understanding of how natural filtration truly operates in vivo in whales and other organisms. The hope is that both endeavors (i.e., traditional and reciprocal biomimetics) can guide valuable applications, such as removal of macro- or microscopic plastic debris (Law 2017; MacLeod et al. 2021) polluting bodies of marine and freshwater.

5.2 Basics of Filtration

As described earlier, filtration is defined as separation of solid physical matter suspended within a circulating fluid (liquid or gaseous) by a semipermeable porous filter. Although the filter is generally a solid structure, filtration (i.e., separation and sequestration of the solid) is enhanced by dynamic flow phenomena or filtration schemes including the speed and direction of flow relative to the filter. The filtered solid (or retentate) can be retained if its accumulation is desired, or discarded if the objective is to purify passed fluid (the filtrate, also called permeate). Both the retention and elimination strategies are common in nature, where they occur at multiple scales ranging from microscopic structures of organisms to ecosystem-level arrangements. Animal kidneys or similar excretory organs use specialized cells (e.g., podocytes) to remove wastes to be discarded from circulating body fluids, whereas livers and similar digestive organs likewise use specialized sinusoids and cells (e.g., hepatocytes) to remove nutrients to be retained. In the same way, but at vastly differing scales and Reynolds numbers (defined below), wetlands filter and sequester particulate matter from environmental watersheds (Werth 2019).

Filters are as common in the human-engineered world as they are in nature. Our world is filled with porous structures and phenomena. From the fuel, air, and water filters in automobiles to coffee pots, tea bags, and strainers or colanders in kitchens, we depend on porosity (abbreviated as phi, ϕ) to get through every day via porous structures that engineers or industrial designers plan and fashion from natural and man-made materials. Non-living (abiotic) filters can be simple aspects of the natural environment, such as soils, through which a different medium (especially air or water) can flow, in addition to naturally porous structures that comprise parts of living organisms, including cells, tissues, and organs. Obviously, a major difference between human-engineered porosity and natural porosity is that porous structures and processes designed by humans are the consequence of careful foresight involving planning toward a goal. In contrast, natural filters evolve non-teleologically through a stochastic, stepwise process of natural selection. Natural filters may achieve an optimal function, or operate in non-optimal fashion that still ensures species survival, but both evolutionary paths occur without goal-directed foresight.

Related to this fundamental difference between man-made and naturally porous structures and processes, another important distinction can be made between *functional porosity* and *non-functional porosity*. Soils, made of differentially accumulated particles including sands, silts, and gravels, are well known for being porous, but soil porosity—although important for soil scientists, soil-dwelling organisms, and geologic processes—is a type of non-functional, *incidental porosity*: in other words, the soil was neither created nor did it evolve with the express purpose of having a porous function. Contrast this with other filters described above, such as kidneys and baleen plates, as well as with engineered filters such as in a coffee pot: these *functional structures* exist to be porous. Porosity plays a vital, essential role (rather than an incidental role) in these structures' existence.

In most types of filtration, the filtrate, or material to be filtered, flows directly (= straight) into the filtration medium or membrane—i.e., the filter itself. This is called *dead-end* or *throughput filtration* (*TPF*). An example involves pouring pasta noodles, along with boiled water, into a kitchen sieve or colander. The filter retains the noodles while draining the water. Although *sieving* is, like other forms of throughput filtration, simple and direct (French 1971), a basic problem arises

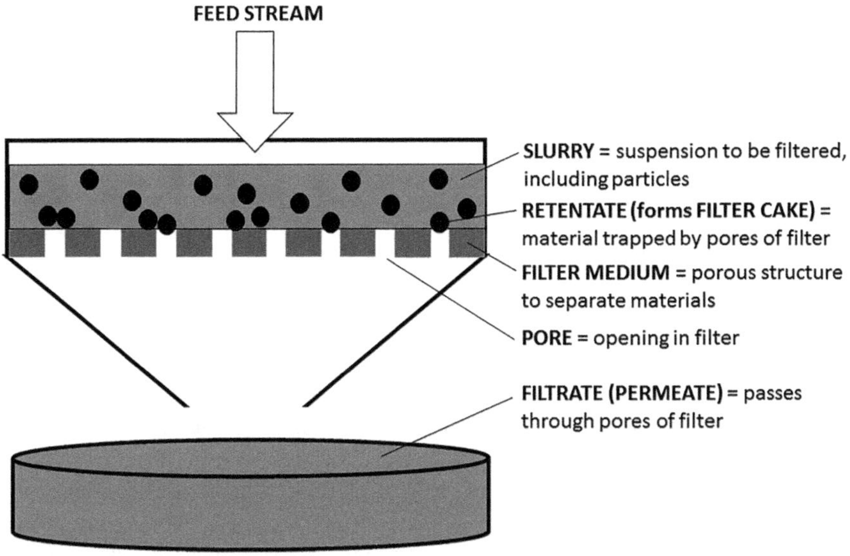

Fig. 5.1 Basic principles of filtration showing how key filter parameters affect flow output and particle retention

from the filter's perpendicular orientation relative to flow: *retentate*, like cooked pasta noodles, accumulates on and covers the filter surface. This potentially *fouls* (i.e., clogs or "cakes") the filter, impeding flow and thereby reducing or wholly preventing further filtration.

Because many filtration processes involve a *slurry* of suspended matter, the accumulated retentate (left behind by elimination of permeate; Fig. 5.1) creates a *filter cake* of retained matter (Blatt et al. 1970). Depending on the type of insoluble retentate, the cake may be solid or semisolid. In situations where filtration is an extensive, ongoing process, such as with industrial purification of large volumes of juice, wine, or other beverages (Starbard 2009; Tamime 2013; Jain and De 2019), the clogged filter will need to be periodically replaced, or else cleaned mechanically or via *backflushing*. Not only does the need to sporadically clean the filter temporarily halt filtration, but it also obviously decreases TPF's effectiveness and potentially wastes large volumes of material (e.g., the unfiltered beverage stream), which spills around rather than through the filter (Fig. 5.2).

A plain yet elegant solution—employed both in industrial filtration as well as biological systems, as will be explained—is to reorient the filter so it is tangential rather than perpendicular to the flow stream (Fig. 5.2). In *tangential filtration*, as turned by a lateral pressure gradient above and across the filtering surface (Fig. 5.2), the slurry passes partially or wholly parallel to the filter surface, hence this strategy is often described as *cross-flow filtration* or *CFF* (Baker et al. 1985; Murkes and Carlsson 1988; Lu and Ju 1989; Lu et al. 1993; Song and Elimelech 1995; Zeman and Zydney 1996; Sanderson et al. 2001; Ripperger and Altmann 2002; Potvin and

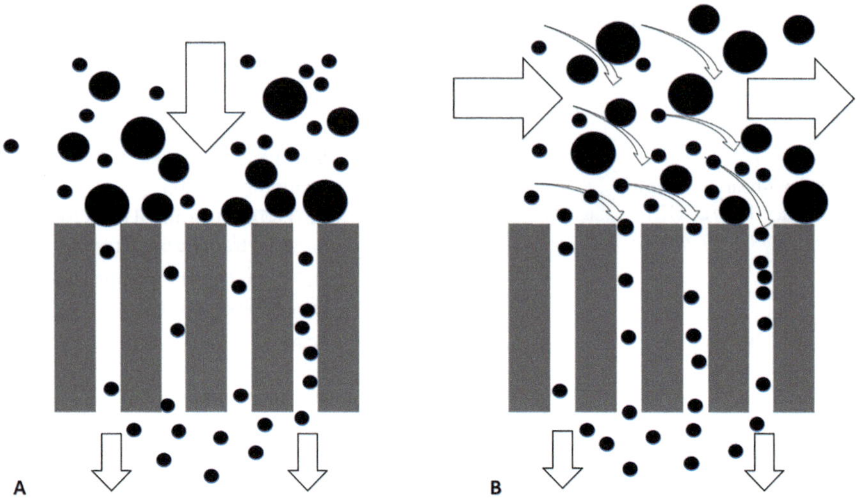

Fig. 5.2 Throughput and cross-flow filtration differ primarily by the orientation of the filter surface relative to the flow stream. (**a**) Dead-end or throughput filtration (TPF), (**b**) tangential or cross-flow filtration (CFF)

Werth 2017; Sanderson 2024). A prime benefit of CFF relative to TPF is that the flow propels retentate mostly along the filter surface prior to eventual capture. In many cases, this occurs without the retentate ever directly contacting the filter surface, or with the retentate partially coating the surface with a thin layer, thereby eliminating clogging, obviating the periodic need to clean the filter, and increasing filtration efficiency (Belfort et al. 1994; Vogel and Todaro 1997; Bott et al. 2000; Sibanda et al. 2001).

Other benefits of CFF include possible fractionation of filtered particles according to size and a generally softer mass of accumulated retentate, which remains in the flow stream for further processing or easy removal from the system. On the other hand, because CFF is less direct, the filter surface may be larger, with greater overall area (Trakumas et al. 2001). Like TPF, CFF is found in man-made as well as natural situations requiring filtration or purification; indeed, engineers often draw biomimetic inspiration from nature. There are multiple means of filtering, as described by Rubenstein and Koehl (1977) and LaBarbera (1984), and as shown in Fig. 5.3.

Another fundamental distinction can be drawn between filters and other porous structures that have *fixed porosity*, like a net or kitchen strainer, versus those with *variable porosity*. Variable porosity, in which the porosity of a particular structure or process changes in response to short-term changes in environmental input, is not the same as *varying porosity* (Fig. 5.4), which simply occurs when pores within a structure vary in size. However, varying porosity can involve varied nets or soils whose porosity changes over extended periods of time, for example due to long-term

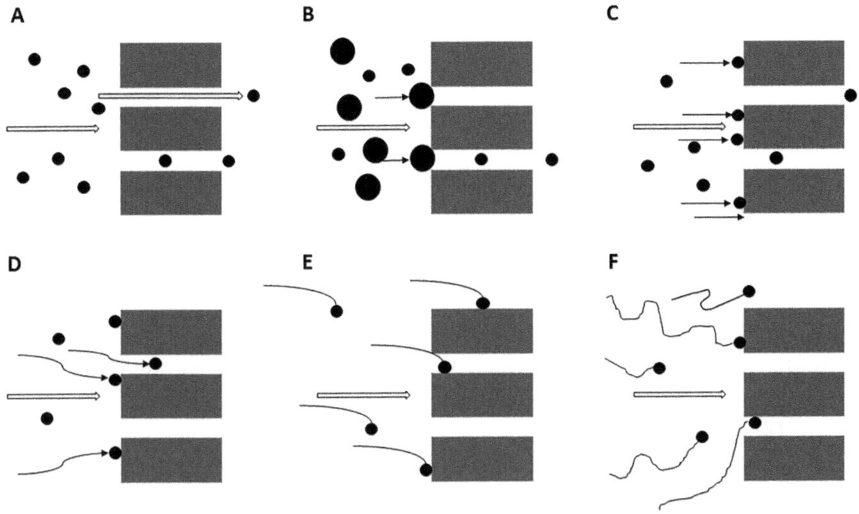

Fig. 5.3 Natural/biological and man-made filters both use multiple means to capture particles/retentate. (**a**) Flow through pore, (**b**) sieving, (**c**) direct interception, (**d**) inertial impaction, (**e**) gravitational deposition, (**f**) motile (diffusing) particle deposition. (Illustration scheme modified from Rubenstein and Koehl (1977))

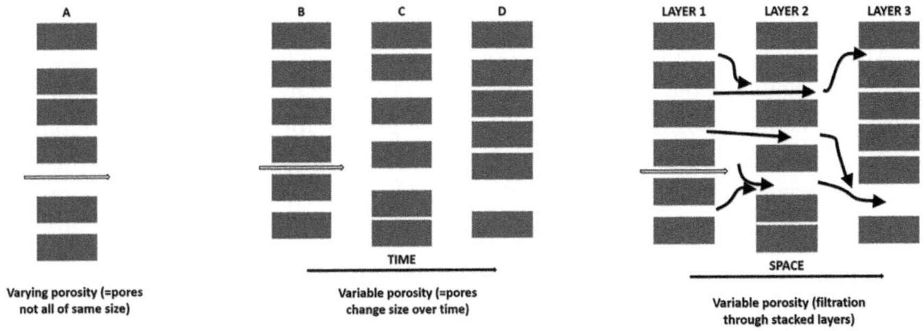

Fig. 5.4 Comparison of varying, variable (dynamic), and layered porosity (redrawn from Werth 2019), with flows shown after their turn toward the filter surface as indicated in Fig. 5.2

weathering or breakdown of lithic components, or due to influx and accumulation of new components (e.g., sand from distant erosion).

Because variable porosity typically involves a rapid, responsive change caused by local conditions, it can also be called *dynamic porosity*. Note that the rate, degree, and direction of change (with porosity either increasing or decreasing) may itself be functional, as an engineered or evolved property of the system. Alternatively, the dynamic change may simply involve an invariant response to fundamental (universal and unalterable) physical or chemical laws. For example, porosity may vary according to changes in the speed or force of flow, or of the density or viscosity of the medium passing through it. Note that in oilfield drilling and other geological settings, dynamic filtration means something different: it refers to a filtering process whereby the slurry (of suspended solids and liquid or liquids) being filtered is recirculated over the filtering medium, such that the filter cake residue is eroded at the same time as it is deposited. If the fluid's shear rate at the filter cake's surface remains constant, this process reaches a steady state.

Variable porosity of many biological filters (Werth 2019; Werth and Potvin 2024), whether controlled or simply reacting in response to prevailing conditions, enables whole organisms and organ systems to alter filtrate flow and retentate capture, avoid clogging, and maintain structural and operational integrity over long spans of time, even years or decades. Muscle and other contractile tissues can be used to govern pore size and filtration rate or efficiency.

Entire wetlands such as marshes or bogs can rightly be interpreted as fundamentally porous filters, as can tiny, microscopic layers of cell membranes (Werth 2019). Hence the scope and scale of porous filtering structures varies dramatically from kilometers (10^3 m) to nanometers (10^{-9} m). Likewise, the flow may involve tremendously different flow regimes as measured by *Reynolds number* (R_e), a dimensionless ratio of inertial to viscous forces, and therefore a measure or description of the turbulence or laminarity of the flow. ($R_e = [\rho \times V \times L]/\mu$, where ρ is the density of the fluid, V is the flow velocity, L is a characteristic length of the flow geometry, and μ is the dynamic viscosity of the fluid.) Such a flow may also involve *tortuous* movements within the filtration medium (e.g., in the layered filter in the right panel of Fig. 5.4).

Much study of porosity and filtration, and many practical applications, relate to geology, whether in the contexts of soil compaction and hydration; drilling for oil or other petroleum deposits; and suitability of the ground for construction, agriculture, or other use. Although porosity's key importance in geology, industrial design, and manufacturing must be acknowledged, emphasis will be placed here on the biological uses of baleen-based biofiltration as inspiration to engineered filtration systems.

Again, the presence of varying porosity can refer to differential porosity, as in *semipermeable* membranes, alone and when linked to form epithelial barriers as often manifested in osmosis. Semipermeability refers to the fact that only certain molecules pass through the membrane based on size, charge, polarity, etc. Semipermeable membranes are found, for example, in cell membranes, dialysis tubing, and reverse osmosis systems. However, this is not the same thing as variable porosity unless changes occur, such as opening channel proteins or pores, such that the structure's porosity changes over time in response to environmental stimuli (Fig. 5.4).

Note too that engineered structures and processes, such as the reverse osmosis method used for water desalination plants (which make fresh water from seawater), can involve artificial porosity which mimics the natural porosity of living cells or other abiotic structures. Designing and manufacturing artificial biomimetic structures inspired by natural structures and processes is a highly successful means of adopting tried-and-true principles to solve human problems (Vogel 1996, 2013), such as the increasingly apparent quandary of widespread microplastic pollution throughout ocean basins and other bodies of water (Baraza et al. 2022; Jambeck et al. 2015; Law 2017; MacLeod et al. 2021). As will be shown, a compelling twist on traditional biomimetics involves creation of structures and materials which in fact do not exist in nature or otherwise occur naturally. This allows investigators both to create naturally inspired yet fundamentally novel structures; as well as to study, on a deeper level, natural processes that underlie basic activities, such as filtration within whale mouths, by extending the range of conditions beyond those found in nature.

5.3 Baleen Histology and Material Properties

As with bone and some other connective tissues, baleen can be viewed and studied on three distinct levels (Werth and Potvin 2024). First, baleen is a structural *material* with its own unique construction and material properties (Werth 2017); in this way baleen is as unique as kinds of wood, glass, metal, or concrete and other organic and inorganic materials. Second, baleen can be considered a growing biological *tissue* like muscle, cartilage, or skin. Third, baleen exists as the equivalent of an *organ* like a femur, heart, or pancreas. Although baleen is—as material, tissue, and organ—generally the same in each extant mysticete species, there are obvious species-specific differences in size and shape, growth rate and pattern, pigmentation, and other parameters that often have ecological or life-history significance.

Each whale's full complement of baleen consists of bilaterally paired racks suspended from the left and right margins of the palate, with each rack comprising approximately 300 individual plates of baleen tissue. Each plate (Fig. 5.5) is roughly shaped like a skewed right triangle (Lambertsen et al. 2005). Because each rack is therefore a layered, comb-like structure, each individual baleen plate is also sometimes called a lamina (Young 2012; Loch et al. 2020). The precise number of plates varies by species (ranging from about 160–360 plates per rack), and to a lesser degree by a whale's overall body size, which in turn depends on age and sex (Werth et al. 2018a). However, virtually all adult mysticetes have approximately 300 baleen plates suspended from each side of the rostrum. There is no baleen attached to the lower jaw; and although early fossil mysticetes had rudimentary dentition and embryonic mysticetes briefly have small tooth germs (anlagen) that never erupt, there is never any upper or lower dentition in extant mysticetes.

Baleen plate size varies by species. Plates are generally about 3-mm thick (anteroposteriorly), ~10–20-cm wide (mediolaterally) for most of their length, and 20–40-cm wide at their dorsal-most origin where they emerge from and are anchored within palatal gingiva. Baleen plates are generally ~50 cm long (dorsoventrally), but range from ~35–75 cm in length for most species, with the major exception of bowhead (*Balaena mysticetus*) and right whales (*Eubalaena* spp.), both of Family Balaenidae. Plates can exceed 4 m in length in exceptionally large balaenid specimens (Werth 2000). Plate dimensions vary by position along the rack, with the longest plates found near the center of a rack (Werth et al. 2018a, 2020). In addition to major plates, most whales have smaller, medial minor

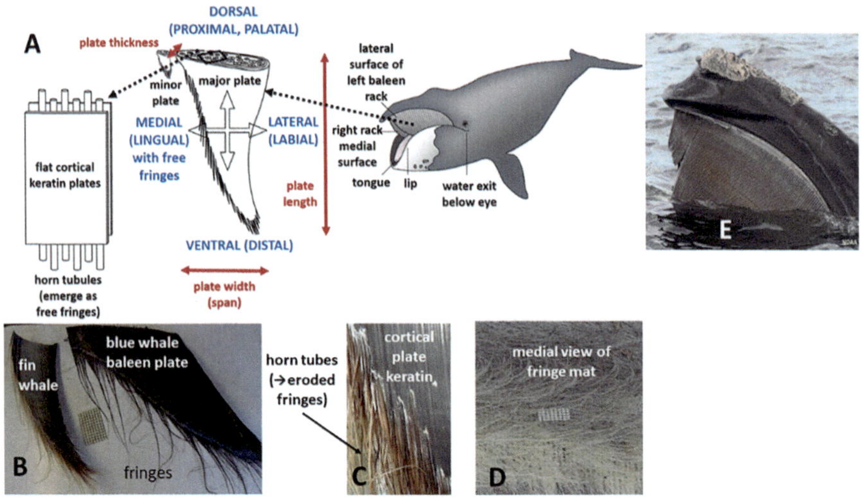

Fig. 5.5 Baleen is a complex oral metafilter, arranged in comb-like fashion, comprising different alpha-keratinous elements arranged sandwich-style with flat, fingernail-like plates enclosing hollow hair-like horn tubules (**a**). Each element has its own filter function along with unique structural and biomechanical properties. Paired racks of baleen tissue (**a**) each consist of ~300 triangular major plates (**b**) that erode along their interior (medial or lingual [= next to the tongue]) surfaces (**c**) to reveal the horn tubes as fringes (bristles; **d**). An entire rack (**e**, image courtesy NOAA), major/minor plate, or fringe can separately or collectively serve as a filter element. (Modified from Werth and Potvin 2024)

plates along each rack. A major plate can have one or multiple adjacent minor plates, all of which resemble fragments broken from the medial edge of a major plate (Williamson 1973; Pivorunas 1976). As their name suggests, minor plates are considerably smaller than major plates (Fig. 5.1). Relative to their reduced size, minor plates likely also play a greatly reduced functional role in filtration, although this has not been adequately addressed. Recent studies indicate that baleen's structural intricacy has been underestimated (Young et al. 2015; Jensen et al. 2017; Vandenberg et al. 2023; Potvin and Werth 2024).

Like most tissues, baleen is an ever-growing substance that grows via basal cell proliferation throughout life (Werth 2017). This is true of other keratinous tissues such as mammalian hair, nails, and claws (which, like baleen, consist entirely of dead cells once erupted), but not true of most teeth and most other vertebrate filtering materials (Marshall et al. 1991; Wang et al. 2016). In contrast, human-engineered artificial filters are static rather than ever-growing structures. Baleen begins growing early in neonatal life, prior to weaning, and continues until the animal's death (George et al. 2016). In an elegantly simple turnover that balances baleen tissue generation and loss, new filtering baleen material arises to replace eroded baleen tissue that becomes frayed and shed during filtration (Werth et al. 2020, 2021). This ever-changing nature of filter growth and loss aptly relates baleen's dynamic structure with its dynamic function. Although baleen growth rates vary somewhat by species and age, in most whales the plates grow about 12–18 cm/year (Sumich 2001; Werth et al. 2021). Bowhead and right whales, which have much longer plates and thus substantially larger baleen filters, are again the exception; balaenid plates typically grow 20–27 cm/year in adults and can exceed 30 cm/year in juveniles (Werth et al. 2020). In most mysticete species, baleen plates grow disproportionately faster in the first 2–3 years of life (Werth et al. 2021). This has been explained as young whales needing to prioritize post-weaning growth of their oral filter to obtain calories needed to sustain growth of other body parts, such that skeletal growth is often delayed or downplayed until the filter attains its full adult proportions (Lubetkin et al. 2008; George et al. 2016; Fortune et al. 2012).

As mentioned earlier, embryonic mysticetes normally bear multiple temporary tooth germs that do not persist but are absorbed during fetal development (Slijper 1962; Peredo et al. 2017), so that they serve as temporary vestiges or atavisms. As these tooth germs fade, they are replaced by arcs of tissue, running along both sides of the rostrum, that generate the alpha-keratin components comprising baleen (Fudge et al. 2009; Szewciw et al. 2010; Thewissen et al. 2017). Alpha-keratin is a strong, insoluble, elastic structural protein of mammalian integumentary tissues (Feughelman 1997; Greenberg and Fudge 2012). Essentially, baleen tissue comprises two primary paired components that are almost exactly identical to human hair and fingernails. These elements are arrayed as a sandwich (Fig. 5.5), with two flat outer layers of nail-like sheets (Forslind 1970) surrounding a core of long, hollow "horn tubules" that dangle down ventrally from the palate (Werth 2000). The interior of every plate consists of an array of one to two dozen variably sized horn tubules within a matrix of intertubular keratin (Werth 2017). The intertubular keratin matrix binds all components together and acts as an important, if much lesser third ingredient (Werth et al. 2018b) for the total baleen "sandwich" shown in Fig. 5.5.

In this way, baleen's construction is highly reminiscent of a fiberglass composite material, with long, narrow fibers interspersed between alternating layers of flat sheets (Greenberg and Fudge 2012). Together, these components provide baleen with unparalleled flexibility and rigidity, combining to create a highly pliant and bendable, yet stiff and resilient material (McKittrick et al. 2012) that absorbs intensely high forces and pressures yet recoils to its original form, all the while resisting crack propagation (Werth 2013; Werth et al. 2018b). Interestingly, this same keratinous composite construction is also found in rhinoceros "horns," which are not at all like true mammalian horns but instead are simply made of matted hair-like fibers interspersed with alpha-keratin sheets. As with baleen, rhino horns' material construction renders them extremely strong yet pliant, so that they are not brittle and do not crack or deform even as they are subjected to remarkably strong loading forces (Benyus 1997).

Like rhino horns, baleen highly resists breaking due to its valuable combination of flexibility and strength (Werth et al. 2018b). However, unlike rhino horns, baleen plates are custom-"designed," via natural selection, to degrade and erode, but only along the vertical (palatal to ventral) axis, especially along the medial (= lingual) edge (Fig. 5.5). Indeed, baleen's composite construction, with two flat outer sheets sandwiching hollow interior straw-like tubules that run solely vertically (Fig. 5.5), is key to its structure and function. This anisotropic design not only enables but indeed facilitates the erosion of each whale's baleen into a tattered curtain of exposed horn tubules that are revealed, like the frayed border of a worn fabric, as hair-like fringes dangling at the exposed, eroded edges of each baleen plate on its medial (internal) side facing the tongue (Werth 2017). These fringes, also called baleen bristles, not only resemble hairs but in fact are created and grow in virtually the same way as human scalp and body hairs (Fraser et al. 1972, 1976; Hearle 2000; Feughelman 2002), as tubes of alpha-keratin that emerge from a follicle where dermal and epidermal generative cells interact (Thewissen et al. 2017).

Because these hair-like fringes run solely dorsoventrally, from a plate's gingival origin to its ventral-most vertex, transverse (mediolateral) cracking of any baleen plate is prohibited even as longitudinal cracking is promoted. Not only does this make baleen a strong, resilient material, but even more importantly it also creates a more elaborate, less coarse filter that consists not only of the flattened baleen plates (major and minor) but also the much smaller, finer, more flexible fringes (Pfeiffer 1992; Young 2012; Jensen et al. 2017; Loch et al. 2020; Vandenberg et al. 2023), as described in the following section on baleen's overall filtering arrangement.

Like other mammalian alpha-keratin tissues including hair, claws, and nails, baleen's only living cells lie in the unerupted generative portion below the exposed surface of the body. All exposed, erupted baleen is, like hair and nails, composed entirely of now-dead but formerly living cells. As such, there are neither nerves or blood vessels of any kind in exposed, erupted baleen, nor any nervous or vascular connections. Baleen might possess some mechanosensory ability due to its anchoring deep within the so-called *zwischensubstanz*, the remarkably tough, pliant, rubbery, gum-like gingival tissue from which baleen emerges and which surrounds and embeds all plates in the palate. However, visibly erupted baleen possesses no sensory reception, as it is entirely dead. In the unlikely event that a baleen plate breaks, it will continue growing from the palate, but the broken portion will not be replaced until the plate retains its former size. Fortunately, the gingival zwischensubstanz further resists shear forces and propagation of baleen cracking or other mechanical damage, enabling the baleen filter to withstand powerful flows as it separates food from water (Pinto and Shadwick 2013; Werth et al. 2019c).

Baleen's flexibility is tempered by the inclusion of mineral salts, particularly calcium, which act to stiffen the plates via deposition in the intertubular and surface keratin proteins (Pautard 1963; Szewciw et al. 2010). Mineral content appears to vary by species and diet, ensuring that baleen of whales that habitually feed on tiny prey is more flexible, creating an overall less porous filter (Werth et al. 2018a). Other studies on baleen's physicochemical properties have demonstrated that baleen is generally hydrophilic (enabling penetration of water, which in turn facilitates flexibility and prevents breakage; Werth et al. 2016a) and largely oleophobic (resisting penetration of oil, which is advantageous for species that live near natural submarine petroleum seeps or man-made oil drilling platforms; Werth et al. 2019a). Baleen has, however, been found to be susceptible to ocean acidification caused by carbon emissions (Werth and Whaley 2019).

Baleen's pliant resilience has long been appreciated by indigenous cultures (Werth et al. 2018b). It is capable of withstanding strong compressive, tensile, and shear forces (Werth 2013, 2019). Calcification potentially stiffens baleen into the hardest of all alpha-keratin tissues (Pautard 1963; Szewciw et al. 2010), but baleen remains highly pliable when hydrated in seawater (Werth et al. 2016a). As a tough yet flexible, non-degradable tissue, baleen has been used by natives for artwork, baskets, implements, armor, and weapons (Moffat et al. 2008; Dubner 2023). During the era of industrial whaling, baleen was a highly valuable commodity, peaking in 1853 with over 5.6 million pounds, mostly from right whales, sold in US ports for almost $2,000,000 (Stevenson 1907). Fringes were made into brushes; plates were used for corset stays, skirt hoops, umbrella ribs, buggy whips, and numerous other flexible items (Lee 1998). Prior to the invention of plastic polymers, baleen served as the plastic of its day, such that whaling ports were among the world's most commercially productive locales. Together with whale oil, baleen made whaling the first true American industry (Dubner 2023).

Baleen's commodity value derived in large part from its exceptional flexibility; its biological value for filtering also arises from this flexibility (Stevenson 1907). Baleen plates bend along their entire length (Fig. 5.6) in response to the forces acting upon them during filtration (Werth et al. 2018b), which enables baleen to withstand strong forces without cracking, thereby providing great longevity. Three-point bending tests of dried and hydrated baleen specimens using an Instron E1000 ElectroPuls or Mark-10 ES30 universal testing machine enabled calculation of uniaxial flexural stress, strain, and modulus in samples of baleen plates and fringes from different whale species and distinct locations within a rack and plate (Werth and Potvin 2024). This enabled creation of a loading (force-deformation) curve (Fig. 5.6), with determination of elastic stiffness and yield and failure points (McKittrick et al. 2012; Werth et al. 2018b). Such experiments showed that baleen can withstand forces >1 kN without breaking (Werth et al. 2018b), and once again at levels that vary along a rack.

These biomechanical experiments (Werth 2013; Werth et al. 2016a, 2018b) also demonstrated the importance of studying baleen under natural conditions: namely, when baleen is fully saturated with water, rather than dried out, the way it is typically stored and displayed in museum collections and exhibits (Werth et al. 2019c). Baleen in water, the way it perpetually exists in vivo, absorbs and holds much water: 32.35% by weight (Werth et al. 2016a). Mechanical materials testing confirmed that dried baleen is stiff but readily shatters, whereas hydrated baleen (like other hydrated keratins; Kitchener and Vincent 1987; Feughelman 1997; Taylor et al. 2004) is much less stiff and bends with little applied force, yet rarely breaks (Werth et al. 2016a, 2018b). This is similar to the difference between dried and fully hydrated pasta noodles (made of dough from unleavened wheat flour), and, further, demonstrates the obvious importance of studying biological materials under natural conditions. Contact angle and wettability tests (Werth et al. 2016a, 2019a) also confirmed that hydrated baleen is highly hydrophilic and oleophobic (i.e., oil resistant or oil-shedding). This is important given that small schooling fish and zooplankton have bodies filled with oil droplets to store energy and maintain neutral

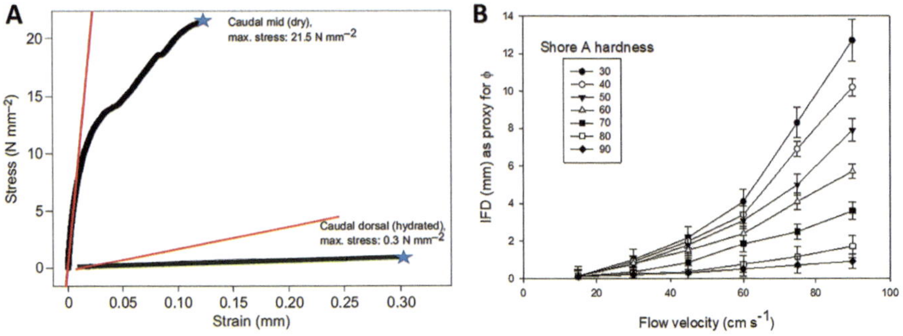

Fig. 5.6 Mechanical testing results (**a**) showing maximal stress recorded from three-point bending tests of right whale baleen, showing elastic flexibility of wet baleen and brittle fracturing of dried baleen, which failed at 12-mm displacement (strain). (**b**) Shows how porosity, as indicated by interfringe distance (IFD), varies by flow speed and Shore hardness, an industrial measure of rubber material softness (measured here in variably hydrated baleen plates). (Data from Werth (2013) and Werth et al. (2016a, 2018b, 2020))

buoyancy (St. Aubin et al. 1984; Michaud and Taggart 2007); because baleen is oleophobic, oily prey are less likely to adhere to or clog the oral filter. Modern imaging techniques such as scanning and atomic force microscopy aid in resolving the microscopic three-dimensional ultrastructure underlying baleen filtration (Werth et al. 2018a, b). Our view of the ways in which baleen's ultrastructure influence and determine its porosity, stiffness, and other functional properties is slowly coming into focus.

To summarize, baleen evolved and filter exceptionally well due to its anisotropically arrayed, composite constituent material. Like human-designed artificial filters, baleen is a non-living structure made solely of dead material, with no nerves, blood vessels, or other elements either exposed or hidden within. Nonetheless, it is an ever-growing tissue that proliferates and propagates (from its basal attachment to the palate) over the course of each whale's lifetime, which can span well over a century (George et al. 2016; Werth and Potvin 2024).

5.4 Mysticete Overall Filter Form

5.4.1 The Balaenid Metafilter

The baleen material described in the previous section gives rise to a three-tiered, low-maintenance and robust filtration system composed of the following elements (Fig. 5.7): (1) two racks of about 300 baleen plates, arranged in comb-like format along the palatial margins and eroding on the medial side of each rack, to produce long strands of fringes or bristles (Fig. 5.5); (2) a plurality of fringes forming thick mats that yield low-porosity surfaces which retain/capture solids and prey (Werth

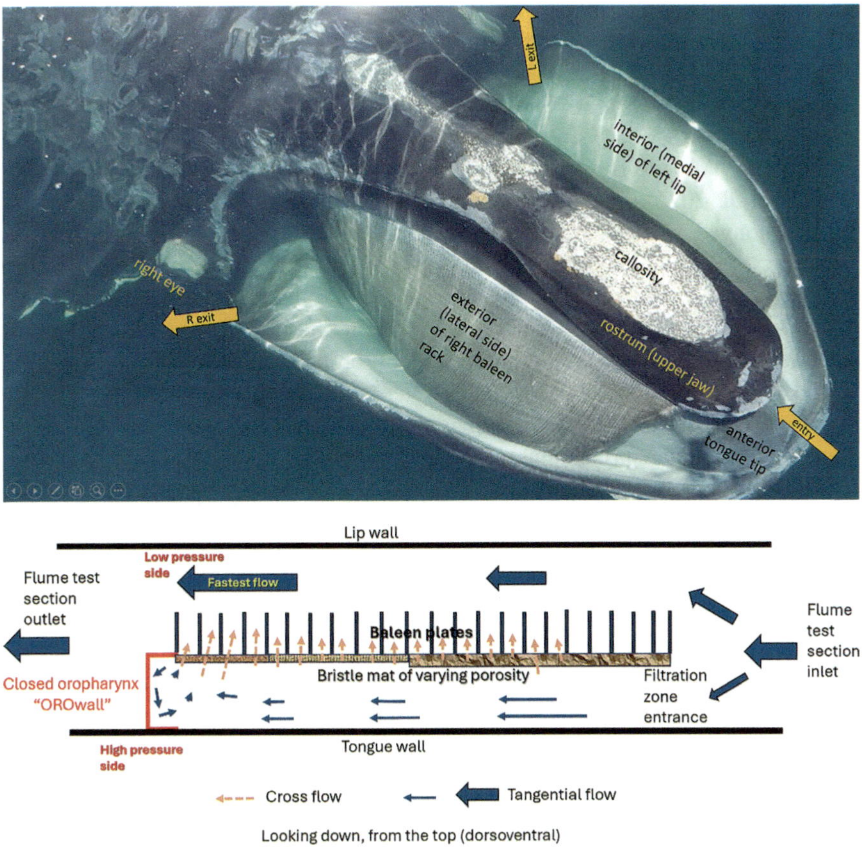

Fig. 5.7 Top: Drone-based photograph of a feeding right whale (EGNo. 4425; NMFS permit #633–1763-0). Bottom: Simplified model, again in dorsoventral view, of a representative mysticete filter highlighting its three salient components

and Potvin 2024; Potvin and Werth 2024); and (3) the closed oropharyngeal wall (or "OROwall," at the entrance to the throat) that generates the higher pressure on the medial side which "pushes" the prey/water toward the filter surface (Fig. 5.2b).

Such a three-tiered structure, encompassing a zero-porosity OROwall and porous filter surface parallel to the incoming flow, remains sufficiently robust to generate CFF regardless of filter surface porosity, baleen plate spacing, or incoming flow speed (albeit of varying capture efficiency). Interestingly, such robustness may have ensured cross-flow's appearance early in mysticete evolution. This design may also prove salutary where large gaps along a rack are accidentally created, for example, from entrapment or entanglement with fishing gear (Moore and van der Hoop 2012; van der Hoop et al. 2015, 2016; Lysiak et al. 2018).

Although much of the following analysis relates to continuous filtration in balaenids (Werth and Potvin 2016; Potvin and Werth 2017, 2024), key aspects presumably relate to all baleen filtration, including intermittent filtration in lunge-feeding groove-throated whales (rorquals; Lambertsen 1983; Orton and Brodie 1987; Simon et al. 2009, 2012; Cade et al. 2016; Werth and Ito 2017; Kahane-Rapport and Goldbogen 2018, Kahane-Rapport et al. 2020; Shadwick et al. 2019; Werth and Potvin 2024) and benthic suction feeding in gray whales (Caraveo-Patiño et al. 2007; Torres et al. 2023; Hildebrand et al. 2022; Colson et al. 2024; Webber et al. 2024). Because balaenids employ steady-state, continuous filtration over prolonged periods, with foraging bouts ranging up to 12–15 min between swallows of accumulated prey (Werth and Potvin 2016, 2024), their filtration is both more easily studied and replicated under controlled laboratory conditions, but also more relevant to commercial or other human filtering applications. In contrast, the brief and powerful intermittent filtration of rorquals and gray whales, in which a single engulfed mouthful of prey-laden water is filtered at a time, offers a dynamic spectacle but is both far more challenging to replicate and also more difficult to translate to artificial applications.

5.4.2 *The Bristle Mat*

Generally, human-designed filters are structurally and functionally more streamlined and simplistic than natural filters, given that they result from foresight and planning rather than generations of blind variation and natural selection. Due to their evolutionary heritage, natural filters may be needlessly complex, cumbersome, or not entirely efficient. On the other hand, they might also be remarkably efficient, given that they, unlike human-designed filters, may have had millions of years of honing, depending on their age.

Although natural filters may be “purpose-built,” they often involve structures co-opted from some other original purpose. Gill rakers used to collect particles for filter feeding in numerous bony and cartilaginous fish species offer a prime example. Rakers originally served to protect the respiratory system from damage to delicate branchial gill filaments caused by contact with food within the pharynx (Brainerd 2001; Sanderson 2024). The widely varied, efficient filtration systems of numerous fish, shark, and ray species display diverse forms and functions, ranging from long gill rakers of basking sharks, to ridged gill pads of mobulid (manta and devil) rays, to tiered gills of small filtering fish such as anchovies (Sanderson 2024). In contrast, whale baleen is a complexly evolved filter that features an entirely novel tissue (Werth and Potvin 2024). Its sole function, now and presumably since its initial inception, is filtration (Werth 2001). There are also key functional differences between fish and whale filtration; notably, flow across gill filters is typically unidirectional, given the opening(s) near gills, whereas water that enters the cetacean mouth must, if not swallowed, be expelled through the same opening, generally leading to more complex flow regimes (although this is not to imply that gill-related filtration in all fishes involves simple or straightforward flow; Sanderson 2024).

Not only does baleen's composite construction confer tremendous strength with flexibility, but it is also essential in creating the filter's complete form, which in turn determines overall function. In brief, each whale's baleen operates on multiple levels as a combined three-stage "metafilter," with differential components arising independently but acting cooperatively. The paired racks suspended from each side of the palate make up the overall filter, but as outlined in the previous section, each rack secondarily comprises several hundred individual plates that grow and erode independently. Water flows around and between these triangular plates. Finally, a tertiary level forms by cracking of plates, exposing the interior, hollow horn tubules, which emerge as hair-like fringes along the medial (inner, tongue-facing) side of every plate, especially at the heavily worn lower vertex of each plate, which is the plate's oldest portion. Like bristles of a broom, these fringes offer a much finer-scale porous mesh than the plates and racks alone.

Depending on the species of whale (based on its foraging ecology), these eroded fringes may be exceptionally, long, thin, and fine, or else short, rough, and coarse. Although there is variation along this continuum, baleen of most species exists at one or the other of the extreme endpoints. Fine, hair-like fringes (and very long baleen plates, reaching or exceeding 4 m from the palate to exposed tip) occur in bowhead and right whales, whereas coarse fringes and much shorter plates are found in gray whales and groove-throated rorquals such as humpback and blue whales. These stark differences are tied to diet (Tomilin 1954; Werth 2000; Croll et al. 2008; Marshall and Goldbogen 2015; Marshall and Pyenson 2019; Savoca et al. 2021). Bowhead and right whales, and the pygmy right whale (Fordyce and Marx 2013; Werth et al. 2018a), skim feed primarily on very small (1–10 mm, typically the size of rice grains) copepods and similar zooplankton; rorquals generally lunge feed on schools of larger (10–50 mm) shrimp-like krill and 8–20 cm forage fish.

Several routine filtering stresses combine to facilitate baleen's erosion into fringes (Werth et al. 2016b). These include prey accumulation, seawater flow, and mechanical abrasion from adjacent impinging oral tissues, particularly the tongue and lips (Werth 2001, 2007). Not only do these forces produce the filter's final structure, but they also contribute to the erosive loss of distal-most fringes, which often appear in whale stomach contents and feces (Werth et al. 2020). Erosive loss is balanced by an equivalent proximal growth of new plate and tubular material (Ruud 1940), such that the ever-growing baleen filter maintains a near-constant size and shape despite constant turnover from continued wear and tear of feeding (Werth et al. 2021). In this way, each plate represents several years of growth, depending on each species' growth rate and typical plate length. In rorquals, each plate includes keratin from about 3–5 years; in balaenids, plates include 7–13+ years of keratin. Erosive loss of baleen is presumably similar in all mysticetes despite differing in key details. Filtration forces and flows are greater in rorquals and gray whales than in balaenids, but are of much longer (10–30×) duration in balaenids.

Although individual fringes can capture prey, together the numerous fringes of adjacent plates, especially in balaenid whales, unite and interact to form a mesh-like mat covering the rack's entire medial (inner) surface facing the tongue. This mat has the coarse texture of a plastic-mesh kitchen scouring pad in rorquals, but in balaenids it more closely resembles a soft mat or high-pile rug of long, thin hairs (Potvin and Werth 2024). As mentioned above, balaenid fringes are not only thinner, finer, and softer than those of rorquals, but also considerably longer (averaging 9.4 vs. 4.6 cm; Werth et al. 2018a, 2020), and more than twice as dense, with 52 fringes per cm of baleen edge in balaenids vs. 21 fringes/cm in rorquals (Werth and Potvin 2024). These differences contribute to the divergent mat textures, and therefore to differing porosities and styles of filtration (Potvin and Werth 2017, 2024).

Mysticetes are, like many animals, opportunistic and resourceful foragers whose diets are frequently broad (Nemoto 1959, 1970; Slijper 1962; Gaskin 1982). However, in mysticete taxa whose diet habitually includes very small prey items (e.g., copepods), the eroded fringes are measurably longer, finer (i.e., with smaller diameter), and denser (i.e., more eroded fringes per cm of plate) than the coarser fringes of species that typically feed on schooling fish or large zooplankton such as 4–10 cm krill (Werth 2000, 2001). Whale species with fine filters include the balaenid (bowhead and right) whales and the sei whale, *Balaenoptera borealis*, of the Family Balaenopteridae (Kawamura 1974, 1980; Brodie and Vikingsson 2009; Horwood 2017; Werth et al. 2018a; Segre et al. 2021).

Because baleen grows continually at a generally predictable rate (Werth et al. 2021), and because it, like mammalian hair and nails, therefore encompasses a "snapshot" of a few years of a whale's life, baleen has proven useful in physiological research (Werth et al. 2020). This is even more valuable considering how well baleen (again, like other keratinous tissues) retains both endogenous and exogenous substances within a whale's body, most notably hormones, isotopes, and seawater contaminants (Caraveo-Patiño et al. 2007; Pomerleau et al. 2018). Therefore, baleen has become highly advantageous for biologists whose studies focus not on filtration but instead on endocrinological, isotopic, and toxicological or pollutant research.

The best way to conceive of the total baleen filter is not as a collection of serial major and minor plates, eroded fringes, and racks, but instead as a composite, multi-level structure, which can be termed *a metafilter* (Werth and Potvin 2024). This comprises multiple components with differing morphological and biomechanical properties (Pivorunas 1976; McKittrick et al. 2012; Loch et al. 2020). Each hierarchical component serves as its own kind of filter, but together they work as an integrative system. Together, the larger plates and smaller eroded fringes produce a dual-phase (i.e., plate plus fringe) filter, with a combined surface area much larger than would be contributed by the flat plates alone (Werth et al. 2018a). This surface area includes the smooth planar surfaces of baleen plates, the tattered region of eroded baleen fringes along each individual plate, and the combined dimensions of each exposed hair-like fringe. Calculations of overall filter area range from roughly 2–4 m^2 in most whales to >6 m^2 in large balaenids, if considering only the surface area of the medial mat of eroded fringe hairs (Kawamura 1974; Werth et al. 2018a). If planar surfaces of all plates and dimensions of individual fringes are also included, the total surface area ranges from ~30 m^2 in smaller whales to >150 m^2 in large balaenids (Werth et al. 2018a).

5.4.3 The Oropharyngeal Wall

A seldom recognized feature of continuous filtration in balaenid whales is the presence of the closed oropharyngeal wall (OROwall), which, besides preventing swallowed seawater during feeding, creates the crucial higher pressure area next to the tongue that turns flow onto and then through the baleen fringe mat comprising the filtration surface (see Figs. 5.2b and 5.7b). This simple and robust mechanism does not require the presence of steady vorticity along a well-defined structure, as with gill rakers (Witkop et al. 2023). Generally, this finite and low porosity balaenid filter, working in tandem with the closed and therefore zero-porosity adjacent OROwall, drives sufficiently strong cross-flows over a significant area of the filter surface to coat the latter with a thin layer of captured particles. The surface coat is particularly important for easy retrieval of the captured prey via scraping by the tongue, head shaking, or flushing via backwash flow (Werth 2001). By directing flow, the OROwall also accelerates flow to reduce pressure between the wall and the baleen filter's outer (labial) surface. As further discussed below in Sect. 5.6, leaky OROwalls and high-porosity filtration surfaces might permit more prey and particles to escape but may also generate stronger cross-flows in the filter's posterior region.

As noted above, the brief yet powerful sporadic bouts of intermittent filtration in rorquals and gray whales pose a difficult challenge to recreate under controlled laboratory conditions, and (given their dynamically variable nature) to describe in fixed, steady terms. Nonetheless, it is clear, unsurprisingly, that their filtration operates via a different strategy, with all flow originating from the expanded center of the oral cavity, medial to the baleen (i.e., on its lingual side), where a single mouthful of engulfed, pre-laden water is briefly held while the oral cavity is compressed via contraction of musculature, primarily in the oral floor, to expel water while captured prey items are retained by the baleen filter before they are swallowed (Werth 2000; Werth and Potvin 2024). The OROwall, which plays a key role in continuous balaenid filtration, is unimportant in intermittent mysticete filtration. Rorquals and gray whales likewise lack crucial anatomical structures, such as the large lower lip and gutter-like orolabial sulcus it creates outside (= lateral to) each baleen rack, and the subrostral gap through which prey-laden water enters the anterior of the mouth (Fig. 5.7). The ventral groove blubber of rorquals, and corresponding thick gular musculature of gray whales, contracts to expel water laterally through the baleen filter, presumably in a largely throughput scheme, although the potential extent to which filtration in these non-balaenid mysticetes is characterized by tangential filtration is unclear (Werth and Ito 2017). In any case, water flows rapidly through the filter in a different direction (at least partly), and through a notably different means (Werth and Potvin 2024).

To conclude, baleen filtration acts on micro- to macroscopic scales, with each component contributing its own function to the entire filtration system (Pivorunas 1976; Sekiguchi et al. 1992; Vandenberg et al. 2023). Depending on the size of filtered items (1-mm copepods to 20-cm fish) and parameters of filtration (e.g., flow speed and direction), different component elements are involved. Additionally, adjacent oral structures influence flow through and within the mouth, and hence play important roles in baleen filtration. These include the tongue and lips (medial and lateral to the baleen racks, respectively), and the oropharyngeal opening, posterior to the racks, where the mouth transitions to the pharyngeal tube.

5.5 Baleen's Filtering Process Variations

In addition to the baleen filter's material and structural complexity, a third key element of variability has been made apparent by controlled laboratory experiments on baleen tissue (e.g., Werth 2013; Werth and Potvin 2016; Potvin and Werth 2017, 2024), confirming that processes of baleen filtration vary according to operating conditions. This has been corroborated by aerial drone recordings (Werth et al. 2019b) and data from whales with archival and telemetric biologging data tags temporarily affixed to their bodies (Goldbogen et al. 2006, 2011, 2012a, b, 2015, 2017a). Tag data reveal that whales control body movements (roll, pitch, yaw), open and close the mouth for specific durations, swim at certain speeds, and feed at various depths during filtration, all of which have implications for how the baleen filter operates (Goldbogen et al. 2007, 2013, 2017b, 2023).

Most notably, these parameters involve flow direction, speed, and volume. Further, because of baleen's structural complexity (Werth 2017), parameters vary by location within the filter (anteroposteriorly and mediolaterally within a single rack and dorsoventrally along each plate (Potvin and Werth 2024). The end result is that baleen's porosity, a ratio of mat "see-through" void area to overall filter surface area (Matyka et al. 2008; Vogel 2013; Potvin and Werth 2024), is flow-dependent and region-dependent (Fig. 5.4; Werth 2013; Potvin and Werth 2024). Unlike a simple kitchen colander or other strainer, baleen does not demonstrate fixed porosity (Werth 2019). Rather, its porosity depends on the oral region of the rack being filtered—which appears to shift from the beginning to end of each bout of filtration—as well as, crucially, the direction of water flow relative to the plates. In short, as will be explained, much intraoral filtration via baleen appears to be primarily tangential rather than throughput.

Perhaps less dramatic than tagged whales but equally instructive are controlled laboratory experiments using either actual baleen tissue (Mayo et al. 2001; Werth and Potvin 2016; Potvin and Werth 2024) or simulated tissue (Jensen et al. 2017; Vandenberg et al. 2023) or models (Sect. 5.6 below) in circulating flow tanks or flumes. Because of the large size of actual baleen plates, small sections are usually cut from plates and assembled to create "mini-racks," although giant flow tanks used to test full-sized vessel hulls have also been used (Werth et al. 2018b). Baleen can then be observed as it responds to moving water, either as water circulates past stationary baleen (Werth 2013) or as baleen is towed or otherwise propelled through water (Werth et al. 2018b). In this way, changes in baleen form and "posture," such as bending and spacing between plates and fringes, can be observed, recorded, and analyzed. Additionally, dyes or other materials can be added to water to better visualize hydrodynamics of baleen-water interactions. Small particles such as plastic beads or genuine prey items (e.g., copepods or other zooplankton; Fields and Yen 1997; Werth 2012) can be introduced to record actual filtration events via cameras placed directly underwater in the flow stream or by outside viewing ports, ideally with ruled grids or other scales to indicate and quantify flow regimes (Werth 2013). One way to measure such interactions is by analyzing rates at which baleen plates and fringes capture plastic or prey particles in the water stream (Werth et al. 2024); another is to determine the extent to which baleen porosity changes as plates and fringes move within the flow (Fig. 5.5). Note that "capture" typically means that particles are caught by the baleen filter (Shimeta and Jumars 1991; Potvin and Werth 2024; Werth et al. 2024). Capture could also include ingested particles flowing directly into the oropharynx without contacting the filter. The distance between the free tips of eroded fringes or of another location along fringes (e.g., 1 cm or 10 cm from where it attaches to a plate), defined as *interfringe distance* (IFD), has provided valuable data about porosity and other filter parameters as variables such as flow speed are experimentally manipulated (Werth 2013, 2019; Potvin and Werth 2024; Werth and Potvin 2024).

Experiments show that the direction of water flow, determined by the angle of attack of plate orientation relative to incurrent flow, alters baleen's filtration ability, as revealed by particle and pigment movement and capture (Werth 2013). Recent research (Potvin and Werth 2024; Werth and Potvin 2024) has shown that flow direction varies throughout baleen filtration (1) by whale species, as well as (2) by location within the rack, and (3) throughout the overall time course of an individual bout of filtration. Our latest results are presented in the next section (5.6), following general considerations of how baleen might conceivably inspire biomimetic filters.

In addition to variation in filtration according to the flow angle of attack (the angle at which the baleen plates meet oncoming flow, hence affecting lift and drag), flow speed also leads to differences in baleen's filtering abilities as indicated by mat porosity (Fig. 5.5), particle capture rate (Fig. 5.6b), and IFD, which is a major determinant of porosity (Fig. 5.8; Werth 2013; Potvin and Werth 2024). As flow

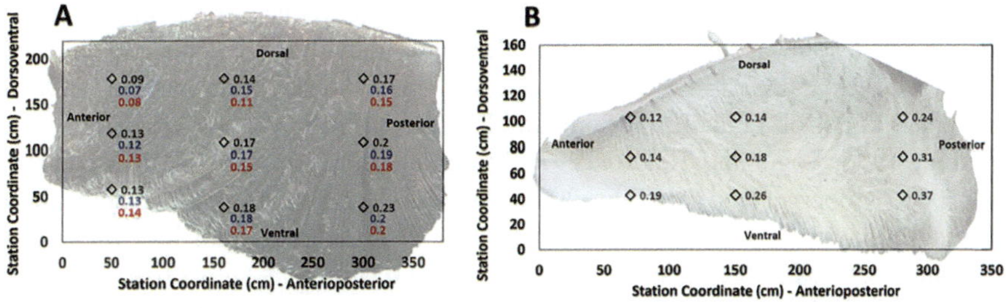

Fig. 5.8 Regional porosity of typical bowhead (**a**) and fin (**b**) whale baleen mats (based on data from Potvin and Werth 2024). The diamonds indicate approximate locations on the rack, and labels correspond to porosities measured at flume speeds of 0.75 m/s (black print), 0.50 m/s (blue), and 0.25 m/s (red)

speed and volume flow rate increase, fringes are generally pushed farther apart, leading to higher IFD (Fig. 5.5b). As noted above, flow speed and volume vary by location and timing within a single-filtration event, especially in rorqual lunge feeding but also in balaenid skimming (Mayo and Marx 1990; van der Hoop et al. 2019) and gray whale benthic suction feeding (Werth 2019). Experiments have revealed the crucial importance of how free baleen fringes interact between adjacent plates (Werth 2013; Werth et al. 2018b). With single plates tested in a flume, it appears that higher flows, particularly above 60 cm/s, lead to diminished particle capture and higher IFD, meaning the filter is more porous (Werth 2013). However, with multiple plates tested together in a "mini-rack" formation (Fig. 5.5), baleen fringes swirling in the flow actually overlap and intermingle to form a denser, less porous mat (Potvin and Werth 2017, 2024). Apart from the overall lesson that basic experimental design is a key consideration for experimental simulations, the more pertinent conclusion is that IFD is a variable parameter that depends on interaction between multiple plates. Each free, eroded horn tube fringe erupts from a single keratinous plate, but fringes from multiple plates work together to create the structurally and functionally important fringe mat (Werth 2013; Potvin and Werth 2017, 2024; Werth et al. 2018a). Porosity and other measures of filtration depend not only on baleen's physical arrangement and dimensions (e.g., length and density of fringes and plates), but also on flow parameters, notably the speed and direction of water flow.

Ongoing experimentation reveals that baleen filtration, particularly in balaenids, apparently operates largely via tangential filtration (Potvin and Werth 2024). Traditional accounts depict baleen as a sieve (Slijper 1962; Pivorunas 1979; Werth 2001) which works merely via separating prey by size with a passive strainer (Rubenstein and Koehl 1977). According to this dead-end or throughput scheme, prey-laden water flows directly into the filter at a more-or-less perpendicular angle, thereby collecting and retaining items larger than the gaps (pores) of the filter, which in this case consist of gaps between baleen plates and/or between fringes (Pivorunas 1976; LaBarbera 1984). Finally, a clear advantage of CFF over TPF may be allowing the use of large filters without incurring dramatic energy losses to drag (whales) or pumping power (engineered systems).

Incurrent flow striking the filter at a low angle reduces or even precludes problems of solid accumulation, filter clogging, and filter cleaning (Baker et al. 1985; Murkes and Carlsson 1988; Lu and Ju 1989; Mei 1992; Lu and Ju 1989; Lu et al. 1993; Belfort et al. 1994; Song and Elimelech 1995; Vogel 1996; Zeman and Zydney 1996; Vogel and Todaro 1997; Bott et al. 2000; Brainerd 2001; Sibanda et al. 2001; Ripperger and Altmann 2002; Humphries 2009; Peng and Dabiri 2009; Espinosa-Gayosso et al. 2012; Makabe et al. 2021). Apart from less filter wear or damage, separation of solid and fluid is more efficient with cross-flow than throughput filtration in terms of delayed or absent clogging, and thus more fluid processed per unit time (Potvin and Werth 2024). For whale feeding, all these benefits would apply (Potvin and Werth 2017), with the further benefit that the solid retentate—that is, the collected prey items, which are the desired outcome of whale filtration, unlike filtration of commercial beverages or other products—accumulate at or near the excurrent flow output (Werth and Potvin 2024). In the case of the whale mouth, this is right at the entrance of the oropharyngeal opening, where prey are swallowed. Accumulation of a bulk slurry of small prey items that have been efficiently separated from undesirable seawater is an optimal form of filtration (Zhu et al. 2020a, b, 2021, 2023). Cross-flow filtration feeding has been well-documented in bony fish (Langeland and Nost 1995; Goodrich et al. 2000; Brainerd 2001; Cheer et al. 2001, 2012; Sanderson et al. 2001, 2016; Smith and Sanderson 2008; Paig-Tran et al. 2011; Witkop et al. 2023), sharks (Sims 2008; Motta et al. 2010; Wegner 2015), and rays (Paig-Tran et al. 2013; Paig-Tran and Summers 2014; Divi et al. 2018).

The very large (up to 1-m high and wide and 4+ m long), firm muscular tongue of all balaenid species also channels flow, dividing it and directing it into either baleen rack (Werth 2000; Werth and Crompton 2023). Lambertsen et al. (1989) speculated that this intraoral morphology might create a Bernoulli effect, as flow speed increases due to narrowing of the pipe-like path, and that this might further generate a Venturi effect to pull water from the center of the mouth, along the tongue, through baleen. These hydrodynamic effects could also slightly diminish pressures where water enters at the anterior of the mouth, perhaps not sufficiently to generate truly notable subambient suction pressures, but at least enough to obviate an anterior compressive bow wave that might physically disperse small prey and/or warn them of an approaching whale (Fish et al. 2013). Flow experiments (Werth 2004, 2013; Werth and Sformo 2020) using pressure transducers and videorecording of particles and prey (Werth 2012) have confirmed that these limited yet real pressure effects in laboratory settings (Fig. 5.7), and tantalizing photographs and video recordings of whales foraging at sea (Werth and Potvin 2016; Werth and Ito 2017; Werth and Sformo 2020; Werth and Crompton 2023; Werth and Marshall 2023) have likewise supported conclusions of Bernoulli and Venturi effects.

Although cross-flow filtration has been best considered in balaenid (bowhead and right) whales, it conceivably applies to all mysticetes (Potvin and Werth 2017; Werth and Ito 2017), including intermittently feeding rorqual and gray whales, although likely to a lesser extent. We must note that most recent accounts of baleen filtration (e.g., Hamann and Blanke 2022; Vandenberg et al. 2023) rely on the notion of throughput filtration in rorquals. However, ongoing flow experiments suggest that at least some flow during rorqual feeding, particularly during latter stages of filtration and water expulsion, involves varying flow pathways that might indicate some cross-flow rather than throughput filtration (Werth and Potvin 2016), and we expect this to be addressed in future publications. Intraoral helical vortices like those envisioned within the balaenid mouth (Potvin and Werth 2017) could form in the mouth of all whales during filtration (Zhu et al. 2023). This is especially true considering the growing realization that multiple distinct phases of flow are apparent within a single mysticete feeding event (Werth and Potvin 2024). Vortex formation could be important for baleen-inspired filters used to remove microplastic particles and other small debris items from water, as will be discussed.

5.6 Baleen Biomimicry and Application to Microplastic Removal

5.6.1 General Biomimicry

As the above sections make clear, baleen's unique structure and function, both more complex than traditionally envisioned, enable highly effective filtration for suspension feeding in toothless whales. Although this may seem to be an odd and small-niche market for human designers to show interest in, aspects of baleen filtration likely apply to other biological filters. Ideally, these can better enable engineers and industrial designers to create artificial filters that are stronger and longer-lasting, as well as more efficient (e.g., less prone to clogging).

Scientists who study filter feeding in varied fishes (bony and cartilaginous, tiny and gigantic, marine and freshwater) have documented nature's variously precise filtering structures, and their equally unique flow mechanisms, that could inspire creation of novel artificial filters. These include flattened, elongated, and stratified gill features used for ricochetal filtration in manta rays (Divi et al. 2018; Kahane-Rapport et al. 2025) and various mechanisms including vortical cross-step filtration, inertial lift, and shear-induced migration in shad and other schooling forage fish (Sanderson et al. 2001, 2016; Witkop et al. 2023). Other filtration studies currently underway, and as yet unpublished, involve close focus on the gill rakers of basking sharks, pores of sponges, and appendages of zooplankton, in addition to ongoing work on baleen filtration.

There are many potential commercial and industrial avenues for biomimetic filtration and separation technology at meso- and microfluidic scales, including cleaning of potable water supplies and expansion of such supplies via desalination of seawater, wastewater treatment, crop irrigation, and oil-spill remediation. Other uses range from obvious, long-standing processes such as filtration of varied beverages and foods, to potential removal of environmental microplastics from fresh and seawater (Reisser et al. 2015; Auta et al. 2017; Law 2017; MacLeod et al. 2021; Harris et al. 2023; Mutuku et al. 2024)—the latter currently more a dream than a reality, but definitely a looming challenge on the horizon.

Before outlining ways that baleen might aid biomimetically inspired filtration, it is worth pointing out that biomimicry is a two-way street. An opposite course, where technology teaches us about nature, may be particularly fruitful. One can just as easily seek ideas and motivation from man-made filters to better understand exactly how baleen works in vivo, something that remains poorly understood due to numerous manifest challenges. Specifically, creation of artificially 3D-printed baleen models (Mironov et al. 2003; Jensen et al. 2017) with unrealistically high or low stiffness or other material properties (Mironov et al. 2003; Yeong et al. 2004) can help researchers to figure out what does and does not happen, or could potentially happen, in actual whale mouths. This approach, in which biology learns from novel technological applications and findings in a positive feedback loop, could be called reverse or reciprocal biomimetics (Werth 2019; Braun et al. 2020).

Fig. 5.9 Three-dimensional scanning and 3D printing of baleen models with varied biomechanical or material properties (e.g., stiffness) can aid in studying baleen function as well as creation of naturally inspired biomimetic filters

This is similar to the dichotomy between the "bottom–up" or "biology push" versus the "top–down" or "technology pull" approaches (Speck and Speck 2008), the former commencing with basic science research and the latter with a technological problem to be addressed (Knippers and Speck 2012). The many prominent hurdles that limit research on whales (legal, logistical, and fiscal) are impediments to carrying out basic experiments beyond observing structures in or testing tissues from dead whales. Nonetheless, testing of 3-D printed models of whale baleen with exaggerated, unrealistic material properties (Fig. 5.9) can help to reveal details of baleen function (Werth 2019). This reversal of conventional biomimetics, in combination with a spiraling, ratcheting strategy of both conventional and reciprocal biomimetics (Werth and Potvin 2024), holds great promise. As we better understand natural systems, we can then design and manufacture more effective synthetic filters.

It must also be emphasized that biomimetic imitation of baleen might focus solely on filtering function and might not be structural in nature. Attempts to produce detailed 3D-printed copies of actual whale baleen (Fig. 5.9) have been reasonably, although not entirely, successful for several reasons. First, the hair-like fringe mat is difficult both to scan and then to create with additive printing or sintering. Second, even when such details can be created, they are often connected, during production, to a background matrix, and it is difficult to free the delicate fringes from the attached matrix. Attempts to remove them via direct cutting, melting, or water jetting have proven less than satisfactory (Werth 2019). Third, attempts to replicate baleen's composite interior structure have also fallen short relative to properties of the biological tissue.

A venture called Baleen Filters (baleen.com), founded in 1998 but apparently now defunct, was a proprietary limited private company based in Adelaide, Australia, that designed, built, and sold industrial wastewater filtration systems. Unlike traditional municipal systems built to handle wastewater sludge, which typically rely on sequential flotation and settling pools, the Baleen Filters system was also a multi-stage system yet depended on rinsing water sprays. The water sprayed on filters of sequentially diminishing porosity both wiped the accumulated filtrate from collection screens and swept it away for collection and removal. As the company advertised, this scheme offered a cost-effective separation system that was simple and environmentally friendly, occupied less space than typical systems, used little energy, and worked efficiently. The company touted its product as working not only for municipal wastewater filtration, but also for food and beverage processing applications in various sizes and configurations. It is not clear why the company is no longer operating. The general process was well-devised. It is also unclear why the outfit was named "Baleen" apart from the very general fact that whale feeding uses filtration. There seems to have been neither special inspiration from nor any effort to emulate baleen tissue or processes by which any mysticete species separate retained prey from expelled water.

Despite the apparent shortcomings of this commercial venture, it would seem preferable to attempt to create biomimetic filters that mimic baleen's superior filtering abilities without trying to replicate its precise structure, which would be difficult and unnecessary. As explained in the previous section, the authors of this chapter are currently engaged in flow experiments with models that seek to copy fundamental principles of mysticete intraoral filtration without using actual baleen. Instead, plastic structures reproduce important outlines of the oral space while perforated or otherwise porous structures replicate essential processes of tangential filtration fulfilled by baleen, particularly in balaenid whales that use continuous filtration.

5.6.2 Microplastics and Recent Results

Large-scale removal of micro- and nanoplastic debris (<5 mm and <1 μm, respectively) from oceans, lakes, and rivers poses a major technological challenge (Baraza et al. 2022; Jambeck et al. 2015; Werth et al. 2024). Recent findings bring dawning awareness of both the magnitude of this environmental problem and the adverse consequences for organisms, virtually all of which experience contamination. A recent study (Ten Hietbrink et al. 2025) estimated that the North Atlantic Ocean alone holds at least 27 million metric tons of microplastics—more than the combined weight of all wild land mammals. Micro- and nanoplastic accumulation have been documented in bodies of humans and other terrestrial and aquatic species (Bucci et al. 2020; Vo and Pham 2021; Jandang et al. 2024). Nanoplastic particles are small enough to cross cell walls, and they easily accumulate in trophic webs (Auta et al. 2017). Precise health effects remain undetermined, but plastics have been shown to cause organ damage and potential poisoning via harmful toxins (Cole et al. 2011; Hwang et al. 2020; Prata et al. 2020; Nihart et al. 2025; Marfella et al. 2024; Huang et al. 2025; Makwana et al. 2025).

The mysticete oral filter, particularly of continuously filtering balaenid (right and bowhead) whales (Werth and Sformo 2020), offers an obvious bio-inspired solution to this problem, given the typical 0.5–15 mm size of whales' planktonic prey (Werth 2012). Over the 25 million years of baleen evolution, whales' filter feeding strategy and oral anatomy have offered many advantages, including filtration area scalability (to several square meters; Werth et al. 2018a), modulation via tongue position (Werth and Crompton 2023), and ease of captured particulate removal (Werth 2001), obviating the need for filter replacement. Moreover, the balaenid filtration system represents a low-power operation involving minimal contraction of oral musculature as well as slow, fluke-based locomotion through prey patches (Werth and Potvin 2016; Potvin and Werth 2017). Of particular biomimetic interest here are the mysticete CFF architectures for microplastic removal from ephemeral and concentrated sources, such as laundry wastewater and rainstorm runoff (Jambeck et al. 2015), as well as from continuous but unconcentrated sources as found in streams and rivers (Hurley et al. 2018; Brown et al. 2023). Another compelling application is the use of these bio-inspired architectures in small-scale detectors used in the laboratory or in the field (Morales-Caselles et al. 2021).

Should baleen-inspired biomimetic filtration be adopted to remove microplastics from the sea, it could be deployed at coastal locations (as with the Australia-based Baleen Filters sewage treatment) or in deeper water, including pelagic sites far out to sea (Zhao et al. 2022). Apart from major oceanic gyres that concentrate plastics in the central North Pacific and North Atlantic Oceans (Gibbs et al. 2019), other "hotspots" for plastic pollution include the Mediterranean and Yellow Seas (Guerrini et al. 2019; Gregorietti et al. 2021), and in general heavily polluted coastal zones of South and East Asia. Another logical place to deploy biomimetic filters would be at river mouths, where high concentrations of microplastics enter the ocean, but these and other solid pollutants could foul and clog the filters, overwhelming them. Sadly, plastic debris is now widely distributed throughout all major ocean regions due to constant entry streams as well as rapid transport, long lifespan, and negligible removal (GESAMP 2015, 2019).

One might wisely ask if widespread use of biomimetically inspired filtration to alleviate the problem of aquatic microplastic pollution could conceivably, and counterintuitively, create another predicament, namely of limiting the food supply of whales and other aquatic organisms in an ironic twist (Werth et al. 2024). Because this and other filtration mechanisms are based on porosity, any items exceeding the pore size would be removed from water, including not only microplastics and related anthropogenic debris but also phytoplankton and algae, zooplankton, fish eggs and larvae, and organic detritus and other large nutrients (Reisser et al. 2015; Law 2017). Perhaps strategies to deploy biomimetic filters on natural water bodies and on a large scale should be limited or considered carefully, at least initially (Gallo et al. 2018; Morales-Caselles et al. 2021; Brown et al. 2023). However, these risks are likely minimal, especially when one considers that filtration is one of the oldest and most successful feeding methods for aquatic organisms of all kinds, and that natural bodies of water, both sea- and freshwater, are already fully stocked with sessile and mobile creatures, both planktonic and nektonic (i.e., free swimming) that derive sustenance via filtration (Hamann and Blanke 2022), including vast schools of sardines, anchovies, and other small fish, not to mention countless arthropods, poriferans, cnidarians, echinoderms, molluscs, vertebrates, and other chordates; both solitary and colonial, that collectively filter many nutrients from water, but which are poorly suited to handle microplastic pollution, or which die or suffer other health consequences from filtering it (Fossi et al. 2020). Even freshwater streams and brackish bays are regularly cleaned by filtering mussels or copepods. As ecologists have noted, baleen whales pose no risk to other filter feeders (Choy et al. 2019; Corkeron 2025); rather, they spread nutrients by vertical and horizontal transport (Pearson et al. 2023).

Herein we describe our initial attempts at constructing and testing a novel, balaenid-inspired CFF system with a view of scaling it to applications for the environmental sciences as well as for use in the laboratory. Most discussion here will center on variations, within a single filter design, of the five most important architectural elements, and their effects on particle capture efficiency. These elements include oropharyngeal wall (OROwall) depth, filtration surface porosity, bristle curtain structure, and baleen rack angle to the incoming flow. The importance of these factors has been studied in custom-built water tunnels (AKA flow tanks or flumes; Vogel and LaBarbera 1978), as well as explored in Computational Fluid Dynamics simulations as described below.

5.6.3 Capture Efficiency Scaling

A prime metric for filtration design is the *capture efficiency* (ε), defined as the ratio of the number n_{cap} of particles captured by a filter, after travel over "integration" distance D_T, a measure analogous to the length of a prey field swept by a feeding balaenid, divided by the number n_{inj} of injected particles entering the "filtration zone":

$$\varepsilon\left(D_T\right) = \frac{n_{\text{cap}}\left(D_T\right)}{n_{\text{inj}}\left(D_T\right)} \tag{5.1}$$

Note that this efficiency also depends on the relative void size ratio (SR) of the mean mat void diameter-to-the mean particle diameter, as well as their corresponding spread (standard deviation) vs. diameter. Such metrics should allow efficiency predictions upon scaling to larger or smaller filter surface porosity.

For whales, the filtration zone corresponds to the volume bordered laterally by the medial sides of the baleen racks, and dorsoventrally by the palate and tongue floor (Fig. 5.7). Such definition excludes the volume of water (and prey) contained in the gutter-like labial (= next to the lip) channel defined by the distal side of each rack and medial side of each lip (Fig. 5.7). In the case of a microplastic removal device, and in reference to Fig. 5.7, the zone's volume would correspond to the cuboid defined by the filtration surface and "tongue" wall, the OROwall, and the "entrance." Cases in which all particles entering the zone are captured by the filter will yield value ~1, and where all particles pass through, one finds $\varepsilon \sim 0$.

Distance D_T is connected to a whale's feeding duration T while swimming through a patch and calculated by

$$D_T = V_{\text{zone}} T \tag{5.2}$$

Parameter V_{zone} corresponds to the speed of the water-prey slurry entering the filtration zone. In other words, D_T parametrizes the actual number of particles moving along a streamline that enters the filtration zone and interacts with the filter, either being captured or passing through. Thus, more particles are thus bound to interact if the integration duration (T) or flow speed (V_{zone}) are increased (with the other factor held constant). Moreover, and as explained below, flow speed V_{zone} depends on the porosity of the filter (e.g., lower value at lower filter surface porosity and higher value at higher filter surface porosity) and as such will not generally be equal to the whale's swimming speed through the patch, except when filter porosity is very large.

Transposing Eq. (5.1) in a form fit for scaling is achieved with the use of filter area HL (Fig. 5.10), slurry volume HD_TW (with H, L and W, as filter height, length and width, respectively), captured particle surface density σ_{cap} (units m^{-2}), and the volumetric density ρ_{inj} of injected particles as follows:

$$\varepsilon(D_T) = \frac{\sigma_{\text{cap}}(D_T)\,HL}{\rho_{\text{inj}} H D_T W} = \frac{\sigma_{\text{cap}}(D_T)\,L}{\rho_{\text{inj}} W D_T} \tag{5.3}$$

The use of densities is particularly useful in the comparative assessment of filter architecture details vs. environmental conditions. In cases where the volumetric density of particles entering the filtration zone is uniform over "feeding" (or particle encounter) duration T, the value of ρ_{inj} becomes an independent variable characterized by environmental conditions, thereby allowing the factorization of D_T out of n_{inj}, following the definition $n_{\text{inj}}(D_T) \equiv \rho_{\text{inj}}\, HWD_T$ and the D_T-independent

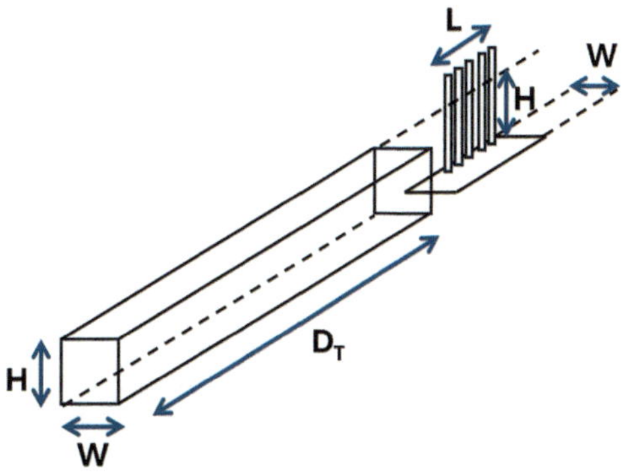

Fig. 5.10 Schematics of the volume of feed stream about to interact with a filter over integration distance D_T. The filter sketch is shown with its "baleen" plates but without its low-porosity mat

density ρ_{inj}. (Generally, and in non-uniform prey patches, density ρ_{inj} would remain D_T-dependent.) On the other hand, and in contrast to actual filter shapes and geometries found in living animals (Fig. 5.7a), Eq. (5.3) works under the assumption of a filtration volume shaped more-or-less like a rectangular cuboid (Figs. 5.7b and 5.10).

Equation (5.3) lends itself to scaling predictions, for example in cases where the integration distance (D_T) and injected densities (ρ_{inj}) are the same for different whale body or environmental microplastic collector size scales. It would follow that filters that are isometrically scaled smaller by half would collect prey/particles with the same (capture) efficiency as the larger filters, given the ratio $\varepsilon_{\text{small}}/\varepsilon_{\text{large}} = \sigma_{\text{cap small}}/\sigma_{\text{cap large}}$ (as derived from Eq. 5.3). In other words, finding the same particle densities σ_{cap} over the filters of both collectors would correspond to the same efficiency. On the other hand, the same filter collecting over different distances D_T (say $D_T^{(1)} > D_T^{(2)}$) may see a leveling off, decrease, or increase in efficiency with increasing values of D_T, depending on the filter's collection modalities (given the proportionality $\varepsilon \propto \sigma_{\text{cap}}(D_T)/D_T$). For example, and with an "empty" filter operating where ρ_{inj} is fixed, the efficiency should remain constant with the capture density $\sigma_{\text{cap}}(D_T)$ expected to increase at the same rate as D_T. In the case of extreme leaky filters, on the other hand, one would expect $\sigma_{\text{cap}}(D_T) \sim 0$ and a continually decreasing efficiency over time as $\varepsilon \sim \sigma_{\text{cap}}/D_T \sim$ "*small number*"/D_T. With highly efficient collectors, accumulation of particles over larger D_T ends up blocking more of the filter's pores, with the results that $\varepsilon \to 1$. As further particles get collected, there is a point where caking occurs, the pressure in the filter zone increases, and the entering flow comes to a full stop ($V_{\text{zone}} \sim 0$) and $D_T \sim 0$, even when the collection time continues to increase. Finally, and where the relative void size ratio SR is less than unity, scaling to larger or smaller LH filters (Fig. 5.10) at same SR should yield similar efficiencies at similar values of D_T/L and ρ_{inj}—that is, where the target particles/prey are larger in the field than in the laboratory.

In the case of whales, another interesting question involves ways that different whales meet the same-D_T constraint, either by swimming faster and/or taking more time to collect prey. Fortunately, answers to this question and all others above can be found in flume settings where the explicit D_T-dependence of the efficiency itself can be directly measured.

5.6.4 Flow Tank Construction

Two custom-built flow tanks based on a design by Vogel and LaBarbera (1978) were constructed for the explicit study of baleen hydrodynamics at Hampden-Sydney College (HSC), and for study of subscale filter models at Saint Louis University (SLU) (Fig. 5.11).

Both the HSC and SLU flumes were built with 6-inch diameter PVC piping supported by a frame of standard 2″ × 4″ wooden boards. The test sections (HSC dimensions: 70 cm length, cross-sectional area 345 cm^2, volume 14.8 l;

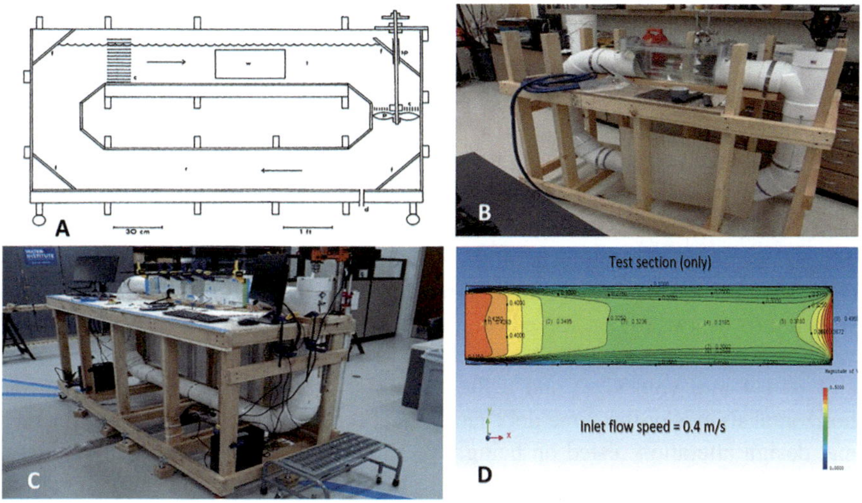

Fig. 5.11 Custom-built flow tanks for balaenid filtration study: per early design of Vogel and LaBarbera (1978) (**a**), at Hampden-Sydney College (**b**) and Saint Louis University (**c**), and Computational Fluid Dynamics simulation of the SLU test section without the filter (**d**)

SLU dimensions: 116 cm, 323 cm^2), were made of shaped 0.25″-thick plexiglass sheets. Both flumes use the same type of drill press motor capable of five drill RPM's (740, 1100, 1530, 2100, or 3140 RPM with SLU's WEN drill press motor Model 4208T), to drive a 0.61 m-long shaft capped with either 3- or 4″ diameter impellers to double the number of flow speed options. Finally, both flumes include turbulence diffusers to dissipate impeller-generated and pipe-based secondary flow turbulence; the SLU flow tank uses standard nylon-based door mesh with square voids of 1.4-mm width, supported by large-bore (2 cm) straws installed at the test section's inlet and outlet. With the exception of the impellers, all materials, tools, and accessories were procured at local hardware stores. Given that the filtration flows in a whale's filter (Fig. 5.7) are channeled flows rather than open flows, as with aircraft, wall proximity was an intentionally built-in rather than avoided feature. In the same vein and given the varying widths of the balaenid filter along the anteroposterior axis, flume flow uniformity was not expected to occur across the filter, as confirmed in CFD simulations (Fig. 5.11d). CFD simulations were carried out with the commercial CFD software package SC/Tetra (Hexagon AB), using RANS, SST k-ω turbulence modeling, no-slip boundary conditions on the longitudinal walls (and free-slip on the surface) with over 20 million tetrahedral mesh elements. Further details are provided by Kennedy (2021).

5.6.5 Filter Architecture

Our bio-inspired microplastic removal system incorporates, albeit in modelled form, all elements found in a whale's oral apparatus, namely, a structural rack of blade-like plates, a filtration mat surface, and a closed oropharyngeal wall (Potvin and Werth 2017, 2024) (Fig. 5.12). This architecture gives rise to prototypes defined by their rack angle (with respect to the incoming flow), filter surface porosity, and OROwall dimensions. Given the curved comb-like plate structure observed in balaenid whales (Fig. 5.7), flume racks incorporating three angles (0°, 9°, and 13°) off the incoming stream direction were considered (Fig. 5.12). In contrast, the bristle/fringe mats of varying porosities were built by combining stacks of standard nylon door mesh (1.4 × 1.4 mm void size) and a flat curtain of "bristle" built by removing the horizontal strands of the same door mesh sheet (Fig. 5.12). The motivation for such mat design alteration rested on being able to strictly control the porosity value of the filtration surface. This was achieved by stacking layers of door mesh at various relative angles (Fig. 5.12). As explained below, a curtain of bristle was added to the stack to enhance particle capture, particularly on the anterior end of the filter. The bristle curtain's far greater porosity than the underlying mesh stacks make the latter the main driver of the overall porosity of the bristle curtain—mesh stack pairing.

5.6.6 Results and Lessons Learned

Capture efficiency of the balaenid-inspired filter shown in Fig. 5.12 (and variations) was obtained from photogrammetric analysis of captured 0.8–1.0 mm polystyrene particles released in the flume prior to capture by a filter prototype (Fig. 5.13). Typically, and at any rack angle with respect to flow, as porosity of the filter surface (bristle curtain plus mesh stack) increased, there was greater clustering of captured particles on the rack's posterior end (i.e., closer to the OROwall). As confirmed by CFD simulations, this clustering is the result of significantly faster cross-flows through the mesh and "baleen" plate channels nearer the OROwall (Fig. 5.13d). Flume tests showed that reducing mesh stack porosity to 10%–20% results in more particles being captured over the filter's anterior end, but in amounts that still show a captured density increasing toward the OROwall.

Another important finding involves the bristle curtain's role in enhancing capture via particle entanglement with the bristle mat. As recently discussed by Werth et al. (2024), particles do become snagged by bristles all over a rack. What transpires here is the importance of such snagging over regions of the filter surface characterized by reduced suction (and weaker cross-flows), as is the case over the anterior end in our prototypes. Snagging is important also for canceling the effect of particle drag in the anteroposterior flow (Potvin et al. 2009, 2012, 2020, 2021). Most importantly, this snagging effect is important enough to create a near-uniform particle capture

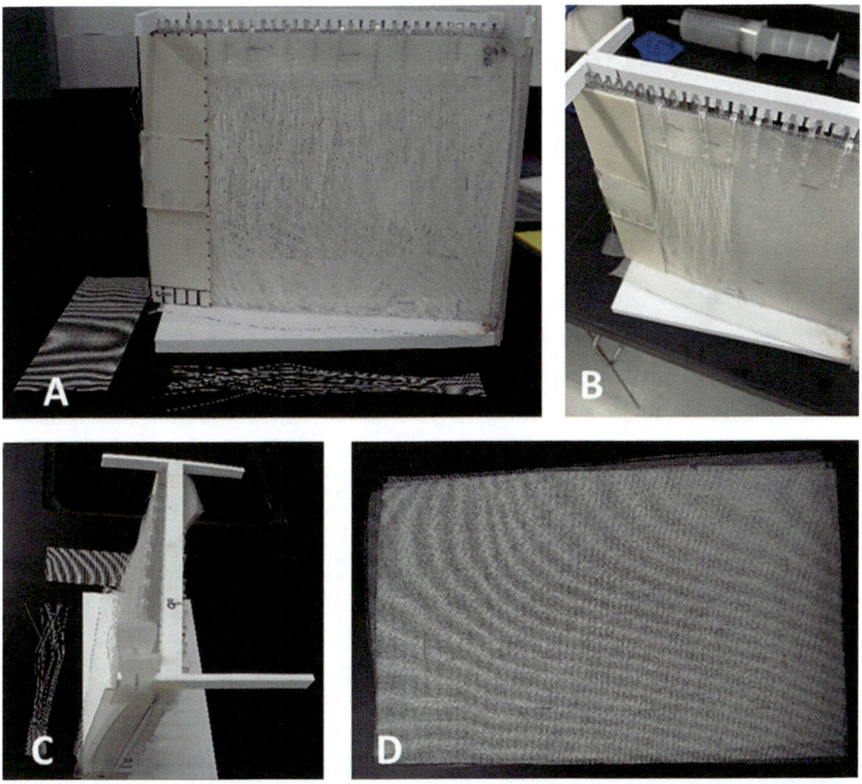

Fig. 5.12 Filter prototype photograph, showing (**a**) lateral and medial views of the filtration surface and particle velocity reference plate (on the left); (**b**) skewed dorsoventral view showing the rack, bristle curtain, basic nylon door mesh and bristle sample constructed from the latter; (**c**) another view of a third of bristle curtain installed; and (**d**) sample mesh stack without the bristle curtain (20% porosity). In frame (**a**), the rack makes a 9° angle with respect to the incoming flow. The frame also shows a bristle curtain section made from the vertical strands of a standard nylon door mesh, with several sections superposed to achieve higher "bristle" density

distribution over the entire filter area (Fig. 5.13) when mesh stack porosity reaches the 10%–20% levels observed in mysticete whales (Potvin and Werth 2024).

Note that such capture density uniformity over the entire filter area is further established with implementation of rack angles relative to the incoming flow. In both whale and flume, such angles cause particles entering the filtration zone nearest to the filter surface to interact with the filter's anterior zone, in contrast to particles entering furthest to the surface, which generally interact with the posterior region of the filter surface (near the OROwall).

Flume experiments also reveal the importance of the zero-porosity OROwall (Fig. 5.7) in generating the lateral pressure gradient needed to create cross-flows through the filter mesh, as well as (when present) the bristle curtain. Remarkably, the presence of a non-porous OROwall guarantees the presence of cross-flows

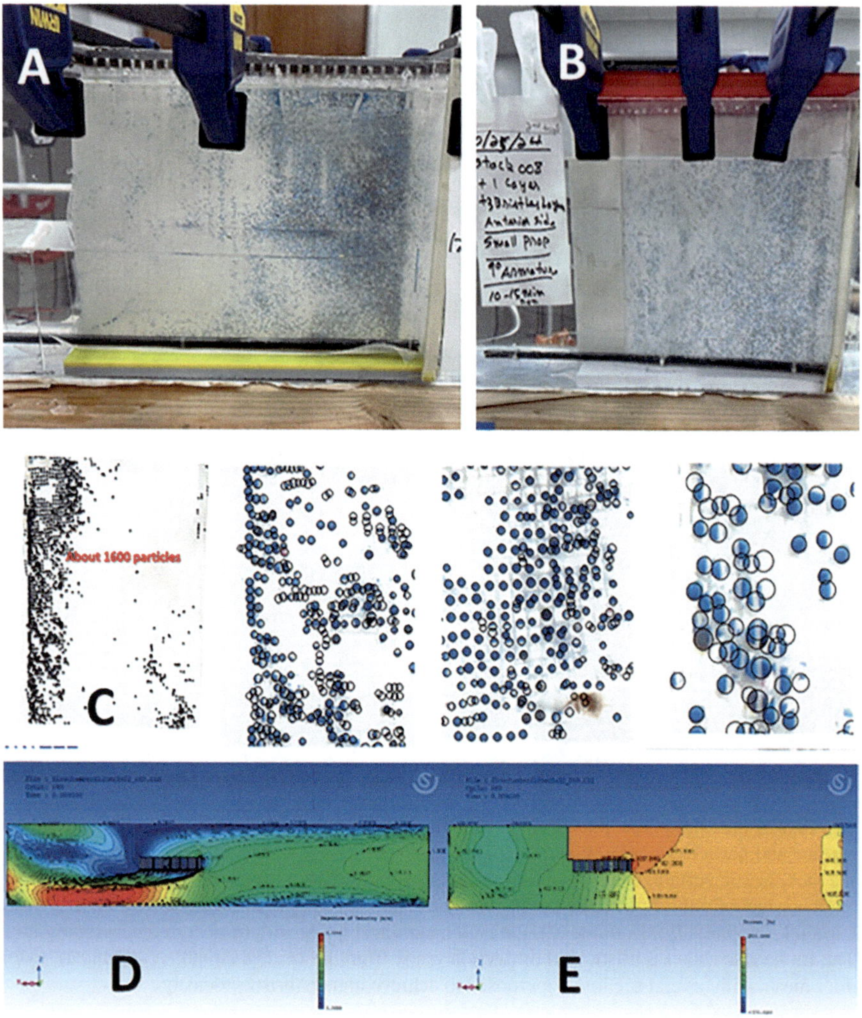

Fig. 5.13 Filters in the tank after collecting particles for >5 min: (**a**) high-porosity filter without a bristle curtain (0° angle with respect to the incoming flow), and (**b**) low-porosity filter with a bristle curtain (and 9° angle with respect to the incoming flow). (**c**) Captured particle counting example using a home-grown MATLAB-based pixel analysis program. (**d, e**) CFD simulation of a 30% porosity mesh overlapping a 35-blade ("baleen") rack showing variations in the crossflow speed (**d**) and pressure (**e**) over individual blades

regardless of filter surface porosity—or, presumably, in the absence of baleen, through teeth. Using a porous OROwall would most likely also establish cross-flows through the filter surface, as long as its porosity is less than the filter surface's.

Flume tests and CFD simulations also point to a somewhat different role for the "baleen" plates than described in the past (Werth 2013). Baleen porosity measurements of about 30% when including plates alone (i.e., not incorporating a bristle mat; Werth et al. 2018a; Potvin and Werth 2017) exceed the 10%–20% porosity of the bristle mats themselves (Potvin and Werth 2024). For the combined "metafilter" of living whales, this suggests dual structural and functional roles for the bristly fringes, both supporting the mat and producing new fringes, rather than the sole role in basic filtration previously ascribed (Werth 2000). In whales, this dual role is crucial in generating the thick bristle mat, as well as keeping it in place within the oral cavity against of the push of the tongue (during mechanical cleaning/clearing of accumulated prey) and against higher pressure in the medial regions of the mouth near the oropharynx. In engineered systems, on the other hand, differing (low-drag) filter support architectures can be devised to implement the same goal. Finally, and with regard to filtration efficiency, one would expect intra-plate spacing to have little effect as long as the spacing yields higher porosities than through the filter surface itself (i.e., bristle mat in whales, bristle curtain and mesh stack in our experimental flume prototypes). However, this may not hold true in architectures featuring plates/blades that encounter the feed stream *prior to* the low-porosity mat.

Photogrammetry analysis of filter images suggest capture efficiencies in the range of 20%–80%, depending mostly on porosity (the lower the porosity, the higher the efficiency) and rack angle to the flow (superficially). However, another crucial factor is the speed of flow entering a whale's oral cavity or test section in a flume, which in turns yield the flow speed V_{zone} of either system's filtration section and ultimately the filtration (or integration) distance D_T (Eq. 5.2). But faster flows entail higher particle capture rates and higher capture densities σ_{cap}, which at the same collection durations (T) yield quicker prey/microplastic collection. Moreover, and starting with clean filters (i.e., away from caking) would per Eq. (5.3) yield collection carried out at near constant efficiency. All of these effects combine to create CFF architecture that can be scaled to the large size seen in whales (tens of m^2 of area) or to smaller dimensions (cm^2 of area) that could be used in devices to detect microplastics.

5.6.7 Conclusions

So how might baleen inspire novel filters and separation processes? To be sure, varied fish gill-inspired mechanisms mentioned above all offer potentially useful biomimetic strategies that closely follow natural processes (Sanderson 2024), and yet are wholly unrelated to whale baleen. As explained at the outset of this chapter, natural selection rarely creates new structures and never steps back to ponder, with foresight, future applications; instead, it tinkers with and exploits existing structures in a piecemeal, jerry-rigged, step-wise fashion. Evolution, it must be emphasized, does not generate perfect solutions: it produces "winners" that survive merely by being slightly better than whatever the competition has to offer. The clear benefit of biomimetic strategies, in contrast, is that they offer purpose-built, forward-looking outcomes. Whether inspired by or wholly unrelated to natural solutions, man-made systems seldom achieve perfection either, but they usually arise much more rapidly, plainly, and effectively than nature's solutions. Like Velcro, baleen works remarkably well, and given its robust filtering architecture insensitive to fine scale details, it was undoubtedly honed by millions of years of evolution, with wrong turns and dead ends along the way.

What aspects of baleen might engineers emulate? First, baleen's extreme resilience is a chief highlight, offering unparalleled strength with flexibility and enabling whales' oral filters to withstand tremendous forces over a century or more. The axial, anisotropic strength is conferred by the composite nature of baleen's microscopic construction. Certainly, these fine structural details and principles can be copied by human designers.

Although baleen's composite nature is crucial, steady processes of growth, erosion, and replacement are also key factors in its overall function. The overall multi-level composition of the total baleen filter (encompassing racks, plates, and fringes) arises from the inexorably slow yet steady growth of baleen tissue, which originates with ever-growing germinative layers buried deep within the palatal gingiva, and which leads eventually to cracking of plates along their longitudinal axis and erosion at their oldest, most exposed and vulnerable erupted margins, exposes the interior horn tubules. These hollow tubules, now freed of the other two alpha keratin elements from which baleen is made (namely the nail-like surface sheets and interstitial, intertubular keratin), emerge as the hair-like fringes, which are essential for creating the porous interior mat of each baleen rack.

It is theoretically possible, if unlikely that an artificial filter might grow like baleen, with eroded and shed fibers gradually replaced by new fibers. Alternatively, it would be easier simply to gradually or sporadically add new components or levels of a filter (rather than wholly replacing the entire filter) as parts of it wear or break off, as happens with baleen. More likely in a commercial copy is basic multi-stage filtration, emulating baleen's tiered scheme of flow between plates as well as through the less porous fringe mat.

Application of biologically inspired designs to build better structures and materials for human use can apply many useful findings from recent and ongoing baleen research. Specifically, baleen's composite keratinous material, the geometry of the mysticete oral metafilter and its fluid dynamics, the precise mechanisms of particle flow and collection, the pathways of water entry and effluence, the forces and pressure gradients that baleen encounters during filtration, and methods of clearing the filter are all key points that could be exploited via biomimetics. Bioinspired structures and systems based on improved knowledge of baleen function could have many conceivable applications proving valuable for a range of technical and industrial design concerns, such as how to improve filter flow and efficiency while limiting breakage and clogging, and how to deal with real-world environmental issues such as clearing waters of microplastics and other waste (Hung et al. 2012;

Zhong et al. 2012; Bach et al. 2015; Divi et al. 2018; Schroeder et al. 2019; Witkop et al. 2023).

Acknowledgments The authors thank numerous colleagues and research assistants for ideas, inspiration, and sweat, including SLU students O. Bante, E. Cabigas, S. Liang, and R. Immekus who helped in flume construction. We also thank the SLU School of Science Engineering, the SLU WATER Institute, and Hampden-Sydney College for financial support.

Author Contributions AW and JP designed and conducted experiments and analyzed data, with help from AC, JG, WH, and DW. DW and WH designed and tested the particle-counting algorithm that the rest of the collaborative used. CFD simulations were carried by JP and JG. All figures were prepared by AW and JP and are the original property of the authors. All authors contributed to the article and reviewed and approved the final submitted version.

Conflict of Interest The authors declare that the research was conducted in the absence of any commercial or financial relationships that could be construed as a potential conflict of interest.

References

Adelmann B, Schwiddessen T, Götzendorfer B, Hellmann R (2022) Evaluation of SLS 3D-printed filter structures based on bionic manta structures. Materials 15:8454. https://doi.org/10.3390/ma15238454

Auta HS, Emenike CU, Fauziah SH (2017) Distribution and importance of microplastics in the marine environment: a review of the sources, fate, effects, and potential solutions. Environ Int 102:165–176

Bach D, Schmich F, Masselter T, Speck T (2015) A review of selected pumping systems in nature and engineering—potential biomimetic concepts for improving displacement pumps and pulsation damping. Bioinspir Biomim 10:051001

Baker RJ, Fane AG, Fell CJD, Yoo BH (1985) Factors affecting flux in crossflow filtration. Desalination 53:81–93

Baraza T, Hernandez NF, Sebok JN, Wu CL, Hasenmueller EA, Knouft JH (2022) Integrating land cover, point source pollution, and watershed hydrologic processes data to understand the distribution of microplastics in riverbed sediments. Environ Pollut 311:119852

Belfort D, Davis RH, Zydney AL (1994) The behavior of suspensions and macromolecular solutions in crossflow microfiltration. J Membr Sci 96:1–58

Benyus J (1997) Biomimicry: innovation inspired by nature. William Morrow, New York

Berta A, Sumich JL, Kovacs K (2015) Marine mammals: evolutionary biology, 3rd edn. Elsevier, San Diego

Berta A, Lanzetti A, Ekdale EG, Deméré TA (2016) From teeth to baleen and raptorial to bulk filter feeding in mysticete cetaceans: the role of paleontological, genetic, and geochemical data in feeding evolution and ecology. Integr Comp Biol 56:1271–1284

Bianciardi A, Cascini G (2022) A bio-inspired approach for boosting innovation in the separation technology sector. Proc Inst Mech Eng C J Mech Eng Sci 236:4533–4550. https://doi.org/10.1177/09544062211052128

Bisconti M, Pellegrino L, Carnevale G (2023) The chronology of mysticete diversification (Mammalia, Cetacea, Mysticeti): body size, morphological evolution and global change. Earth-Sci Rev 239:e104373

Blatt W, Dravid A, Michael AS, Nelson L (1970) Solute polarization and cake formation in membrane ultrafiltration: causes, consequences and control techniques. In: Flinn JE (ed) Membrane science and technology. Plenum, New York, pp 47–91

Boessenecker RW, Fordyce RE (2015) Anatomy, feeding ecology, and ontogeny of a transitional baleen whale: a new genus and species of Eomysticetidae (Mammalia: Cetacea) from the Oligocene of New Zealand. PeerJ 3:1129

Bosio G, Collareta A, DiCelma C, Lambert O, Marx FG, de Muizon C, Gioncada A, Gariboldi K, Malinverno E, Varas Malca R, Urbina M, Biannucci G (2021) Taphonomy of marine vertebrates of the Pisco formation (Miocene, Peru): insights into the origin of an outstanding fossil-lagerstätte. PLoS One 16(7):e0254395

Bott R, Langeloh TH, Ehrfeld E (2000) Dynamic cross flow filtration. Chem Eng J 80:245–249

Brainerd EL (2001) Caught in the crossflow. Nature 412:387–388

Braun J-M, Manoonpong P, Xiong X (2020) Biology-inspired engineering and engineering-inspired biology. Front Neurorobot 14:e614683

Brodie P, Vikingsson G (2009) On the feeding mechanisms of the sei whale (*Balaenoptera borealis*). J Northwest Atl Fish Sci 42:49–54

Brown E, MacDonald A, Allen S, Allen D (2023) The potential for a plastic recycling facility to release microplastic pollution and possible filtration remediation effectiveness. J Hazard Mater Adv 10:100309

Bucci K, Tulio M, Rochman CM (2020) What is known and unknown about the effects of plastic pollution: a meta-analysis and systematic review. Ecol Appl 30:e02044

Burdick JG, Burdick BJ (2017) Achieving Flight: The Life and Times of John J. Montgomery. Archway, Bloomington

Cade DE, Friedlaender AS, Calambokidis J, Goldbogen JA (2016) Kinematic diversity in rorqual whale feeding mechanisms. Curr Biol 26:2617–2624

Caraveo-Patiño J, Hobson KA, Soto LA (2007) Feeding ecology of gray whales inferred from stable-carbon and nitrogen isotopic analysis of baleen plates. Hydrobiology 586:17–25

Cheer AY, Ogami Y, Sanderson SL (2001) Computational fluid dynamics in the oral cavity of ram suspension-feeding fishes. J Theor Biol 210:463–474

Cheer A, Cheung S, Hung T-C, Piedrahita RH, Sanderson SL (2012) Computational fluid dynamics of fish gill rakers during crossflow filtration. Bull Math Biol 74:981–1000

Choy CA, Robison BH, Gagne TO, Erwin B, Firl E, Halden RU, Hamilton JA, Katija K, Lisin SE, Rolsky C (2019) The vertical distribution and biological transport of marine microplastics across the epipelagic and mesopelagic water column. Sci Rep 9:7843

Clark AS, San-Miguel A (2021) A bioinspired, passive microfluidic lobe filtration system. Lab Chip 21:3762. https://doi.org/10.1039/D1LC00449B

Cohen YH, Reich Y, Greenberg S (2014) Biomimetics: structure–function patterns approach. J Mech Des 136:111108-1

Cole M, Lindeque P, Halsband C, Galloway TS (2011) Microplastics as contaminants in the marine environment: a review. Mar Pollut Bull 62(12):2588–2597

Colson KM, Pirotta E, New L, Cade DE, Calambokidis J, Bierlich KC, Bird CN, Ajó AF, Hildebrand L, Trites AW, Torres LG (2024) Using accelerometry tags to quantify gray whale foraging behavior. Mar Mamm Sci 41(2):e13210

Corkeron P (2025) Whales still aren't 'eating all the fish' [commentary]. Mongabay 2025:06

Croll DA, Tershy BR, Newton KM (2008) Filter feeding. In: Würsig B, Thewissen JGM, Kovacs K (eds) Encyclopedia of marine mammals, 3rd edn. Elsevier, New York, pp 429–433

Deméré TA, McGowen MR, Berta A, Gatesy J (2008) Morphological and molecular evidence for a stepwise evolutionary transition from teeth to baleen in mysticete whales. Syst Biol 57:15–37

Divi RV, Strother JA, Paig-Tran EWM (2018) Manta rays feed using ricochet separation, a novel nonclogging filtration mechanism. Sci Adv 4:e9533

Dou Y, Tian D, Sun Z, Liu Q, Zhang N, Kim JH, Jiang L, Dou SX (2017) Fish gill inspired crossflow for efficient and continuous collection of spilled oil. ACS Nano 11:2477–2485. https://doi.org/10.1021/acsnano.6b07918

Dubner S (2023) The first great American industry. Freakonomics 549, 12 July 2023. https://freakonomics.com/podcast/the-first-great-american-industry/

Ekdale EG, Deméré TA (2021) Neurovascular evidence for a co-occurrence of teeth and baleen in an Oligocene mysticete and the transition to filter-feeding in baleen whales. Zool J Linnean Soc 194(2):395–415

Espinosa-Gayosso A, Ghisalberti M, Ivey GN, Jones NL (2012) Particle capture and low-Reynolds-number flow around a circular cylinder. J Fluid Mech 710:362–378

Espinosa-Gayosso A, Ghisalberti M, Shimeta J, Ivey GN (2021) On predicting particle capture rates in aquatic ecosystems. PLoS One 16:e0261400. https://doi.org/10.1371/journal.pone.0261400
Feughelman M (1997) The relation between the mechanical properties of wool and hair fibres and the keratin-water system. In: Feughelman M (ed) Mechanical properties and structure of alpha-keratin fibres. Univ. of New South Wales Press, Sydney, pp 16–27
Feughelman M (2002) Natural protein fibers. J Appl Polym Sci 83:489–507
Fields DM, Yen J (1997) The escape behavior of marine copepods in response to a quantifiable fluid mechanical disturbance. J Plankton Res 19:1289–1304
Fish FE (2020) Biomimetics and the application of the leading-edge tubercles of the humpback whale flipper. In: New D, Ng B (eds) Flow control through bio-inspired leading-edge tubercles. Springer, Cham, pp 1–39. https://doi.org/10.1007/978-3-030-23792-9_1
Fish FE, Weber PW, Murray MM, Howle LE (2011) The tubercles on humpback whale flippers: application of bio-inspired technology. Integr Comp Biol 51(1):203–213. https://doi.org/10.1093/icb/icr016
Fish FE, Goetz KT, Rugh DJ, Brattström LV (2013) Hydrodynamic patterns associated with echelon formation swimming by feeding bowhead whales (*Balaena mysticetus*). Mar Mamm Sci 29(4):e498–e507
Fitzgerald EMG (2010) The morphology and systematics of *Mammalodon colliveri*, a toothed mysticete from the Oligocene of Australia. Zool J Linnean Soc 158:367–476
Fordyce RE (1980) Whale evolution and Oligocene southern ocean environments. Palaeogeogr Palaeoclimatol Palaeoecol 31:319–336
Fordyce RE, Barnes LG (1994) The evolutionary history of whales and dolphins. Annu Rev Earth Planet Sci 22:419–455
Fordyce RE, Marx FG (2013) The pygmy right whale *Caperea marginata*: the last of the cetotheres. Proc R Soc B Biol Sci 280:2645
Fordyce RE, Marx FG (2018) Gigantism precedes filter feeding in baleen whale evolution. Curr Biol 28:1–7
Forslind B (1970) Biophysical studies of normal nail. Acta Derm Venereol 50:161–168
Fortune SME, Trites AW, Perryman WL, Moore MJ, Pettis HM, Lynn MS (2012) Growth and rapid early development of North Atlantic right whales (*Eubalaena glacialis*). J Mammal 93:1342–1354
Fossi MC, Baini M, Simmonds MP (2020) Cetaceans as ocean health indicators of marine litter impact at global scale. Front Environ Sci 8:586627
Fraser RD, MacRae TP, Rogers GE (1972) Keratins: their composition, structure, and biosynthesis. Charles C. Thomas, Springfield
Fraser RDB, MacRae TP, Suzuki E (1976) Structure of keratin microfibril. J Mol Biol 108:435–452
French DH (1971) An experiment in water-sieving. Anat Stud 21:59–64
Fudge DS, Szewciw LJ, Schwalb AN (2009) Morphology and development of blue whale baleen: an annotated translation of Tycho Tullberg's classic 1883 paper. Aquat Mamm 35:226–252
Gallo F, Fossi C, Weber R, Santillo D, Sousa J, Ingram I, Nadal A, Romano D (2018) Marine litter plastics and microplastics and their toxic chemicals components: the need for urgent preventive measures. Environ Sci Eur 30:13
Gaskin DE (1982) The ecology of whales and dolphins. Heinemann, New York
Gatesy J, O'Leary MA (2001) Deciphering whale origins with molecules and fossils. Trends Ecol Evol 16:562–571
Gatesy J, Geisler JH, Chang J, Buell C, Berta A, Meredith RW, Springer MS, McGowen MR (2013) A phylogenetic blueprint for a modern whale. Mol Phylogenet Evol 66:479–506
Gatesy J, Ekdale EG, Deméré TA, Lanzetti A, Randall J, Beta A, El Adli JJ, Springer MS, McGowen MR (2022) Anatomical, ontogenetic, and genomic homologies guide reconstructions of the teeth-to-baleen transition in mysticete whales. J Mamm Evol 29:891–930
Geisler JH, Boessenecker RW, Brown M, Beatty BL (2017) The origin of filter feeding in whales. Curr Biol 2017:2036–2042

George JC, Stimmelmayr R, Suydam R, Usip S, Givens G, Sformo T, Thewissen JGM (2016) Severe bone loss as part of the life history strategy of bowhead whales. PLoS One 11:e0156753

George JC, Lubetkin SC, Zeh JE, Thewissen JGM, Wetzel D, Givens GH (2021) Age estimation. In: George JC, Thewissen JGM (eds) The bowhead whale, *Balaena mysticetus*: biology and human interactions. Academic Press, San Diego, pp 309–322

GESAMP (2015) Sources, fate and effects of microplastics in the marine environment: a global assessment. In: Kershaw PJ (ed) IMO/FAO/UNESCO-IOC/UNIDO/WMO/IAEA/UN/UNEP/UNDP Joint Group of Experts on the scientific aspects of marine environmental protection. GESAMP, London

GESAMP/IMO/FAO/UNESCO-IOC/UNIDO/WMO/IAEA/UN/UNEP/UNDP/ISA Joint Group of Experts on the Scientific Aspects of Marine Environmental Protection (2019) Guidelines for the monitoring and assessment of plastic litter and microplastics in the ocean. In: Kershaw PJ, Turra A, Galgani F (eds) GESAMP reports & studies series, vol 99. GESAMP, London, 130p. ISSN 1020-4873

Gibbs SE, Kent S, Slat B, Morales D, Fouda L, Reisser J (2019) Cetacean sightings within the great Pacific Garbage Patch. Mar Biodivers 49:2021–2027

Gioncada A, Collareta A, Gariboldi K, Lambert O, DiCelma C, Bonaccorsi E, Urbina M, Bianucci G (2016) Inside baleen: exceptional microstructure preservation in a late Miocene whale skeleton from Peru. Geology 44:839–842

Goel G, Hélix-Nielsen C, Upadhyaya HM, Goel S (2021) A bibliometric study on biomimetic and bioinspired membranes for water filtration. npj Clean Water 4:41. https://doi.org/10.1038/s41545-021-00131-4

Goldbogen JA (2010) The ultimate mouthful: lunge-feeding in rorqual whales. Am Sci 98:124–131

Goldbogen JA, Madsen PT (2018) The evolution of foraging capacity and gigantism in cetaceans. J Exp Biol 221:jeb166033. https://doi.org/10.1242/jeb.166033

Goldbogen JA, Calambokidis J, Shadwick RE, Oleson E, McDonald MA, Hildebrand JA (2006) Kinematics of foraging dives and lunge-feeding in fin whales. J Exp Biol 209:1231–1244

Goldbogen JA, Pyenson ND, Shadwick RE (2007) Big gulps require high drag for fin whale lunge feeding. Mar Ecol Prog Ser 349:289–301

Goldbogen JA, Calambokidis J, Oleson E, Potvin J, Pyenson ND, Schorr G, Shadwick RE (2011) Mechanics, hydrodynamics and energetics of blue whale lunge feeding: efficiency dependence on krill density. J Exp Biol 214:131–146

Goldbogen JA, Calambokidis J, Croll D, McKenna MF, Potvin J, Pyenson ND, Schoor G, Shadwick RE, Tershy BR (2012a) Scaling of lunge feeding performance in rorqual whales: mass-specific energy expenditure increases with body size and progressively limits diving capacity. Funct Ecol 26:216–226

Goldbogen JA, Calambokidis J, Friedlaender AS, Francis J, DeRuiter SL, Stimpert AK, Falcone E, Southall BL (2012b) Underwater acrobatics by the world's largest predator: 360 degrees rolling manoeuvres by lunge-feeding blue whales. Biol Lett 9:0986

Goldbogen JA, Friedlaender AS, Calambokidis J, McKenna MF, Simon M, Nowacek DP (2013) Integrative approaches to the study of baleen whale diving behavior, feeding performance, and foraging ecology. Bioscience 63:90–100

Goldbogen JA, Hazen EL, Friedlaender AS, Calambokidis J, DeRuiters SL, Stimpert AK, Southall BL (2015) Prey density and distribution drive the three-dimensional foraging strategies of the largest filter feeder. Funct Ecol 29:951–961

Goldbogen JA, Cade DE, Boersma AT, Calambokidis J, Kahane-Rapport SR, Segre PS, Stimpert AK, Friedlaender AS (2017a) Using digital tags with integrated video and inertial sensors to study moving morphology and associated function in large aquatic vertebrates. Anat Rec 300:1935–1941

Goldbogen JA, Cade D, Calambokidis JA, Friedlaender AS, Potvin J, Segre PS, Werth AJ (2017b) How baleen whales feed: the biomechanics of engulfment and filtration. Annu Rev Mar Sci 9:367–386

Goldbogen JA, Pyenson ND, Madsen PT (2023) How whales dive, feast, and fast: the ecophysiological limits of foraging in the evolution of Cetacea. Annu Rev Ecol Evol Syst 54:307–325

Goldstone L (2015) Birdmen: the Wright Brothers, Glenn Curtiss, and the battle to control the skies. Ballantine, New York

Goodrich JS, Sanderson SL, Batjakas IE, Kaufman LS (2000) Branchial arches of suspension-feeding *Oreochromis esculentus*: sieve or sticky filter? J Fish Biol 56:858–875

Greenberg DA, Fudge DS (2012) Regulation of hard alpha-keratin mechanics via control of intermediate filament hydration: matrix squeeze revisited. Proc R Soc B 280:e2158

Gregorietti M, Atzori F, Carosso L, Frau F, Pellegrino G, Sarà G, Arcangeli A (2021) Cetacean presence and distribution in the Central Mediterranean Sea and potential risks deriving from plastic pollution. Mar Pollut Bull 173:112943

Guerrini F, Lorenzo M, Renato C (2019) Modeling plastics exposure for the marine biota: risk maps for fin whales in the Pelagos sanctuary (North-Western Mediterranean). Front Mar Sci 6:00299

Hamann L, Blanke A (2022) Suspension feeders: diversity, principles of particle separation and biomimetic potential. J R Soc Interface 19:e0741. https://doi.org/10.1098/rsif.2021.0741

Harris PT, Maes T, Raubenheimer K, Walsh JPA (2023) A marine plastic cloud—global mass balance assessment of oceanic plastic pollution. Cont Shelf Res 255:104947

Harwood CS, Fogel GB (2012) Quest for flight: John J. Montgomery and the dawn of aviation in the West. University of Oklahoma Press, Norman

Hearle JWS (2000) A critical review of the structural mechanics of wool and hair fibres. Int J Biol Macromol 27:123–138

Hildebrand L, Sullivan FA, Orben RA, Derville S, Torres LG (2022) Trade-offs in prey quantity and quality in gray whale foraging. Mar Ecol Prog Ser 695:189–201

Hocking DP, Marx FG, Fitzgerald EMG, Evans AR (2017a) Ancient whales did not filter feed with their teeth. Biol Lett 13:20170348

Hocking DP, Marx FG, Park T, Fitzgerald EMG, Evans AR (2017b) A behavioural framework for the evolution of feeding in predatory aquatic mammals. Proc R Soc Lond B Biol Sci 284:e2016750

Horwood J (2017) Sei whale. In: Würsig B, Thewissen JGM, Kovacs K (eds) Encyclopedia of marine mammals, 3rd edn. Academic Press, San Diego, pp 845–847

Huang H, Hou J, Li M, Wei F, Liao Y, Xi B (2025) Microplastics in the bloodstream can induce cerebral thrombosis by causing cell obstruction and lead to neurobehavioral abnormalities. Sci Adv 11(4):e8243

Humphries S (2009) Filter feeders and plankton increase particle encounter rates through flow regime control. Proc Natl Acad Sci 106:7882–7887

Hung T-C, Piedrahita RH, Cheer A (2012) Bio-inspired particle separator design based on the food retention mechanism by suspension-feeding fish. Bioinspir Biomim 7:046003

Hurley R, Woodard J, Rothwell JJ (2018) Microplastic contamination of river beds significantly reduced by catchment-wide flooding. Nat Geosci 11:251–257

Hwang J, Choi D, Han S, Jung SY, Choi J, Hong J (2020) Potential toxicity of polystyrene microplastic particles. Sci Rep 10(1):1–12

Jain A, De S (2019) Processing of beverages by membranes. In: Grumezescu AM, Hoban AM (eds) Processing and sustainability of beverages. Woodhead Publishing, Cambridge, pp 517–560

Jambeck JR, Geyer R, Wilcox C, Siegler TR, Perryman M, Andrady A, Narayan R, Law KL (2015) Plastic waste inputs from land into the ocean. Science 347(6223):768–771

Jandang S, Alfonso MB, Nakano H, Phinchan N, Darumas U, Viyakarn V, Chavanich S, Isobo A (2024) Possible sink of missing ocean plastic: accumulation patterns in reef-building corals in the Gulf of Thailand. Sci Total Environ 954:e176210

Jensen MM, Saladrigas AH, Goldbogen JA (2017) Comparative three-dimensional morphology of baleen: cross-sectional profiles and volume measurements using CT images. Anat Rec 300:1942–1952

Johari H, Henoch C, Custodio D, Levshin A (2007) Effects of leading-edge protuberances on airfoil performance. Am Inst Aeronaut Astronaut J 45(11):2634–2642

Jorgensen CB (1966) Biology of suspension feeding. Pergamon Press, Oxford

Kahane-Rapport SR, Goldbogen JA (2018) Allometric scaling of morphology and engulfment capacity in rorqual whales. J Morphol 279(9):1256–1268

Kahane-Rapport SR, Savoca MS, Cade DE, Segre PS, Bierlich KC, Calambokidis J, Dale JA, Fahlbusch AS, Friedlaender AS, Johnston DW, Werth AJ, Goldbogen JA (2020) Lunge filter feeding biomechanics constrain rorqual foraging ecology across scale. J Exp Biol 223:jeb224196. https://doi.org/10.1242/jeb.224196

Kahane-Rapport SR, Teeple J, Liao JC, Paig-Tran EWM, Strother JA (2025) Filter feeding in devil rays is highly sensitive to morphology. Proc R Soc B 292:20242037. https://doi.org/10.1098/rspb.2024.2037

Kawamura A (1974) Food and feeding ecology in the southern sei whale. Sci Rep Whales Res Inst 26:25–144

Kawamura A (1980) A review of food of balaenopterid whales. Sci Rep Whales Res Inst 32:155–198

Kennedy JH (2021) Hydrodynamics of rorquals. PhD thesis, Saint Louis University, St. Louis

Kitchener A, Vincent JFV (1987) Composite theory and the effect of water on the stiffness of horn keratin. J Mater Sci 22:1385–1389

Knippers J, Speck T (2012) Design and construction principles in nature and architecture. Bioinspir Biomim 7:015002

LaBarbera M (1984) Feeding currents and particle capture mechanisms in suspension feeding animals. Am Zool 24:71–84

Lambert O, Martínez-Cáceres M, Bianucci G, DiCelma C, Salas-Gismondi R, Steurbaut E, Urbina M, de Muizon C (2017) Earliest mysticete from the Late Eocene of Peru sheds new light on the origin of baleen whales. Curr Biol 27:1535–1541

Lambertsen RH (1983) Internal mechanism of rorqual feeding. J Mammal 64:76–88

Lambertsen RH, Hintz RJ, Lancaster WC, Hirons A, Kreiton KJ, Moor C (1989) Characterization of the functional morphology of the mouth of the bowhead whale, *Balaena mysticetus*, with special emphasis on the feeding and filtration mechanisms. Report to the Department of Wildlife Management. North Slope Borough, Barrow

Lambertsen RH, Rasmussen KJ, Lancaster WC, Hintz RJ (2005) Functional morphology of the mouth of the bowhead whale and its implications for conservation. J Mammal 86(2):342–352

Langeland A, Nost T (1995) Gill raker structure and selective predation on zooplankton by particulate feeding fish. J Fish Biol 47:719–732

Law KL (2017) Plastics in the marine environment. Annu Rev Mar Sci 9:205–229

Lee M (1998) Baleen basketry of the North Alaskan Eskimo. University of Washington Press, Seattle

Liao Y, Lyu J, Zhang Y, Hong Y, Ding S, Wu Z, Liu H, Wu J (2024) Specialized spatially-arranged non-circular fibers enhance filtration performance of African shrimp (*Atya gabonensis*). Bioinspir Biomim 19(3):6013. https://doi.org/10.1088/1748-3190/ad3b57

Loch C, Viegas SV, Waddell JN, Kemper C, Cook RB, Werth AJ (2020) Structure and properties of baleen in the Southern right (*Eubalaena australis*) and Pygmy right whales (*Caperea marginata*). J Mech Behav Biomed Mater 110:e103939

Lu WM, Ju SC (1989) Selective particle deposition in crossflow filtration. Sep Sci Technol 24:517–540

Lu WM, Hwang KJ, Ju SC (1993) Studies on the mechanism of cross-flow filtration. Chem Eng Sci 48:863–872

Lubetkin SC, Zeh JE, Rosa C, George JC (2008) Age estimation for young bowhead whales (*Balaena mysticetus*) using annual baleen growth increments. Can J Zool 86:525–538

Lysiak NSJ, Trumble SJ, Knowlton AR, Moore MJ (2018) Fishing gear entanglement on a North Atlantic right whale (*Eubalaena glacialis*) using stable isotopes, steroid and thyroid hormones in baleen. Front Mar Sci 5:e168

MacLeod M, Arp HPH, Tekman MB, Jahnke A (2021) The global threat from plastic pollution. Science 373:61–65

Magna RL, Gabler M, Reichert S, Schwinn T, Waimer F, Menges A, Knippers J (2013) From nature to fabrication: biomimetic design principles for the production of complex spatial structures. Int J Space Struct 28(1):27–39. https://doi.org/10.1260/0266-3511.28.1.27
Makabe R, Akamatsu K, Tatsumi R, Koike O, Nakao S (2021) Numerical simulations of lift force and drag force on a particle in cross-flow microfiltration of colloidal suspensions to understand limiting flux. J Membr Sci 621:118998
Makwana B, Khadke S, Kumar A, Nasir K, Wadhera R, Shah R, Shethe S, Kong Y, Navas-Acien A, Adamkiewicz G, Rajagopalan S, Al-Kindo S, Dani SS, Moffat-Bruce S, Ganatra S (2025) Marine microplastic levels and the prevalence of cardiometabolic diseases in US coastline counties. J Am Heart Assoc 14:e039891
Marfella R, Prattichizzo F, Sardu C, Fulgenzi G, Graciotti L, Spadoni T, D'Onofrio N, Paolisso G (2024) Microplastics and nanoplastics in atheromas and cardiovascular events. N Engl J Med 390:900–910
Marshall CD, Goldbogen JA (2015) Marine mammal feeding mechanisms. In: Castellini MA, Mellish JA (eds) Marine mammal physiology: requisites for ocean living. CRC, Boca Raton, pp 95–117
Marshall CD, Pyenson ND (2019) Feeding in aquatic mammals: an evolutionary and functional approach. In: Bels V, Whishaw IQ (eds) Feeding in vertebrates: anatomy, biomechanics, evolution. Springer, New York, pp 743–785
Marshall RC, Orwin DFG, Gillespie JM (1991) Structure and biochemistry of mammalian hard keratin. Electron Microsc Rev 4:47–83
Marx FG, Fordyce RE (2015) Baleen boom and bust: a synthesis of mysticete phylogeny, diversity and disparity. R Soc Open Sci 2:1–14
Marx FG, Hocking DP, Park T, Ziegler T, Evans AR, Fitzgerald EMG (2016a) Suction feeding preceded filtering in baleen whale evolution. Mem Museum Vic 75:71–82
Marx F, Lambert O, Uhen MD (2016b) Cetacean paleobiology. Wiley-Blackwell, London
Marx FG, Collareta A, Gioncada A, Post K, Lambert O, Bonaccorsi E, Urbina M, Bianucci G (2017) How whales used to filter: exceptionally preserved baleen in a Miocene cetotheriid. J Anat 231:212–220
Masselter T, Schaumann U, Kampowski T, Ulrich K, Thielen M, Bold G, Speck T (2023) Improvement of a microfiber filter for domestic washing machines. Bioinspir Biomim 18:016017. https://doi.org/10.1088/1748-3190/acaba2
Matyka M, Khalili A, Kozal Z (2008) Tortuosity–porosity relation in the porous media flow. Phys Rev E 78:e026306
Mayo CA, Marx MK (1990) Surface foraging behaviour of the North Atlantic right whale, *Eubalaena glacialis*, and associated zooplankton characteristics. Can J Zool 68:2214–2220
Mayo CA, Letcher BH, Scott S (2001) Zooplankton filtering efficiency of the baleen of a North Atlantic right whale, *Eubalaena glacialis*. J Cetacean Res Manag 2:225–229
McKittrick J, Chen PY, Bodde SG, Yan W, Novitskaya EE, Meyers MA (2012) The structure, functions, and mechanical properties of keratin. J Miner Met Mater Soc 64:449–468
Mei R (1992) An approximate expression for the shear lift on a spherical particle at finite Reynolds number. Int J Multiphase Flow 18:145–147
Michaud J, Taggart CT (2007) Lipid and gross energy content of North Atlantic right whale food, *Calanus finmarchicus*, in the Bay of Fundy. Endanger Species Res 2007:77–94
Mironov V, Boland T, Trusk G, Forgacs G, Markwald RR (2003) Organ printing: computer-aided jet-based 3D tissue engineering. Trends Biotechnol 21:157–161
Moffat R, Spriggs J, O'Connor S (2008) The use of baleen for arms, armour, and heraldic crests in medieval Britain. Antiq J 88:207–215
Moore MJ, van der Hoop J (2012) The painful side of trap and fixed net fisheries: chronic entanglement of large whales. J Mar Sci 2012:230653. https://doi.org/10.1155/2012/230653
Morales-Caselles C, Viejo J, Marti E, González-Fernández D, Pragnell-Raasch H, González-Gordillo JI, Montero E, Arroyo GM, Hanke G, Salvo VS, Basurko OC, Mallos N, Lebreton L, Echevarria F, van Emmerik T, Duarte CM, Gálvez JA, van Sebille E, Galgani F, Garcia CM,

Ross PS, Bartual A, Ioakeimidis C, Markalain G, Isobe A, Cózar A (2021) An inshore-offshore sorting system revealed from global classification of ocean litter. Nat Sustainability 4:484–493
Motta PJ, Maslanka M, Hueter RE, Davis RL, de la Parra R, Mulvany SL, Habegger ML, Strother JA, Mara KR, Gardiner JM, Tyminski JP, Zeigler LD (2010) Feeding anatomy, filter-feeding rate, and diet of whale sharks *Rhincodon typus* during surface ram filter feeding off the Yucatan Peninsula, Mexico. Zoology 113:199–212
Murkes J, Carlsson CG (1988) Crossflow filtration: theory and practice. Wiley, New York
Mutuku J, Yanotti M, Tocock M, MacDonald DH (2024) The abundance of microplastics in the world's oceans: a systematic review. Oceans 5(3):398–428
Nemoto T (1959) Food of baleen whales with reference to whale movements. Sci Rep Whales Res Inst 14:149–241
Nemoto T (1970) Feeding pattern of baleen whales in the ocean. In: Steele JH (ed) Marine food chains. Oliver & Boyd, Edinburgh, pp 241–252
Nihart AJ, Garcia MA, El Hayek E, Liu R, Olewine M, Kingston JD, Castillo EF, Gullapalli RR, Howard T, Bleske B, Scott J, Gonzalez-Estrella J, Gross JM, Spilde M, Adolphi NL, Gallego DF, Jarrell HS, Dvorscak G, Zuluaga-Ruiz ME, West AB, Campen MJ (2025) Bioaccumulation of microplastics in decedent human brains. Nat Med 31:1114–1119
Orton LS, Brodie PF (1987) Engulfing mechanisms of fin whales. Can J Zool 65:2898–2907
Paig-Tran EWM, Summers AP (2014) Comparison of the structure and composition of the branchial filters in suspension feeding elasmobranchs. Anat Rec 297:701–715
Paig-Tran EWM, Bizzarro JJ, Strother JA, Summers AP (2011) Bottles as models: predicting the effects of varying swimming speed and morphology on size selectivity and filtering efficiency in fishes. J Exp Biol 214:1643–1654
Paig-Tran EWM, Kleinteich T, Summers AP (2013) The filter pads and filtration mechanisms of the devil rays: variation at macro and microscopic scales. J Morphol 274:1026–1043
Pautard FGE (1963) Mineralization of keratin and its comparison with the enamel matrix. Nature 199:531–535
Pearson H, Savoca MS, Costa DP, Lomas MW, Molina R, Pershing AJ, Smith CR, Villaseñor-Derbez JC, Wing SR, Roman J (2023) Whales in the carbon cycle: can recovery remove carbon dioxide? Trends Ecol Evol 38(3):238–249. https://doi.org/10.1016/j.tree.2024.06.009
Peng J, Dabiri JO (2009) Transport of inertial particles by Lagrangian coherent structures: application to predator–prey interaction in jellyfish feeding. J Fluid Mech 623:75–84
Peredo CM, Pyenson ND, Boersma AT (2017) Decoupling tooth loss from the evolution of baleen in whales. Front Mar Sci 4:1–11
Pfeiffer C (1992) Cellular structure of terminal baleen in various mysticete species. Aquat Mamm 18:67–73
Pinto SJD, Shadwick RE (2013) Material and structural properties of fin whale (*Balaenoptera physalus*) zwischensubstanz. J Morphol 274:947–955
Pivorunas A (1976) A mathematical consideration of the function of baleen plates and their fringes. Sci Rep Whales Res Inst 28:37–55
Pivorunas A (1979) The feeding mechanisms of baleen whales. Am Sci 67:432–440
Pomerleau C, Matthews CJD, Gobeil C, Stern GA, Ferguson SH, Macdonald RW (2018) Mercury and stable isotope cycles in baleen plates are consistent with year-round feeding in two bowhead whale (*Balaena mysticetus*) populations. Polar Biol 41:1–13
Potvin J, Werth AJ (2017) Oral cavity hydrodynamics and drag production in balaenid whale suspension feeding. PLoS One 12:5220
Potvin J, Werth AJ (2024) Suffused: baleen fringe mat porosity and hydrodynamics in balaenid and balaenopterid whales. Biol J Linn Soc 143(4):blae030. https://doi.org/10.1093/biolinnean/blae030
Potvin J, Goldbogen JA, Shadwick RE (2009) Passive versus active engulfment: verdict from trajectory simulations of lunge-feeding fin whales *Balaenoptera physalus*. J R Soc Interface 6:1005–1025

Potvin J, Goldbogen JA, Shadwick RE (2012) Metabolic expenditures of lunge feeding rorquals across scale: implications for the evolution of filter feeding and the limits to maximum body size. PLoS One 7:44854

Potvin J, Cade DE, Werth AJ, Shadwick RE, Goldbogen JA (2020) A perfectly inelastic collision: bulk prey engulfment by baleen whales and dynamical implications for the world's largest cetaceans. Am J Phys 2020:e1771

Potvin J, Cade DE, Werth AJ, Shadwick RE, Goldbogen JA (2021) Rorqual lunge-feeding energetics near and away from the kinematic threshold of optimal efficiency. Integr Comp Biol 3(1):obab005

Prata JC, da Costa JP, Lopes I, Duarte AC, Rocha-Santos T (2020) Environmental exposure to microplastics: an overview on possible human health effects. Sci Total Environ 702:134455

Pyenson ND, Vermeij GJ (2016) The rise of ocean giants: maximum body size in Cenozoic marine mammals as an indicator for productivity in the Pacific and Atlantic Oceans. Biol Lett 12:20160186

Reisser J, Slat B, DuPlessis K, Epp M, Proietti M, DeSonneville J, Becker T, Pattiaratchi C (2015) The vertical distribution of buoyant plastics at sea: an observational study in the North Atlantic Gyre. Biogeosciences 12:1249–1256

Ripperger S, Altmann J (2002) Crossflow microfiltration—state of the art. Sep Purif Technol 26:19–31

Rubenstein DI, Koehl MAR (1977) The mechanisms of filter feeding: some theoretical considerations. Am Nat 111:981–994

Ruud JT (1940) The surface structure of the baleen plates and a possible clue to age in whales. Hvalradets Skr 23:1–24

Sanderson SL (2024) Particle separation mechanisms in suspension-feeding fishes: key questions and future directions. Front Mar Sci 11:1331164. https://doi.org/10.3389/fmars.2024.1331164

Sanderson SL, Wassersug RL (1990) Suspension feeding vertebrates. Sci Am 262:96–101

Sanderson SL, Wassersug R (1993) Convergent and alternative designs for vertebrate suspension feeding. In: Hanken J, Hall BK (eds) The skull volume 3: functional and evolutionary mechanisms. University of Chicago Press, Chicago, pp 37–112

Sanderson SL, Cheer AY, Goodrich JS, Graziano JD, Callan WT (2001) Crossflow filtration in suspension-feeding fishes. Nature 412:439–441

Sanderson SL, Roberts E, Lineburg J, Brooks H (2016) Fish mouths as engineering structures for vortical cross-step filtration. Nat Commun 7:e11092

Savoca MS, Czapanskiy MF, Kahane-Rapport SR, Gough WT, Fahlbusch JA, Bierlich KC, Segre PS, DiClemente J, Penry GS, Wiley DN, Calambokidis J, Nowacek DP, Johnston DW, Pyenson ND, Friedlaender AS, Hazen EL, Goldbogen JA (2021) Baleen whale prey consumption based on high-resolution foraging measurements. Nature 599:85–90

Schroeder A, Marshall L, Trease B, Becker A, Sanderson SL (2019) Development of helical, fish-inspired cross-step filter for collecting harmful algae. Bioinspir Biomim 14:056008. https://doi.org/10.1088/1748-3190/ab2d13

Segre PS, Weir CR, Stanworth A, Cartwright S, Friedlaender AS, Goldbogen JA (2021) Biomechanically distinct filter-feeding behaviors distinguish sei whales as a functional intermediate and ecologically flexible species. J Exp Biol 224:e238873

Sekiguchi K, Best PB, Kaczmaruk BZ (1992) New information on the feeding habits and baleen morphology of the pygmy right whale *Caperea marginata*. Mar Mamm Sci 8:288–293

Shadwick RE, Potvin J, Goldbogen JA (2019) Lunge feeding in rorqual whales. Physiology 34(6):409–418

Shimeta J, Jumars PA (1991) Physical mechanisms and rates of particle capture by suspension-feeders. Oceanogr Mar Biol Annu Rev 29:191–257

Sibanda V, Greenwood RW, Seville JPK (2001) Particle separation from gases using cross-flow filtration. Powder Technol 118:193–202

Simon M, Johnson M, Tyack P, Madsen PT (2009) Behaviour and kinematics of ram filtration in bowhead whales (*Balaena mysticetus*). Proc R Soc B 276:3819–3828

Simon M, Johnson M, Madsen PT (2012) Keeping momentum with a mouthful of water: behavior and kinematics of humpback whale lunge feeding. J Exp Biol 215:3786–3798

Sims DW (2008) Sieving a living: a review of the biology, ecology and conservation status of the plankton-feeding basking shark *Cetorhinus maximus*. Adv Mar Biol 54:171–220

Slater GJ, Goldbogen JA, Pyenson ND (2017) Independent evolution of baleen whale gigantism linked to Plio-Pleistocene Ocean dynamics. Proc R Soc B 284:20170546

Slijper EJ (1962) Whales. Basic Books, New York

Smith JC, Sanderson SL (2008) Intra-oral flow patterns and speeds in a suspension-feeding fish with gill rakers removed *versus* intact. Biol Bull 215:309–318

Song L, Elimelech M (1995) Theory of concentration polarization in cross-flow filtration. J Chem Soc Faraday Trans 91:3389–3398

Speck T, Speck O (2008) Process sequences in biomimetic research. In: Brebbia CA (ed) Design and nature IV. WIT transactions on ecology and the environment, vol 114. WIT Press, Southampton, pp 3–11. https://doi.org/10.2495/DN080011

St. Aubin DJ, Stinson RH, Geraci JR (1984) Aspects of the structure and composition of baleen, and some effects of exposure to petroleum hydrocarbons. Can J Zool 62:193–198

Starbard N (2009) Beverage microfiltration. Wiley/Blackwell, Hoboken

Stevenson CH (1907) Whalebone: its production and utilization. Bur Fish Doc 626:1–12

Sumich JL (2001) Growth of baleen in a rehabilitating gray whale calf. Aquat Mamm 27:234–238

Sun F, Zhang X, Xue T, Cheng P, Yu T (2024) The performance testing and analysis of common new filter materials: a case of four filter materials. Materials 17(12):2802. https://doi.org/10.3390/ma17122802

Sutherland K (2005) The A-Z of filtration. Filtr Sep 42(10):40

Szewciw LJ, de Kerkhove DG, Grime GW, Fudge DS (2010) Calcification provides mechanical reinforcement to whale baleen alpha keratin. Proc R Soc B 277:2597–2605

Tamime AY (2013) Membrane processing: dairy and beverage applications. Wiley/Blackwell, Hoboken

Taylor AM, Bonser RHC, Farrent JW (2004) The influence of hydration on the tensile and compressive properties of avian keratinous tissues. J Mater Sci 39:939–942

Ten Hietbrink S, Materić D, Holzinger R, Groeskamp S, Niemann H (2025) Nanoplastic concentrations across the North Atlantic. Nature 643:412–416

Thewissen JGM (1998) The emergence of whales: evolutionary patterns in the origins of Cetacea. Plenum, New York

Thewissen JGM, Bajpai S (2001) Whale origins as a poster child for macroevolution. Bioscience 51:1037–1049

Thewissen JGM, Cooper LN, George JC, Bajpai S (2009) From land to water: the origin of whales, dolphins, and porpoises. Evol: Educ Outreach 2:272–288

Thewissen JGM, Hieronymus TL, George JC, Suydam R, Stimmelmayr R, McBurney D (2017) Evolutionary aspects of the development of teeth and baleen in the bowhead whale. J Anat 230:549–566

Tomilin AG (1954) Adaptive types of the order Cetacea. Zool Zh 33:677–691

Torres LG, Brander SM, Parker JI, Bloom EM, Norman R, Van Brocklin JE, Lasdin KS, Hildebrand L (2023) Zoop to poop: assessment of microparticle loads in gray whale zooplankton prey and fecal matter reveal high daily consumption rates. Front Mar Sci 10:e1201078

Trakumas S, Willeke K, Reponen T, Grinshpun SA, Friedman W (2001) Comparison of filter bag, cyclonic, and wet dust collection methods in vacuum cleaners. Am Ind Hyg Assoc J 62:573–583

Tsai C-H, Fordyce RE (2015) The earliest gulp-feeding mysticete from the Oligocene of New Zealand. J Mamm Evol 22:535–560

Uhen MD (2010) The origin(s) of whales. Annu Rev Earth Planet Sci 38:189–219

van der Hoop JM, Corkeron P, Kenney J, Landry S, Morin D, Smith J, Moore MJ (2015) Drag from fishing gear entangling North Atlantic right whales. Mar Mamm Sci 32:619–642

van der Hoop JM, Corkeron P, Moore MJ (2016) Entanglement is a costly life-history stage in large whales. Ecol Evol 7(1):92–106. https://doi.org/10.1002/ece3.2615

van der Hoop JM, Nousek-McGregor AE, Nowacek DP, Parks SE, Tyack P, Madsen PT (2019) Foraging rates of ram-filtering North Atlantic right whales. Funct Ecol 33:1290–1306
Vandenberg ML, Cohen KE, Rubin RD, Goldbogen JA, Summer AP, Paig-Tran EWM, Kahane-Rapport SR (2023) Formation of a fringe: a look inside baleen morphology using a multimodal visual approach. J Morphol 284(4):e21574
Vo HC, Pham MH (2021) Ecotoxicological effects of microplastics on aquatic organisms: a review. Environ Sci Pollut Res 28:44716–44725
Vogel S (1996) Life in moving fluids, 2nd edn. Princeton University Press, Princeton
Vogel S (2013) Comparative biomechanics: life's physical world, 2nd edn. Princeton University Press, Princeton
Vogel S, LaBarbera M (1978) Simple flow tanks for research and teaching. Bioscience 28(10):638–643
Vogel HC, Todaro CM (1997) Cross flow filtration. In: Bhave RR (ed) Fermentation and biochemical engineering handbook. Elsevier (Noyes), New York, pp 271–322
Wang B, Yan W, McKittrick J, Meyers MA (2016) Keratin: structure, mechanical properties, occurrence in biological organisms, and efforts at bioinspiration. Prog Mater Sci 76:229–318
Webber MA, Keener W, Markowitz TM, Chamberlin D, Allen D, Lane RS, Slaathaug JM, Rodriguez PN, George K, O'Hern JE (2024) Fish feeding and rapid foraging behavior switching by gray whales (*Eschrichtius robustus*) in California. Aquat Mamm 50(2):132–151
Wegner NC (2015) Elasmobranch gill structure. Fish Physiol 34(A):101–151
Werth AJ (2000) Marine mammals. In: Schwenk K (ed) Feeding: form, function and evolution in tetrapod vertebrates. Academic, New York, pp 475–514
Werth AJ (2001) How do mysticetes remove prey trapped in baleen? Bull Museum Comp Zool 156(1):189–203
Werth AJ (2004) Models of hydrodynamic flow in the bowhead whale filter feeding apparatus. J Exp Biol 207(20):3569–3580
Werth AJ (2007) Adaptations of the cetacean hyolingual apparatus for aquatic feeding and thermoregulation. Anat Rec 290:546–568
Werth AJ (2012) Hydrodynamic and sensory factors governing response of copepods to simulated predation by baleen whales. Int J Ecol 2012:208913
Werth AJ (2013) Flow-dependent porosity of baleen from the bowhead whale (*Balaena mysticetus*). J Exp Biol 216:1152–1159
Werth AJ (2017) Baleen. In: Würsig B, Thewissen JGM, Kovacs K (eds) Encyclopedia of marine mammals, 3rd edn. Elsevier, San Diego, pp 60–61
Werth AJ (2019) Variable porosity of throughput and tangential filtration in biological and 3D printed systems. In: Petrova V (ed) Advances in engineering research, volume 29: porosity—properties and measurement. Nova Science Publishers, Hauppauge, pp 37–93
Werth AJ, Crompton AW (2023) Cetacean tongue mobility and function: a comparative review. J Anat 243:343–373
Werth AJ, Ito H (2017) Sling, scoop, squirter: anatomical features facilitating prey transport, concentration, and swallowing in rorqual whales (Mammalia: Mysticeti). Anat Rec 300(11):2070–2086
Werth AJ, Marshall CD (2023) Convergent evolution of secondarily aquatic feeding in mammals. In: Bels V, Legreneur P (eds) Convergent evolution. Springer, Cham, pp 183–220
Werth AJ, Potvin J (2016) Baleen hydrodynamics and morphology of cross-flow filtration in balaenid whale suspension feeding. PLoS One 11:e0150106
Werth AJ, Potvin J (2024) Dynamic filtration in baleen whales: recent discoveries and emerging trends. Front Mar Sci 11:1347497. https://doi.org/10.3389/fmars.2024.1347497
Werth AJ, Sformo TL (2020) Anatomy and function of feeding. In: George JC, Thewissen JGM (eds) The bowhead whale, *Balaena mysticetus*: biology and human interactions. Academic, San Diego, pp 213–223
Werth AJ, Whaley HR (2019) Ocean acidification's potential effects on keratin protein in cetacean baleen and other integumentary tissue. Ann Ecol Environ Sci 3(2):21–28

Werth AJ, Harriss RW, Rosario MV, George JC, Sformo TL (2016a) Hydration affects the physical and mechanical properties of baleen tissue. R Soc Open Sci 3(10):160591
Werth AJ, Straley J, Shadwick R (2016b) Baleen wear reveals intraoral water flow patterns of mysticete filter feeding. J Morphol 277(4):453–471
Werth AJ, Potvin J, Shadwick RE, Jensen MM, Cade DE, Goldbogen JA (2018a) Filtration area scaling and evolution in mysticetes: trophic niche partitioning and the curious cases of the sei and pygmy right whales. Biol J Linn Soc 125(2):264–279
Werth AJ, Rita Espada D, Rosario MV, Moore MJ, Sformo TL (2018b) How do baleen whales stow their filter: a comparative biological analysis. J Exp Biol 221(23):189233
Werth AJ, Blakeney SM, Cothren AI (2019a) Oil adsorption does not structurally or functionally alter whale baleen. R Soc Open Sci 6(5):182194
Werth AJ, Kosma MA, Chenoweth EM, Straley JM (2019b) New views of humpback whale flow dynamics and morphology during prey engulfment. Mar Mamm Sci 35(4):1556–1578
Werth AJ, van de Graaf P, DesJardins R (2019c) Preparation of full baleen racks for long-term exhibition and research. Aquat Mamm 45(5):500–506
Werth AJ, Sformo TL, Lysiak NS, Rita Espada D, George JC (2020) Baleen turnover and gut transit in mysticete whales and its environmental implications. Polar Biol 43(6):707–723
Werth AJ, Sformo TL, Lysiak NS, Rita Espada D, George JC (2021) Differential baleen growth and its consequences. Polar Biol 44(6):1227–1228
Werth AJ, Kahane-Rapport SR, Potvin J, Goldbogen JA, Savoca MS (2024) Baleen–plastic interactions reveal high risk to all filter-feeding whales from clogging, ingestion, and entanglement. Oceans 5(1):48–70
Williamson GR (1973) Counting and measuring baleen and ventral grooves of whales. Sci Rep Whales Res Inst 24:279–292
Witkop EM, Van Wassenbergh S, Heideman PD, Sanderson SL (2023) Biomimetic models of fish gill rakers as lateral displacement arrays for particle separation. Bioinspir Biomim 18(5):056009
Xu Z, Mao X, Gu Y, Chen X, Kuang W, Wang R, Shao X (2023) Analysis of hydrodynamic filtration performance in a cross-step filter for drip irrigation based on the CFD-DEM coupling method. Biosyst Eng 232:114–128. https://doi.org/10.1016/j.biosystemseng.2023.07.004
Yeong W-Y, Chua C-K, Leong K-F, Chandrasekaran M (2004) Rapid prototyping in tissue engineering: challenges and potentials. Trends Biotechnol 22:643–652
Young S (2012) The comparative anatomy of baleen: evolutionary and ecological implications. MSc thesis, San Diego State University, San Diego
Young S, Deméré TE, Ekdale EG, Berta A, Zellmer N (2015) Morphometrics and structure of complete baleen racks in gray whales (*Eschrichtius robustus*) from the Eastern North Pacific Ocean. Anat Rec 298:703–719
Zeman LJ, Zydney AL (1996) Microfiltration and ultrafiltration: principles and applications. Dekker, New York
Zhang X, Ma J, Zheng J, Dai R, Wang X, Wang Z (2022) Recent advances in nature-inspired antifouling membranes for water purification. Chem Eng J 432:134425. https://doi.org/10.1016/j.cej.2021.134425
Zhao S, Zettler ER, Bos RP, Lin P, Amaral-Zettler LA, Mincer TJ (2022) Large quantities of small microplastics permeate the surface ocean to abyssal depths in the South Atlantic Gyre. Glob Chang Biol 28:2991–3006
Zhong PS, Chung T-S, Jeyaseelan K, Armugan A (2012) Aquaporin-embedded biomimetic membranes for nanofiltration. J Membr Sci 407–408:27–33
Zhu Y, Hu D, Yang G (2020a) Theoretical analysis of the hydrodynamic filtering system in the balaenid whales suspension feeding. Bioinspir Biomim 16(2):e26006
Zhu Y, Yang G, Zhuang C, Li C, Hu D (2020b) Oral cavity flow distribution and pressure drop in balaenid whales feeding: a theoretical analysis. Bioinspir Biomim 15(3):e036004
Zhu Y, Hu D, Li C, Zhuang C, Yang G (2021) CFD-DEM simulation of the hydrodynamic filtration performance in balaenid whale feeding. Sci Total Environ 787:147696
Zhu Y, Hu D, Guo Y, Ding H, Yang G (2023) Bio-inspired filter design based on vortex control mechanism of parallel groove structure. J Bionic Eng 20:338–348

Chapter 6
Comparative Study of Two Caudal Fin Morphologies of the Shortfin Mako Shark (*Isurus oxyrinchus*) During Locomotion

Matthew Rodrigez, Erin E. Hackett, and Roi Gurka

Abstract Many aquatic organisms locomote using body-caudal-fin oscillatory motions. Within elasmobranchs, the shortfin mako is considered the most hydromechanically efficient species. They have evolved numerous morphological adaptations to enhance their hydromechanical efficiency. For this study, we considered the lunate caudal fin and thunniform oscillations shortfin Makos employ to propel themselves. Most elasmobranchs have a heterocercal caudal fin paired with carangiform oscillations. This study investigated the hydrodynamics of the shortfin mako using a flexible scale model with a robust oscillating caudal fin that emulates this species' natural thunniform body motions. Here, the shortfin mako's lunate caudal fin was replaced with the heterocercal caudal fin common to other sharks; a comparison between the wake characteristics of two morphologies was performed to shed light on the effects caudal fin morphology has on hydrodynamics when paired with thunniform oscillations. Experiments were conducted in a recirculating water flume using particle image velocimetry (PIV) to measure the wake velocity field formed behind each oscillating caudal fin. The velocity fields were used to estimate the sectional drag formed during steady forward swimming, as well as the near-wake turbulence characteristics. Results indicate that the different morphologies generated similar wake characteristics, yet there was a notable discrepancy in the energetics and size of the wake, where the lunate tail had a narrower, more energetic wake, especially in terms of rotational fluid motion. Due to these features, the wake also decayed more slowly as the distance from the caudal fin increased.

Keywords Shark's hydrodynamics · Mako shark · Locomotion · Turbulence

---

M. Rodrigez · E. E. Hackett
Marine Science, Coastal Carolina University, Conway, SC, USA

R. Gurka (✉)
Physics and Engineering Science, Coastal Carolina University, Conway, SC, USA
e-mail: rgurka@coastal.edu

V. Bels et al. (eds.), *Biomimicry – A Functional Approach*, Biologically-Inspired Systems 20, https://doi.org/10.1007/978-3-032-11949-0_6

6.1 Background

Movement through an aquatic system is affected by changes in abiotic factors (temperature, currents, salinity, etc.) that have led sharks and other fishes to adapt in order to survive (locate prey, migrate, avoid predation, etc.). The field of biomimicry derives inspiration from the natural world for solutions that may be applied to technologies [autonomous underwater vehicles (AUVs), surface vessels, airplanes, etc.] to improve them and mitigate obstacles. For example, different swimming modes (i.e., anquilliform, carangiform, thunniform) utilized by fishes to move throughout their environments may inspire the design of future biomimetic aquatic technologies (Suleman and Crawford 2008). These different swimming modes are paramount to the success of different pelagic organisms and can be influenced by their morphological adaptations (e.g., fins, scales, mucus). For example, the important effects the caudal fin morphology has on a fast, open ocean-dwelling shark species using the thunniform swimming mode for optimizing hydrodynamic efficiency during long migrations (Lauder 2000; Wilga and Lauder 2000; Chopra 1976; Sfakiotakis et al. 1999; Bernal et al. 2001; Graham et al. 1990). Shortfin makos are rare among sharks in that they employ thunniform locomotion in combination with a lunate caudal fin where both the upper and lower lobes are nearly symmetrical. Their propulsion relies primarily on tail oscillations, and this unique combination of movement and morphology presents promising applications for bioinspired autonomous underwater vehicles. In this study, we investigate the hydrodynamic differences between symmetrical and asymmetrical (i.e., blacktip shark caudal fin) caudal fin shapes, both operating under thunniform mode of locomotion. By isolating shape as the variable, this comparison aims to clarify the role of caudal fin symmetry in aquatic propulsion efficiency and maneuverability.

6.1.1 *Fish Propulsion*

The types of undulations performed by swimming fishes using body-caudal-fin (BCL) motion fall into four main categories of modes: anguilliform, subcarangiform, carangiform, and thunniform (Breder 1926; Lighthill 1970; Webb 1978; Lauder and Tytell 2005; Sfakiotakis et al. 1999), which are illustrated in Fig. 6.1. Each of these modes corresponds to differing degrees of lateral extension over a defined body length, with anguilliform and thunniform having the most and least, respectively. By undulating, fishes generate a propulsive wake. This wake may be further augmented by the fish when its median fins are aligned in an angle with respect to the mean flow direction; thus, the fish imparts momentum into the flow more efficiently and reduces the size of the wake (Gray 1953; Lighthill 1970). Other morphological adaptations (e.g., median fins), such as innate behaviors (e.g., schooling) and secretion of materials (e.g., mucus), are important to understand how fish species swim in various modes and at varying efficiencies (Fish 1998).

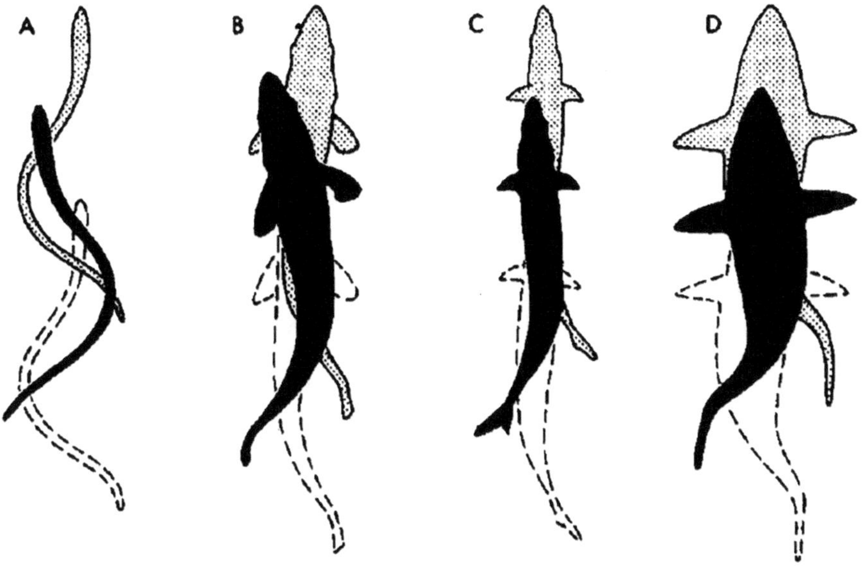

Fig. 6.1 Depiction of the four most common fish swimming modes from (**a**) anguilliform, (**b**) subcarangiform, (**c**) carangiform, and (**d**) thunniform. The black silhouette (dorsal view) is superimposed on successive positions one-half tail beat earlier (broken outline) and one-half tail beat later (stippled). (Lindsey 1978)

Morphological adaptations, including caudal fin shape, body depth, along with kinematic differences between different swimming modes, have implications for hydromechanical efficiency, speed, and thrust generation. Relationships between parameters such as oscillation frequency and oscillation amplitude have been linked with hydrodynamic characteristics such as thrust and efficiency (Xia et al. 2015). Direct relationships between forward velocity and oscillation frequency have been observed, and inverse relationships between oscillation amplitude and efficiency have been reported (Xia et al. 2015; Lighthill 1970; Webb et al. 1984). The oscillation frequency of the caudal fin has been shown to have the greatest effect on hydromechanical performance (Akanyeti et al. 2017; Tytell 2004; Webb and Weihs 2015) and is the largest contributing factor to increased thrust.

Thrust and acceleration are generated to allow for forward propulsion, as well as the capture of prey, and evasion of predators. Certain fishes such as tunas and trout may also mitigate flow separation by controlling body-generated vortices and moving the separation point farther posterior on their bodies. Shifting the separation point posteriorly decreases the effect of pressure drag on the body (Çengel and Cimbala 2006). A previously conducted study concluded that as fish undulate their bodies generate vortices that are then passed posteriorly (Zhu et al. 2002). These body-generated vortices may be augmented by the fish to interact constructively or destructively with the vortices generated by the caudal fin. Pairing same-sign body-generated vortices with caudally generated vortices results in constructive interference of the vortices at the wake and eventually leads to an increase in thrust. While pairing opposite-sign vortices from the body and caudal fin may result in destructive interference that increases propulsive efficiency but decreases thrust generation (Zhu et al. 2002; Yang and Su 2011). This negative interference was shown to be the case for tuna utilizing the thunniform mode of swimming and for the giant danio utilizing the carangiform mode (Zhu et al. 2002). It has also been shown that certain fish with flexible median fins also may augment their swimming by utilizing the vortices shed from the trailing edge of these fins. At higher swimming speeds, the vortex street behind the dorsal fin of a bluegill sunfish combines constructively with the wake behind the body, and this enhances its overall thrust generation (Drucker and Lauder 2001; Tytell 2006).

Different modes of undulation can also affect swimming efficiency (Chopra and Kambe 1977; Lighthill 1960; Schouveiler et al. 2005). For those swimming organisms utilizing a mode with greater lateral oscillations, such as the carangiform or anguilliform modes, swimming efficiency decreases due to the effect of recoil on the body (Lighthill 1969). The force that a swimming organism exerts on the surrounding fluid is equal and opposite to the force that the surrounding fluid exerts back onto the swimming organism, assuming steady state (zero acceleration). Therefore, an organism that moves its body more while swimming expends more energy to move and counteract recoil, thereby reducing efficiency. It has been demonstrated that those organisms performing carangiform oscillations generate higher thrust than thunniform swimmers at the cost of hydromechanical efficiency (Li et al. 2012; Chopra 1976). Hydromechanical efficiency of a flapping foil is the ratio of the work used for propulsion by the total amount of work expended through the flapping motion (Lighthill 1970; Sfakiotakis et al. 1999).

For thunniform swimmers, body oscillations are restricted to approximately the posterior one-third of the body with very little lateral head movement. This reduction in overall body movement helps increase the hydromechanics efficiency by decreasing the effect of recoil. High swimming efficiency does come at the cost of thrust generation as thunniform swimmers must beat their caudal fins faster and at greater amplitudes in order to accelerate (Xia et al. 2015). Body shape also has a significant effect on swimming efficiency. Thunniform swimmers have a fusiform body shape, which includes a reduction in depth of the body's cross section and a narrow caudal peduncle (Syme and Shadwick 2011). The reduction in depth of the cross section is presumably to reduce the recoil experienced by the body as it oscillates. Because of this narrowing, the anterior section of the body has a significant drag-reducing effect while not negatively affecting thrust generation (Lighthill 1970).

6.1.2 Shark Propulsion

The evolutionary history of sharks has seen them radiate and adapt to occupy nearly every oceanic habitat, from sedentary species on the sandy bottom, to those that constantly swim in the open ocean and never see the bottom (Sorenson et al. 2014). Most shark species have a tapered body with one or more dorsal fins, paired pectoral and anal fins (Thomson and Simanek 1977). One of the main differences distinguishing sharks from bony fish is their inability to directly control their buoyancy (Gleiss et al. 2017). Sharks compensate in part for their negative buoyancy with an oil-filled liver or fatty tissues (Cohen and Cleary 2010). Despite this, sharks must generate lift (Moss 1984; Flammang et al. 2011). Some sharks utilize ram ventilation (Graham et al. 1990) as they do not have branchial muscles strong enough to actively draw water over their gills and thus must mechanically force it by forward swimming. By utilizing ram ventilation, sharks can move at cruising speed without altering metabolic costs. Most sharks utilize carangiform or sub-carangiform swimming modes, while more specialized species adapted for high-performance swimming utilize thunniform. Using different morphological adaptations such as the caudal fin upper (epichordal) lobe, caudal fin lower (hypochordal) lobe, median fin shape, or their dermal denticles may increase their hydrodynamic efficiency.

6.1.3 Lamnids and Scombrids

Thunniform swimmers, which include scombrid fishes, cetaceans, and lamnid sharks, are thought to be the most efficient swimming organisms in the oceans (Liu et al. 2017) due to their need for constant forward motion during long-distance migrations. Because of this similarity, lamnids have convergently evolved to be ecologically and morphologically similar to scombrids (Syme and Shadwick 2011; Shadwick 2005). Shared morphological adaptations include internalized red muscle, narrow necking toward the caudal peduncle, endothermy, thunniform oscillations, and caudal fin geometry. Scombrid fishes are considered to be a prey item of certain lamnid sharks, and this selective pressure may be another reason these distantly related families have convergently evolved. These fishes embark on long-distance migrations, that is, they must maximize their swimming efficiency, reaching the fastest cruising speed possible while expending a minimal amount of energy (Trump and Leggett 1980).

Many studies have been conducted to investigate the hydrodynamic characteristics of swimming fish (Triantafyllou et al. 1993; Wang et al. 2010; Wolfgang et al. 1999; Cohen and Cleary 2010). In the specific case of thunniform swimming, the majority of studies on hydrodynamic characteristics have been relegated to studies involving tuna, including numerical simulations, controlled experiments using oscillating flat plates with lunate caudal fin geometries, scale physical models, and water tunnel tests with live individuals. The choice of tuna is due to the comparative ease of using them as a model species rather than lamnids like the shortfin mako (Yang and Su 2011; Hong and Chang-an 2005; Wang et al. 2010; Takagi et al. 2013; Dewar and Graham 1994). This choice primarily stems from the logistical difficulties involved with shortfin makos as a model species, including capture, transport, husbandry, and cost. Simulations of tuna have, for example, quantified vorticity distributions in the wake, as well as observed the interactions of constructive and destructive interference between bodily shed vortices and caudal fin-shed vortices (Gopalkrishnan et al. 1994; Zhu et al. 2002). Xia et al. (2015) determined that the resulting wake generated by tuna consists of a series of disconnected sickle-shaped vortices consistent with observations in nature (Anderson 1996). They also suggested that thunniform swimmers achieve high cruising speeds with high hydromechanical efficiency. While these characteristics have yet to be confirmed using a lamnid shark as the model species instead of tunas, the ecological and morphological similarities between the two distantly related species suggest that lamnids would generate similar characteristics.

6.1.4 Caudal Fin Morphology

Caudal fin geometry plays a role in aquatic organisms locomoting efficiently through their environment (Sambilay 1990). There is a wide variety of caudal fin geometries observed in nature, suggesting that it may serve a different hydrodynamic purpose for various species. One of the useful geometrical characteristics of the caudal fin geometry is the aspect ratio (*AR*), defined as $A_R = {}^{W}/_{c}$, where W is the span of the caudal fin and the mean chord length is c. Previous studies have found that varying the *AR* will change propulsive performance and wake patterns (Raspa et al. 2014; Yeh and Alexeev 2016). A shark's caudal fin serves important functions such as propulsion and generating enough lift to counteract a shark's naturally negatively buoyant body (Lighthill 1975). Slower-moving species that prioritize burst acceleration and maneuverability usually have lower aspect ratio caudal fins. Lower aspect ratio caudal fins generate greater thrust when accelerating or maneuvering, which is important for evading predators or capturing prey (Liu and Dong 2016; Cheng and Murillo 1984; Flammang and Lauder 2009) but generally operate at a lower hydromechanical efficiency (Lee et al. 2017). Faster-moving species that prioritize sustained swimming and efficiency generally have high aspect ratio caudal fins (Lighthill 1969; Chopra 1974; Sagong et al. 2013; Liu and Dong 2016). High aspect ratio caudal fins are normally seen on those species that also employ the thunniform mode of swimming. This pairing, while efficient, has been shown to generate slightly less thrust than lower aspect ratio caudal fins (Lee et al. 2017).

The fastest moving and the most hydromechanically efficient sharks all belong to the family Lamnidae, which have high aspect ratio lunate caudal fins (Shadwick 2005). Lunate caudal fins have been described as a two-dimensional airfoil (Lighthill 1969), with a rounded leading edge and a sharp trailing edge. The epichordal (upper) and hypochordal (lower) lobes are considered to be approximately the same shape and size, similar to that of scombrid fishes. A lunate caudal fin that exhibits a high aspect ratio fin has been shown to generate low levels of thrust compared to ones with lower aspect ratios, presumably due to the pairing with the thunniform mode of locomotion. This pairing elicits leading-edge separation (Chopra 1976) because of the strong leading-edge suction acting on the fin. Leading-edge separation greatly reduces thrust and efficiency. To generate high thrust, Chopra (1976) argues that lunate-tailed organisms need to swim at a high angle of attack. During steady forward locomotion, the oscillating caudal fin generates a forward thrust force, which acts on the center of mass of the shark (Thomson and Simanek 1977). Through each lateral oscillation, force is simultaneously directed horizontally and upward, counteracting (Bushnell and Moore 1991) and gravity. Regardless of caudal fin morphology or swimming mode, the thrust force consists of two components: (i) the along-body force associated with the thrust directed horizontally during each lateral caudal fin beat and is dependent on the angle of the caudal fin with respect to the oncoming flow (Thomson and Simanek 1977) and (ii) the transverse force that is the thrust directed upward (normal to the body) as a result of the caudal fin rotating about the vertical spanwise plane. The specialized lunate caudal fin morphology (in combination with other characteristics) allows high-performance pelagic predators to achieve success. This morphology has convergently evolved in the majority of swimming organisms that use the thunniform swimming mode. The shape of the caudal fin is not the only important factor but the kinematics of the fin are also important (Wilga and Lauder 2000). An oscillating caudal fin may change orientation such that the trailing edge shape varies throughout a tail beat cycle. Certain species, such as the bluegill sunfish, possess a flexible caudal fin made of individual fin rays connected by tissue that can be directly controlled to change the trailing edge shape and thus augment certain hydrodynamic performance. While other species, such as the bluefin tuna, have a rigid caudal fin with a rounded leading edge and sharp trailing edge that cannot be directly controlled. A rigid caudal fin with a rounded leading edge allows the fish to take advantage of water moving sharply around it, generating a leading-edge suction force parallel to the chord of the fin (Sane 2003), while having a sharp trailing edge promotes vortex shedding (Lighthill 1969). A vortex sheet transfers momentum more efficiently than one that is unorganized and may lead to a narrower wake, thereby reducing the drag force acting on the fin (Han et al. 2017; Xia et al. 2015). Liu and Dong (2016) demonstrated that tunas generate more concentrated vorticity in their wake than trout. This concentrated vorticity may be a reason that tunas have been found to be more hydromechanically efficient (0.25) than those with a lower aspect ratio caudal fin such as trout (0.16).

6.1.5 Shortfin Mako Shark

The shortfin mako (*Isurus oxyrinchus*) is a large, pelagic, migratory, and purportedly the fastest shark species found in all the tropical seas. They are capable of bursts up to 19 ms^{-1} (Afroz et al. 2016; Compagno 2001), but are normally found cruising at approximately 0.64 ms^{-1} to conserve energy (Sepulveda et al. 2004) with a

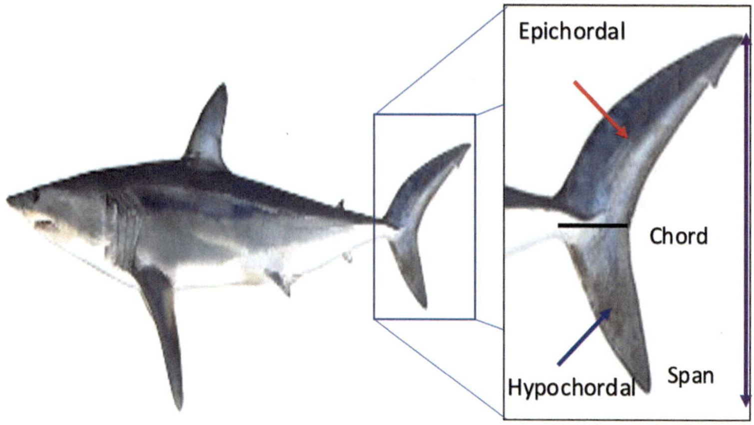

Fig. 6.2 Shortfin mako shark with its caudal fin enlarged to show the hypochordal (lower) lobe (blue arrow), epichordal (upper) lobe (red arrow), chord (black line), and span (purple double-sided arrow). (Source: CSIRO National Fish Collection. License: CC-BY-Attribution)

tail oscillation frequency of 1 Hz (Sepulveda et al. 2007). Shortfin makos are an exclusively pelagic species that never stop swimming for their entire lives. Shortfin makos also feed on fast-moving teleost fish such as tuna, bluefish, and billfish. They are more ecologically and morphologically similar to scombrids than other members of the lamnid family (Syme and Shadwick 2011). Due to their need for constant forward motion, long-distance migrations, and their similarities to scombrids, shortfin makos are thought to be extremely hydromechanically efficient. Shortfin makos have several physiological evolutionary adaptations that are utilized to achieve these high speeds and efficiency of movement. The adaptations include ridged dermal denticles, internalized red muscle, large gill slits, a torpedo-shaped body, lunate caudal fin (Fig. 6.2), and thunniform swimming mode (Du Clos et al. 2018; Sepulveda et al. 2004; Carey and Teal 1969). The increased cruising speed is further facilitated by shortfin makos being endothermic while most other shark species are ectothermic (Harding et al. 2021). These adaptations allow shortfin makos to have a lower caudal fin beat frequency compared to other shark species but still swim at comparable cruising speeds (Graham et al. 1990); this unique pairing of a lunate caudal fin with thunniform oscillations could be related to the shortfin makos' rapid swimming (Stevens 2008). Shortfin makos do however increase their caudal fin oscillation frequency and lateral head movement when accelerating, an action typical of predation and predator avoidance. This increased lateral motion may affect their hydrodynamic performance and is not considered in this study, which focuses on swimming at a constant speed.

We investigate one of the unique features of the shortfin mako, which is the pairing of its unique lunate caudal fin morphology with its thunniform swimming mode. We use a flexible model of the shortfin mako that is able to replicate the thunniform swimming mode using a mechanically generated motion. This oscillating scale model is fit with two different interchangeable caudal fin morphologies to study the effect caudal fin morphology has on the hydromechanics when paired with its natural thunniform swimming mode. We have measured and compared the near-wake characteristics behind an oscillating shortfin mako model with two distinct caudal fin morphologies to shed light on the effect caudal fin morphology has on the swimming efficiency of shortfin makos when paired with thunniform oscillations.

6.2 Experiments

We investigated a high aspect ratio lunate caudal fin and compared it with a lower aspect ratio heterocercal caudal fin both oscillating at a predetermined frequency. The comparison aims to evaluate the lunate caudal fin morphology and its importance to the hydrodynamic performance achieved by the shortfin mako shark while performing thunniform oscillations. We choose heterocercal fin morphology for comparison given that many shark species pair carangiform oscillations with a lower aspect ratio heterocercal caudal fin. Heterocercal caudal fins are characterized by having a larger epichordal lobe and smaller hypochordal lobe. This pairing of oscillations and caudal fin morphology has been shown to generate greater thrust with less hydromechanic efficiency than thunniform swimmers utilizing a lunate caudal fin (Lee et al. 2017). The lower aspect ratio heterocercal caudal fin has a higher burst potential (e.g., for predator evasion or prey capture) but still allows for long-distance migration (Turner 2020; Reyier et al. 2008). One difference between the lunate and heterocercal caudal fin morphologies, other than the aspect ratio, is its flexibility. Both lobes of the lunate caudal fin are known to be rigid, while the heterocercal caudal fin lobes are more flexible (Crofts et al. 2019). This dissimilarity in flexibility may attribute to the differences in hydromechanical efficiency between the two morphologies due to the increased recoil experienced by the heterocercal morphology (Wilga and Lauder 2002).

These two caudal fin morphologies were studied in a set of laboratory flume experiments where the flow field was measured using particle image velocimetry (PIV) (see schematic configuration in Fig. 6.3). The flow field was measured in two planes at three locations in the near wake of both caudal fin morphologies. The measured flow field is used to estimate turbulence characteristics and drag from momentum deficits present in the wake behind the model (Goett 1939). Results are presented as phase averages to account for the unsteady motion of the model.

6.2.1 *Shortfin Mako Model*

The model used in this study was based on a 3D drawing (TurboSquid.com), shown in Fig. 6.4a, and was rendered using SolidWorks. The model total length (TL) was 50 cm, which equates to a 6:1 scale model of a 3 m adult shortfin mako. A negative

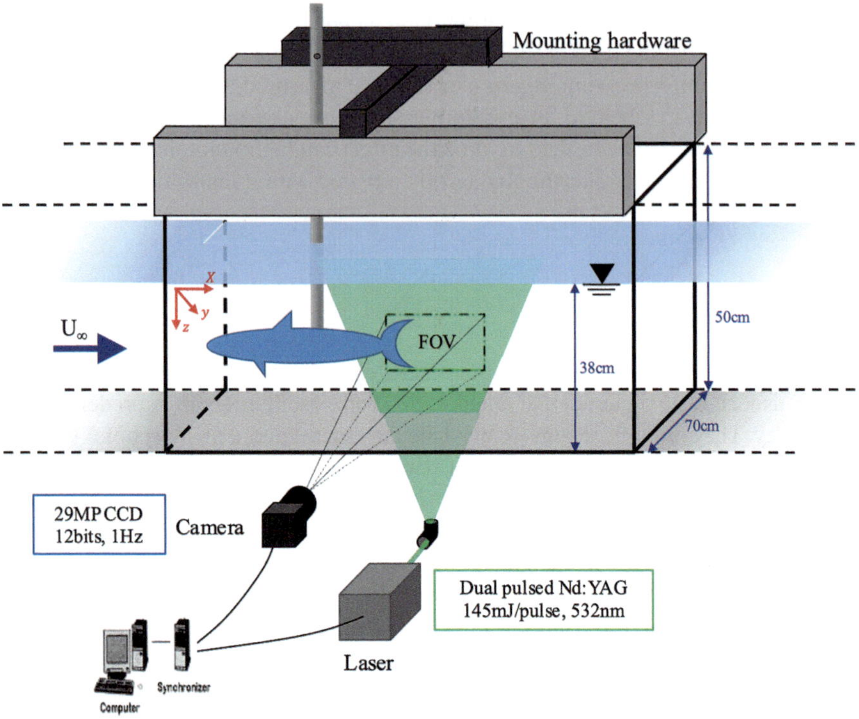

Fig. 6.3 Schematic configuration of the experiential setup

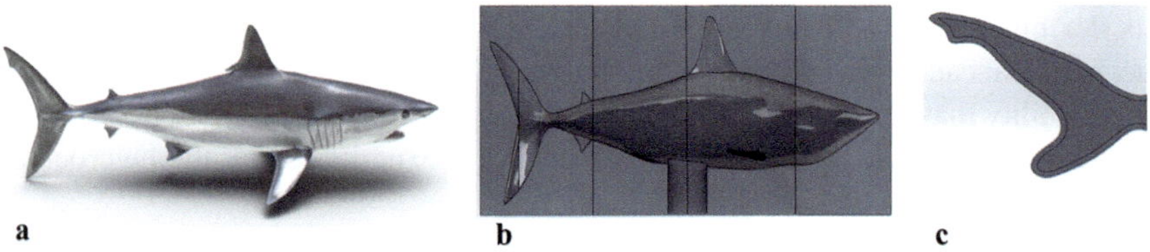

Fig. 6.4 (**a**) Shortfin mako 3D drawing (TurboSquid.com) used to create the mold. (**b**) Schematic of the shortfin mako mold, that is, the negative of the shortfin mako drawing in (**a**), where the vertical black lines indicate the different sections of the mold that were printed separately. (**c**) The heterocercal caudal fin morphology investigated in this experiment, modeled after a blacktip shark (*Carcharhinus limbatus*)

of this shortfin mako model was used to generate a mold. The exterior shell was 3D printed using Onyx plastic filament (Markforged Onyx One). A mix of carbon and nylon served as the printing material for the mold, with pieces printed individually and then combined to form the entire mold, as shown in Fig. 6.4b.

The mold for the alternate heterocercal morphology was printed separately and attached to the existing model once the original lunate morphology was removed; this alternate morphology was derived from an image of a blacktip shark

(*Carcharhinus limbatus*) and is shown in Fig. 6.4c. Recall that we chose to add the heterocercal caudal fin for comparison because it is a common morphology for most other shark species. To add flexibility to the model, dragon skin platinum cure silicone mixed with paint thinner was cast into the mold and allowed to cure. Paint thinner was added to the silicone mixture to generate a material that closely mimicked the flexibility of a living shark. We assume that bodily flexibility of shark skin is generally similar across species; thus, a target Young's modulus was taken to be 1.0×10^3 Pa (Epps et al. 2009; Turner 2020). A mixture of 80% silicone and 20% paint thinner generated a material with a Young's modulus of 1.1×10^3 Pa, which was deemed similar enough. It is worth mentioning that the caudal fin flexibility is not entirely consistent across morphologies in nature with the lunate morphology known to be a little more rigid than the heterocercal morphology. Previous research (Wilga and Lauder 2002; Crofts et al. 2019) emphasizes the role of caudal fin flexibility for hydrodynamic performance. However, we did not account for this difference in the current study due to a lack of data on the differing material properties of the two caudal fins. Thus, both caudal fin morphologies have approximately the same flexibility so that the focus of the study remains on the morphological differences only.

Lateral caudal fin oscillations closely resembling the thunniform swimming mode were generated by a small servo motor embedded inside a protective housing within the mold before the silicone was poured. The servo was controlled via an Arduino microcontroller. The caudal fin beat frequency was set to 0.4 Hz to obtain a Strouhal number within an optimal range (Triantafyllou et al. 1993), which for fish is between 0.25 and 0.45 (Taylor et al. 2003). A support plate was also placed within the model attached to the back (tailward side) of the protective housing around the motor. This plate served as an anchoring point for the model, as well as for directing the torque produced by the motor posteriorly. The rotor arms of the servo motor imparted torque on a vertical swim plate located in the posterior end of the model, causing the caudal to oscillate laterally. This lateral motion was facilitated by attaching 2 C-shaped inserts to the inside of the mold before pouring the silicone. These inserts created a natural pivot point for the model, mimicking that of thunniform swimmers (see Fig. 6.5). The model was positioned in the center of the flume to avoid any interference from the walls and surface of the flume using a threaded anchoring rod. Any turbulence produced by the mounting structure was assumed to be small enough not to interfere with the wake produced by the model (Adaramola et al. 2006).

6.2.2 *Particle Image Velocimetry Measurements*

Experiments took place in a recirculating open-channel flow facility (flume) with an overall length of 15 m and a cross section of 0.7 m × 0.5 m (Fig. 6.6a). A centrifugal pump generates flow rates up to 3 m^3/min. A 5 m long glass section in the middle of the flume allows for visualization of the flow. At the exit of the upstream

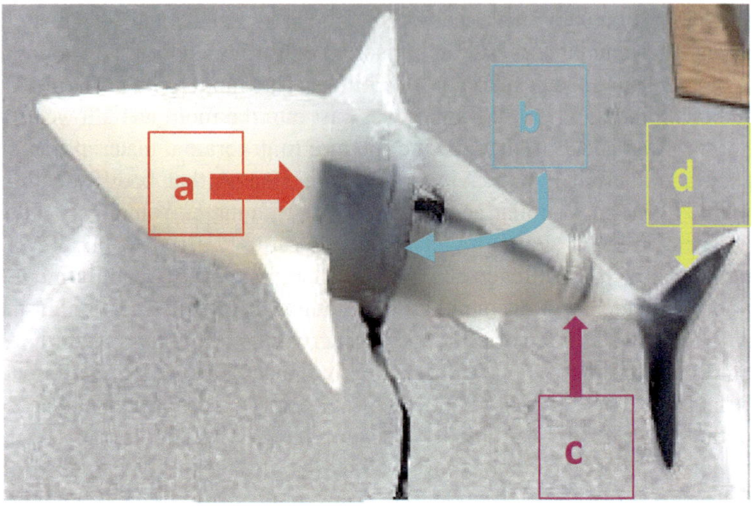

Fig. 6.5 The silicone shark model with the motorized lunate caudal fin. (**a**) Protective housing surrounding the servo motor that activates the motion. (**b**) Support plate that anchors the model. (**c**) Swim plate that allows the caudal fin to oscillate in a controlled manner. (**d**) Vertical plate inserted within the lunate caudal fin

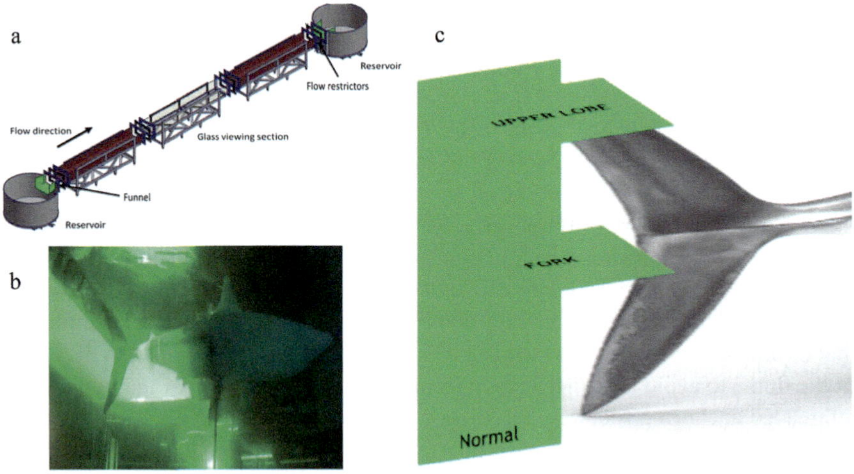

Fig. 6.6 (**a**) Isometric rendering of the flume. (**b**) The shortfin mako model positioned in the flume with the laser illuminating the streamwise-spanwise (x, y) plane behind the fork of the lunate caudal. (**c**) The PIV light sheet configurations used in this study

reservoir, a funnel and set of screens serve to straighten the flow as it enters the trough, while flow restrictors (a series of vertical pipes) at the exit of the trough serve to maintain a constant water level across the trough and prevent wave reflections from the downstream reservoir reverberating back into the flume. Water temperature in the flume was maintained by the surrounding ambient room temperature and was normally 20–22 °C and water height in the flume was 38 cm to ensure the model was fully submerged. The streamwise (flow) velocity was 0.20 m/s, resulting in a Reynolds number of 10^5 based on the model length and a Strouhal number of 0.36. It is noteworthy to mention that common Reynolds numbers based on sharks' lengths are in the order of 10^7. However, at these ranges, the flow is inherently turbulent; thus, while the Reynolds number herein is low compared to nature, the underlying flow mechanisms (i.e., inertia, viscosity, and unsteadiness; Gurka et al. 2022) are presumably similar.

Flow measurements were obtained using particle image velocimetry (PIV) (Adrian 1991; Raffel et al. 2018). Microscopic (11 μm diameter), neutrally buoyant, and chemically inert glass sphere beads were used as seeding particles (Melling 1997). A dual pulsed laser, a series of optical lenses, and a mirror were used to create a thin light sheet (~1 mm thick) that illuminated the flow and tracer particles. The laser was a high-powered Nd-YAG laser, emitting a light with power up to 140 mJ/pulse at a wavelength of 532 nm. A CCD double-exposure camera with 29-million-pixel resolution, double-exposure frame rate maximum of 1 Hz, and a dynamic bit depth of 12 was used to capture images of the illuminated particles. The light sheet illuminated either the streamwise-spanwise (x, y) plane or the streamwise-normal plane (Fig. 6.6c). The former setup was used to measure two-dimensional velocities (u, v) at two vertically separated locations for each morphology; specifically, near the top of the upper lobe and at the fork (Fig. 6.6c). Measurements at the upper lobe were taken 1 cm down from the tip of the lobe for each morphology. To illuminate the streamwise-normal plane, the laser was placed below the flume and the beam directed horizontally at a 45° mirror so that it reflected upward at a 90° angle, then through the same series of optics. This light sheet illuminated the streamwise-normal plane (x, z) directly posterior to the caudal fin measuring the u and w components of the velocity field. A total of 1200 image pairs were collected in each experiment to ensure statistical convergence. An image of the setup for the streamwise-spanwise plane is shown in Fig. 6.6b.

The analysis of the PIV images was performed on 64×64 pixel^2 interrogation windows with 50% overlap between adjacent windows (Willert and Gharib 1991). The tailbeat cycle was barely resolved given the 0.4 Hz caudal fin oscillation and the 1 Hz PIV frame rate; thus, we coarsely divide the PIV velocity vector maps into three phases: the leftmost and rightmost excursions of the caudal fin and the mid-phase. We present results for the first two phases (leftmost and rightmost) for data taken in the streamwise-spanwise plane, while the mid-phase data is used for data taken in the span plane. Division of the data into these phases was done visually for all image pairs, where any images not clearly at the leftmost, rightmost, or mid-position were removed from the data analysis. Flow quantities were subsequently phase-averaged over the set of PIV frames corresponding to each "left," "right," and "mid" phase, and with phase-averaging denoted with a caret (^) above the variable.

6.3 Results

We compare the general wake characteristics behind each of the caudal fin morphologies. With this understanding of the mean flow, we compare turbulence characteristics to examine whether the relatively unique high aspect ratio caudal fin morphology of the shortfin mako (relative to other shark species) interacts differently with the flow field.

6.3.1 Wake Mean Flow Characteristics

We show only the characteristics of the left phase of each morphology in the streamwise-spanwise plane given the similarity (symmetry) in results between the rightmost and leftmost phases. Results from the streamwise-normal (or span) plane are for the phase of the tailbeat cycle where the caudal fin is directly behind the model. Figure 6.7 presents contour map of the magnitude of the phase-averaged streamwise velocity, $\hat{u}$. In each measured PIV plane, a velocity deficit appears in the wake for both morphologies. The lowest velocity can be seen at small x-positions, directly behind the caudal fin. The velocity field does not recover back to the free stream velocity magnitude at any measured x-position within the wake in the FOV (field of view), but it does generally increase as the streamwise distance from the caudal fin increases. Measurements taken at the upper lobe for each morphology show the smallest velocity decrease and smallest spanwise wake width, with the velocity only decreasing ~0.04 m/s. These observations are possibly due to the small local length, where local length refers to the distance between the leading and trailing edges of the fin for each PIV measurement plane (for both morphologies). The local length is the smallest at the upper lobe of the lunate morphology, at 0.012 m. The overall similarity in the size and velocity magnitude for each PIV plane suggests that the wake is not significantly affected by the change in caudal fin morphology; however, the wake formed at the fork appears to be slightly more organized with strong shear at the edges for the lunate tail compared to the other morphology where the wake is weaker (lower deficit) but has broader signature in the spanwise direction and recovers at closer x-positions. This faster wake recovery may be associated with the caudal fin morphology and may present a "by-product" advantage for avoiding detection by predators that is needed by slower swimming sharks.

The phase-averaged vorticity ($\hat{\omega_z}$ and $\hat{\omega_y}$) shown in Fig. 6.8 presents a similar picture of the wake dynamics. The presence of two regions with vorticity in opposite directions directly behind the caudal fin shows that the oscillating caudal fin is generating two distinct shear layers (Williamson and Roshko 1988), suggesting that both morphologies could be propulsive. The upper lobe of each morphology generates smaller regions of concentrated vorticity than behind the fork, possibly due to the small local length again. The shear layers at this location are largest in

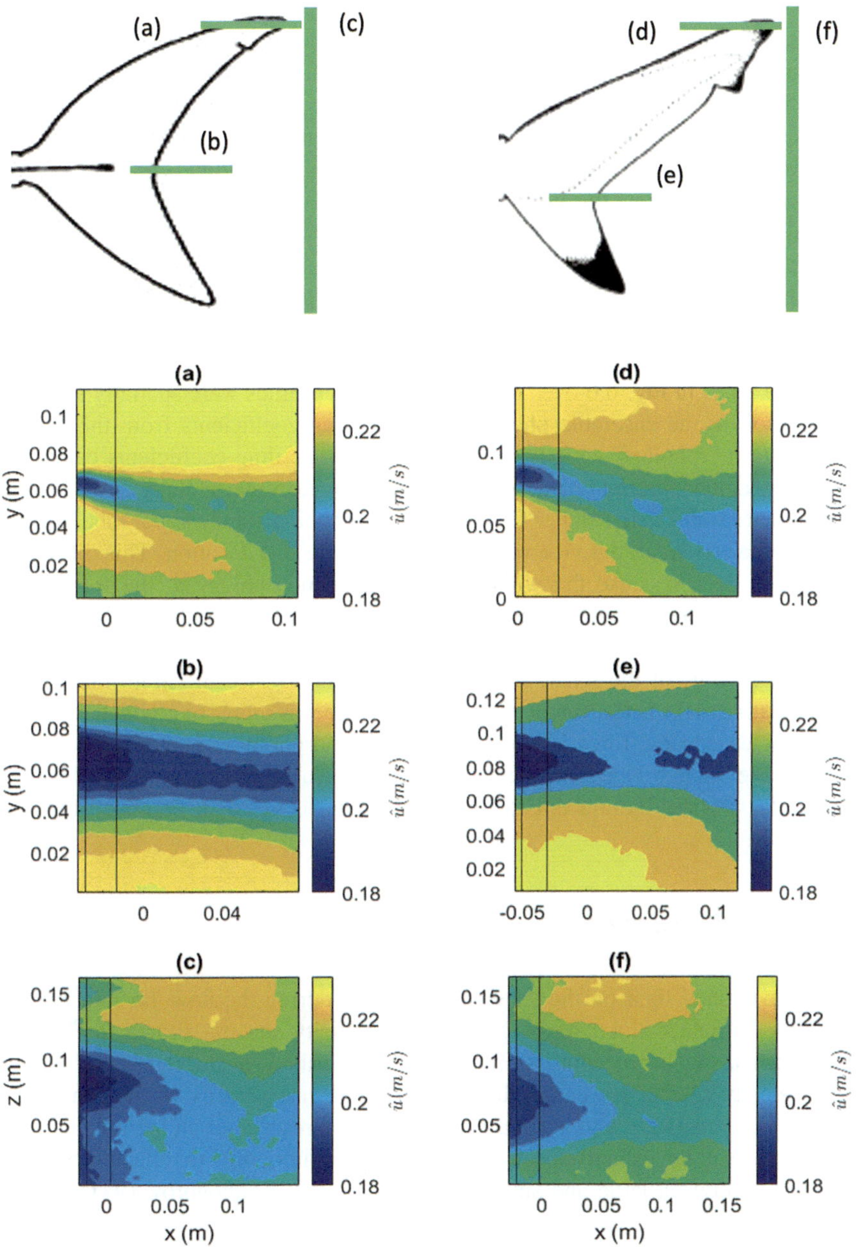

Fig. 6.7 Contours of the left and mid-phase of the streamwise velocity component for the streamwise-spanwise planes and the streamwise-normal plane, respectively, for each morphology. Subplots (**a**, **b**, **c**) show the upper (streamwise-spanwise), fork (streamwise-spanwise), and normal (streamwise-normal) measurement planes for the lunate morphology, and subplots (**d**, **e**, **f**) show the upper, fork, and normal planes of the heterocercal morphology. The panel at the top shows the PIV measurement plane locations for reference (not to scale), labeled by their corresponding subfigure letters. The vertical black lines mark the columns where the sectional drag was calculated, shown in Table 6.1

terms of magnitude but in a more concentrated region, indicating a strong local interaction with the flow. In comparison, at the fork, the concentrated vorticity is similar between morphologies but weaker for the heterocercal morphology, along with the shear layers being spread over a slightly larger region. Also, in both morphologies, the vertical vorticity is much larger than the spanwise vorticity. While the overall picture is similar for both morphologies, consistent with Fig. 6.7, the wake structure behind the lunate tail appears more coherent and organized, and extends a further distance downstream from the caudal fin, especially at the fork and upper lobe.

The sectional drag coefficient for each phase, morphology, and PIV measurement plane was estimated using the wake velocity deficit (Goett 1939). The momentum deficit profiles used for these calculations are extracted from the region bordered by the black lines in Fig. 6.6 over which the velocity profiles were spatially averaged in addition to the ensemble phase averaging. Drag coefficients from the fork of the lunate morphology were the highest at 0.02. The drag coefficients calculated in the streamwise-normal plane were the same for each morphology at 0.01. Drag coefficient results were smaller for the lunate morphology at the upper lobe location than the heterocercal, but vice versa measured at the fork location. This difference may have been because the width of the lunate morphology at the fork in the spanwise direction was greater than that of the heterocercal morphology at the same location, 0.013 and 0.007 m, respectively. This additional width may have altered the trailing edge of the lunate morphology such that it became more like a bluff body relative to the heterocercal morphology, which was more streamlined, resulting in an increased drag force acting on the lunate morphology. The drag coefficient results (see Table 6.1) are consistent with those from previously conducted studies on similar species (Tamura and Takagi 2009; Sagong et al. 2013; Takagi et al. 2013).

6.3.2 Wake Turbulence

In order to shed light on the state of turbulence during the flow–fin interactions, we have estimated fluctuating velocity and vorticity spectra, shown in Fig. 6.9, computed using a procedure similar to that described in Hackett et al. (2009). The velocity spectra are of the fluctuating streamwise velocity component in the streamwise, spanwise, and normal directions. The presence of an inertial subrange is evident in all the velocity spectra (as defined by Kolmogorov to reflect the transfer of energy between large and small turbulence scales; Kolmogorov 1941) in each measured PIV plane, as each spectrum reasonably follows the $-5/3$ scaling law (Saddoughi and Veeravalli 1994); however, there does not appear to be a large separation in scales, likely a consequence of the relatively low Reynolds number used in this study (Pope 2000), but the presence of an inertial subrange indicates that there is some scale separation between larger energy containing eddies and dissipation-scale eddies. The velocity spectra for the streamwise-spanwise PIV planes at the fork and upper lobe show turbulent kinetic energy of the flow is up

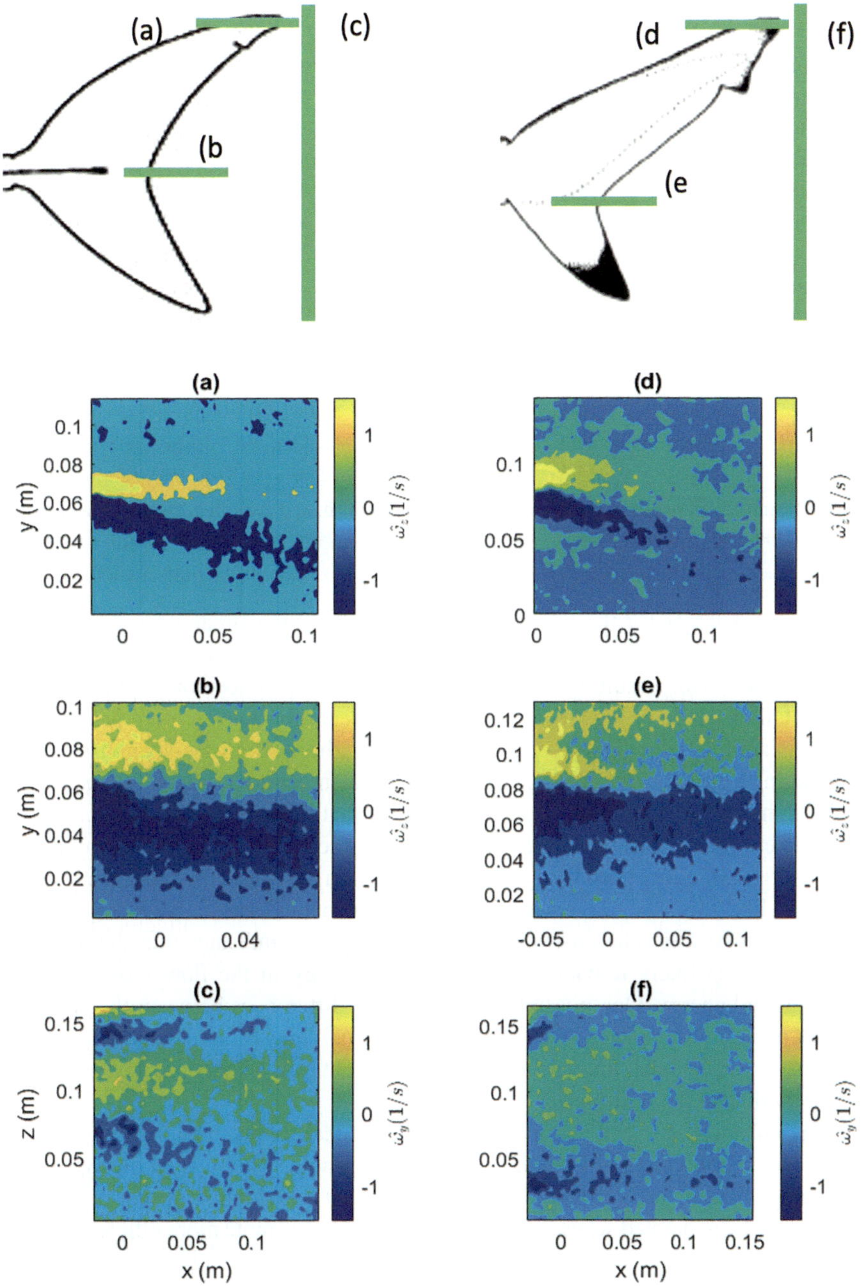

Fig. 6.8 Contours of the left and mid-phase for the normal and spanwise vorticity components, respectively, for each morphology. Subplots (**a**, **b**, **c**) show the upper (streamwise-spanwise), fork (streamwise-spanwise), and streamwise-normal PIV measurement planes of the lunate morphology, and subplots (**d**, **e**, **f**) show the upper, fork, and streamwise-normal PIV measurement planes of the heterocercal morphology. The panels at the top show the PIV measurement plane locations for reference (not to scale), labeled by their corresponding subfigure letter

Table 6.1 Measured PIV planes and morphologies along with associated fin parameters: phase, local Re, free stream velocity, and the drag coefficient (C_D)

Morphology	Location	Phase	Local length (m)	Local Re	U_∞ (m/s)	C_D
Lunate	Upper	Left	0.012	2.76×10^3	0.23	0.010
Lunate	Upper	Right	0.012	2.76×10^3	0.23	0.006
Lunate	Fork	Left	0.040	9.24×10^3	0.23	0.020
Lunate	Fork	Right	0.040	9.28×10^3	0.23	0.020
Lunate	Span	Mid	0.040	8.80×10^3	0.22	0.010
Hetero	Upper	Left	0.026	5.72×10^3	0.22	0.006
Hetero	Upper	Right	0.026	5.72×10^3	0.22	0.006
Hetero	Fork	Left	0.050	1.15×10^4	0.23	0.010
Hetero	Fork	Right	0.050	1.13×10^4	0.23	0.010
Hetero	Span	Mid	0.050	1.12×10^4	0.22	0.010
Lunate[a]	Span	Mid	0.040	3.00×10^4	0.75	0.010
Lunate[b]	Span	Mid	0.040	2.40×10^4	0.60	0.010
Lunate[c]	Span	Mid	0.10	2.00×10^6	2.00	0.009

Note: Results from previous experiments using tuna, sailfish, and swordfish are also listed in the table. Tamura and Takagi (2009)[a], Takagi et al. (2013)[b], Sagong et al. (2013)[c]

to two times larger for the lunate tail, while results in the streamwise-normal PIV plane show more similarity in turbulence levels between the two tail morphologies at all scales. This result is consistent with the more concentrated wake of the lunate tail observed in Figs. 6.7 and 6.8. This observation is even more pronounced when examining fluctuating vorticity where for the streamwise-spanwise planes fluctuating vertical vorticity is nearly an order of magnitude larger for the lunate tail, as shown in Fig. 6.9d–f. It is worth noting that these differences in enstrophy (kinetic energy of vorticity) hold across the entire scale range measured, indicating that energy differences are likely stemming from larger-scale (more energetic) flow features like the shear layers observed to form behind the tails in Fig. 6.7. A comparison between the enstrophy and energy spectra indicates that the different morphology is likely impacting the rotational energy in the flow more than the translational energy. One possible reason for the differences between the two PIV planes (streamwise-spanwise vs. streamwise-normal) may be associated with phase, that is, as the caudal fin reaches is leftmost/rightmost excursion it sheds more vorticity than at the mid-phase.

The enhanced spectral energy of the lunate caudal fin presumably suggests that the wake is more organized behind the lunate morphology because of the symmetry of the epichordal and hypochordal lobes, consistent with similar observations with respect to the generated wake; such organization may promote more efficient energy transfer within the wake in terms of propulsion and lift. The slightly larger drag of the higher aspect ratio (lunate) tail previously discussed is consistent with prior studies, but the more organized and more energetic wake may enhance either propulsion or lift as they are related to such flow structures, resulting in an overall net increase in hydrodynamic performance (lift/drag) despite slightly higher drag

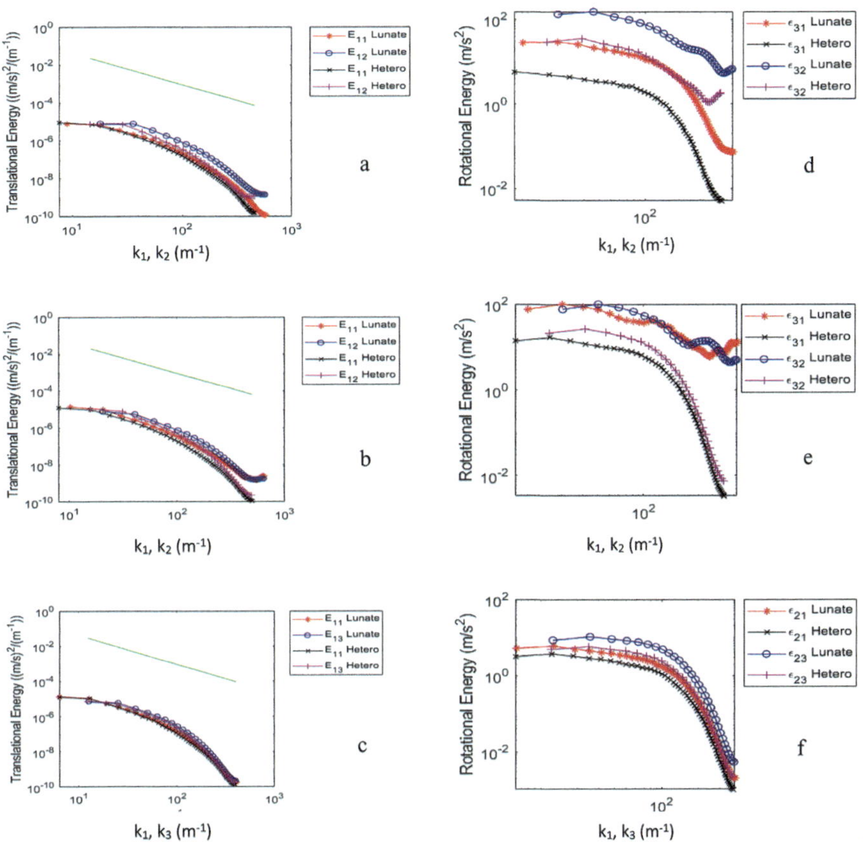

Fig. 6.9 (**a–c**) Fluctuating velocity spectra and (**d–f**) fluctuating vorticity spectra. (**a**) Upper lobe: streamwise velocity spectra in the streamwise (E_{11}) and spanwise (E_{12}) directions for both lunate and heterocercal. (**b**) Fork: streamwise velocity spectra in the streamwise (E_{11}) and spanwise (E_{12}) directions for both lunate and heterocercal. (**c**) Normal: streamwise velocity spectra in the streamwise (E_{11}) and normal (E_{13}) directions for both lunate and heterocercal. (**d**) Upper lobe: normal vorticity spectra in the streamwise (ε_{31}) and spanwise (ε_{32}) directions for both lunate and heterocercal. (**e**) Fork: normal vorticity spectra in the streamwise (ε_{31}) and spanwise (ε_{32}) directions for both lunate and heterocercal. (**c**) Normal: spanwise vorticity spectra in the streamwise (ε_{21}) and normal (ε_{23}) directions for both lunate and heterocercal. (**a–c**) The y-axes depict the transitional energy that corresponds to the turbulent kinetic energy. (**d–f**) The y-axes depict the rotational energy, which corresponds to the enstrophy (turbulent rotational kinetic energy). k_1, k_2, and k_3 are the wavenumbers in the streamwise, spanwise, and normal directions, respectively. The green line represents the $k^{-5/3}$ power law for turbulence in the inertial range

coefficients. This interpretation is consistent with previous studies, in which the vortices generated by an oscillating lunate caudal fin were found to be more concentrated and contain more energy than other tested morphologies (Liu and Dong 2016; Han et al. 2017). These results also indicate that the trailing edge shape may be the dominant factor when considering the amount of rotational kinetic energy imparted to the flow.

6.4 Summary

We set out to further our understanding of shark hydromechanics, specifically the pairing of thunniform oscillations and caudal fin morphology of the shortfin mako shark. To this end, we (i) evaluated the wake characteristics behind an oscillating shortfin mako shark with two separate caudal fin morphologies in three different measurement planes and (ii) investigated the effect caudal fin morphology has on drag. It appears that the general distribution of flow features is similar between the tails (lunate and heterocercal). The lunate exerts slightly higher drag given its larger velocity deficit magnitude, yet its wake appears somewhat more organized as demonstrated by the presences of strong concentrated positive and negative vorticity regions generating more rotational kinetic energy, which causes slower wake recovery. This result is also supported by measuring higher values in the turbulent kinetic and rotational (enstrophy) energies. It is worth mentioning that the drag estimated is based on sectional profile drag and does not account for unsteady contribution to the drag given the undulatory motion of the fin that may alter the drag values (see, e.g., Ben-Gida et al. 2013). This characteristic could potentially provide benefits in terms of propulsion and lift efficiency. In addition, the wake appears to decay faster for the heterocercal tail, which might be advantageous for predator avoidance in slower swimming sharks but not applicable for the mako.

Other hydrodynamic characteristics such as acceleration, lift, or stability may play a significantly more beneficial role in the evolution of the lunate morphology (than the heterocercal). For example, this possibility is supported by previous research conducted using bottlenose dolphins, which also perform thunniform oscillations utilizing a lunate caudal fin albeit rotated 90° to that of the shortfin mako, which found that the limited lateral oscillations of the anterior part of the body increased stability while cruising (Fish et al. 2000). This finding may indicate that those organisms utilizing other swimming modes are more unstable because of the increased amount of lateral movement at the anterior of the organism. Another possible selective pressure not examined during this experiment was the flexibility of each caudal fin morphology. The heterocercal morphology was more rigid in this study than what can be expected in nature (relative to the lunate morphology) and may have altered the hydrodynamics of the heterocercal morphology such that it became more similar to the lunate morphology in terms of hydrodynamic performance than would be valid in the natural environment. Future studies should explore these differing aspects of the caudal fins to provide more context for the results presented herein.

References

Adaramola MS, Akinlade OG, Sumner D, Bergstrom DJ, Schenstead AJ (2006) Turbulent wake of a finite circular cylinder of small aspect ratio. J Fluids Struct 22(6–7):919–928

Adrian RJ (1991) Particle-imaging techniques for experimental fluid mechanics. Annu Rev Fluid Mech 23(1):261–304

Afroz F, Lang A, Habegger ML, Motta P, Hueter R (2016) Experimental study of laminar and turbulent boundary layer separation control of shark skin. Bioinspir Biomim 12(1):016009

Akanyeti O, Putney J, Yanagitsuru YR, Lauder GV, Stewart WJ, Liao JC (2017) Accelerating fishes increase propulsive efficiency by modulating vortex ring geometry. Proc Natl Acad Sci 114(52):13828–13833

Anderson, J.M. (1996). Vorticity control for efficient propulsion. MIT, Cambridge

Ben-Gida H, Kirchhefer A, Taylor ZJ, Bezner-Kerr W, Guglielmo CG, Kopp GA, Gurka R (2013) Estimation of unsteady aerodynamics in the wake of a freely flying European starling (Sturnus vulgaris). PLoS One 8(11):e80086

Bernal D, Dickson KA, Shadwick RE, Graham JB (2001) Analysis of the evolutionary convergence for high performance swimming in lamnid sharks and tunas. Comp Biochem Physiol A: Mol Integr Physiol 129(2–3):695–726

Breder CM (1926) The locomotion of fishes. Zoologica 4:159–291

Bushnell DM, Moore KJ (1991) Drag reduction in nature. Annu Rev Fluid Mech 23(1):65–79

Carey FG, Teal JM (1969) Mako and porbeagle: warm-bodied sharks. Comp Biochem Physiol 28(1):199–204

Çengel YA, Cimbala JM (2006) Fluid mechanics: fundamentals and applications. McGraw-HillHigher Education, Boston

Cheng HK, Murillo LE (1984) Lunate-tail swimming propulsion as a problem of curved lifting line in unsteady flow. Part 1. Asymptotic theory. J Fluid Mech 143:327–350

Chopra MG (1974) Hydromechanics of lunate-tail swimming propulsion. J Fluid Mech 64(2):375–392

Chopra MG (1976) Large amplitude lunate-tail theory of fish locomotion. J Fluid Mech 74(1):161–182

Chopra MG, Kambe T (1977) Hydromechanics of lunate-tail swimming propulsion. Part 2. J Fluid Mech 79(1):49–69

Cohen RCZ, Cleary PW (2010) Computational studies of the locomotion of dolphins and sharks using smoothed particle hydrodynamics. In: 6th World Congress of Biomechanics (WCB 2010), Singapore

Compagno LJ (2001) Sharks of the world: an annotated and illustrated catalogue of shark species known to date (No. 1). Food and Agriculture Organization

Crofts SB, Shehata R, Flammang B (2019) Flexibility of heterocercal tails: what can the functional morphology of shark tails tell us about ichthyosaur swimming? Integr Org Biol 1(1):obz002

Dewar H, Graham J (1994) Studies of tropical tuna swimming performance in a large water tunnel-kinematics. J Exp Biol 192(1):45–59

Drucker EG, Lauder GV (2001) Locomotor function of the dorsal fin in teleost fishes: experimental analysis of wake forces in sunfish. J Exp Biol 204(17):2943–2958

Du Clos KT, Lang A, Devey S, Motta PJ, Habegger ML, Gemmell BJ (2018) Passive bristling of mako shark scales in reversing flows. J R Soc Interface 15(147):20180473

Epps BP, Alvarado PV, Youcef-Toumi K, Techet AH (2009) Swimming performance of a biomimetic compliant fish-like robot. Exp Fluids 47(6):927–939

Fish FE (1998) Imaginative solutions by marine organisms for drag reduction. In: Proceedings of the international symposium on seawater drag reduction. Rhode Island, Newport, pp 443–450

Fish FE, Peacock JE, Rohr J (2000) Phase relationships between body components of odontocete cetaceans in relation to stability and propulsive mechanisms. In: Kato N, Suzuki Y (eds) Proceedings 1st international symposium on aqua bio-mechanisms/international, seminar on aqua bio-mechanisms. Tokai University Pacific Center, Honolulu, pp 57–60

Flammang BE, Lauder GV (2009) Caudal fin shape modulation and control during acceleration, braking and backing maneuvers in bluegill sunfish, *Lepomis macrochirus*. J Exp Biol 212(2):277–286

Flammang BE, Lauder GV, Troolin DR, Strand T (2011) Volumetric imaging of shark tail hydrodynamics reveals a three-dimensional dual-ring vortex wake structure. Proc R Soc B Biol Sci 278(1725):3670–3678

Gleiss AC, Potvin J, Goldbogen JA (2017) Physical trade-offs shape the evolution of buoyancy control in sharks. Proc R Soc B Biol Sci 284(1866):20171345

Goett HJ (1939) Experimental investigation of the momentum method for determining profile drag. In: NACA-TR-660. US Government Printing Office

Gopalkrishnan R, Triantafyllou MS, Triantafyllou GS, Barrett DS (1994) Active vorticity control in a shear flow using a flapping foil. J Fluid Mech 274:1–21

Graham JB, Dewar H, Lai NC, Lowell WR, Arce SM (1990) Aspects of shark swimming performance determined using a large water tunnel. J Exp Biol 151(1):175–192

Gray J (1953) Undulatory propulsion. J Cell Sci 3(28):551–578

Gurka R, Nafi AS, Weihs D (2022) On an adaptation of the Reynolds number, applicable to body-caudal-fin aquatic locomotion. Front Mar Sci 9:914214

Hackett EE, Luznik L, Katz J, Osborn TR (2009) Effect of finite spatial resolution on the turbulent energy spectrum measured in the coastal ocean bottom boundary layer. J Atmos Ocean Technol 26(12):2610–2625

Han P, Liu G, Dong H (2017) Proper orthogonal decomposition analysis on the hydrodynamic effects of the propulsor geometries in flapping propulsion. 55th AIAA Aerospace Sciences Meeting, 1708

Harding L, Jackson A, Barnett A, Donohue I, Halsey L, Huveneers C, Meyer C, Papastamatiou Y, Semmens JM, Spencer E, Watanabe Y, Payne N (2021) Endothermy makes fishes faster but does not expand their thermal niche. Funct Ecol 35:1–9

Hong C, Chang-an Z (2005) Modeling the dynamics of biomimetic underwater robot fish. 2005 IEEE International Conference on Robotics and Biomimetics-ROBIO, pp 478–483

Kolmogorov A (1941) Dissipation of energy in locally isotropic turbulence. Proc USSR Acad Sci 31:538–540

Lauder GV (2000) Function of the caudal fin during locomotion in fishes: kinematics, flow visualization, and evolutionary patterns. Am Zool 40(1):101–122

Lauder GV, Tytell ED (2005) Hydrodynamics of undulatory propulsion. Fish Physiol 23:425–468

Lee J, Park YJ, Cho KJ, Kim D, Kim HY (2017) Hydrodynamic advantages of a low aspect-ratio flapping foil. J Fluids Struct 71:70–77

Li GJ, Luodin ZHU, Lu XY (2012) Numerical studies on locomotion performance of fish-like tail fins. J Hydrodyn Ser B 24(4):488–495

Lighthill MJ (1960) Note on the swimming of slender fish. J Fluid Mech 9(2):305–317

Lighthill MJ (1969) Hydromechanics of aquatic animal propulsion. Annu Rev Fluid Mech 1(1):413–446

Lighthill MJ (1970) Aquatic animal propulsion of high hydromechanical efficiency. J Fluid Mech 44(2):265–301

Lighthill MJ (1975) Mathematical Biofluiddynamics. Society for Industrial and Applied Mathematics

Lindsey CC (1978) Form, function, and locomotory habits in fish. In: Fish physiology. Academic Press

Liu G, Dong H (2016) Effects of tail geometries on the performance and wake pattern in flapping propulsion, ASME 2016 Fluids Engineering Division Summer Meeting. ASME Digital Collection

Liu G, Ren Y, Dong H, Akanyeti O, Liao JC, Lauder GV (2017) Computational analysis of vortex dynamics and performance enhancement due to body-fin and fin-fin interactions in fish-like locomotion. J Fluid Mech 829:65–88

Melling A (1997) Tracer particles and seeding for particle image velocimetry. Meas Sci Technol 8(12):1406

Moss SA (1984) Sharks: an introduction for the amateur naturalist. Prentice Hall
Pope SB (2000) Turbulent flows. Cambridge University Press
Raffel M, Willert CE, Scarano F, Kähler CJ, Wereley ST, Kompenhans J (2018) Particle image velocimetry: a practical guide. Springer
Raspa V, Ramananarivo S, Thiria B, Godoy-Diana R (2014) Vortex-induced drag and the role of aspect ratio in undulatory swimmers. Phys Fluids 26(4):041701
Reyier EA, Adams DH, Lowers RH (2008) First evidence of a high density nursery ground for the lemon shark, *Negaprion brevirostris*, near Cape Canaveral, Florida. Fla Sci 71(2):134–148
Saddoughi SG, Veeravalli SV (1994) Local isotropy in turbulent boundary layers at high Reynolds number. J Fluid Mech 268:333–372
Sagong W, Jeon WP, Choi H (2013) Hydrodynamic characteristics of the sailfish (*Istiophorus platypterus*) and swordfish (*Xiphias gladius*) in gliding postures at their cruise speeds. PLoS One 8(12):e81323
Sambilay VC (1990) Interrelationships between swimming speed, caudal fin aspect ratio and body length of fishes. Fishbyte, The WorldFish Center 8(3):16–20
Sane SP (2003) The aerodynamics of insect flight. J Exp Biol 206(23):4191–4208
Schouveiler L, Hover FS, Triantafyllou MS (2005) Performance of flapping foil propulsion. J Fluids Struct 20(7):949–959
Sepulveda CA, Kohin S, Chan C, Vetter R, Graham JB (2004) Movement patterns, depth preferences, and stomach temperatures of free-swimming juvenile mako sharks, *Isurus oxyrinchus*, in the Southern California bight. Mar Biol 145(1):191–199
Sepulveda CA, Graham JB, Bernal D (2007) Aerobic metabolic rates of swimming juvenile mako sharks, Isurus oxyrinchus. Mar Biol 152(5):1087–1094
Sfakiotakis M, Lane DM, Davies JBC (1999) Review of fish swimming modes for aquatic locomotion. IEEE J Ocean Eng 24(2):237–252
Shadwick RE (2005) How tunas and Lamnid sharks swim: an evolutionary convergence: these fishes diverged millions of years ago, but selection pressures have brought them very similar biomechanical schemes for movement. Am Sci 93(6):524–531
Sorenson L, Santini F, Alfaro ME (2014) The effect of habitat on modern shark diversification. J Evol Biol 27(8):1536–1548
Stevens JD (2008) The biology and ecology of the shortfin mako shark, Isurus oxyrinchus. In: Sharks of the open ocean: biology, fisheries and conservation, pp 87–94
Suleman A, Crawford C (2008) Studies on hydrodynamic propulsion of a biomimetic tuna. In: Underwater Vehicles, pp 459–486
Syme DA, Shadwick RE (2011) Red muscle function in stiff-bodied swimmers: there and almost back again. Philos Trans R Soc B: Biol Sci 366(1570):1507–1515
Takagi T, Tamura Y, Weihs D (2013) Hydrodynamics and energy-saving swimming techniques of Pacific bluefin tuna. J Theor Biol 336:158–172
Tamura YC, Takagi T (2009) Morphological features and functions of bluefin tuna change with growth. Fish Sci 75(3):567–575
Taylor GK, Nudds RL, Thomas AL (2003) Flying and swimming animals cruise at a Strouhal number tuned for high power efficiency. Nature 425(6959):707–711
Thomson KS, Simanek DE (1977) Body form and locomotion in sharks. Am Zool 17(2):343–354
Triantafyllou GS, Triantafyllou MS, Grosenbaugh MA (1993) Optimal thrust development in oscillating foils with application to fish propulsion. J Fluids Struct 7(2):205–224
Trump CL, Leggett WC (1980) Optimum swimming speeds in fish: the problem of currents. Can J Fish Aquat Sci 37(7):1086–1092
Turner VL (2020) Investigating dorsal fin hydrodynamic function of the lemon shark, Negaprion brevirostris. Thesis. CCU, Conway
Tytell ED (2004) The hydrodynamics of eel swimming II. Effect of swimming speed. J Exp Biol 207(19):3265–3279
Tytell ED (2006) Median fin function in bluegill sunfish Lepomis macrochirus: streamwise vortex structure during steady swimming. J Exp Biol 209(8):1516–1534

Wang T, Wen L, Liang J, Wu G (2010) Fuzzy vorticity control of a biomimetic robotic fish using a flapping lunate tail. J Bionic Eng 7(1):56–65
Webb PW (1978) Fast-start performance and body form in seven species of teleost fish. J Exp Biol 74(1):211–226
Webb PW, Weihs D (2015) Stability versus maneuvering: challenges for stability during swimming by fishes. Integr Comp Biol 55(4):753–764
Webb PW, Kostecki PT, Stevens ED (1984) The effect of size and swimming speed on locomotor kinematics of rainbow trout. J Exp Biol 109(1):77–95
Wilga CD, Lauder GV (2000) Three-dimensional kinematics and wake structure of the pectoral fins during locomotion in leopard sharks, *Triakis semifasciata*. J Exp Biol 203(15):2261–2278
Wilga CD, Lauder GV (2002) Function of the heterocercal tail in sharks: quantitative wake dynamics during steady horizontal swimming and vertical maneuvering. J Exp Biol 205(16):2365–2374
Willert CE, Gharib M (1991) Digital particle image velocimetry. Exp Fluids 10(4):181–193
Williamson CH, Roshko A (1988) Vortex formation in the wake of an oscillating cylinder. J Fluids Struct 2(4):355–381
Wolfgang MJ, Anderson JM, Grosenbaugh MA, Yue DK, Triantafyllou MS (1999) Near-body flow dynamics in swimming fish. J Exp Biol 202(17):2303–2327
Xia D, Chen W, Liu J, Wu Z, Cao Y (2015) The three-dimensional hydrodynamics of thunniform swimming under self-propulsion. Ocean Eng 110:1–14
Yang L, Su YM (2011) CFD simulation of flow features and vorticity structures in tuna-like swimming. China Ocean Eng 25(1):73–82
Yeh PD, Alexeev A (2016) Effect of aspect ratio in free-swimming plunging flexible plates. Comput Fluids 124:220–225
Zhu Q, Wolfgang MJ, Yue DKP, Triantafyllou MS (2002) Three-dimensional flow structures and vorticity control in fish-like swimming. J Fluid Mech 468:1–28

Chapter 7
Strategies for Improving Swimming Efficiency: From Individual Fish to Schools

Liang Li and Li-Ming Chao

Abstract Swimming organisms provide examples of efficient movement through strategies such as reducing drag with specialized surface structures, taking advantage of tail flexibility, optimizing their swimming kinematics, and harnessing energy from flow patterns created by nearby swimmers. Understanding the strategies behind these high-efficiency swimming behaviors is invaluable across biological studies, fluid mechanics, and engineering. In this chapter, we present several primary strategies that are used either by individual fish or by fish schools to enhance swimming performance, especially those identified based on studies utilizing cutting-edge methods like theoretical analyses, computational fluid dynamics (CFD) simulations, and bioinspired robotics. These studies not only offer valuable hypotheses for biological research but also inspire engineers in the design of efficient underwater vehicles.

Keywords Swimming efficiency · Drag reduction · Embodied Intelligence · Fish school · Hydrodynamic interactions

7.1 Introduction

Energy is the cornerstone of biological systems, as it drives nearly all processes necessary for life. Over billions of years, nature has evolved both sophisticated mechanisms for acquiring energy and highly efficient strategies for its utilization

L. Li (✉) · L.-M. Chao (✉)
Department of Collective Behaviour, Max Planck Institute of Animal Behavior, Konstanz, Germany

Centre for the Advanced Study of Collective Behaviour, University of Konstanz, Konstanz, Germany

Department of Biology, University of Konstanz, Konstanz, Germany

Department of Computer and Information Science, University of Konstanz, Konstanz, Germany
e-mail: lli@ab.mpg.de; chaoliming92@gmail.com

V. Bels et al. (eds.), *Biomimicry – A Functional Approach*, Biologically-Inspired Systems 20, https://doi.org/10.1007/978-3-032-11949-0_7

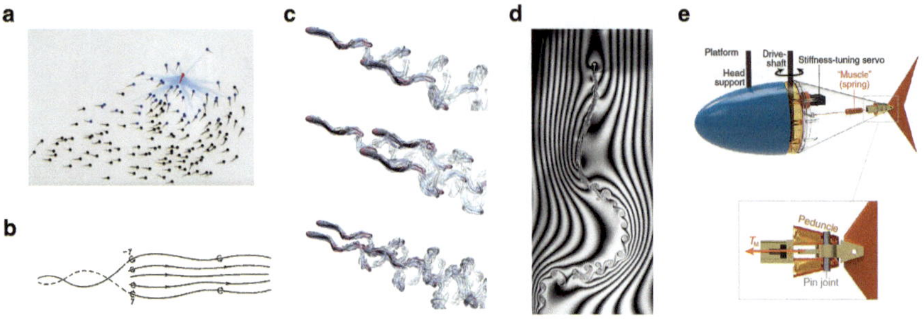

Fig. 7.1 A diverse array of tools used to explore and understand the strategies of improving swimming efficiency. (**a**) Biological observations; (**b**) Mathematical analysis; (**c**) Computational fluid dynamics (CFD) simulations; (**d**) Experimental fluid dynamics (EFD); (**e**) Bioinspired robotics as a physical model. Figures reproduced from Rosenthal et al. (2015), Pitcher et al. (1976), Wu (1961), Verma et al. (2018), Zhang et al. (2000), Zhong et al. (2021)

(Kooijman 2000). In aquatic ecosystems, animals often exhibit incredible behaviors while conserving limited energy resources. Examples include the long migration of fish from the Norwegian Sea to their home rivers in Norway (Hansen et al. 1993), European eels crossing vast distances across the Atlantic Ocean (Wright et al. 2022), and species such as the sailfish (*Istiophorus platypterus*) and shortfin mako shark (*Isurus oxyrinchus*), known for their remarkable speed and endurance—some reaching velocities over 70 km/h (Sagong et al. 2013; Compagno et al. 2001). All these examples suggest that aquatic creatures should have evolved highly effective strategies to perform these behaviors. Investigating these strategies can significantly enhance our understanding of the fundamental principles underpinning swimming behavior. Furthermore, such insights may inspire innovative designs for human-engineered underwater vehicles, which, despite ongoing technological advancements, still struggle with efficiency improving. Researchers are thus increasingly studying the biomechanics of fish locomotion, hoping to translate biological principles into improved hydrodynamic performance for underwater vehicles (Du et al. 2015).

Lots of tools have been developed to explore and understand the mechanisms underlying fish swimming (Fig. 7.1), with a focus on energy saving and efficiency enhancement. Early approaches primarily relied on observational biology and hypothesis generation (Fig. 7.1a). For instance, studies of drag-reducing morphological adaptations in fish, such as streamlined body shapes and specialized surface textures, directly inspired engineers to design more efficient marine vessels (Soyener 1993). Subsequently, fluid dynamic theory provided a robust framework for analyzing these phenomena (Fig. 7.1b). Seminal work by Sir James Lighthill and Wu, for example, established theoretical models that link swimming modes, body kinematics, and collective behaviors (e.g., schooling) to energy optimization in aquatic organisms (Wu 1961; Lighthill 1970), offering critical insights into how biological systems achieve efficiency in complex fluid environments. Advancements in computational power and engineering tools have since revolutionized the field (Fig. 7.1c). High-resolution studies comparing respiratory metrics in schooling versus solitary fish have quantified the energy-saving benefits of collective swimming (Zhang and Lauder 2023), while computational fluid dynamics (CFD) simulations have elucidated the fluid-structure interactions governing propulsion (Kern and Koumoutsakos 2006). Experimental platforms (Fig. 7.1d) have validated hypotheses about energy-efficient swimming strategies, providing empirical insights into undulatory motion, vortex dynamics, and self-organized schooling (Triantafyllou et al. 2000; Lauder et al. 2007). Additionally, bioinspired robotics (Fig. 7.1e) has bridged theory and practice. These robotic systems not only replicate biological principles but also inform the design of next-generation underwater vehicles.

Within these tools, researchers have identified remarkable strategies that aquatic organisms use to enhance swimming efficiency during underwater locomotion. Some strategies have clearly defined mechanisms supported by both biological observations and robotic validations. For example, the "vortex phase matching" mechanism, whereby follower fish adjust their swimming patterns to align with vortices generated by leading fish, has been well documented and tested in both biological and robotic contexts (Li et al. 2020). Other strategies, although well understood mechanistically and demonstrated in robotic systems, remain difficult to verify directly in biological organisms due to their inherent complexity. An example is the use of simple proprioceptive sensing to save energy independently of lateral line sensing or vision, as confirmed in robotic experiments but difficult to isolate in biological fish due to integrated movement systems (Li et al. 2021a). Lastly, certain observed phenomena in biological systems clearly indicate energy-saving capabilities, but their underlying mechanisms remain elusive. For instance, aquatic organisms appear to "harvest" energy effectively from turbulence flow (Zhang et al. 2024), highlighting a remarkable biological capability not yet fully understand the fundamental mechanisms.

Naturally, all these strategies could inspire innovations in bioinspired engineering. For instance, engineered surfaces featuring bioinspired textures have been demonstrated to reduce drag in underwater vehicles (Golovin et al. 2016). Moreover, some of these mechanisms can be extended beyond aquatic locomotion to address broader engineering challenges. For instance, fish schooling patterns have even informed novel wind farm array designs aimed at optimizing energy capture (Whittlesey et al. 2010).

In this chapter, we review strategies that improve swimming efficiency in both solitary and school, delving into their underlying mechanisms and potential applications. We organize these strategies across scales—from individual locomotion to collective schooling dynamics—and categorize them by mode of energy use, spanning passive adaptations to active biomechanical control. Finally, we conclude by highlighting promising future research directions, particularly those enabled by emerging technologies such as advanced computational modeling, bioinspired robotics, and high-resolution fluid dynamics visualization.

7.2 Individual Swimming

Fish have developed highly efficient swimming capabilities over millions of years, employing both passive and active mechanisms to optimize energy utilization (Fish and Lauder 2006). Using their bodies and appendages to propel water, fish generate hydrodynamic thrust while simultaneously minimizing resistance through morphological adaptations, synchronized kinematics, and strategic behaviors. At the individual level, passive mechanisms for enhancing swimming efficiency rely on morphological features that shape flow patterns around the body, thereby reducing drag (Alexander 1967; Bushnell and Moore 1991). In contrast, active mechanisms involve coordinated muscular movements that dynamically interact with the surrounding fluid, enabling fish to finely adjust their swimming efficiency in response to changing environmental conditions (Smits 2019).

7.2.1 *Passive Mechanisms*

Aquatic swimmers have developed several purely passive strategies to reduce drag and enhance swimming efficiency, including optimized body morphology, skin denticles, mucus secretion, and prominent bumps on fins (Abdulbari et al. 2015).

One of nature's primary strategies for minimizing drag is the optimization of body morphology (Lauder and Drucker 2004). Most fish, for example, exhibit a fusiform body shape (Fig. 7.2a), characterized by an elongated, teardrop-like form with a rounded leading edge that reaches maximum thickness before gradually

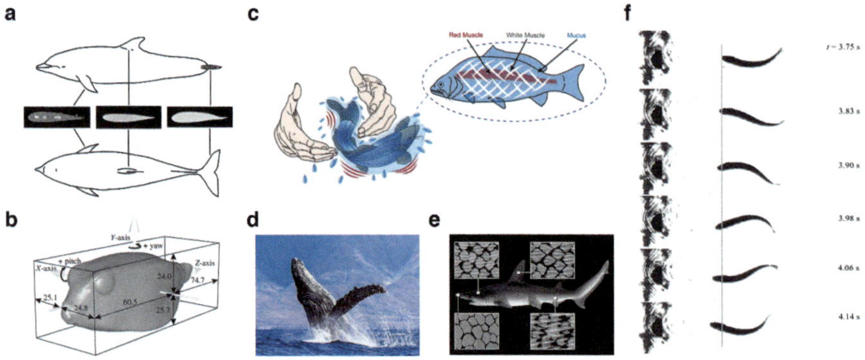

Fig. 7.2 (**a**) Lateral and dorsal body shapes, along with cross-sectional profiles of control surfaces (flippers, dorsal fin, and flukes) of a bottlenose dolphin; (**b**) Morphologies and drag coefficient of a yellow boxfish (*Ostracion cubicus*); (**c**) Modulus adaptive switchable lubrication when fish being caught; (**d**) Bumps employed by whales; and (**e**) denticles employed by sharks for reducing drag; (**f**) Experiments shown even a dead fish can swim in Kármán vortex streets. Figures reproduced from Fish (2006), Van Wassenbergh et al. (2015), Zhang et al. (2022b), Lauder et al. (2016), and Beal et al. (2006)

tapering toward the tail (Hertel 1966; Fish 2006). This streamlined shape helps delay flow separation as close to the trailing edge as possible, thereby minimizing the wake and reducing pressure drag. Certain species, such as the boxfish (Fig. 7.2b), deviate from this typical morphology. Researchers have hypothesized that its unconventional shape may offer benefits in drag reduction and stability, though some studies suggest it provides no significant drag reduction compared to more generalized fish morphologies (Van Wassenbergh et al. 2015).

Fish can reduce friction drag through mucus secreted by their bodies while swimming (Shephard 1994). The mucus layer acts as a lubricant (Fig. 7.2c), smoothing the body's surface and reducing turbulence at the boundary layer, which minimizes frictional resistance. Reports indicate that this mucus system can enable fish to achieve up to 65.9% drag reduction (Rosen and Cornford 1971).

Some swimmers, such as whales, have evolved large bumps, or tubercles, along the leading edges of their flippers (Fig. 7.2d), which generate vortices that stabilize the boundary layer, preventing flow separation and reducing drag (Miklosovic et al. 2004). Similarly, sharks minimize friction drag through microscopic denticles on their skin (Fig. 7.2e). Studies using 3D-printed replicas of real shark denticles, obtained via 3D scanning, have demonstrated that these structures can increase swimming speed by 6.6% and reduce the energy cost of transport by 5.9% (Dean and Bhushan 2010; Wen et al. 2014; Graybill and Xu 2024). Two potential mechanisms explain how denticles contribute to drag reduction. First, denticles may strategically lift streamwise vortices away from the surface, reducing the extent of surface exposure to high shear stress and minimizing drag (Mawignon et al. 2023). Second, they may restrict crossflow velocity fluctuations by channeling the flow, thereby reducing momentum transfer from vortex interactions (Martin and Bhushan 2014).

Finally, fish could potentially passively swim in Kármán vortex streets to reduce their energy expenditure. Comparative studies involving live and dead trout subjected to such flow conditions indicate that live fish can naturally align their bodies with the vortices generated behind a cylinder (Liao et al. 2003; Beal et al. 2006; Liao 2007). This passive alignment allows their bodies to bend naturally with the flow, thereby generating thrust without extensive muscular effort (Fig. 7.2f). Consequently, fish benefit from significant energy savings. However, determining whether fish actively or passively adjusts their movements in natural environments, where flow patterns are complex, such as around stones generating Kármán vortices, remains challenging. Nonetheless, this energy-saving strategy could inspire engineers to develop artificial systems that move in similar environments, such as multiple ships moving in specific formations (He et al. 2022).

7.2.2 *Active Mechanisms*

Fish have evolved a range of active strategies involving body and fin movements to save energy and/or enhance swimming efficiency in aquatic environments. These include optimizing body undulation and tail flapping to reduce drag and

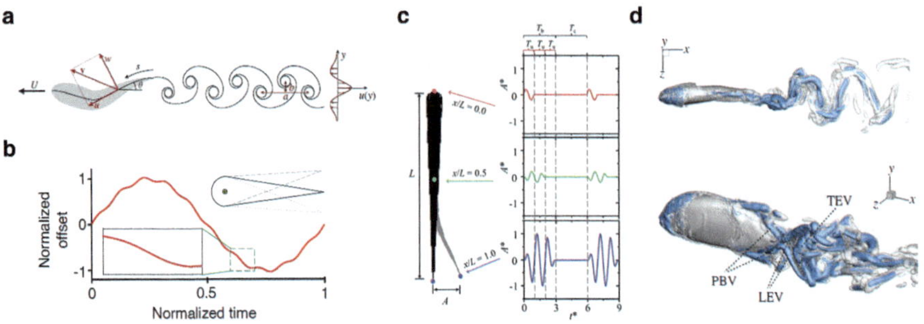

Fig. 7.3 (**a**) Strouhal number, defined as the ratio between wake width and length of a reversed Kármán vortex street, is usually used to estimate the swimming performance; (**b**) oscillating tailbeats with superimposed rhythmic perturbations of high frequency and low amplitude; (**c**) burst-and-coast strategies used by the swimming animals; (**d**) wake structures illustrating fin-fin and body-fin interactions. Figures reproduced from Eloy (2012), Chao et al. (2024c), Chao et al. (2024a), and Liu et al. (2017)

improve thrust (Triantafyllou and Triantafyllou 1995; Barrett et al. 1999), adopting intermittent swimming behaviors such as burst-and-coast to minimize energy expenditure (Weihs 1974; Videler and Weihs 1982), and leveraging fin-fin interactions to amplify propulsion while lowering drag (Liu et al. 2017; Drucker and Lauder 2001). Additionally, fish actively position themselves near surfaces to exploit ground effects, boosting efficiency without extra energy costs (Weihs 1973; Webb 1993). These strategies, supported by theoretical, experimental, and computational studies, highlight how fish dynamically adjust their movements to optimize hydrodynamic performance and reduce power consumption in complex flow conditions (Moored et al. 2012; Gleiss et al. 2011; Bale et al. 2014).

Fish actively employ body undulation and tail flapping to reduce drag and enhance swimming efficiency, leveraging precise kinematic control to optimize energy use. Body undulation delays boundary-layer separation, maintaining laminar flow longer to minimize drag, as observed in dolphins (Floryan et al. 2018), and prevents the formation of energy-wasting leading-edge vortices (Wang 2000). Tail flapping, when tuned to an optimal Strouhal number ($St = \frac{fA}{U}$, where f is flapping frequency, A is amplitude, and U is speed), operates within the range $0.2 < St < 0.4$, producing reversed Kármán vortex street that maximizes thrust while reducing drag (Fig. 7.3a) (Taylor et al. 2003; Anderson et al. 1998; Read et al. 2003; Hover et al. 2004). This synchronization with the flow field enhances propulsive efficiency (Moored et al. 2012; Zhang 2017), with lift forces counteracting lift-induced drag (Triantafyllou and Triantafyllou 1995). Particularly, the optima in propulsive efficiency occur when the driving frequency of a flapping fin matches the resonant frequency of the jet profile (Moored et al. 2012). Robotic models demonstrate that active undulation requires less power than towing a rigid body at the same speed (Barrett et al. 1999), and recent studies by Chao et al. show that high-frequency, low-amplitude tailbeat perturbations further boost thrust and efficiency by reducing phase lag with the fluid response (Fig. 7.3b) (Chao et al. 2024c,b).

Fish actively adopt swimming behavior strategies like burst-and-coast to save energy by alternating between intense movement and passive gliding phases (Fig. 7.3c) (Gleiss et al. 2011; Amichay et al. 2024). In the burst phase, rapid tail flapping generates high thrust, while the coast phase allows gliding with reduced drag compared to continuous swimming, which increases form and lift-induced drag (Lighthill 1971). By adjusting the duty cycle (DC)—the fraction of the locomotor cycle spent actively bursting—fish can optimize energy use. With simplified dynamic models, Weihs and Videler theoretically showed that intermittent swimming—when performed with an appropriate duty cycle (DC)—can reduce energy expenditure by up to 50% compared to steady swimming at the same speed (Weihs 1974; Videler and Weihs 1982). These predictions have been supported by computational fluid dynamics (CFD) studies, which demonstrate that burst-and-coast swimming can lead to a 40–50% reduction in cost of transport (CoT)—defined as the energy required to move a unit body mass over a unit distance (Akoz and Moored 2018; Dai et al. 2018; Kang et al. 2023; Yang et al. 2024). Further supporting evidence comes from experiments on koi carp (*Cyprinus carpio koi*), where digital particle image velocimetry (DPIV) and vortex-ring modeling revealed that burst-and-coast swimming can save up to 45% more energy than continuous swimming. This benefit is attributed to a better match between instantaneous swimming speed and hydrodynamic drag during the glide phases (Wu et al. 2007). While burst-and-coast swimming is widely observed and associated with energy savings, the underlying mechanisms remain not fully understood. Several hypotheses have been proposed to explain how this strategy reduces energetic cost. One classical explanation is the Bone-Lighthill hypothesis (Lighthill 1971), which suggests that continuous flapping motion increases skin-friction drag due to boundary-layer thinning along the fish's body. In contrast, during gliding phases, the boundary layer can redevelop, reducing drag. As a result, intermittently reducing flapping, i.e., shortening the active propulsion period, can lower overall power expenditure (Ehrenstein and Eloy 2013). Another explanation is the inviscid "Garrick" mechanism, investigated by Akoz and Moored (Akoz and Moored 2018), which highlights how burst-and-coast swimming alters the balance between added-mass thrust and vortex-induced drag. In unsteady flapping, certain phases, such as deceleration or motion reversal, can cause the fluid to do negative work on the swimmer. By inserting glide phases at the right moment, fish can avoid these inefficient intervals. For instance, gliding after a burst allows shed vortices to dissipate, preventing interference with subsequent tailbeats and reducing energy loss. Through this modulation of thrust and drag timing, burst-and-coast swimming enables significant energy savings.

Fish could enhance swimming efficiency through active fin-fin interactions, coordinating the movements of dorsal, pectoral, pelvic, anal, and caudal fins to optimize thrust and reduce drag. Live fish studies demonstrate that these interactions create synergistic hydrodynamic effects, amplifying propulsion beyond the sum of individual fin contributions (Drucker and Lauder 2001; Tytell 2006). CFD simulations of tuna show that dorsal-caudal fin interactions, tuned to specific distances, Strouhal numbers, and Reynolds numbers, increase thrust while conserving energy (Zhang et al. 2022a; Guo et al. 2023). Liu et al.'s jackfish model revealed that vortices shed by dorsal and anal fins are captured by the caudal fin, enhancing thrust and minimizing drag (Fig. 7.3d) (Liu et al. 2017). Bioinspired robotic platforms confirm that coordinated fin movements improve propulsion efficiency (Zhong et al. 2019; Matthews and Lauder 2021), with body-fin interactions conditioning the flow to reduce energy costs (Liu et al. 2017). These active strategies highlight how fish dynamically adjust fin kinematics to achieve efficient locomotion in diverse aquatic conditions.

7.3 Swimming in a School

The hydrodynamic effects of group locomotion can significantly influence how individuals arrange themselves within a school (Marras et al. 2015). A long-standing hypothesis posits that fish reduce their energy expenditure by swimming collectively. Daniel Weihs was the first to propose a "diamond formation" for fish schools, suggesting that this arrangement minimizes the group's energy cost (Weihs 1973). While direct observations of such a precise diamond pattern in natural schools remain inconclusive (Partridge and Pitcher 1979), Weihs's seminal work has spurred extensive research on energy savings in fish schools—particularly because the vortices shed by one fish can provide usable kinetic energy for its neighbors, thereby reducing overall power consumption or boosting thrust for improved efficiency.

7.3.1 *Passive Mechanisms*

One of the most famous conjecture proposed by Sir James Lighthill is that fish could potentially spontaneously form schools through passive hydrodynamic interactions without actively controlling their body movements to achieve specific positions (Lighthill 1975). Unlike active strategies requiring deliberate undulation or positioning, this passive mechanism suggests that fish naturally align into energy-saving configurations solely due to the surrounding flow dynamics.

Simulations using the immersed boundary method show that a downstream filament in a tandem pair reduces energy expenditure by aligning with vortex cores shed by the upstream filament, leveraging lateral flows to lower propulsion costs by about 80% (Fig. 7.4a) (Peskin 2002; Zhu et al. 2014). This energy saving arises as the follower drifts into the wake's low-pressure zones, a process confirmed across tandem self-propelled bodies where vortex interactions consistently reduce the downstream swimmer's power needs (Peng et al. 2018a; Lin et al. 2019; Heydari and Kanso 2021; Lin et al. 2021; Dal Jeong et al. 2023; Chao et al. 2023). Experimental studies with flapping foils in a rotating tank further demonstrate this: Without coordination, the downstream foil exploits the upstream foil's reversed Kármán

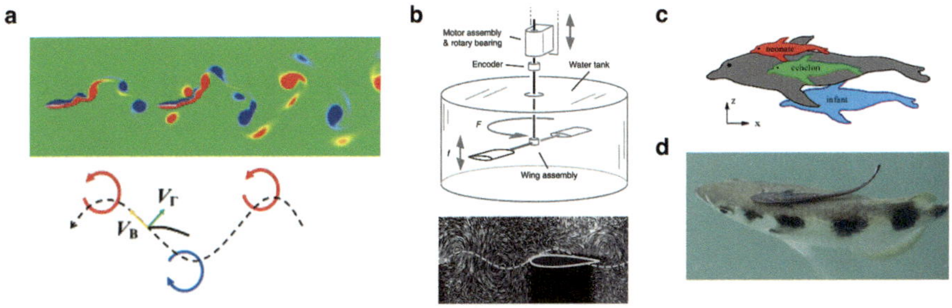

Fig. 7.4 (**a**) Passive self-organization and energy harvesting in two self-propelled tandem filaments; (**b**) The rotating-tank system employed to investigate energy harvesting in two and multiple self-propelled foils, including wake patterns captured via particle image velocimetry (PIV); (**c**) Side view of the swimming positions of a dolphin calf in relation to their mother; (**d**) Hitchhiking with a larger swimmer (Photo by Brendon Samuels, licensed under CC BY-NC 4.0. Source: iNaturalist (Observation ID: 159449954).). Panels (**a**–**c**) reproduced from Zhu et al. (2014), Becker et al. (2015), Ramananarivo et al. (2016), and Shoele and Zhu (2015)

vortex street to cut energy use, spontaneously forming slow or fast modes based on flow-mediated energy exchange (Fig. 7.4b) (Becker et al. 2015; Ramananarivo et al. 2016; Newbolt et al. 2019, 2022, 2024). These findings suggest that passive wake exploitation is a key mechanism driving Lighthill's predicted schooling efficiency.

Similar as the potential passive energy saving within vortices, schooling fish could also passively save energy swimming the Kármán vortices induced with two leading schooling fish shedding reverse Kármán vortices. A potential passive energy-saving strategy in schooling fish involves utilizing vortex-induced propulsion, where fish exploit surrounding vortices to reduce the effort required for swimming. Studies have shown that even dead fish placed behind a cylinder can be propelled by vortex flows, suggesting that live fish may similarly interact with vortices to enhance swimming efficiency (Liao 2004; Beal et al. 2006). While replicating this effect in live fish may be more complex due to their active movement control, the underlying hydrodynamic mechanisms present an intriguing possibility for energy-efficient locomotion.

Calves can benefit from drafting forces by swimming alongside their mothers (Fig. 7.4c). In this case, the larger body size of the mother swimming in front creates a favorable hydrodynamic flow around her body, allowing the calves to passively swim nearby and take advantage of the reduced drag forces without active control. This phenomenon has been particularly studied in dolphin mother-calf pairs, where the calf can gain up to 90% of the thrust required to maintain speed, leveraging the hydrodynamic benefits of swimming close to the mother (Weihs 2004; Noren et al. 2008; Alexander 2004). Similar drafting behavior has been observed in young sharks swimming behind adults, benefiting from the energy-saving effects of hydrodynamic drag reduction. This strategy is particularly important during migrations, where energy saving is essential for survival (Noren et al. 2008).

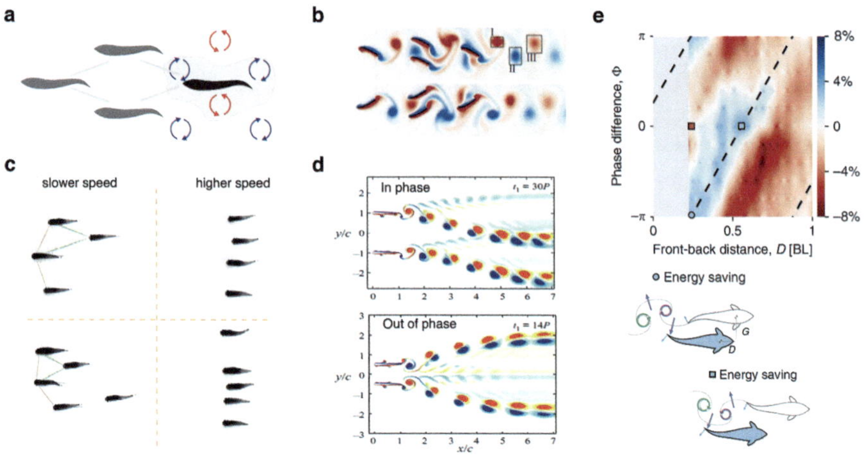

Fig. 7.5 (**a**) The optimal "diamond formation" used by schooling fish to save energy. (**b**) CFD-based simulation of four filaments swimming in a diamond configuration to achieve high swimming efficiency. (**c**) *Hemigrammus bleheri* adopts a simple phalanx pattern to save energy. (**d**) CFD analysis shows that phase difference significantly influences swimming efficiency in a side-by-side configuration. (**e**) Fish reduce energy expenditure through spatiotemporal formations. Figures reproduced from Liao (2022), Park and Sung (2018), Ashraf et al. (2017), Gungor et al. (2022), and Li et al. (2020)

Another strategy of saving energy is known as "hitchhiking," where the aquatic animal such as remora could attach to the body of a big swimmer to passively towed along with the host (Fig. 7.4d). Within this they could save tremendous energy while swimming or migrating. Recent CFD work also reveals the resistance of the remora can generally be halved when attached. Besides, the results have also demonstrated that the drag reduction rate increases with the developed boundary-layer thickness and can be estimated using the boundary-layer thickness ratio and velocity deficit (Xu et al. 2021).

7.3.2 Active Mechanisms

In 1973, Daniel Weihs proposed that fish swimming in a diamond formation could effectively reduce energy consumption compared to swimming alone (Fig. 7.5a) (Weihs 1973). Since then, scientists have been exploring the types of formations in which fish can save energy relative to solitary swimming with various tools, including computational simulations and experiments with physical models. Although energy savings or efficiency improvements have been observed through computational simulations, physical experiments, and biological studies across various formations—including in-line (tandem), side-by-side (phalanx), staggered, rectangular, and diamond arrangements (Hemelrijk et al. 2015; Peng et al. 2018a,b;

Heydari et al. 2024)—the underlying mechanisms remain largely unclear. Two primary hydrodynamic mechanisms have been proposed to explain these observed energy savings in fish schools. The first is known as the vortex interaction mechanism. When fish swim in specific spatial arrangements, the vortices shed by each individual can interact constructively, forming coherent patterns—sometimes resembling Kármán vortices. These structured vortex fields can generate regions of favorable flow for trailing fish, such as low-pressure zones or forward-directed jets, effectively reducing the thrust needed to maintain speed (Deng and Shao 2006; Kelly et al. 2023; Park and Sung 2018). The second mechanism is often referred to as the channeling effect. When fish swim side-by-side or in densely packed arrays, the flow between their bodies is constrained. According to principles related to Bernoulli's theory, the velocity of the water in these narrow channels increases, leading to a drop in pressure between individuals. This pressure differential can reduce lateral drag and enhance thrust, making it easier for each fish to propel itself forward. This mechanism is particularly relevant in side-by-side, rectangular, and lattice formations and has been observed in both numerical simulations and experimental studies (Fig. 7.5b) (Hemelrijk et al. 2015). However, verifying whether fish employ one or both mechanisms during schooling swimming is challenging, even though we observe that fish tend to swim in a phalanx formation (side-by-side), achieving energy savings (Fig. 7.5c) (Ashraf et al. 2017).

Beyond maintaining specific spatial distributions, it is also hypothetisized that fish in a school could also regulate the phase difference of their tailbeats to achieve energy-saving and efficiency improvement (Fig. 7.5d). For instance, when two fish swim side-by-side, their relative tailbeat phase plays a crucial role in determining their hydrodynamic efficiency. Studies have shown that in-phase synchronization, where both fish move their tails in the same phase, allows them to save energy. In contrast, swimming out of phase, where their tails move in opposite directions, results in increased energy expenditure with enhancement of thrust/speed (Dong and Lu 2007; Heydari et al. 2024). This trade-off between energy conservation and propulsion suggests that fish may dynamically adjust their phase relationships based on demands.

It is natural to hypothesize that both spatial and temporal factors are important for enhancing swimming efficiency and energy savings. Using innovative robotic and live-fish experiments, (Li et al. 2020) investigated spatiotemporal formation control in fish schools, finding that downstream-positioned fish dynamically adjust their distance (spatial) and phase difference (temporal) relative to the leading fish. Flow visualization indicated the involvement of a mechanism termed "vortex phase matching," in which followers adjust their tailbeat according to the induced flow from reverse Kármán vortices shed by the leader (Fig. 7.5e). Consequently, followers need to adjust their body undulations to achieve a phase difference linearly correlated with distance, thereby continuously saving energy. This phenomenon has also been observed in biological systems, serving as indirect verification of the proposed mechanism. More recently, Thandiackal and Lauder developed a robot-fish interaction platform incorporating a flapping foil designed to replicate the wake generated by a swimming fish (Thandiackal and Lauder 2023). They similarly observed this linear correlation between phase difference and distance: Trout swimming behind the foil within the artificially induced wake exhibited a reduced tailbeat frequency and synchronized their body undulations with the foil's oscillatory motion. Furthermore, this mechanism has been extended to three-dimensional scenarios, demonstrating that fish swimming at different heights can also benefit energetically from the vortex phase-matching mechanism in 3D (Li et al. 2021b).

In addition to direct hydrodynamic interactions, fish can achieve energy savings through behavioral strategies. For example, storks have been observed to follow others in thermals at the behavioral level, thereby conserving energy by exploiting the rising air currents used by individuals ahead of them (Flack et al. 2018). Analogously, fish can exploit ocean currents and complex hydrodynamic landscapes to reduce energy expenditure without relying solely on direct interactions with neighbors (Di Santo and Goerig 2025). For instance, fish may utilize updraft zones generated by tidal flows along submarine slopes, effectively "surfing" these currents to maintain position with minimal active swimming effort (Papastamatiou et al. 2021). In coral reef environments, fish also take advantage of the habitat's structural complexity, seeking refuge in low-velocity flow zones to reduce drag and save energy (Johansen et al. 2008).

7.4 Future Studies

Strategies for energy saving or efficiency improvement in aquatic swimmers have attracted attention for centuries across a wide range of scientific disciplines, including biology, physics, engineering, and mathematics. Despite advances in analytical tools such as mathematical modeling, computational simulations, and robotics, the underlying strategies used by aquatic organisms to reduce energy expenditure remain largely unclear. Some hypotheses have been proposed purely from mathematical or computational models that require very specific conditions—conditions that biological systems may not always be capable of achieving or maintaining. For example, the well-known diamond formation has been theoretically shown to save energy from a fluid dynamics perspective (Weihs 1973). However, in nature, fish rarely swim in such stable or precisely maintained formations (Partridge and Pitcher 1979), raising questions about the biological feasibility of this strategy. Nonetheless, these models are still highly valuable for engineering applications, where underwater vehicles can be precisely controlled to maintain energy-efficient formations. In other cases, biologists have uncovered surprising evidence that biological systems can indeed save energy in highly complex environments—scenarios that engineers might not have previously considered feasible. For instance, schooling fish have been shown to reduce energy expenditure when swimming in turbulent flows (Zhang et al. 2024), based on measurements of power consumption by individuals swimming alone versus in groups within the same turbulent conditions. Yet, the mechanism by which they achieve this remains a mystery.

One hypothesis suggests that fish may be able to modulate turbulence—possibly organizing their bodies in ways that mimic a honeycomb structure (Zorlutuna et al. 2012), thereby promoting laminar flow conditions within the group. All these highlight the importance of deep interdisciplinary collaboration between biology and engineering. Biological systems may reveal unexpected strategies that inspire new engineering solutions, while engineering tools such as simulation, robotics, and optimization can help test hypotheses and deepen our understanding of natural phenomena (Li et al. 2020). A reciprocal exchange, where engineers propose mechanisms and biologists validate them experimentally, holds great promise for advancing both fields.

With the advancement of computer science and the increasing accessibility of engineering technologies, there is growing potential to systematically explore and validate the energy-saving strategies employed by biological systems and to translate these strategies into engineering applications. First, integrating high-fidelity fluid simulations and experimental techniques with machine learning approaches can enable the development of data-driven optimization models for fish locomotion (Gazzola et al. 2014, 2016; Verma et al. 2018). These models offer deeper insights into turbulent energy transfer and wake vortex dynamics. Second, recent progress in smart materials and soft actuation technologies (Katzschmann et al. 2018), such as dielectric elastomers that mimic muscle contraction or light-responsive morphing fins, enhances the adaptability of robotic fish in complex flow environments. Third, embedding swarm intelligence into bioinspired robotic systems could replicate the collective swimming behaviors observed in nature (Ioannou 2017). By leveraging distributed sensing and reinforcement learning algorithms, robotic fish swarms may be capable of synchronizing their movements and optimizing energy use through coordinated vortex interactions, thereby significantly improving overall propulsion efficiency.

In addition to understanding the strategies aquatic animals use to save energy or improve efficiency, it is equally critical to explore how animals or physical systems are able to implement these mechanisms. Most hypotheses suggest that fish utilize their lateral line system to sense flow information and respond optimally (Liu et al. 2016). However, in some cases, fish may not rely solely on such sensory input. For instance, studies have shown that even after impairing both the lateral line and vision, fish can still perform vortex phase matching strategies to save energy (Li et al. 2020). This suggests that detailed sensing of the surrounding flow field may not be necessary. Through robotic system as a model, simple proprioceptive sensing, involving low-dimensional force or motion feedback, may be sufficient to guide the locomotion controller toward energy-efficient behavior (Li et al. 2021a). Such findings imply that fish might be using internal feedback, such as body deformation or muscular effort, rather than external flow sensing, to optimize their movement (Sánchez-Rodríguez et al. 2021). Further research is needed to verify this mechanism in biological systems and to explore how similar principles can be applied to the design of energy-efficient robotic or engineering systems.

Finally, biological systems likely employ a combination of strategies to save energy and improve efficiency—either by integrating multiple passive mechanisms or by combining them with active control strategies. In some cases, they may also dynamically switch between these active strategies in response to environmental conditions or task demands. For example, beyond individual and collective locomotion, cross-scale bioinspired design methodologies must be further developed to bridge propulsion strategies across both micro- and macroscales. At the microscale, biomimetic shark-skin-inspired microtextures can be used to reduce hydrodynamic drag (Wen et al. 2014), while at the macroscale, morphologically optimized, adjustable-span pectoral fins may enhance maneuverability and propulsion control (Liu et al. 2017). Schooling fish often exhibit dynamic reconfigurations in formation, which may indicate that they are switching among different energy-saving strategies depending on context (Marras et al. 2015). A more detailed exploration of when and how biological systems select specific strategies to save energy or enhance efficiency is needed. Understanding these potential strategy-switching behaviors is a critical direction for future bioinspired research and could lead to more adaptable and efficient engineering systems.

References

Abdulbari HA, Hassan DM, Zulkefli BYH (2015) Bio-inspired passive drag reduction techniques: a review. ChemBioEng Rev 2(3):185–203. https://doi.org/10.1002/cben.201400033

Akoz E, Moored KW (2018) Unsteady propulsion by an intermittent swimming gait. J Fluid Mech 834:149–172. https://doi.org/10.1017/jfm.2017.731

Alexander RM (1967) Functional design in fishes. Hutchinson University Library, London. ISBN 0090847709, 9780090847709

Alexander RM (2004) Hitching a lift hydrodynamically-in swimming, flying and cycling. J Biol 3:1–3. https://doi.org/10.1186/jbiol5

Amichay G, Li L, Nagy M, Couzin ID (2024) Revealing the mechanism and function underlying pairwise temporal coupling in collective motion. Nat Commun 15(1):4356. https://doi.org/10.1038/s41467-024-48458-z

Anderson JM, Streitlien K, Barrett D, Triantafyllou MS (1998) Oscillating foils of high propulsive efficiency. J Fluid Mech 360:41–72. https://doi.org/10.1017/S0022112097008392

Ashraf I, Bradshaw H, Ha TT, Halloy J, Godoy-Diana R, Thiria B (2017) Simple phalanx pattern leads to energy saving in cohesive fish schooling. Proc Natl Acad Sci 114(36):9599–9604. https://doi.org/10.1073/pnas.1706503114

Bale R, Hao M, Bhalla APS, Patankar NA (2014) Energy efficiency and allometry of movement of swimming and flying animals. Proc Natl Acad Sci 111(21):7517–7521. https://doi.org/10.1073/pnas.1310544111

Barrett D, Triantafyllou M, Yue D, Grosenbaugh M, Wolfgang M (1999) Drag reduction in fish-like locomotion. J Fluid Mech 392:183–212. https://doi.org/10.1017/S0022112099005455

Beal DN, Hover FS, Triantafyllou MS, Liao JC, Lauder GV (2006) Passive propulsion in vortex wakes. J Fluid Mech 549:385–402. https://doi.org/10.1017/S0022112005007925

Becker AD, Masoud H, Newbolt JW, Shelley M, Ristroph L (2015) Hydrodynamic schooling of flapping swimmers. Nat Commun 6(1):8514. https://doi.org/10.1038/ncomms9514

Bushnell DM, Moore K (1991) Drag reduction in nature. Annu Rev Fluid Mech 23:65–79. https://doi.org/10.1146/annurev.fl.23.010191.000433

Chao LM, Bhalla APS, Li L (2023) Vortex interactions of two burst-and-coast swimmers in a side-by-side arrangement. Theor Comput Fluid Dyn 37(4):505–517. https://doi.org/10.1007/s00162-023-00664-z

Chao LM, Couzin ID, Li L (2024a) On turning maneuverability in self-propelled burst-and-coast swimming. Phys Fluids 36(11):111918. https://doi.org/10.1063/5.0237171
Chao LM, Jia L, Li L (2024b) Tailbeat perturbations improve swimming efficiency in self-propelled flapping foils. J Fluid Mech 984:A46. https://doi.org/10.1017/jfm.2024.262
Chao LM, Jia L, Wang S, Liberzon A, Ravi S, Couzin ID, Li L (2024c) Tailbeat perturbations improve swimming efficiency by reducing the phase lag between body motion and the resulting fluid response. PNAS Nexus 3(3):pgae073. https://doi.org/10.1093/pnasnexus/pgae073
Compagno LJ et al (2001) Sharks of the world: an annotated and illustrated catalogue of shark species known to date, vol 1. Food and Agriculture Organization. ISBN 9251045437, 9789251045435
Dai L, He G, Zhang X, Zhang X (2018) Intermittent locomotion of a fish-like swimmer driven by passive elastic mechanism. Bioinspiration Biomimetics 13(5):056011. https://doi.org/10.1088/1748-3190/aad419
Dal Jeong Y, Kim MJ, Lee JH (2023) Intermittent swimming of two self-propelled flexible fins with laterally constrained heaving motions in a side-by-side configuration. J Fluid Mech 960:A39. https://doi.org/10.1017/jfm.2023.203
Dean B, Bhushan B (2010) Shark-skin surfaces for fluid-drag reduction in turbulent flow: a review. Philos Trans Roy Soc A: Math Phys Eng Sci 368(1929):4775–4806. https://doi.org/10.1098/rsta.2010.0201
Di Santo V, Goerig E (2025) Swimming smarter, not harder: fishes exploit habitat heterogeneity to increase locomotor performance. J Exp Biol 228(Suppl_1):JEB247918. https://doi.org/10.1242/jeb.247918
Dong GJ, Lu XY (2007) Characteristics of flow over traveling wavy foils in a side-by-side arrangement. Phys Fluids 19(5):057107. https://doi.org/10.1063/1.2736083
Drucker EG, Lauder GV (2001) Locomotor function of the dorsal fin in teleost fishes: experimental analysis of wake forces in sunfish. J Exp Biol 204(17):2943–2958. https://doi.org/10.1242/jeb.204.17.2943
Du R, Li Z, Kamal Y, Pablo V (2015) Robot Fish: Bio-inspired Fishlike Underwater Robots. Springer tracts in mechanical engineering. Springer, Berlin. ISBN 3662468700, 9783662468708
Ehrenstein U, Eloy C (2013) Skin friction on a moving wall and its implications for swimming animals. J Fluid Mech 718:321–346. https://doi.org/10.1017/jfm.2012.613
Eloy C (2012) Optimal Strouhal number for swimming animals. J Fluids Struct 30:205–218. https://doi.org/10.1016/j.jfluidstructs.2012.02.008
Fish FE (2006) The myth and reality of gray's paradox: implication of dolphin drag reduction for technology. Bioinspiration Biomimetics 1(2):R17. https://doi.org/10.1088/1748-3182/1/2/R01
Fish FE, Lauder GV (2006) Passive and active flow control by swimming fishes and mammals. Annu Rev Fluid Mech 38(1):193–224. https://doi.org/10.1146/annurev.fluid.38.050304.092201
Flack A, Nagy M, Fiedler W, Couzin ID, Wikelski M (2018) From local collective behavior to global migratory patterns in white storks. Science 360(6391):911–914. https://doi.org/10.1126/science.aap7781
Floryan D, Van Buren T, Smits AJ (2018) Efficient cruising for swimming and flying animals is dictated by fluid drag. Proc Natl Acad Sci 115(32):8116–8118. https://doi.org/10.1073/pnas.1805941115
Gazzola M, Hejazialhosseini B, Koumoutsakos P (2014) Reinforcement learning and wavelet adapted vortex methods for simulations of self-propelled swimmers. SIAM J Sci Comput 36(3):B622–B639. https://doi.org/10.1137/130943078
Gazzola M, Tchieu AA, Alexeev D, De Brauer A, Koumoutsakos P (2016) Learning to school in the presence of hydrodynamic interactions. J Fluid Mech 789:726–749. https://doi.org/10.1017/jfm.2015.686
Gleiss AC, Jorgensen SJ, Liebsch N, Sala JE, Norman B, Hays GC, Quintana F, Grundy E, Campagna C, Trites AW et al (2011) Convergent evolution in locomotory patterns of flying and swimming animals. Nat Commun 2(1):352. https://doi.org/10.1038/ncomms1350

Golovin KB, Gose JW, Perlin M, Ceccio SL, Tuteja A (2016) Bioinspired surfaces for turbulent drag reduction. Philos Trans Roy Soc A: Math Phys Eng Sci 374(2073):20160189. https://doi.org/10.1098/rsta.2016.0189

Graybill MT, Xu NW (2024) Experimental studies of bioinspired shark denticles for drag reduction. Integrative Comp Biol 64(3):742–752. https://doi.org/10.1093/icb/icae086

Gungor A, Khalid MSU, Hemmati A (2022) Classification of vortex patterns of oscillating foils in side-by-side configurations. J Fluid Mech 951:A37. https://doi.org/10.1017/jfm.2022.785

Guo J, Han P, Zhang W, Wang J, Lauder GV, Di Santo V, Dong H (2023) Vortex dynamics and fin-fin interactions resulting in performance enhancement in fish-like propulsion. Phys Rev Fluids 8(7):073101. https://doi.org/10.1103/PhysRevFluids.8.073101

Hansen LP, Jonsson N, Jonsson B (1993) Oceanic migration in homing Atlantic salmon. Anim Behav 45(5):927–941. https://doi.org/10.1006/anbe.1993.1112

He Y, Mou J, Chen L, Zeng Q, Huang Y, Chen P, Zhang S (2022) Will sailing in formation reduce energy consumption? Numerical prediction of resistance for ships in different formation configurations. Appl Energy 312:118695. https://doi.org/10.1016/j.apenergy.2022.118695

Hemelrijk CK, Reid D, Hildenbrandt H, Padding J (2015) The increased efficiency of fish swimming in a school. Fish Fisheries 16(3):511–521. https://doi.org/10.1111/faf.12072

Hertel H (1966) Structure, form, movement. Biology and technology. Reinhold. https://cir.nii.ac.jp/crid/1574231874650636032

Heydari S, Kanso E (2021) School cohesion, speed and efficiency are modulated by the swimmers flapping motion. J Fluid Mech 922:A27. https://doi.org/10.1017/jfm.2021.551

Heydari S, Hang H, Kanso E (2024) Mapping spatial patterns to energetic benefits in groups of flow-coupled swimmers. Elife 13:RP96129. https://doi.org/10.7554/eLife.96129.3

Hover F, Haugsdal Ø, Triantafyllou M (2004) Effect of angle of attack profiles in flapping foil propulsion. J Fluids Struct 19(1):37–47. https://doi.org/10.1016/j.jfluidstructs.2003.10.003

Ioannou CC (2017) Swarm intelligence in fish? the difficulty in demonstrating distributed and self-organised collective intelligence in (some) animal groups. Behav Process 141:141–151. https://doi.org/10.1016/j.beproc.2016.10.005

Deng J, Shao Xm (2006) Hydrodynamics in a diamond-shaped fish school. J Hydrodyn Ser B 18(3):438–442. https://doi.org/10.1016/S1001-6058(06)60090-5

Johansen J, Bellwood D, Fulton C (2008) Coral reef fishes exploit flow refuges in high-flow habitats. Marine Ecol Progress Series 360:219–226. https://doi.org/10.1137/130943078

Kang Ll, Gong Sx, Lu XY, Cui Wc, Fan Dx (2023) Scaling laws for the intermittent swimming performance of a flexible plate at low Reynolds number. J Hydrodyn 35(4):803–810. https://doi.org/10.1007/s42241-023-0059-2

Katzschmann RK, DelPreto J, MacCurdy R, Rus D (2018) Exploration of underwater life with an acoustically controlled soft robotic fish. Sci Robot 3(16):eaar3449. https://doi.org/10.1126/scirobotics.aar3449

Kelly J, Pan Y, Menzer A, Dong H (2023) Hydrodynamics of body–body interactions in dense synchronous elongated fish schools. Phys Fluids 35(4):041906. https://doi.org/10.1063/5.0142950

Kern S, Koumoutsakos P (2006) Simulations of optimized anguilliform swimming. J Exp Biol 209(24):4841–4857. https://doi.org/10.1242/jeb.02526

Kooijman SALM (2000) Dynamic energy and mass budgets in biological systems. Cambridge University Press, Cambridge. ISBN 0521786088, 9780521786089

Lauder GV, Drucker EG (2004) Morphology and experimental hydrodynamics of fish fin control surfaces. IEEE J Oceanic Eng 29(3):556–571. https://doi.org/10.1109/JOE.2004.833219

Lauder GV, Anderson EJ, Tangorra J, Madden PG (2007) Fish biorobotics: kinematics and hydrodynamics of self-propulsion. J Exp Biol 210(16):2767–2780. https://doi.org/10.1242/jeb.000265

Lauder GV, Wainwright DK, Domel AG, Weaver JC, Wen L, Bertoldi K (2016) Structure, biomimetics, and fluid dynamics of fish skin surfaces. Phys Rev Fluids 1(6):060502. https://doi.org/10.1103/PhysRevFluids.1.060502

Li L, Nagy M, Graving JM, Bak-Coleman J, Xie G, Couzin ID (2020) Vortex phase matching as a strategy for schooling in robots and in fish. Nat Commun 11(1):5408. https://doi.org/10.1038/s41467-020-19086-0
Li L, Liu D, Deng J, Lutz MJ, Xie G (2021a) Fish can save energy via proprioceptive sensing. Bioinspiration Biomimetics 16(5):056013. https://doi.org/10.1088/1748-3190/ac165e
Li L, Zheng X, Mao R, Xie G (2021b) Energy saving of schooling robotic fish in three-dimensional formations. IEEE Robot Autom Lett 6(2):1694–1699. https://doi.org/10.1109/LRA.2021.3059629
Liao JC (2004) Neuromuscular control of trout swimming in a vortex street: implications for energy economy during the Karman gait. J Exp Biol 207(20):3495–3506. https://doi.org/10.1242/jeb.01125
Liao JC (2007) A review of fish swimming mechanics and behaviour in altered flows. Philos Trans Roy Soc B: Biol Sci 362(1487):1973–1993. https://doi.org/10.1098/rstb.2007.2082
Liao JC (2022) Fish swimming efficiency. Curr Biol 32(12):R666–R671. https://doi.org/10.1016/j.cub.2022.04.073
Liao JC, Beal DN, Lauder GV, Triantafyllou MS (2003) Fish exploiting vortices decrease muscle activity. Science 302(5650):1566–1569. https://doi.org/10.1126/science.1088295
Lighthill MJ (1970) Aquatic animal propulsion of high hydromechanical efficiency. J Fluid Mech 44(2):265–301. https://doi.org/10.1017/S0022112070001830
Lighthill MJ (1971) Large-amplitude elongated-body theory of fish locomotion. Proc Roy Soc Lond Ser B Biol Sci 179(1055):125–138. https://doi.org/10.1098/rspb.1971.0085
Lighthill SJ (1975) Mathematical biofluid dynamics. SIAM. ISBN 1611970512, 9781611970517
Lin X, Wu J, Zhang T, Yang L (2019) Phase difference effect on collective locomotion of two tandem autopropelled flapping foils. Phys Rev Fluids 4(5):054101. https://doi.org/10.1103/PhysRevFluids.4.054101
Lin X, Wu J, Zhang T, Yang L (2021) Flow-mediated organization of two freely flapping swimmers. J Fluid Mech 912:A37. https://doi.org/10.1017/jfm.2020.1143
Liu G, Wang A, Wang X, Liu P (2016) A review of artificial lateral line in sensor fabrication and bionic applications for robot fish. Appl Bionics Biomech 2016(1):4732703. https://doi.org/10.1155/2016/4732703
Liu G, Ren Y, Dong H, Akanyeti O, Liao JC, Lauder GV (2017) Computational analysis of vortex dynamics and performance enhancement due to body–fin and fin–fin interactions in fish-like locomotion. J Fluid Mech 829:65–88. https://doi.org/10.1017/jfm.2017.533
Marras S, Killen SS, Lindström J, McKenzie DJ, Steffensen JF, Domenici P (2015) Fish swimming in schools save energy regardless of their spatial position. Behav Ecol Sociobiol 69:219–226. https://doi.org/10.1007/s00265-014-1834-4
Martin S, Bhushan B (2014) Fluid flow analysis of a shark-inspired microstructure. J Fluid Mech 756:5–29. https://doi.org/10.1017/jfm.2014.447
Matthews DG, Lauder GV (2021) Fin–fin interactions during locomotion in a simplified biomimetic fish model. Bioinspiration Biomimetics 16(4):046023. https://doi.org/10.1088/1748-3190/ac03a8
Mawignon FJ, Liu J, Qin L, Kouediatouka AN, Ma Z, Lv B, Dong G (2023) The optimization of biomimetic sharkskin riblet for the adaptation of drag reduction. Ocean Eng 275:114135. https://doi.org/10.1016/j.oceaneng.2023.114135
Miklosovic D, Murray M, Howle L, Fish F (2004) Leading-edge tubercles delay stall on humpback whale (megaptera novaeangliae) flippers. Phys fluids 16(5):L39–L42. https://doi.org/10.1063/1.1688341
Moored KW, Dewey PA, Smits A, Haj-Hariri H (2012) Hydrodynamic wake resonance as an underlying principle of efficient unsteady propulsion. J Fluid Mech 708:329–348. https://doi.org/10.1017/jfm.2012.313
Newbolt JW, Zhang J, Ristroph L (2019) Flow interactions between uncoordinated flapping swimmers give rise to group cohesion. Proc Natl Acad Sci 116(7):2419–2424. https://doi.org/10.1073/pnas.1816098116

Newbolt JW, Zhang J, Ristroph L (2022) Lateral flow interactions enhance speed and stabilize formations of flapping swimmers. Phys Rev Fluids 7(6):L061101. https://doi.org/10.1103/PhysRevFluids.7.L061101

Newbolt JW, Lewis N, Bleu M, Wu J, Mavroyiakoumou C, Ramananarivo S, Ristroph L (2024) Flow interactions lead to self-organized flight formations disrupted by self-amplifying waves. Nat Commun 15(1):3462. https://doi.org/10.1038/s41467-024-47525-9

Noren SR, Biedenbach G, Redfern JV, Edwards EF (2008) Hitching a ride: the formation locomotion strategy of dolphin calves. Funct Ecol 22(2):278–283. https://doi.org/10.1111/j.1365-2435.2007.01353.x

Papastamatiou YP, Iosilevskii G, Di Santo V, Huveneers C, Hattab T, Planes S, Ballesta L, Mourier J (2021) Sharks surf the slope: current updrafts reduce energy expenditure for aggregating marine predators. J Anim Ecol 90(10):2302–2314. https://doi.org/10.1111/1365-2656.13536

Park SG, Sung HJ (2018) Hydrodynamics of flexible fins propelled in tandem, diagonal, triangular and diamond configurations. J Fluid Mech 840:154–189. https://doi.org/10.1017/jfm.2018.64

Partridge B, Pitcher T (1979) Evidence against a hydrodynamic function for fish schools. Nature 279(5712):418–419. https://doi.org/10.1038/279418a0

Peng ZR, Huang H, Lu XY (2018a) Collective locomotion of two closely spaced self-propelled flapping plates. J Fluid Mech 849:1068–1095. https://doi.org/10.1017/jfm.2018.447

Peng ZR, Huang H, Lu XY (2018b) Hydrodynamic schooling of multiple self-propelled flapping plates. J Fluid Mech 853:587–600. https://doi.org/10.1017/jfm.2018.634

Peskin CS (2002) The immersed boundary method. Acta Numerica 11:479–517. https://doi.org/10.1017/S0962492902000077

Pitcher TJ, Partridge BL, Wardle C (1976) A blind fish can school. Science 194(4268):963–965. https://doi.org/10.1126/science.982056

Ramananarivo S, Fang F, Oza A, Zhang J, Ristroph L (2016) Flow interactions lead to orderly formations of flapping wings in forward flight. Phys Rev Fluids 1(7):071201. https://doi.org/10.1103/PhysRevFluids.1.071201

Read DA, Hover F, Triantafyllou M (2003) Forces on oscillating foils for propulsion and maneuvering. J Fluids Struct 17(1):163–183. https://doi.org/10.1016/S0889-9746(02)00115-9

Rosen MW, Cornford NE (1971) Fluid friction of fish slimes. Nature 234(5323):49–51. https://doi.org/10.1038/234049a0

Rosenthal SB, Twomey CR, Hartnett AT, Wu HS, Couzin ID (2015) Revealing the hidden networks of interaction in mobile animal groups allows prediction of complex behavioral contagion. Proc Natl Acad Sci 112(15):4690–4695. https://doi.org/10.1073/pnas.1420068112

Sagong W, Jeon WP, Choi H (2013) Hydrodynamic characteristics of the sailfish (Istiophorus platypterus) and swordfish (Xiphias gladius) in gliding postures at their cruise speeds. PLoS One 8(12):e81323. https://doi.org/10.1371/journal.pone.0081323

Sánchez-Rodríguez J, Celestini F, Raufaste C, Argentina M (2021) Proprioceptive mechanism for bioinspired fish swimming. Phys Rev Lett 126(23):234501. https://doi.org/10.1103/PhysRevLett.126.234501

Shephard KL (1994) Functions for fish mucus. Rev Fish Biol Fisheries 4:401–429. https://doi.org/10.1007/BF00042888

Shoele K, Zhu Q (2015) Drafting mechanisms between a dolphin mother and calf. J Theor Biol 382:363–377. https://doi.org/10.1016/j.jtbi.2015.07.017

Smits AJ (2019) Undulatory and oscillatory swimming. J Fluid Mech 874:P1. https://doi.org/10.1017/jfm.2019.284

Soyener JK (1993) Die elisabethanische galeone: Der einfluss von schiffsbau und bewaffnung auf die seegefechtstaktik in der zweiten hälfte des 16. jh. Das Logbuch 29(4):160–165, includes secondary citation of Baker (ca. 1590), "Fragments of ancient Shipwrightry"

Taylor GK, Nudds RL, Thomas AL (2003) Flying and swimming animals cruise at a Strouhal number tuned for high power efficiency. Nature 425(6959):707–711. https://doi.org/10.1038/nature02000

Thandiackal R, Lauder G (2023) In-line swimming dynamics revealed by fish interacting with a robotic mechanism. elife 12:e81392. https://doi.org/10.7554/eLife.81392

Triantafyllou MS, Triantafyllou GS (1995) An efficient swimming machine. Sci Am 272(3):64–70. https://doi.org/10.1038/scientificamerican0395-64

Triantafyllou MS, Triantafyllou G, Yue DK (2000) Hydrodynamics of fishlike swimming. Annu Rev Fluid Mech 32(1):33–53. https://doi.org/10.1146/annurev.fluid.32.1.33

Tytell ED (2006) Median fin function in bluegill sunfish Lepomis macrochirus: streamwise vortex structure during steady swimming. J Exp Biol 209(8):1516–1534. https://doi.org/10.1242/jeb.02154

Van Wassenbergh S, van Manen K, Marcroft TA, Alfaro ME, Stamhuis EJ (2015) Boxfish swimming paradox resolved: forces by the flow of water around the body promote manoeuvrability. J Roy Soc Interface 12(103):20141146. https://doi.org/10.1098/rsif.2014.1146

Verma S, Novati G, Koumoutsakos P (2018) Efficient collective swimming by harnessing vortices through deep reinforcement learning. Proc Natl Acad Sci 115(23):5849–5854. https://doi.org/10.1073/pnas.1800923115

Videler JJ, Weihs D (1982) Energetic advantages of burst-and-coast swimming of fish at high speeds. J Exp Biol 97(1):169–178. https://doi.org/10.1242/jeb.97.1.169

Wang ZJ (2000) Vortex shedding and frequency selection in flapping flight. J Fluid Mech 410:323–341. https://doi.org/10.1017/S0022112099008071

Webb PW (1993) The effect of solid and porous channel walls on steady swimming of steelhead trout Oncorhynchus mykiss. J Exp Biol 178(1):97–108. https://doi.org/10.1242/jeb.178.1.97

Weihs D (1973) Hydromechanics of fish schooling. Nature 241(5387):290–291. https://doi.org/10.1038/241290a0

Weihs D (1974) Energetic advantages of burst swimming of fish. J Theor Biol 48(1):215–229. https://doi.org/10.1016/0022-5193(74)90192-1

Weihs D (2004) The hydrodynamics of dolphin drafting. J Biol 3:1–16. https://doi.org/10.1186/jbiol2

Wen L, Weaver JC, Lauder GV (2014) Biomimetic shark skin: design, fabrication and hydrodynamic function. J Exp Biol 217(10):1656–1666. https://doi.org/10.1242/jeb.097097

Whittlesey RW, Liska S, Dabiri JO (2010) Fish schooling as a basis for vertical axis wind turbine farm design. Bioinspiration Biomimetics 5(3):035005. https://doi.org/10.1088/1748-3182/5/3/035005

Wright RM, Piper AT, Aarestrup K, Azevedo JM, Cowan G, Don A, Gollock M, Rodriguez Ramallo S, Velterop R, Walker A, et al (2022) First direct evidence of adult European eels migrating to their breeding place in the Sargasso Sea. Sci Rep 12(1):15362. https://doi.org/10.1038/s41598-022-19248-8

Wu G, Yang Y, Zeng L (2007) Kinematics, hydrodynamics and energetic advantages of burst-and-coast swimming of koi carps (*Cyprinus carpio koi*). J Exp Biol 210(12):2181–2191. https://doi.org/10.1242/jeb.001842

Wu TYT (1961) Swimming of a waving plate. J Fluid Mech 10(3):321–344. https://doi.org/10.1017/S0022112061000949

Xu Y, Shi W, Arredondo-Galeana A, Mei L, Demirel YK (2021) Understanding of remora's "hitchhiking" behaviour from a hydrodynamic point of view. Sci Rep 11(1):14837. https://doi.org/10.1038/s41598-021-94342-x

Yang D, Wu J, Khedkar K, Chao LM, Bhalla APS (2024) Hydrodynamics and scaling laws for intermittent s-start swimming. J Fluid Mech 984:A2. https://doi.org/10.1017/jfm.2024.103

Zhang J (2017) Footprints of a flapping wing. J Fluid Mech 818:1–4. https://doi.org/10.1017/jfm.2017.173

Zhang J, Childress S, Libchaber A, Shelley M (2000) Flexible filaments in a flowing soap film as a model for one-dimensional flags in a two-dimensional wind. Nature 408(6814):835–839. https://doi.org/10.1038/35048530

Zhang JD, Sung HJ, Huang WX (2022a) Hydrodynamic interaction of dorsal fin and caudal fin in swimming tuna. Bioinspiration Biomimetics 17(6):066004. https://doi.org/10.1088/1748-3190/ac84b8

Zhang Y, Lauder GV (2023) Energetics of collective movement in vertebrates. J Exp Biol 226(20):jeb245617. https://doi.org/10.1242/jeb.245617

Zhang Y, Zhao W, Ma S, Liu H, Wang X, Zhao X, Yu B, Cai M, Zhou F (2022b) Modulus adaptive lubricating prototype inspired by instant muscle hardening mechanism of catfish skin. Nat Commun 13(1):377. https://doi.org/10.1038/s41467-022-28038-9

Zhang Y, Ko H, Calicchia MA, Ni R, Lauder GV (2024) Collective movement of schooling fish reduces the costs of locomotion in turbulent conditions. PLoS Biol 22(6):e3002501. https://doi.org/10.1371/journal.pbio.3002501

Zhong Q, Dong H, Quinn DB (2019) How dorsal fin sharpness affects swimming speed and economy. J Fluid Mech 878:370–385. https://doi.org/10.1017/jfm.2019.612

Zhong Q, Zhu J, Fish FE, Kerr SJ, Downs A, Bart-Smith H, Quinn D (2021) Tunable stiffness enables fast and efficient swimming in fish-like robots. Sci Robot 6(57):eabe4088. https://doi.org/10.1126/scirobotics.abe4088

Zhu X, He G, Zhang X (2014) Flow-mediated interactions between two self-propelled flapping filaments in tandem configuration. Phys Rev Lett 113(23):238105. https://doi.org/10.1103/PhysRevLett.113.238105

Zorlutuna P, Annabi N, Camci-Unal G, Nikkhah M, Cha JM, Nichol JW, Manbachi A, Bae H, Chen S, Khademhosseini A (2012) Microfabricated biomaterials for engineering 3D tissues. Adv Mater 24(14):1782–1804. https://doi.org/10.1002/adma.201104631

Chapter 8
Biomimicry That Is Inspired by Spider Predation Tactic

Frédéric Ysnel and Alain Canard

Abstract Spiders (order Araneae) have developed a range of unique and effective adaptations as a result of their diverse predation strategies. Based on the latter, they can be placed into one of three major functional groups: wanderers, ambush hunters, and web weavers. Here, we detail key biological facets of species in these groups and draw upon this information to present a number of innovative bioinspired research directions. More specifically, we focus on spider vision, locomotion, mechanoreceptors, deceptive signals, and silk. The telescope-like eyes of jumping spiders are particularly effective in detecting stimuli over long distances. Furthermore, the retinas of their large anterior median eyes contain multiple layers of packed photoreceptors that are each sensitive to different wavelengths of light and that can analyse polarised light. Optics researchers have drawn upon jumping spider eye structure to develop several kinds of high-resolution nano- and microcameras. Spider locomotion relies on a hydrostatic skeleton, unique leg angulation patterns, and dry adhesion via scopulae, mechanisms that can all inform the design of spider-like robots. Spiders possess leg mechanoreceptors that are stimulated by airflow, proximal low-frequency sounds, direct touch, olfactory signals, or proprioceptive input, which could inspire the development of application-driven artificial flow-sensing hairs. UV reflectance is a deceptive signal commonly used by spiders and could be employed in novel miniature light-dispersive components. Because of its many unique biochemical and biomechanical properties, spider silk is being used in numerous bioinspired solutions under development, including in novel medical and technological applications.

Keywords Functional groups · Vision · Locomotion · Mechanosensors · Silk properties

F. Ysnel (✉) · A. Canard
Université de Rennes-MNHN – UMR 8067 BOREA-Campus de Beaulieu, Rennes Cedex, France
e-mail: frederic.ysnel@univ-rennes1.fr

V. Bels et al. (eds.), *Biomimicry – A Functional Approach*, Biologically-Inspired Systems 20, https://doi.org/10.1007/978-3-032-11949-0_8

8.1 Introduction

The ancient fossil records dating back approximately 390 million years (Middle Devonian period) reveal the presence of the extinct arachnid genus *Attercopus fimbriunguis* exhibiting one of the essential morphological characteristics specific to the spider, notably the presence of species with opisthosomal spigots already reflecting the ability of the species to produce silk (Selden et al. 2008). Little is generally known about ancient spiders because they are poorly preserved in the fossil record—their exoskeleton was not hardened over the entire body. However, *A. fimbriunguis* and others seem to mark the transition from the presumed common ancestors of arachnids, members of the order Trigonotarbida, to modern-day spiders. Nearly perfectly preserved specimens found in Baltic amber show that spider diversification has likely been underway for more than 100 million years.

The long evolutionary history of spiders has played a key part in the proliferation of the life strategies exhibited by extant species. Spiders are found across all latitudes and continents, with the notable exception of Antarctica. The scientific literature contains in-depth descriptions of spider life and spiders' biological, ecological, and behavioural traits (Nentwig 1987; Wise 1993; Barth 2002; Foelix 2011; Herberstein 2012; Boutry and Blamires 2013; Penney 2013; Eberhard 2020). The vast majority of the world's 53,000 spider species (World Spider Catalog 2025) appear to be zoophagous, and some are more specifically monophagous or araneophagic. They also tend to consume juvenile or adult terrestrial arthropods, with the exception of some large mygalomorph or araneomorph species that feed on small vertebrates. In recent decades, direct and indirect experimental research on spider predation has demonstrated that spiders have an impact on a significant proportion of the secondary consumers found in natural areas (see reviews by Wise 1993; Marc et al. 1993; Nyffeler and Benz 1987). The salticid *Bagheera kipingli* represents an exception to this general rule. It is the only spider known to have a primarily herbivorous diet. It is also known that juveniles of some tropical salticid genera may feed on nectar when prey are scarce (Kuja et al. 2012). Although all spiders are predatory in nature, they display heterogeneity in their predation strategies, which can be broadly classified into three functional groups: pursuit (wanderers), sit-and-wait (ambush hunters), and indirect prey capture (web weavers).

In summary, spiders have developed a range of adaptations that have allowed them to colonise habitats of all types. Their core morphological, ecophysiological and behavioural characteristics can be used as a source of bioinspiration.

8.2 Ingenious Predation Strategies for Bioinspired Technologies

For spiders in all three functional groups, predation involves three main sequential steps (Fig. 8.1): (i) hunting site selection and prey detection; (ii) prey interception

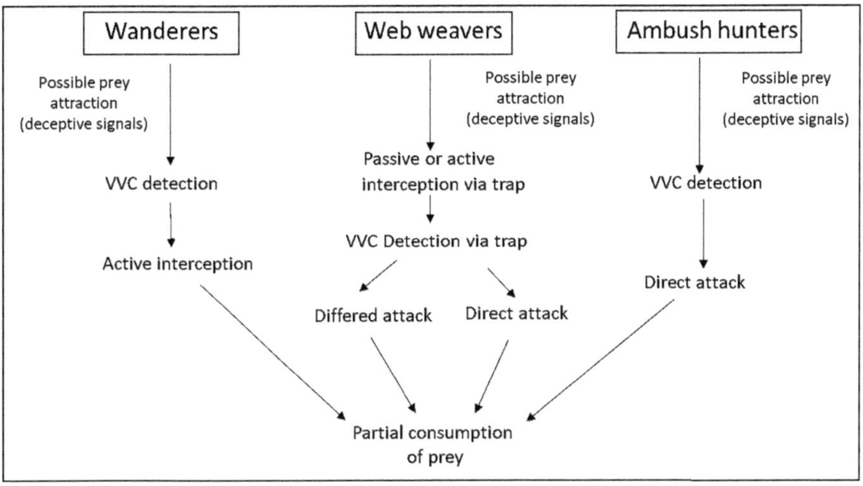

Fig. 8.1 Summary of the main sequential steps involved in spider predation strategies. VVC detection = detection of visual, vibrational, or chemical signals

and immobilisation; and (iii) immediate or delayed prey consumption. More recently, another intermediate step has been identified: actively attracting prey.

About two-thirds of known spider families do not build webs. Rather, they intercept their prey via active hunting or ambush hunting, which may take place on the ground or in herbaceous, shrubby, or arboreal vegetation (including the canopy). Depending on their nychthemeral rhythm of activity (nocturnal or diurnal), wanderers and ambush hunters may differ from web weavers in the functional mechanisms used. These mechanisms involve specific vision systems, locomotor abilities, and silk properties. This chapter focuses on these three facets of spider biology and their potential for informing bioinspired technologies.

8.3 High-Performance Prey Detection Systems

8.3.1 Spider Vision

Spiders display highly diverse visual mechanisms and vision-related behaviours. Better understanding this biological wealth could help drive the development of useful innovations (see review by Winsor et al. 2023 for a recent synthesis of knowledge on spider vision). Spider eyes are always simple ocelli, of which most spiders possess eight. In certain groups, there are only six, although sometimes four or two are seen; certain cave-dwelling species have none. These ocelli are generally arranged in two lines (Fig. 8.2c). They basically fall into one of two groups based on the position of the sensory cells in the retina. The principal eyes [i.e. the anterior

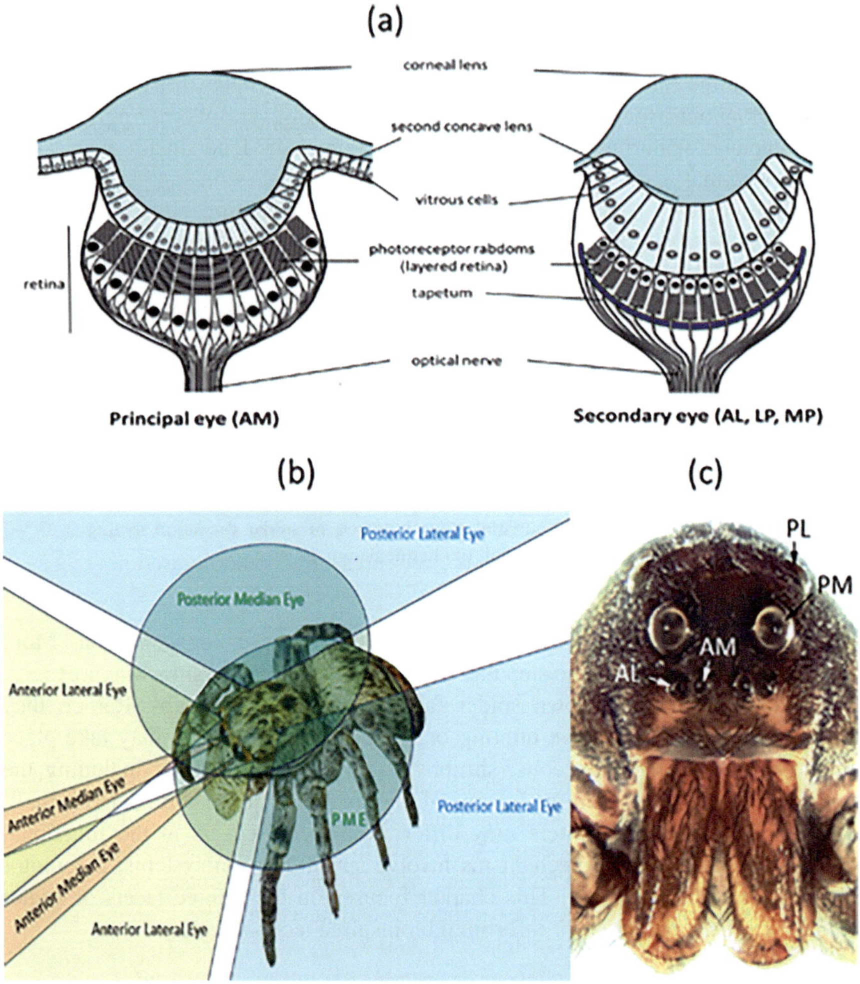

Fig. 8.2 Unique visual optic system of spiders. (**a**) Eye structure of principal and secondary eyes. (**b**) Diagram of the almost 360° field of vision of a jumping spider. (**c**) Ocular group (front view). *AL* anterior lateral eye; *AM* anterior median eye; *PM* posterior median eye; *PL* posterior lateral eye. (Adapted from Foelix 2011)

median (AM) eyes] have rhabdomeres that face the lenses, while the secondary eyes have rhabdomeres that face towards the optic nerve (Fig. 8.2a). As a result, the principal eyes can potentially detect shapes and colours, while the secondary eyes are mainly motion detectors. While the principal eyes provide daytime vision for diurnal wanderers, the secondary eyes have a tapetum—a reflective crystalline structure—that concentrates light and allows the nocturnal perception of prey (Foelix 2011). There is pronounced variation in eye morphology and functionality. For example, spider eyes vary in size, arrangement, anatomy, optical properties, photoreceptor structures, and underlying brain neuromorphology (see review by Winsor et al. 2023). In most web weaver species, vision plays a minor role, while visual acuity is much greater in diurnal and nocturnal wanderers.

The vision of jumping spiders (family *Salticidae*) has been studied far more than the vision of spiders in other families. In jumping spiders, vision is involved in prey detection, prey attack, predator escape, and courtship behaviour. Their eye groups are particularly effective at detecting stimuli over long distances (i.e. their telescope-like eyes); such is the case even for nocturnal species. Lateral muscles move their retinas in all three dimensions, allowing them to significantly expand their field of vision. The angle of vision is larger or smaller depending on the ocelli (Fig. 8.2b). Furthermore, some jumping spiders have eyes that are sensitive to UV wavelengths, enabling them to locate and distinguish their prey, whose body parts strongly reflect light in the UV spectrum (Li and Lim 2005). Jumping spiders have particularly large AM eyes, which are characterised by wide lenses and a long focal length; the retinas of their AM eyes contain multiple layers of packed photoreceptors that are each sensitive to different wavelengths of light. This retinal structure can simultaneously obtain multiple images of the same scene using different levels of defocus, and behavioural evidence suggests that jumping spiders perceive depth using image defocus cues (Guo et al. 2019). The structural organisation of jumping spider eyes also suggests they analyse polarised light, an essential function allowing orientation during nocturnal or diurnal activity or the detection of prey movement. Different patterns of retinal photoreceptor organisation have been seen in other families, such as *Lycosidae* and *Gnaphosidae*, and allow spiders to navigate using polarised light in the sky (Dacke et al. 2001).

8.3.2 Using Air Vibrations and Volatile Compounds to Detect Prey

Some spiders experience the environment through visual cues, but most tend to use tactile, vibratory, and olfactory cues; several studies have highlighted the role of other sensory systems, such as chemoreception and mechanoreception (Nelson 2023). Therefore, prey detection may rely on ocelli as well as on mechanoreceptor setae—trichobothria (e.g. filiform 'hairs'). Trichobothria are elongate exocuticular capillary stalks whose bases are inserted into circular cavities, cuticular cups, or cuticular sockets (Fig. 8.3) that are connected to nerve endings that transmit information to the central nervous system. These lightweight setae move in response to the frictional force of the air; the tilting deflection of an exposed seta causes the embedded section to deflect, which stimulates the neuron. As has been shown in scorpions, for example, trichobothria display maximum mechanical efficiency: mechanical forces are not lost in bending the hair but are rather deflected to the basal area, which is responsible for signal production (Zhang et al. 2020). Trichobothria are generally arranged in straight lines or small clusters on the dorsal

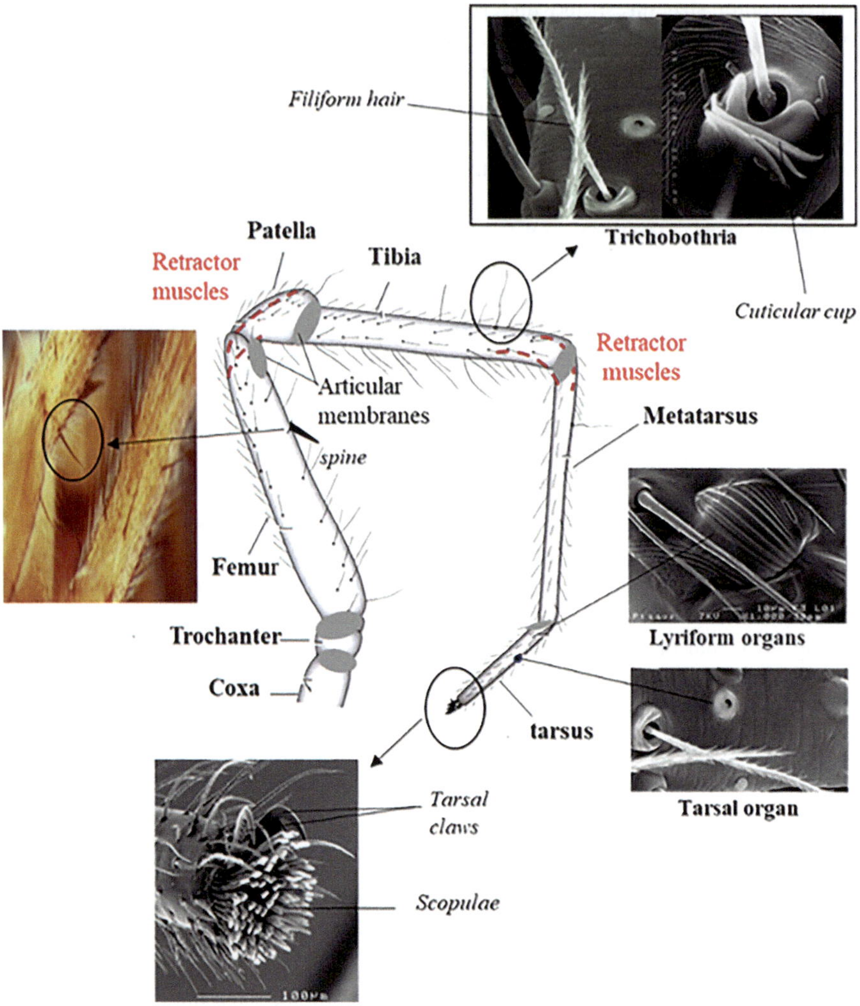

Fig. 8.3 Structure and sensory equipment of a spider walking leg

side of appendages (i.e. femur, tibia, metatarsus, or tarsus) (Fig. 8.3). They are stimulated by airflow, proximal low-frequency sounds, direct touch, olfaction, and proprioceptive input resulting from a spider's own locomotion (Barth 2021). They can be stimulated by airborne vibrations transmitted directly by prey or indirectly by prey provoking web vibrations.

Other mechanosensors called lyriform organs (Fig. 8.3) consist of a series of parallel slit sensilla, which can pick up on airborne sound at the limit of what is physically possible (Barth 2021). The slit sensilla on the legs are like small strain gauges. They consist of tiny skeletal slits that, when stretched by tens of nanometres, trigger nerve impulses. In web weavers, slit sensilla can perceive longitudinal vibrations that propagate through the web on radial threads and can use amplitude changes in the incoming vibrations to determine the source's direction (Ratnam et al. 2023).

Chemical detection systems also have an important role in prey detection. Surprisingly, spiders can pick up on certain volatile compounds emitted by humans or vegetables, which can then stimulate predatory behaviour. Farid et al. (2019) explored how plant essential oils influenced orientation and predation in the wandering spider *Pardosa pseudoannulata* (*Lycosidae*); they concluded that some individual and combined essential oils could have an influence on both processes. The African salticid *Evarcha culicivora* (known as the vampire spider) can effectively target *Anopheles* mosquitoes that have consumed a blood meal. This spider is attracted to the odour of mosquitoes that have fed on blood and the odour of humans; thus, adult and juvenile *E. culicivora* are drawn to human habitations where their mosquito prey are potentially present (Cross and Jackson 2011). Spiders have two types of chemoreceptors on their legs: (i) tarsal organs through which nerve endings are exposed to volatile substances via microscopic pores and (ii) contact chemoreceptors (or taste 'hairs') situated on the distal segments of the palps and legs (see Foelix 2011 or Barth 2021 for a basic description).

8.4 Efficient Locomotory Capabilities of Wandering Spiders

Spiders have four pairs of walking legs, which can vary in size, attached to the cephalothorax. These appendages are made up of seven segments (Fig. 8.3) that are connected by thin membranes, which allows articulation between the different segments. The lateral mobility of the different leg segments varies depending on joint type, and the range of the joint-rotation angle is smallest between the patella and the tibia. In wandering spiders, rapid movement can be achieved by jumping or running, which occurs during hunting or escape behaviour. These differences in joint-rotation angles, combined with the variation in segment size, serve to increase velocity, stability, and impact resistance during movement (Wang et al. 2011). Spider legs function via a particular mechanism involving the longitudinal muscle complex of the hydrostatic skeleton (see review by Landkammer et al. 2016; Barth 2021; Ratnam et al. 2023). The many longitudinal muscles result in a link between leg segments. Consequently, muscle contraction in one leg segment can provoke retraction or extension in an adjacent segment. However, biomechanical research has shown that there are no muscle connections enabling extension at the femur-patella and tibia-metatarsus junctions. In the semi-closed circulatory system of spiders, extension results from increasing the hydrostatic pressure of the haemolymph in the intermuscular gaps of the target segments (i.e. the principle of the hydrostatic skeleton). The abdominal heart generates circulation throughout the body, and the prosomatic muscles selectively increase the pressure of the haemolymph in the legs. The first of the seven segments of the legs—coxa and trochanter—are crucial for the orientation and regulation of this pressure. This increase in pressure causes the articular membranes between the target segments to swell and extend. The performance of this process is highly dependent on the nature of the working fluid, which must correspond to required operating speeds and pressures (Göttler et al. 2021). Measurements of hydrostatic pressure at different levels show variations depending on the appendages or the relative weight of the species (Kropf 2013). In the case of spiders, the legs can autotomize at specific areas (the coxa-trochanter junction) where a sealing process exists (Roth and Roth 1984). A predator grabbing a spider by a leg allows the captured animal to escape by sacrificing that leg. A unique form of locomotion has been observed in the desert spider *Cebrennus rechenbergi* (family *Sparassidae*), also known as the Moroccan flic-flac spider. If threatened, this species can escape by doubling its normal crawling speed and performing forward or backward flips using all of its legs simultaneously, a movement that resembles a gymnast's back handspring (https://www.youtube.com/watch?v=IJhTrPDU oxs&t=12s).

In many spiders, the ventral side of the tarsi may be covered with special microscopic or nanoscopic 'hairs' that form two distinct or contiguous groups: (i) the claw tufts, located in two juxtaposed tufts on the underside of the claws, and (ii) the scopulae, which can extend over the entire dorsal surface of the tarsus (Fig. 8.3). Wandering spiders use their claw tufts and scopulae to adhere to different substrate types as they move across horizontal and vertical planes. The claws mainly allow spiders to cling firmly to microcrevices in substrates or to move along their webs. In contrast, claw tufts and scopulae allow spiders to stick to and climb on many kinds of surfaces, thanks to the intermolecular forces of attraction between the tarsus' nanoscopic 'hairs' and the contact surface (Gasparetto et al. 2008). In the genus *Evarcha* (*Salticidae*), for example, the adhesiveness of the tarsal extremities results from the dichotomous shape of the microscopic 'hairs', which are subdivided into setulae. The ends of the setulae are flattened, forming what are called spatula. These microstructures give rise to adhesive forces that can reach the equivalent of 160 times the weight of the spider (for details, see Kesel et al. 2003, 2004 in Gasparetto et al. 2008). The scopulae of spiders are adapted to safely allow vertical movement over surfaces that vary greatly in roughness. The scopulae of *Evarcha arcuata* in particular do not rely on the presence of humidity, permitting dry adhesion. Spiders display at least equivalent performance to geckos on surfaces with low rugosity and exceed geckos on surfaces with high rugosity (Gasparetto et al. 2009). Spider species in the families *Ctenidae, Trechalaeidae, Pisauridae*, and *Lycosidae* regularly displace themselves across water surfaces using tarsi that are covered in highly hydrophobic microscopic 'hairs' (Foelix 2011; Suter 2013) that also possess micro- and nanoscale irregularities. These two traits mean the spiders' tarsi are extremely water resistant (Suter 2013).

8.5 Attracting Prey via Contrast and Camouflage

Spiders may attract prey during the day or at night using a variety of mechanisms, including deceptive signals. Prey attraction may involve subtle techniques such as the use of visual lures, UV reflectance, or chemical compounds. Some tropical and temperate web weavers (notably in family *Araneae*) employ conspicuous ventral spots or dorsally located abdominal colour patterns (Chuang et al. 2008). Visual lures may be based on the visible colour of the spider's body, as is seen in the genus *Gasteracantha* or *Nephila* (Fig. 8.4 d,h): they work by maximising the contrast in brightness between the spider's body and its hunting sites/immediate environment (Heiling et al. 2005; Vieira et al. 2017). However, it must be noted that, in some species, the sole function of these contrasting colours is to scare away predators (Ximenes and Gawryszewski 2020). Prey are also attracted by the light-coloured

Fig. 8.4 Examples of spider prey capture strategies. (**a**) Silk decoration (stabilimentum) of *Argiope* sp. (*Aranaeidae*). (**b**) Active surface wanderer (*Dolomedes fimbriatus*, *Pisauridae*). (**c**) Anterior median and lateral eyes of the diurnal terrestrial wanderer *Saitis barbipes*, *Salticidae*). (**d**) Body coloration (possible visual lure) of a golden silk orb-weaver (*Nephilidae*). (**e**) Aquatic hunting behaviour of the diving bell spider *Argyroneta aquatica* (*Argyronetidae*). (**f**) Sit and wait predator (*Misumena vatia*, *Thomisidae*). (**g**) Prey carcasses as decorations (*Cyclosa* sp., *Aranaeidae*). (**h**) Body coloration (possible visual lure) of *Gasteracantha* sp. (*Aranaeidae*). (Author's photographs)

stripes of nocturnal wanderers (e.g. the fishing spider *Dolomedes raptor,* Pisauridae) or sit-and-wait spiders (e.g. the brown huntsman spider, *Heteropoda venatoria*) (Lin et al. 2015; Zhang and Chen 2015). On the whole, the contrasting colours or white stripes seen on spider bodies may have a dual function, attracting both prey and mating partners. Ambush hunters in the family *Thomisidae* utilise the colour contrast between their bodies and flower-filled environments to attract pollinators, although they can also position themselves within a floral background that matches their body colour to obscure their presence from potential prey or to hide from predators (Fig. 8.4f).

During the day and at night, several web weaver species may draw in prey using the reflectance spectra of their silk threads and silk decorations (e.g. stabilimenta), a strategy that improves their foraging success. Silk decorations tend to function as visual attractants for insect prey, although they may sometimes purely play a role in predator defence (Nakata 2009). Materials that reflect UV wavelengths (100–400 nm) are attractive to pollinators, which are evolutionarily primed by the fact that UV-reflecting flowers contain more nectar; such materials simultaneously repel or ward off larger animals like birds. *Nephila clavipes*, the renowned golden orb-weaver, is known for its ability to adjust web reflectance to local light conditions and produce pigments that enhance web visibility by increasing the light reflected by its silk. Similarly, *Argiope aethereal* and *Argiope keyserlingi* build a larger number of longer decorations when light conditions are dim versus bright, probably to better attract approaching prey or to advertise a warning to oncoming birds (see Craig et al. 1996; Elgar et al. 1996; Herberstein and Fleisch 2003 in Théry and Casas 2008). In several cases, it has been found that, due to their UV reflectance, webs with stabilimenta intercept more prey than do undecorated webs (Lai et al. 2017). Silk decorations may also lead to the high levels of prey retention seen for giant orb-weaving spiders in the genus *Nephila* (Blamires et al. 2010).

Spiders' bodies may also reflect the same UV or visible wavelengths as flowers, a mechanism that can serve to attract insect pollinators. This technique has specifically been seen in certain European, Indian, and tropical species in the family *Thomisidae*, which hunt on flowers (Ratz et al. 2023). That said, this characteristic is not universal within this family (Nogushi 2021). Spiders in the neotropical genus *Epicadus* (*Thomisidae*) have bodies that mimic the shape of a flower and that reflect UV, meaning they do not need to hunt on flowers (Vieira et al. 2017). Although not exclusively involved in prey attraction, the highly coloured abdomen of the male peacock spider (*Maratus splendens*, Salticidae) reflects the full spectrum of visible light (Stavenga et al. 2016). The resulting iridescent rainbow signal is used as part of its courtship display (see asknature.org, https://youtu.be/kIQMXEARzFE).

Some spiders use chemical compounds to attract prey. For example, the silk of *A. keserlingi* (family *Araneidae*) contains allomones, such as the amide putrescine, a foul-smelling organic compound that is attractive to several categories of common terrestrial prey (Henneken et al. 2017). By decorating their webs with prey carcasses and other materials (e.g. plant detritus), *Nephila clavata* and *Cyclosa mulmeinensis* (family *Araneidae*) may increase their capture rates and prey retention rates (Tan et al. 2010). Such decorations (Fig. 8.4g) may boost web attractiveness via odour emissions. It has been suggested that the degree of attractiveness depends on whether the *Cyclosa* species is social or solitary (Blamires et al. 2010). Female and subadult bolas spiders (genus *Mastophora*; *Araneidae*) synthesise compounds resembling the sex pheromones produced by females of certain moth species (Diaz Jr and Long Jr 2022). These nocturnal spiders dangle a line of thread with a 'bola' at the end: a viscous ball of droplets containing these compounds. Male moths attracted by this chemical cocktail are then snared by the spider, which swings or launches the sticky ball towards its prey (see Nat. Geo. Animals: An alluring spider; https://www.youtube.com/watch?v=j7yU87enMRo). Very recently, research uncovered a unique prey attraction strategy used by the orb-weaving *Araneus ventricosus*: the spider exploits the bioluminescent signals of male fireflies it has captured, turning the fireflies into bait for capturing other males. Trapped male fireflies emit stress signals that closely resemble those of free females, leading more and more males to be drawn into the web (Fu et al. 2024).

8.6 Webs and the Interesting Properties of Their Silk

8.6.1 Web Diversity

Spiders in about one-third of the 134 currently described spider families (World Spider Catalog 2025) weave webs to capture prey. Some atypical species actively deploy their webs. For example, spiders in the tropical genus *Deinopis* behave like Roman gladiators: they carry around a small upright rectangular web that they throw over prey. A common inhabitant of European homes, the spitting spider *Scytodes thoracica* (Scytodidae), has venom glands that, in addition to their usual secretions, produce a sticky substance that the spider can project onto prey a few centimetres away.

However, these few cases aside, spiders tend to use their webs to passively trap prey. Depending on their architecture, passive webs can be grouped into seven broad types: orb, funnel, sheet, lace, tangle, radial, and purse (Fig. 8.5). Within each type, specific patterns in web structure can be used to identify spider family and, sometimes, genus. Compared to wanderers, web weavers have considerably greater hunting efficiency. Based on their architecture, surface area, adhesiveness, and vertical distribution in space, webs will trap prey that vary in quality and quantity. Web weavers may exhibit diurnal or nocturnal hunting activity. Such temporal differences logically decrease the degree of interspecific competition and interguild predation, allowing different spider species to coexist within the same biotope. The seven types of webs clearly differ in their biomechanical characteristics or in the fine-scale chemical composition of their silk threads.

Fig. 8.5 Various ways spiders use silk to trap prey. (**a**) Web complex (colony) of the social funnel web spider *Agelena consociata* (family *Agelenidae*). (**b**) Orb web spider *Larinioides cornutus* (family *Araneidae*) with a warning thread running from its second walking leg to the centre of its web. (**c**) Silk-lined burrow with cork-like trapdoor made of soil, vegetation, and silk constructed by a ctenizid spider (infraorder *Mygalomorphae*). (**d**) Above-ground portion of silken tubes (purse web) created by a subterranean atypid spider (infraorder *Mygalomorphae*). (**e**) Horizontal sheet web with its supporting threads. (**f**) Orb web with its radii, temporary spirals, and sticky spirals (family *Araneidae*). (**g**) Funnel web spider (family *Agelenidae*). (**h**) Three-dimensional web typical of theridiid spiders. (**i**) Small net-shaped web held by the front legs of a net-casting spider (family *Deinopidae*); the net can be stretched wide. (**j**) Spitting spider (*Scytodes thoracica,* family *Scytodidae*) that captures its prey by squirting a sticky, venomous fluid from its chelicerae. (**k**) Tube-dwelling spider (family *Segestridae*). (**l**) A few barely visible silk threads suffice to support a pholcid spider and be used to capture prey. (Author's photographs)

8.6.2 Overview of the Chemical and Mechanical Properties of Spider Silk

A defining characteristic of spiders is that they produce silk from opisthosomal glands. Silk emerges from modified setae—spigots—located on the apical region of the spinnerets, which are modified appendages. Spider silk is a biopolymer primarily consisting of spidroins, large proteins composed of tandem repeats that

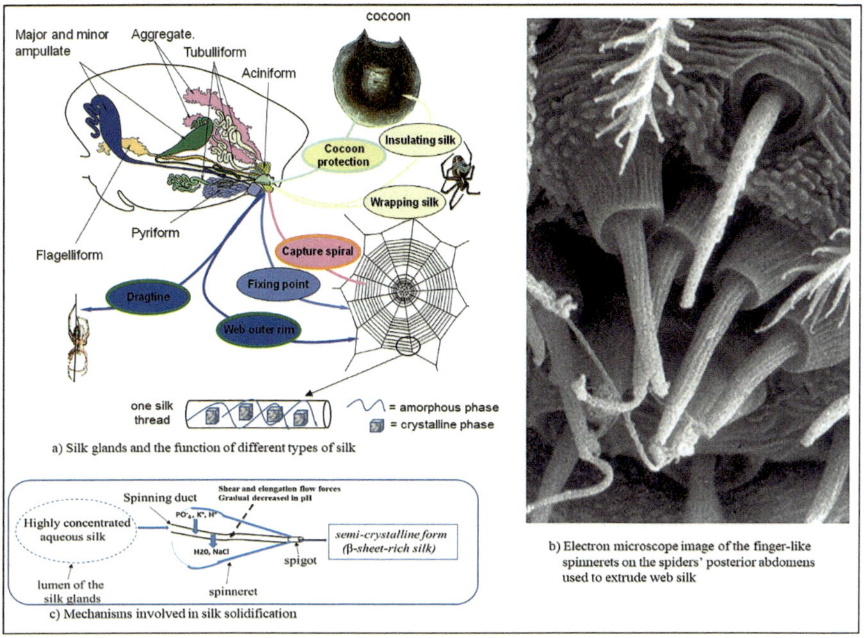

Fig. 8.6 Schematic overview of spider silk functions, production, and structure

are secreted by at least six types of sericigenous glands (Fig. 8.6a). Spidroins are part of hydrophilic scleroproteins or fibrous proteins, which consist of elongated or fibrous polypeptide chains that form filamentous, sheet-like structures with a low level of solubility in water. More than 20 genes encode spidroins, suggesting that many types of fibrous proteins, some as yet unidentified, can be secreted. Spidroins are primarily made up of the amino acids glycine, alanine, and serine. In their secondary structure, the proteins arrange themselves into polyalanine-rich β-sheets, whose crystalline phase confers stiffness; these β-sheets bathe in a non-crystalline matrix, whose alpha-helix-based amorphous phase confers elasticity (Fig. 8.6c). The β-sheets can also interact to form tertiary or quaternary structures. Proteins are organised into domains, where more or less crystalline polypeptides occur in different proportions in various silk varieties, depending on the type of secreting gland. Structural analysis of silk fibres confirms that β-sheets constitute the backbone of the crystalline domains, which are highly interspersed between the helical amorphous regions and other secondary structures of the silk protein. Both the mechanical strength and extensibility of a silk protein are determined by the ratio between the total amount of β-sheet (crystalline) regions and amorphous (non-crystalline) regions within the protein.

Spidroins initially accumulate in liquid form in the lumen of the sericigenous glands. The final physical properties of a silk thread depend on both the chemical nature of the gland's product and on the processes that occur in the spinneret. There may be fine-scale fluctuations in the solidification mechanisms of silk threads depending on the species, but the basic processes that modify liquid silk in its initial state are the following: water resorption, the secretion of potassium phosphate and H+ ions, a drop in pH, and shear flow forces as the liquid silks moves through the gland and spinneret ducts. Therefore, silk threads are structured and solidified by the ion gradient within the silk glands and spinnerets and by the physical forces associated with silk gland and spinneret conduit geometry (Fig. 8.6b).

A silk thread results from the fusion of many subunits, resulting in different properties. Depending on the secretory gland or organ of production, silk threads range from being rubbery and non-adhesive (flagelliform glands) to gluey and fluid (aggregate glands) to highly adhesive (the cribellum) to highly resistant and elastic (ampullate glands). In contrast to other silk types, capture silk is coated with a thin layer of a glycoprotein glue, whose primary function is to snare insects. In cribellate spiders, the core silk thread can be spooled and tightly packed into each of the gluey droplets, thus maintaining the thread and the whole web under tension. Under compression, the drag line completely changes its behaviour, becoming liquid like: it seems to indefinitely shrink while exerting a constant level of tension. This mechanism can be repeated indefinitely, without weakening or breaking the thread (Elettro et al. 2016). Most silk threads have the remarkable ability to collect microdroplets. In humid environments, moist air condenses and nucleates on the surface of spider silk, forming small water droplets. Indeed, spider silk has the ability to soften and decrease in length by up to ~60% upon exposure to conditions of high humidity, a phenomenon known as supercontraction. For a more comprehensive discussion of spider silk structure and differences in silk chemical properties among spider taxa, see the book by Eberhard (2020) and the reviews by Heim et al. (2009) and Li et al. (2022).

The two-phase nature and specific spatial arrangement of spider silk proteins result in high levels of elasticity and resistance, especially when it comes to spider draglines, which display peak torsional damping capacity. The latter is a unique property of dragline silk and arises from reversible interactions in hierarchical protein structure (Emile et al. 2007). Orb-weavers produce draglines with remarkable levels of tensile strength (i.e. maximum pulling force withstood), elongation at break (i.e. ability to withstand deformation before breaking), and toughness (i.e. ability to absorb energy before breaking). Spider dragline silk mechanically outperforms high-performance human-made fibres (Heim et al. 2009). Its values for all three metrics are always higher or much higher than those for materials such as Kevlar 49, ligaments, or silkworm silk (Fig. 8.7). The same is true in comparisons with other fibres, such as high-tensile steel, carbon fibres, and nylon (Heim et al. 2009).

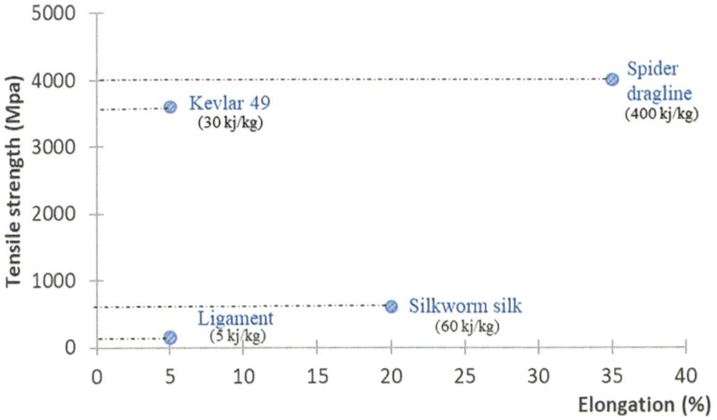

Fig. 8.7 Mechanical properties of spider dragline silk versus human-made fibres. (Adapted from Branković et al. 2024) (MPa: megapascal; in brackets: toughness values)

8.7 Spiders and the Web of Biomimetic Innovations

8.7.1 Spider Eyes as Sources of Bioinspiration for Optics Researchers

In industry settings (e.g. during inspections) and in the field of medical endoscopy, the quality of artificial vision systems—fixed cameras, endoscopes, or telephoto lenses—is generally limited by objective size and the degree of image resolution. To overcome these constraints, optics researchers have turned to solutions based on spider eyes. Tonet et al. (2008) describe a dual-camera robotic vision system inspired by the visual system of salticid spiders. This bioinspired system uses a wide-angle field of view to generate images whose resolution levels are as high as those obtained using a telephoto lens. The system could prove particularly useful for fixed camera surveillance or teleoperation applications (e.g. remote monitoring, minimally invasive surgery). Indeed, it produces images whose resolution levels are 24 times greater than those obtained using traditional commercial endoscopes.

Salticid spiders have been the source of another innovative technology in medical endoscopy, namely a depth sensor. The Metalens depth sensor, developed by Guo et al. (2019), was inspired by these spiders' multilayer retinas, which perceive depth via image defocus. The Metalens can simultaneously capture two microimages with different defocus levels. This technology could be employed in microrobots or small ingestible sensors equipped with numerous nanotechnologies (e.g. sensors, optical signal processing).

8.7.2 Technologies Based on Spider Trichobothria

Over the past decade, spider trichobothria have inspired a number of artificial hair flow sensors with applied uses. The reason is that these miniscule setae display a high level of sensitivity and are capable of detecting small changes in the dynamics of fluids (i.e. air and water). According to the classification system initially proposed by Tao and Yu (2012), capillary flow sensors include thermal, piezoresistive, capacitive, magnetic, and oxygen sensors, as well as their respective robotics applications. A recent review by Hollenbeck et al. (2023) explores this wide range of artificial bioinspired applications in depth.

8.7.3 Modern Societal Challenges and Examples of Bioinspired Solutions Informed by Spider Reflectance

Among the diversity of mechanisms used to attract preys, UV or light reflection are the main sources of inspiration. Digital display technologies are communication tools that allow multimedia content to be displayed on any device that possesses a screen. These technologies may be based on a cathode-ray tube (CRT), a liquid-crystal display (LCD), plasma, or an organic light-emitting diode (OLED). However, they all rely on intense, energy-consuming lights positioned behind the screens, which can lead to headaches and various health issues following extended use. The latter could be reduced by finely separating out the different wavelengths of light, resulting in the emission of iridescent light. Building upon this idea, Hsiung et al. (2017) demonstrated that the design of the peacock spider's abdominal scales could inspire novel, miniature light-dispersive components.

Bird collisions with building windows are one of the main sources of avian mortality. Birds' eyes can detect UV wavelengths, a fact that spiders exploit: their expansive orbital webs or stabilimenta reflect UV light, warning birds away from their webs (Théry and Casas 2008). With a view to reducing bird mortality, private companies have reproduced the reflective properties of spider silk and developed coatings for windows in urban environments. The UV reflectance of these coatings remains invisible to the human eye.

8.7.4 Spider Locomotion: Bioinspired Robots and Hydrophobic Materials

The locomotory mechanisms of spiders have greatly inspired the field of robotics, as well-documented in the first review on this subject by Vidoni and Gasparetto (2011). Many biomimetic approaches have used knowledge of hydrostatic skeleton functioning, leg angulation patterns, and the dry adhesion of scopulae to design spider-like robots. The performance of these robots, notably their agility, adaptability, and robustness, has been enhanced by exploring the use of new materials and incorporating the most recent, highly detailed discoveries about leg actuation patterns. Important applications include search and rescue operations, exploration of hazardous environments, and infrastructure maintenance in remote or inaccessible areas (Bouddih et al. 2024, Nie et al. 2024). In particular, the dry adhesion capacities of scopulae could have a wide range of applications, particularly in the space industry. Under conditions of microgravity, it is important to have a versatile and reliable system for securely affixing all kinds of loose objects (e.g. tools, spare parts, complex robotic devices) and for moving across surfaces (Gasparetto et al. 2009).

Like certain surface-dwelling aquatic insects, spiders possess hydrophobic 'microhairs' on their appendages, which may constitute another source of bioinspiration. Research examining how these setae interact with water could inform the development of hydrophobic materials or sterile micro- and nanosurfaces with major industrial or medical applications.

In addition, the hydrostatic skeleton of spiders could serve as a model for actuator systems (i.e. systems that regulate fluid pressure to act on another system) that are much faster and less energy intensive than conventional actuators or semi-rotary actuators. These bioinspired actuators could have practical applications in robotics or in the medical field, such as in the design of new prostheses (Göttler et al. 2021; Nelson 2023).

The behaviour of the Moroccan flic-flac spider served as the foundation for a reconfigurable rolling–crawling robot, which mimics the spider's unique leg movements (Tan et al. 2016).

8.7.5 Spider Silk: One of the Greatest Natural Sources of Human Inspiration

Compared to other kinds of arthropod silks, spider silk threads have unique and remarkable properties—great strength combined with extensibility—that arise from their biochemical composition and their unusual biomechanical characteristics (Heim et al. 2009; Gu et al. 2020; Li et al. 2022; Cohen et al. 2021; Guo et al. 2022; Branković et al. 2024). Focusing on dragline silk, these properties can be summarised as follows:

- Great mechanical strength, toughness, and extensibility.
- Pronounced torsional energy dissipation and shape-memory behaviour.
- High degree of stability at high temperatures (200 °C).
- Elevated capacity for damping and water collection.
- Excellent performance in vibration transmission and/or attenuation.
- Ability to supercontract.
- Ability to become liquid-like under compression (cribellate spiders).
- High degree of biodegradability.
- High degree of biocompatibility.

The spider represents a unique animal model, and its biology has inspired a large number of innovations with novel biomedical and biotechnological applications (Table 8.1).

8.8 Conclusion

Like other zoophagous predators, spiders employ a range of adaptive mechanisms over the sequential course of predation (Fig. 8.1). These mechanisms are generating a wealth of bioinspired applications (Fig. 8.8) (although the wealth of biological characteristics involved in predation sequences presented in this synthesis are not systematic sources of bioinspiration).

Other sources of bioinspiration regularly emerge from research on how species adapt to challenging environments. For example, the diving bell spider (Fig. 8.4e) can live completely submerged in freshwater, and the hydrophobic 'hairs' that form its physical gill have inspired an innovative CO2 catalyst (Wakerley et al. 2019). Among the range of potential bioinspired solutions, the greatest promise is shown by those drawing on increasingly advanced analyses of the biochemical and biomechanical properties of spider silk fibres. However, steps must be taken to develop large-scale systems capable of producing artificial silk with similar physical properties via genetic engineering (Mi et al. 2023) or chemical synthesis (Li et al. 2022).

Table 8.1 The bioinspired applications of spider silk.

Field of applications	Recent applications (compiled from Blamires et al. 2010; Li et al. 2022; Gu et al. 2020; Babu and Suamte 2024; Branković et al. 2024)
Biomedical	Tissue, bones, and cartilage repairs or engineering Acceptance of implants Spinal and osteorathritis treatments Skin regeneration Vascularisation process Support of growth and proliferation of fibroblast and keratinocyte Corneal epithelial restoration Ischaemic heart disease Clinical sutures, biomedical textiles (thin films), and hydrogels Repair of peripheral nerves Drug delivery system for bioactive molecules (nanomedicine)
Biotechnology	Silk-based food sensors, oxygen sensors, humidity sensors Wearable textiles Fruit and vegetable packaging Catalytic activity Silk-based nanocomposite (high stiffness, strength , toughness) Acoustic applications Optical fibers (light transmission) Shape-memory fibers Thermal and electrical conductance

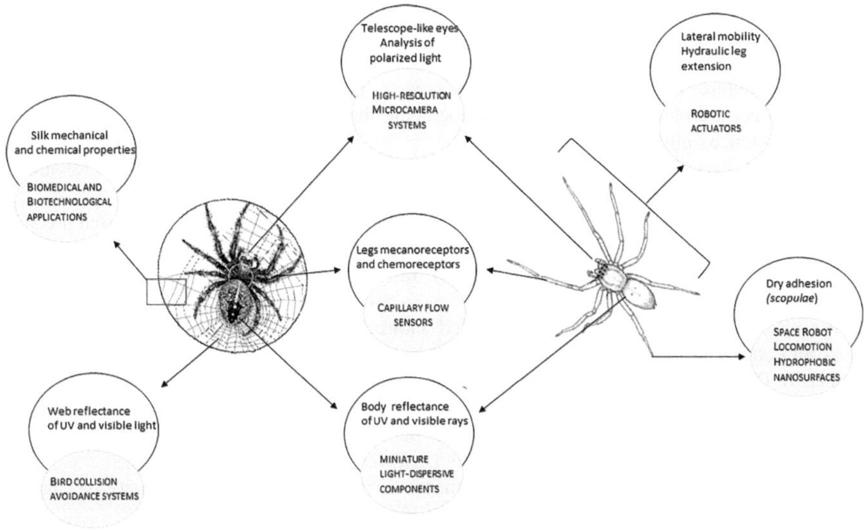

Fig. 8.8 Main types of biomechanisms involved in predation (white circles) and potential bioinspired applications (grey circles)

The influence of spiders extends beyond their silk. For example, the architecture of their webs is reflected in the radar charts used in descriptive statistics and the roofs of certain large buildings. Spiders have also inspired cultural biomimicry, notably the design, creation, and perception of cultural works. In Indonesia, for example, a series of rattan rings resembling a spider web's spiral pattern connects the Earth to the tabernacle of an altar, allowing the soul of the deceased to rise to the surface. Furthermore, web-building species display tremendous ingenuity in how they optimise their use of restricted spaces. For example, they share resources via their different patterns of nychthemeral activity, and they employ the same tool in different ways (i.e. a trap web). Spiders could teach humans how to improve resource sharing and diversification in various socioeconomic domains. Finally, a challenge we are facing is to deploy effective strategies for maintaining spider diversity, which also helps ensure that we maintain the biological wealth they have to offer.

References

Babu PJ, Suamte L (2024) Applications of silk-based biomaterials in biomedicine and biotechnology. Eng Regen 5(1):56–69

Barth FG (2002) A spider world: senses and behavior. Springer, Berlin, Heidelberg

Barth FG (2021) A spider in motion: facets of sensory guidance. J Comp Physiol A 207:239–255

Blamires SJ, Lee YH, Chang CM, Lin, Jou-An Chen JA, Lin TY, Tso IM (2010) Multiple structures interactively influence prey capture efficiency in spider orb webs. Anim Behav 80(6):947–953

Bouddih BR, Budiyono A, Prihatmanto AS (2024) Bio-inspired locomotion and sensing mechanisms for spider robots: enhancing agility, adaptability, and robustness (August 23, 2024). In: 20th international conference on intelligent unmanned systems (ICIUS 2024)

Boutry C, Blamires SJ (2013). Plasticity in spider webs and silk: an overview of current evidence? In: spiders morphology behavior and geographic distribution editor: Marq Santerre edotor, Nova Science Publishers, Inc.

Branković M, Zivic F, Grujovic N, Stojadinovic I, Milenkovic S, Kotorcevic N (2024) Review of spider silk applications in biomedical and tissue engineering. Biomimetics 9:169

Chuang CY, Yang EC, Tso IM (2008) Deceptive color signaling in the night: a nocturnal predator attracts prey with visual lures. Behav Ecol 19(2):237–244

Cohen N, Levin M, Eisenbach CD (2021) On the origin of Supercontraction in spider silk. Biomacromolecules 22:993–1000

Cross FR, Jackson RR (2011) Olfaction-based anthropophily in a mosquito-specialist predator. Biol Lett 7:510–512

Dacke M, Doan TA, O'caroll DC (2001) Polarized light detection in spiders. J Exp Biol 204:2481–2490

Diaz C Jr, Long JH Jr (2022) Behavior and bioadhesives: how bolas spiders, Mastophora hutchinsoni catch moths insects. Insects 13(12):1166

Eberhard W (2020) Spider webs: behavior, function, and evolution. The University of Chicago Press, Chicago

Elettro E, Neukirch S, Vollrath F, Antkowiak A (2016) In-drop capillary spooling of spider capture thread inspires hybrid fibres with mixed solid-liquid mechanical properties. Proc Natl Acad Sci USA 13(22):6143–6147

Emile O, Le Floch A, Vollrath F (2007) Time-resolved torsional relaxation of spider draglines by an optical technique. Phys Rev Lett 98(16):167402

Farid MF, Chakira H, Caia W, Zhaoa J, Hua H (2019) Effect of some plant essential oils on the orientation and predation capacity of the predatory spider Pardosa pesudoannulata. J Asia Pac Entomol 22(3):927–993

Foelix RF (2011) Biology of spiders, 3rd edn. Oxford University Press, Oxford

Fu X, Yu L, Zhou W, Lei C, Jackson RR, Kuntner M, Huang Q, Zhang S, Li D (2024) Spider manipulates and exploits firefly's bioluminescent signals. Curr Biol 34(16):R768–R769

Gasparetto A, Vidoni R, Seidl T. (2008) Kinematic syudy of the spider system in a biomimetic perspective., In: Conference paper, conference: intelligent robots and systems, IROS 2008. IEEE/RSJ international conference: 3077–3082

Gasparetto A, Tobias S, Renato V (2009) A mechanical model for the adhesion of spiders to nominally flat surfaces. J Bionic Eng 6(2):135–142

Göttler C, Elflein K, Siegwart R, Sitti M (2021) spider origami: folding principle of jumping spider leg joints for bioinspired fluidic actuators. Adv Sci 8:2003890

Gu Y, Yu L, Mou J, Wu D, Zhou P, Xu M (2020) Mechanical properties and application analysis of spider silk bionic material. e-Polymers 20:443–457

Guo Q, Shi Z, Huang YW, Alexander E, Qiu CW, Capasso F, Zickler T (2019) Compact single-shot metalens depth sensors inspired by eyes of jumping spiders. Proc Natl Acad Sci USA 116(46):22959–22965

Guo C, Wang C, Huang Q, Wang Z, Gong X, Ramakrishna S (2022) 3D-printed spider-web structures for highly efficient water. Heliyon 8(8):e10007

Heiling AM, Cheng K, Chittka L, Goeth A, Herberstein ME (2005) The role of UV in crab spider signals: effects on perception by prey and predators. J Exp Biol 208:3925–3931

Heim M, Römer L, Sheibel T (2009) Hierarchical structures made of proteins. The complex architecture of spider webs and their constituent silk proteins. Chem Soc Rev 39:156–164

Henneken J, Goodger JQD, Elgar MA (2017) The potential role of web-based putrescine as a prey-attracting allomone. Anim Behav 129:205–210

Herberstein MH (2012) Spider behaviour-flexibility and versatility. Macquarie, Sydney Publisher, Cambridge University

Hollenbeck AC, Grandhi R, Hansen JH, Pankonien AM (2023) Bioinspired artificial hair sensors for flight-by-feel of unmanned aerial vehicles: a review. AIAA J 61:5206–5231
Hsiung BK, Siddique R, Stavenga DG, Otto CJ, Allen MC, Liu Y, Lu YF, Deheyn DD, Shawkey MD, Blackledge TA (2017) Rainbow peacock spiders inspire miniature super-iridescent optics. Nat Commun 8:2278
Kropf C (2013) Hydraulic system of locomotion. In: Nentwig W (ed) Spider Ecophysiology. Springer Verlag, Berlin, Heidelberg, pp 43–56
Kuja JO, Jackson RR, Sune GO, RNH K, Lagat ZO, Carvell GE (2012) Nectar Meals of a Mosquito-Specialist Spider. Psyche 2012:898721. 7 pages
Lai CW, Zhang S, Tso IM (2017) A trap and a lure: dual function of a nocturnal animal construction. Anim Behav 130:159–164
Landkammer S, Winter F, Schneider D, Hornfeck R (2016) Biomimetic spider leg joints: a review from biomechanical research to compliant robotic actuators. Robotics 5(3):15
Li D, Lim MLM (2005) Ultraviolet cues affect the foraging behaviour of jumping spiders. Anim Behav 70:771–776
Li J, Li S, Huang J, Khan AQ, An B, Zhou X, Liu Z, Zhu M (2022) Spider silk-inspired artificial fibers. Adv Sci 9:2103965
Lin TS, Zhang S, Tso IM (2015) A dual function of white coloration in a nocturnal spider Dolomedes raptor (Araneae: Pisauridae). Anim Behav 108:25–32
Marc P, Canard A, Ysnel F (1993) Spiders (Araneae) useful for pest limitation and bioindication. Agric Ecosyst Environ 74:229–273
Mi J, Zhou Y, Ma S, Zhou X, Xu S, Yang Y, Sun Y, Xia Q, Zhu H, Wang S, Tian L, Meng Q (2023) High-strength and ultra-tough whole spider silk fibers spun from transgenic silkworms. Matter 6(10):3661–3683
Nakata K (2009) No be or not to be conspicuous: the effects of prey availability and predator risk on spider's web decoration building. Anim Behav 78(5):1255–1260
Nelson XJ (2023) A road map of jumping spider behavior. J Arachnol 51(2):139–154
Nentwig W (1987) Ecophysiology of spiders. Springer-Verlag, Berlin, Heidelberg, New York, London, Paris & Tokyo
Nie Y, Zhu D, Chen Y, Hu X, Wang L (2024) Design of a Bionic Spider Robot with a two-degrees-of freedom leg structure and body frame. Appl Sci 14:9809
Nogushi D (2021) UV-Vis reflectance in East Asian crab spider, Thomisus labefactus (Araneae: Thomisidae). Rep Grad Sch Eng Nagasaki Univ 51(96):55–59
Nyffeler M, Benz BJ (1987) Spiders in natural pest control: a review. J Appl Entomol 103(1-5):321–339
Penney D (2013) Spider research in the 21th century. Siri Scientific Press
Ratnam VS, Rajeev C, Subbareddy B (2023) A look into how mechanical design and biological systems may work together. Indo-Am J Mech Eng 12(2):22–32
Ratz T, Bourdiol J, Moreau S, Vadnais C, Pierre-Olivier Montiglio PO (2023) The evolution of prey-attraction strategies in spiders: the interplay betwee foraging and predator avoidance. Oecologia 202:669–684
Roth VD, Roth BM (1984) A review of appendotomy in spiders and other arachnids. BullBrarachnol Soc 6(4):137–146
Selden PA, Shear WA, Sutton MD (2008) Fossil evidence for the origin of spider spinnerets, and a proposed arachnid order. Proc Natl Acad Sci USA 105(52):20781–20785
Stavenga DG, Otto JC, Wilts BD (2016) Splendid coloration of the peacock spider Maratus splendens. J R Soc Interface 13(121):20160437
Suter RB (2013) Spider locomotion on the water surface: biomechanics and diversity. J Arachnol 41:93–101
Tan JE, Seah WHS, Li D (2010) Why do orb-weaving spiders (Cyclosa ginnaga) decorate their webs with silk spirals and plant detritus? Anim Behav 79:179–186
Tan N, Mohan RE, Elangovan K (2016) Scorpio: a biomimetic reconfigurable rolling-crawling robot. Int J Adv Robot Syst 13(5)

Tao J, Yu XB (2012) Hair flow sensors: from bio-inspiration to bio-mimicking—a review. Smart Mater Struct 21:113001

Théry M, Casas J (2008) The multiple disguises of spiders: web colour and decorations, body colour and movement. Philos Trans R Soc Lond Ser B Biol Sci 364(1516):471–480

Tonet O, Focacci F, Piccigallo M, Mattei L, Quaglia C, Megali G, Mazzolai B, Dario P (2008) Bioinspired robotic dual-camera system for high-resolution vision. IEEE Trans Robot 24:55–64

Vidoni R, Gasparetto A (2011) Efficient force distribution and leg posture for a bio-inspired spider robot. Robot Auton Syst 59(2):142–150

Vieira C, Ramires EN, Vasconcellos-Neto J, Poppi RJ, Romero GQ (2017) Crab spider lures prey in flowerless neighborhoods. Sci Rep 7:9188

Wakerley D, Lamaison S, Ozanam F, Menguy N, Mercier D, Marcus P, Fontecave M, Mougel V (2019) Bio-inspired hydrophobicity promotes CO_2 reduction on a cu surface. Nat Mater 18:1222–1227

Wang ZY, Wang JT, Ji A, Li H, Dai ZD (2011) Movement behavior of a spider on a horizontal surface. Chin Sci Bull 56(25):2748–2757

Winsor AM, Morehouse NI, Jakob EM (2023) Distributed vision in spiders. In: Buschbeck E, Bok M (eds) Distributed vision, Springer Series in Vision Research. Springer, Cham

Wise DH (1993) Spiders in ecological webs. Cambridge University Press, New York

World Spider Catalog (2025) World Spider Catalog. Version 26. Natural History Museum Bern, online at http://wsc.nmbe.ch. Accessed on {date of access}. https://doi.org/10.24436/2

Ximenes NG, Gawryszewski FM (2020) Conspicuous colours in a polymorphic orb-web spider: evidence of predator avoidance but not prey attraction. AnimBehav 169:35–43

Zhang S, Chen HL (2015) A nocturnal cursorial predator attracts flying prey with a visual lure. Anim Behav 102:119–125

Zhang C, Chen D, Niu S, Zhang J, Meng X, Liu L, Sun, T Wen S, Zhou Y, Shi Y, ZHan Z, Rena L (2020) High-aspect-ratio deflection transducers inspired by the ultra-sensitive cantilever configuration of scorpion trichobothria. J Mater Chem 8(18):6093–6101

Chapter 9
Biologically Inspired Chemical Plume Tracking

Marc Weissburg

Abstract Autonomous agents capable of tracking odor signals have a variety of important applications, including locating people, sources of hazardous or dangerous chemicals, or ordinance. The size and scale of these agents, in air or water, are roughly equivalent to organisms such as insects, crustaceans, fish, and mammals that track complex odor plumes to their source with a level of reliability and efficiency that far exceeds available technology. Like agents, organisms that track odor plumes are computationally limited. Consequently, translating strategies from organisms into human devices has been suggested as a way to develop better-tracking agents. Here, I review the biological strategies that underlie effective chemosensory navigation, which range from morphological adaptations that condition incoming signals, the use of simple but salient signal properties individually and in combination for guidance, to behavioral adaptations to maintain or facilitate recontact with plumes, and ways to encode information. I examine how these strategies are translated into agents tracking plumes in both air and water. The successful biomimetic design of a variety of tracking agents and algorithms clearly shows the advantages and utility of using biological principles. Despite this success, however, current efforts are focused on a very small set of potential strategies and mechanisms often incompletely translated into technology. Many applications are narrowly geared to a specific set of conditions that often are relatively simple or representative of a very narrow range of environments. The barriers are often less technological than ideological and reflect differing disciplinary approaches and incomplete appreciation or knowledge of biology. Considerable progress is possible even with current technology.

Keywords Functional groups · Vision · Locomotion · Mechanosensors · Chemosensors

---

M. Weissburg (✉)
School of Biological Sciences, Georgia Institute of Technology, Atlanta, GA, USA
e-mail: marc.weissburg@biology.gatech.edu

V. Bels et al. (eds.), *Biomimicry – A Functional Approach*, Biologically-Inspired Systems 20, https://doi.org/10.1007/978-3-032-11949-0_9

9.1 Introduction

Many animals rely on chemical cues for identifying and locating mates, food, dwelling sites, or other essential resources, and the ability to follow scent trails or plumes has been finely honed by evolution in a variety of animal groups. The canonical example of plume tracking is represented by moths, where a 1 cm male can locate a female from a distance of 100 m (or more) by following her pheromones (Cardé 2021). But a variety of other insects such as ants (Sharpe and Webb 1998), flies (Duistermars et al. 2009), and many crustaceans, particularly lobsters (Moore et al. 1991) and crabs (Weissburg and Zimmer-Faust 1994), also display impressive scent tracking. This ability is found even in tiny crustaceans such as the 1 mm planktonic male copepod, which can locate a female from 100 to 1000 body lengths away (Yen et al. 1998). Among vertebrates, canines are acknowledged masters of chemical trail following using both airborne and surface-bound chemical cues over large distances. Snakes display a similar ability, tracking prey and conspecifics using deposited and airborne scents (Shine et al. 2003). Sharks are commonly referred to as "swimming noses" and hunt prey using chemical scent (Hodgson and Mathewson 1971).

Technology for tracking odors is an advantageous solution for a number of human challenges from locating dangerous or toxic substances (e.g., ordinance or pollutants) to identifying gas leaks or finding people hidden after a disaster such as a flood, earthquake, or explosion (Prakash et al. 2024). The difficulty of this challenge, and our relative unfamiliarity with using scent in this way, has led to the idea that understanding chemical tracking in animals can be the basis for odor-tracking devices and processes. Current interest in this idea is associated with the development of autonomous air and water vehicles that operate at similar scales to animals, but which (like animals) are computationally and energetically constrained. This places a premium on finding robust, efficient, and relatively simple solutions such as those organisms are thought to embody. Motivated by both technological applications and fundamental understanding of chemosensory tracking, there is now a substantial accumulation of studies on artificial chemical tracking approaches in terrestrial, aerial, and aquatic environments that have been inspired by biology.

This chapter reviews progress in biologically inspired design (BID) of chemical tracking technology, with a focus on autonomous agents. Given the consensus definition of BID as translating biological mechanisms into human technology as solutions for a given problem (Shu et al. 2011), our focus is on devices or processes that incorporate features from biological systems to provide functions necessary for chemical tracking. Important functions comprise both behavioral strategies and strategies used to encode, or respond to, information in chemical cues, as well as other organismal properties that account for the unrivaled ability of animals to track scent. Following a review of some fundamental biological mechanisms and properties important for chemical tracking, we examine how these features have been incorporated in human devices and identify current shortcomings, promising approaches, and potential future directions.

9.2 Chemical Tracking in Animals

Organisms across a range of sizes, from bacteria (Dusenbery 1998) to whale sharks and whales (Dove 2015), use chemical cues. This size span (11 orders of magnitude) means there are substantial changes in the physical environment across this range that affect what strategies are viable and useful. In particular, animals in the μm to mm range live in a low Reynolds (Re) number regime where chemical transport is mainly diffusive, whereas beyond this size the distribution of chemical cues becomes more strongly affected by fluid flow and turbulence (Dusenbery 1992). Low Re conditions are characterized by smooth concentration gradients and concentrations are predictable as functions of time, distance from the source, and the diffusion coefficient (Webster and Weissburg 2009). High Re environments are dominated by advective transport and turbulent diffusion, resulting in chemical plumes that are unpredictable in space and time. Odor signals in these environments are composed of a series of chemical filaments intermingled with unscented fluid (imagine a smoke plume) that remain coherent for some distance downstream from the source until dispersed (Webster and Weissburg 2009). Thus, a particular location might experience large fluctuations in odor intensity through time, and chemical concentrations at adjacent locations may differ widely (Webster and Weissburg 2001). Average concentrations are predictable; in the downstream direction concentrations decline with a power law function based on distance to the source, whereas concentrations across the width of the plume are roughly Gaussian (Webster et al. 2001; Rahman and Webster 2005; Webster and Weissburg 2009). However, these gradients are a statistical abstraction, and it may take considerable sampling time for patterns to emerge (Webster and Weissburg 2001).

The most relevant human problem scenarios for chemical tracking occur in turbulent conditions where plumes are stochastic and unpredictable, and the bulk of both biological studies and engineering applications are focused on this scenario. Investigations have uncovered some of the biological mechanisms for chemical tracking in low Re environments, and early researchers have produced devices that can effectively operate in smooth gradients (Kowadlo and Russell 2008). Many of the biological strategies identified for low Re chemical tracking rely on sensing gradients that are regular and easily predicted in space and time and so are ineffective in turbulent flow. The mechanisms underlying the impressive odor-tracking abilities in turbulent environments are indicated in Table 9.1.

Potential strategies for following turbulent plumes are dependent on the time and space scales of organismal sampling abilities compared to the scales of variation of the plume (Weissburg 2000). Organisms that are small compared to the plume cannot rely on spatial sampling to extract information regarding their location relative to the plume; their sensor spans are too small for spatial comparisons to be meaningful. Larger organisms can potentially span the plume so that comparisons of sensors located "within" vs. "outside" the plume provide information on the direction to the plume centerline, or even, detect differences between upstream and downstream sensors.

Table 9.1 Important mechanisms and properties for plume tracking in animals

Strategy	Description	Benefits
Morphological filtering	Take advantage of sensor placement and sensory organ morphology to condition incoming signals	Increase signal reliability and detection Reduce computational complexity
Odor-gated response to wind (amenotaxis) or water flow (rheotaxis)	Odor detection above a critical threshold evokes up current movement	Eliminates the need to follow odor concentration gradients
Internally generated counter turns	Odor detection above threshold evokes an internally generated pattern of opposing turns	Does not require encoding spatial properties of odor plume that may not be possible for small animals
Use spatial asymmetry	Odor filament properties (instantaneous concentration; frequency) are compared across sensory appendages, within an appendage or through time, to find the plume "edge"	Simple detection scheme to maintain contact with plume or determine location relative to the plume, particularly for animals with good spatial sampling
Surge-cast; surge-stop	Move quickly upstream when an above threshold odor filament is detected but gradually move cross-stream, or stop, if odor is not detected	Produces rapid and consistent progress as long as detection frequency is high enough, while allowing a response to temporary signal absence, without requiring encoding of concentration
Dynamic rules	Critical detection thresholds or frequency ranges determined on the fly by local conditions	Allows a simple rule to be used flexibly in a variety of conditions
State-dependent responses	Stimulus history modifies the response to acute signals	Tunes responses to provide appropriate context-dependent behavior

The ability to sample in time depends on sensor sampling frequency and animal movement rate (Weissburg 2000). Chemosensor sampling frequency seems to be in the range of 1–100 Hz (Gomez et al. 1994; Szyszka et al. 2014), although this information comes mainly from insects and crustaceans. Even at the upper end of this range, the movement rates of these animals are sufficiently rapid so as to prevent animals from obtaining reliable estimates of the local average concentration at a given point in their trajectory. Thus, gradient tracking methods (i.e., strategies based on moving toward higher average concentrations) used by organisms in low Re environments generally do not work in turbulent plumes. The one caveat here is that slowly moving animals may sample sufficiently finely in time to acquire information on average concentrations (Weissburg 2000). Gastropods, a group comprised of shelled and shell-less snails and relatives, move at rates of mm s^{-1}, and their behavior is not strongly influenced by turbulence, suggesting some sort of time-

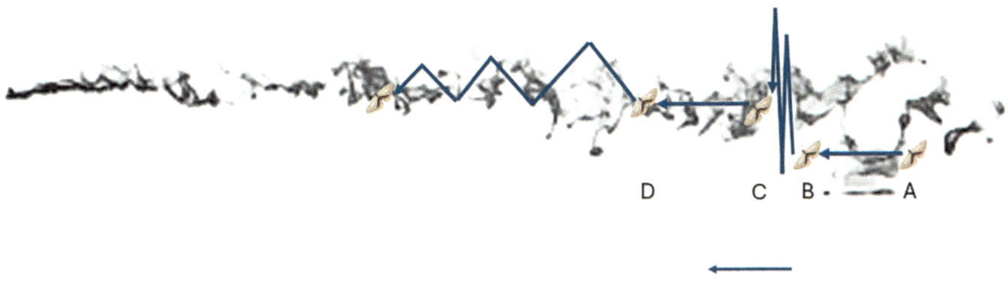

Fig. 9.1 Illustration of typical behavioral strategies used by moths to track chemical plumes. The moth detects odor (**a**) and surges forward using visually guided response to wind direction (optomotor amenotaxis) and then fails to detect the plume again during the interval at which point the behavior transitions to cross-wind casting (**b**). Contact with the plume initiates another upwind surge (**c**), and regular counterturns at a set angle relative to the wind direction (**d**). Changes in the frequency of odor detection, or other properties decrease flight speed and the probability of counterturning and prevents the moth from straying beyond the plume as it narrows to the source. This is a qualitative description; surge-cast and counter-turning may be expressions of the same underlying mechanism in different stimulus patterns and may or may not be observed in the same tracking event

averaging strategy (Alansari et al. 2025; Ferner and Weissburg 2005). This raises an important point about the biological solution sets used in engineered devices; they are contingent on the perceived problem specification that is geared toward rapid movement and minimizing tracking time. Other sorts of solutions may be possible if such constraints are relaxed.

Animals with coarse temporal sampling abilities use perception of individual odorant filaments to guide behavior. Because coarse sampling of odor filaments in a turbulent plume is insufficient to resolve direction to the source, animals use flow as a general cue, which is gated by odor perception. The tendency of macroscopic animals to move upstream when detecting odor is displayed in all animal groups studied, from fish (Montgomery et al. 2000; Baker et al. 2002) and sharks (Hodgson and Mathewson 1971), to crustaceans (McLeese 1973; Weissburg and Zimmer-Faust 1993, 1994), gastropods (Wyeth 2019; Yu et al. 2019), mollusks (Wyeth et al. 2006), echinoderms (Swenson and McClintock 1998), and insects (Vickers 2000; Cardé 2021).

The canonical and most well-studied systems are moths that use pheromone plumes for mate location, with a general picture emerging based on a variety of studies on different species (Vickers 2000; Cardé 2021; Wechsler and Bhandawat 2023). Movements toward an odor source are the result of two related, but different motor programs referred to as cast-surge and counterturning (Fig. 9.1). Investigators have often emphasized one or the other mechanism based primarily on behavioral studies, some of which show moths progressing to a source via a regular pattern of opposing turns that move the animal obliquely to the source, whereas other studies show periods of cross-wind casting interspersed with upwind surges toward the source (Cardé 2021, Wechsler and Bhandawat 2023). The prevailing belief is that both the reflexive surge-cast and the endogenously generated counterturning programs are required, and that differences in behavior result from different patterns of odor arrival (Wechsler and Bhandawat 2023).

The surge-cast motor program results in upstream movement triggered by the detection of individual odor filaments by sensors on insect antennae, which elicits an upstream movement surge (Mafra-Neto and Carde 1994, 1995 Vickers and Baker 1994). Upstream surges terminate in casting after some interval during which no further odor is detected. Moths cannot accurately perceive wind direction in flight and use visual cues to determine the upstream heading (odor-gated amenotaxis) (Kennedy and Marsh 1974; Willis and Arbas 1998).

Superimposed on the surge-cast program are endogenously specified counterturns (Kanzaki et al. 1992; Kanzaki 1998) that alternate with straight flight. Counterturns occur at regular intervals, suggesting they may be independent of odor arrival, and the end result is that moths move upstream in a zig-zag fashion. Turning periodically reduces the probability the searcher will stray from the plume and lose contact.

Although both the surge-cast and counter-turn programs represent rules where odor gates a fixed behavior, the rules themselves appear dynamic and/or state dependent. This is a necessary precondition for tracking in turbulent plumes with stochastic variation in arrival frequency, concentration, filament size, etc. As discussed more fully below, odor contact frequency appears to modulate behavior so as to minimize the probability that a searcher will move outside of the plume, which narrows as the source is approached (Mafra-Neto and Carde 1994; Vickers and Baker 1994; Baker and Haynes 1987; Willis and Arbas 1991; Willis and Baker 1994). Odorant detection appears to be a binary encoding scheme where odor must be above a particular threshold. Thus, progress toward the source is set by the frequency of arriving filaments above the detection threshold; moths move in an increasingly cross wind direction in the absence of odor to regain contact with the plume and may engage in looping or spiral flight patterns without renewed odor contact.

Recent work using a variety of techniques some quite novel has examined odor-guided navigation in *Drosophila*. Flies moving through a plume in a wind tunnel show short (1–several seconds) contact with odor that induces upwind surges and the loss of contact with odor produces 3D casting, with movements relative to the wind being visually guided (van Breugel and Dickinson 2014). Stupski and van Breugel 2024 use optogenetic techniques to activate odor responses in freely moving flies and report similar findings, although they document decreased movement speed in response to odor (not unlike moths in constant or near constant plumes), and a downward circular casting upon odor loss (Stupski and van Breugel 2024). Both investigations report increased salience of visual cues upon plume loss, and flies respond to the loss of odor in still air by downward flight and visually guided proximal search (Stupski and van Breugel 2024).

The behaviors of flies are functionally similar to that of moths, although mechanisms may be different. The observations of flies in windborne plumes (van Breugel and Dickinson 2014) were interpreted as suggesting flies do not possess and internal counter-turn generator, based on the constancy of the interval between odor loss and the beginning of turning, which is inconsistent with observations in moths (see below). Instead, they posit turning as a reflexive behavior triggered with delay based on odor loss. Stupski (Stupski and van Breugel 2024) also shows the transition to downward searching with a very similar delay in their optogenetically activated flies. Further, the delay to casting is longer than the delay to initiating upwind flight (van Breugel and Dickinson 2014; Stupski and van Breugel 2024), which prevents flies from casting so quickly that they fly out of the plume before reiterative contact with odor. This seems to serve the same function as the dynamic responses to odor frequency shown by moths (see below), although it is obviously less sensitive to the particular plume dynamics (unless these delay periods are dynamic, which remains untested). Lastly, the visually guided amenotaxis involves two rapid turns to use optic flow to determine wind direction in a manner different from that thought to be used in moths (Stupski and van Breugel 2024).

Walking flies examined in a well-controlled wind tunnel also display behaviors similar to surge-cast, with some important differences (Álvarez-Salvado et al. 2018; Demir et al. 2020). Flies in a constant plume move faster upstream and in straighter trajectories in response to odor, whereas odor loss resulted in increased turning and decreased velocity characteristic of local search. There is no evidence of an internal counter turn generator; odor loss triggers the local search behavior with a latency of several seconds. Interestingly, these responses are contingent on odor concentration, with higher concentrations inducing more rapid upstream movement in response to odor and increased turning without odor recontact (Álvarez-Salvado et al. 2018). These findings are in marked contrast to behaviors displayed by most insects (including flies navigating through airborne plumes) and also to those of noninsect arthropods and further study seems necessary. Perhaps these unusual responses are related to the role of walking during olfactory navigation in flies, which typically occurs very near the source. Wind flow is critical, whereas vision is not; flies used in these trials were from a genetically blind strain and velocity changes and suppressed turning rate to odor contact are preserved in the absence of wind but are not directed toward the upstream source (Álvarez-Salvado et al. 2018). Demir et al. used an artificially pulsed plume and also showed flies respond to odor stimulation with bouts of upstream walking as a result of upwind turn bias, but did not detect changes in walking speed or turn frequency or magnitude.

Other insects may use other mechanisms that have not been well investigated and are still unresolved (Cardé 2021), but which also may represent alternate strategies to the endogenously generated cast-surge paradigm. Some insects may move so rapidly as to prevent odor-gated amenotaxis and the perception of odor results using visual cues to set a fixed heading. Once the insect takes flight, this heading is maintained for as long as odor is perceived. Odor loss causes the insect to land and repeat the process.

Some of the general features displayed by moths occur in crustaceans (Fig. 9.2), which also have been well studied (Weissburg 2000). Similarities in the mechanisms used by these two groups reflect robust and effective tracking mechanisms, whereas differences are assumed to be a product of size differences (crustaceans are large relative to their plumes) and the particular exigencies of walking vs. flying.

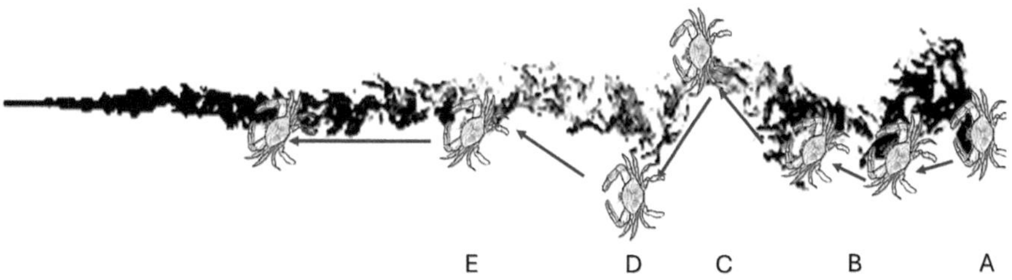

Fig. 9.2 Illustration of typical behavioral strategies used by crustaceans to track chemical plumes. The crustacean detects the odor plume and uses the asymmetry of stimulation to turn toward the plume (**a**) combined with a flow response to move upstream and tracks along the plume edge (**b**, **c**). The tracker transits far enough from the plume center to lose contact and waits to reacquire a signal (**d**), and then moves upstream (**e**). Increasing frequency of odor contact causes the crab to slow and possibly evokes near-source searching

Blue crabs and lobsters, like insects, surge upstream in response to odorant filament detection by their antennae/antennules, and this response also is best explained by a binary encoding scheme as in moths (Page et al. 2011a) combined with a response to flow (rheotaxis). Shaving the chemosensory sensilla on the antennules results in slowed movement and poor upstream progress toward the odor source in a turbulent plume (Keller et al. 2003). The odor-gated rheotaxis is thought to be encoded by mechanosensors that are distributed on many body surfaces and appendages, including the antennae/antennules (Derby and Weissburg 2014) Also unlike moths, crustaceans do not utilize casting or spiral search as an immediate mechanism to relocate the plume. Instead, they tend to stop upon signal loss and resume movement when they reacquire contact if the period is short enough (e.g., less than several seconds), although prolonged signal loss eventually results in cross-stream movement (Weissburg and Zimmer-Faust 1993, 1994; Page et al. 2011a). Similar behaviors are seen in fish, suggesting that this is a ubiquitous response for animals that can hold station during tracking (Montgomery et al. 2000).

Macroscopic crustaceans have a variety of chemosensory organs, with populations of sensors on legs, claws, and occasionally other body surfaces complimenting the antennae/antennules, which are considered the major chemosensory organs (Derby and Weissburg 2014). The bilateral arrangement of sensors (one claw, four legs, and one antennae and antennule on each body side) allows for the encoding of left–right differences in odor properties, and together these sensory appendages sample both in front and at the rear of the body, providing considerable potential for detecting spatial patterns in different directions. The relatively large size of blue crabs compared to the scale of the plume allows for spatial sensing as a way to locate themselves relative to the plume, which is difficult for moths. Crustaceans seem to combine a cast-surge-like mechanism to move upstream in response to flow with spatial sensing to direct lateral movements.

At least in blue crabs, chemosensory input from claws and/or legs provides important spatial information, allowing the animal to locate itself relative to the plume. The large span and spatial coverage allow these sensors to encode spatial patterns in odor signal properties that help the animal locate itself relative to the plume. Crabs tend to straddle the (time averaged) "edge" of plumes during tracking (Zimmer-Faust et al. 1995) and turn toward the side of body encountering stronger odorant concentrations (Page et al. 2011b). Deactivating chemosensors on the legs and claws of blue crabs results in highly meandering trajectories during plume tracking, suggesting a deficit in spatial sensing (Keller et al. 2003).

The need for and utility of bilateral comparison for spatial localization during tracking are demonstrated by other organisms. Lobsters and crayfish use some sort of bilateral comparison for plume tracking (Reeder and Ache 1980; Moore et al. 1991; Moore and Grills 1999), which is facilitated by the reach and mobility of their antennae, but again, it is unclear what specific property conveys information. Lobsters appear to encode filament arrival frequency, which presumably could be used to track plume edges (Michaelis et al. 2020). Sharks can detect very small differences (100 ms) in the arrival time of odorants between their left and right nares, which house chemosensors (Gardiner and Atema 2010), and turn toward the side of earlier arrival. Walking silk moths also encode arrival time differences of as little as 50 ms, yet turn toward the side of later arrival (Takasaki et al. 2012), a curious result with an unclear explanation. Walking silk moths also will turn toward the side of the more strongly stimulated antennae, at least when concentration differences exceed a certain threshold. Both mechanisms are thought to have some role in either setting the initial turn direction when the plume is detected or resetting the direction if the moth is at the plume edge since moths seem to use the endogenous counterturning mechanism while tracking (Takasaki et al. 2012). This observation highlights that different phases of the overall tracking task may require different mechanisms.

Other walking (cockroaches) and flying insects (flies) have been reported to employ bilateral comparison during at least some phases of tracking (Bell and Tobin 1982; Willis and Avondet 2005; Duistermars et al. 2009), but again, the specific stimulus properties remain unresolved. Note that any general principles regarding necessary thresholds for either concentration differences or time lags remain unestablished. A priori, processing speed as constrained by action potential generation, conduction velocity, and chemical second messenger transduction are limiting factors, but other processes occurring at receptors also might be determinant; for instance, it takes roughly 150–300 ms for individual chemosensory neuron responses in lobsters to encode concentration differences, and differences become less robust as concentrations become more similar (Gomez and Atema 1996). Interestingly, insect olfactory neurons have much more rapid dynamic responses and can encode peak odor concentrations within a few milliseconds (Szyszka et al. 2014). This suggests the speed of animal sensors are matched to environmental stimulus properties to accurately encode the characteristics of transient odor bursts.

The different sensory organs mediating blue crab orientation capitalize on the local hydrodynamic environment to condition incoming chemical signals to increase their ability to reliably encode critical information (Keller et al. 2003). The antennules and antennae are located high enough above the ground so as to reduce boundary layer shear that would otherwise diffuse odor filaments (Webster and Weissburg 2009). These sensors consequently receive a sharp and distinct signal. In contrast, the chemosensors on legs are within the boundary layer region where shear stretches incoming odor filaments, increasing spatial spread and reducing intermittence (Webster and Weissburg 2009), which is desirable for reliably encoding spatial signals. These effects can be thought of as "morphological filtering" (referred to as embodiment by de Croon et al. (2022)) that can reduce the reliance on postprocessing mechanisms to increase the simplicity and robustness of plume following strategies.

Organisms commonly rely on morphological filtering to condition incoming signals. Moth antennae project in front of the organism so they are not affected by flow disruptions caused by the body and wings. Walking moths create air currents by fanning their wings during plume tracking (Mishima and Kanzaki 1998) to facilitate odor encounters and increase odor penetrance at sensors in the densely packed, hairy antennae (Loudon and Koehl 2000). Similarly, chemosensors of mate-tracking copepods are located within the animal's self-generated propulsive flow that pulls odor-laden water across the sensors, which decreases the fluid boundary layer and enhances odor signal strength (Fields and Yen 1993).

Morphological "conditioning" can enhance useful aspects of signal structure but represents important design tradeoffs. Lobsters flick their antennules rapidly through odor plumes at a velocity that allows the tightly packed sensory hairs to become "leaky," enabling them to catch a distinct parcel of odor-laden fluid (Koehl et al. 2001), serving the same function as moth wing fanning. However, fluid dynamics at these scales result in a tradeoff between maximizing odorant capture and preserving the fine-scale properties of individual filaments (Schuech et al. 2011). Thus, the reliance on particular aspects of odor signal properties may reflect both physical constraints as well as the adaptive value of robust and computationally simple decision-making algorithms.

One significant but incompletely addressed question is how context dependence in rules governing odor responses enables effective strategies. For instance, frequency of odorant arrival seems a critical signal feature in regulating moth turning and flight speed as moths approach the source. As discussed below, the upwind surge of male moths' transitions to a cross-wind casting behavior in the event that odor is not detected. But is the critical time interval constant, or in some way dependent on the history of recent encounters? Prior filament contact also reduces the latency to surge in blue crabs (see below). Do other odor properties such as concentration modulate this interval or is arrival frequency the only significant parameter? Are threshold concentrations required to elicit or modify behaviors fixed or contingent? Such questions are difficult to answer without visualizing behavior while simultaneously collecting independent information on the properties of the odor signal evoking tracking. The previously mentioned ambiguity regarding whether cast-surge and endogenous counter-turning represent alternative tracking strategies is a manifestation of the difficulty of reaching firm conclusions without having simultaneous access to information on animal movements and the sensory inputs.

Experiments where both animal behavior and odor signal structure are collected simultaneously are rare given the technical complexity of the task. The chemicals that animals respond to in many situations are either unknown, not commercially available, or cannot be detected with commercially available sensors at the appropriate time and space scales. A very few experiments with moths used excised antennae as independent sensors (e.g., Vickers and Baker 1994), which are attached to free-flying moths so the incoming pheromonal signal can be determined simultaneously with the evoked behavior. The resulting data sets are too small to clearly disambiguate the role of different stimulus properties or preceding context, although offer tantalizing clues, discussed below. Optogenetic techniques offer an interesting possibility to precisely control temporal stimulus dynamics (Stupski and van Breugel 2024) but is obviously limited to a few organisms. How organisms respond to a ubiquitous and general stimulation of olfactory transduction, as opposed to more spatially discrete signals, is also unclear.

Simultaneous odor and behavioral measurements are more experimentally tractable in water than air (but see Jayaram et al. 2022 for a notable exception in airborne plumes). Attractant solutions can be prepared with a very small amount of nontoxic fluorescent dye that emits light when illuminated by a laser, which allows for visualizing odor signals around a freely moving animal whose behavior also is recorded (Mead et al. 2003; Dickman et al. 2009). Again, such experiments are technically challenging and so data is scarce, but experiments with blue crabs suggest while strategies remain the same, the rules themselves are dynamic. For instance, blue crabs use the same surge-stop strategy in conditions where the instantaneous odor concentrations in arriving filaments are substantially different (Page et al. 2011a). For this to be successful, the threshold concentrations that elicit upstream movement must be context dependent; a fixed threshold would either be too high to elicit movement in conditions of low instantaneous concentrations or so low that it would produce upstream movement to spurious signals in plumes where concentrations are high. The performance of tracking animals is best explained using a threshold rule of $C_{\mathrm{Amax}}/2$, where C_{Amax} is the highest instantaneous concentration reaching the anntennules of a particular animal during tracking (Page et al. 2011a). Different animals, even when under the same nominal conditions, experience different stimulus patterns and C_{Amax} varies accordingly.

The response to suprathreshold spikes also seems to reflect a dynamic rule contingent on stimulus history. Blue crabs make the most consistent progress toward a source when the recent frequency of above-threshold filament encounters is around the modal filament frequency experienced by the animal during tracking (Page et al. 2011a). Here, the time lag between filament arrival and surge decreases as prior filament encounters appear to prime the animal to move upstream more quickly. Similar responses occur in flying moths, where a bout of arriving filaments fails to elicit upwind surge early in tracking, but a series of arriving filaments with a similar temporal profile induce upwind surges later on in a track when the animal had been moving consistently upwind (Vickers and Baker 1994). This suggests that patterns of stimulus history that are not salient at one point during tracking become so later.

Prior contact with suprathreshold filaments amplifies the response to subsequent stimuli within a certain frequency range, but beyond this, high-contact frequencies depress movement. Interturn frequencies decrease as moths approach the source, suggesting a change in the endogenous counterturn clock (Willis and Arbas 1991) that reduces the tendency of animals to move beyond the plume as it narrows. Flying moths reduce their flight speed and make less progress toward the source in homogeneous plumes (Kennedy et al. 1980; Willis and Baker 1984; Baker et al. 1985), again supporting arguments for state-dependent responses to odor filament contact that suppress upstream movement. What is unclear is whether these critical frequencies are dynamic in the way suggested for crustaceans above.

Evidence from walking moths suggests the probability of counterturning in response to odor detection also is dynamic and decreases with upstream progress (Shigaki et al. 2020). Using fuzzy logic to model the output of neural circuits that mediate the surge in a freely walking moth, they conclude that the probability of an upstream surge is frequency dependent, and high frequencies reduce the probability that a given odor detection event elicits a surge, which is consistent with behavioral evidence suggesting intermittence is important for sustained upwind flight. Filament encounter frequencies higher than the modal value cause blue crabs to decelerate, which would prevent a crab from overshooting the target (Page et al. 2011a). Both insect (Szyszka et al. 2014) and crustacean (Gomez et al. 1994) olfactory receptors can follow odor pulses with sufficient fidelity to use the frequency of odor arrival as a signal property.

Interestingly, the role of high-frequency odor contacts seems to differ in flies, which have been reported to maintain upwind movement even in constant plumes when walking or flying (Álvarez-Salvado et al. 2018). Higher frequency odor contact sustains upwind progress, increasing speed (the opposite of what occurs in moths) and suppressing turning (Álvarez-Salvado et al. 2018). It is unclear whether this represents a different strategy used by other animals (flies spend very little time in contact with plumes in these studies) or limitations on our ability to resolve stimulus dynamics clearly. Further, the dynamic responses displayed by walking flies do not seem to occur in flies navigating through airborne plumes (van Breugel and Dickinson 2014; Stupski and van Breugel 2024), which also seems to require further study and explanation.

Many animals display distinct behaviors that are evoked close to the target and which may involve other sensory modes. Crabs and lobsters start probing the substrate with claws and/or legs when near the source, potentially using chemical and mechanosensory cues. Moths and other insects use visual cues for the final stages of target location (Cardé 2021; Wechsler and Bhandawat 2023). It is unclear what mediates the switch to near-source behaviors, but presumably the mechanisms are similar to those underlying the state-dependent responses discussed above.

9.3 Translating Biologically Inspired Strategies into Technology

Devices and processes for bioinspired autonomous tracking comprise a range of products and approaches. As the previous discussion makes clear, there are a number of design targets for incorporating biologically inspired solutions: important information carrying parameters (binary concentration thresholds, intermittence, odor arrival frequency); strategies (follow a plume edge, move upstream using flow); rules to implement strategies (turn to the side of the stronger stimulus); and higher-order decision making algorithms encompassing several rules. Studies often focus on one or two of these different design targets that may be examined using physical experiments incorporating robots and sensors, computation and simulation, or occasionally both (Table 9.2).

The most common implementation of BID approaches to tracking is to employ the particular behavioral strategies ubiquitous in animals such as cast-surge or endogenous counter-turning. Regardless of other features, incorporating these basic behaviors is sufficient for investigators to identify their efforts as bioinspired, which seems problematic. Most efforts here focus on plume tracking in air and use land-based vehicles, although tracking studies with flying drones are becoming increasingly common. A second, sometimes complimentary approach is using simulated odor dynamics to examine the schemes for encoding information (i.e., useful signal properties) or decision-making algorithms. Here, the lack of a robotic platform allows the investigation to focus on how chemical signal information is extracted and processed for efficient navigation without the need for motion control or accommodating other constraints (e.g., power consumption, processing capability, sensor integration) associated with physical instantiation.

Ishida and colleagues (Ishida et al. 1994) implemented odor-gated amenotaxis in a wheeled robot with four metal oxide gas (MOX) sensors and four thermistor-based flow sensors on a mobile platform to investigate the counterturn and cast-surge behavioral mechanisms. Wind direction was determined by matching the pattern of air sensor responses to an internally stored library of sensor outputs from 0° to 360° at 45° intervals, and the prevailing direction of travel was 60° relative to the calculated wind direction. The current direction was maintained as long as the stage was in the plume as defined by all gas sensor readings being above a fixed threshold. The platform turned opposite to its current direction of travel if the sensor readings fell beneath this threshold but backtracked to its original location if sensor readings continued to be below threshold. This condition was taken to indicate the stage had moved outside of the plume, and retreat served as an analog of casting allowing the stage to relocate the plume. The test area was a small (roughly $1 \times 0.7 \times 0.35$m; l $\times$ w $\times$ h) wind tunnel.

This implementation of basic biological features resulted in a consistent location of the odor source, with tracks displaying an alternating turning pattern. Although successfully translating a biological strategy into hardware and showing its utility, this study also illustrates a number of difficulties and limitations that generally apply to other similar investigations. First, while MOX sensors are inexpensive, ubiquitous, and easy to work with, they take many seconds to reach peak output and even longer to return to baseline (Eu and Yap 2018; Anderson et al. 2020), which limits both movement and response speed, and diminishes the ability of robotic agents to detect odor bursts. As noted, both insect and crustacean olfactory receptors are capable of much more rapid responses and therefore more accurately encode properties of individual odor encounters. MOX sensors may be a poor choice when

Table 9.2 Summary of biologically inspired studies on agent tracking. *Strategies* refers to motor programs or tracking strategies used by organisms, whereas *Rules* describe the implementation of those strategies. *Platform* indicates the type of autonomous agent or the nature of the simulation

Source	Sensors	Strategies	Rules	Platform	Notes/findings
Anderson et al. (2020)	1 chemical sensor (insect antennae)	Surge-cast counterturning	Binary odor detection, steer relative to wind, fixed thresholds, angles, counterturn angles, no stopping rule	Areal drone with wind vanes and insect antennae in a wind tunnel	Comparison of MOX vs. antennal sensor output in the same plume; sensor positioned below rotor downdraft
Belanger and Willis (1996)	Single sensor, flow direction resumed	Surge-cast, counterturning	Binary odor detection, steer relative to wind, no stopping rule	Simulation using wind tunnel data	Tested different combinations of odor-gated amenotaxis, counterturning and surge-case
Edwards et al. (2005)	1 chemical sensor; flow direct presumed	Surge-cast, counterturning	Concentration-dependent flight speeds and angles, fixed counterturning rules, no stopping rule	2D linear drive robotic gantry in wind tunnel	Discrete and pre-set concentration thresholds for modulating flight behavior. Compared counterturning rules using either fixed interval between detection or detection of increased concentration
Eu and Yap (2018)	4 chemical, 1 wind sensor	Surge-cast, conterturning and spiral casting	Binary detection relative to previous concentration + transient signal properties, steer relative to wind, fixed casting angles and fuzzy logical weighting, no stopping rule	Simulation using CFD software + drone tests	Tested different combinations of casting rules and parameters in a quadrotor
Farrell et al. (2005)	1 chemical, 1 flow sensor	Surge-cast, counterturning	Binary detection, steer to flow, source declaration, fixed heading angles, detection thresholds, casting behavior	AUV in coastal environment	Parameters tuned based on initial data

Grasso et al. (2000)	2 chemical sensors	Movement based on signal comparison across two sensors	Forward movement if paired sensors register equal concentrations above fixed threshold, turn toward stronger signal; move backwards if concentrations below fixed threshold	Wheeled robot in a water tunnel	Sampling and movement rates, other parameters based on organism; performance compared with organism, experiments examined impacts of sensor distance, initial robot orientation and distance
Harvey et al. (2008)	1 chemical and 1 wind sensor	Surge-cast, counterturning	Binary odor detection, steer to wind, concentration-dependent countering angles, fixed thresholds,	Wheeled robot in a wind tunnel	Used several different wind speeds, non-biological movement rules also tested
Ishida et al. (1994)	4 chemical and 4 wind sensors	Surge-cast, counterturning	Binary odor detection, steer relative to wind, fixed counter turn angle, fixed heading upwind, no stopping rule	Wheeled robot	Wind direction determined by comparing wind sensor outputs to internally stored data
Ishida et al. (1996)	1 chemical sensor	Upwind movement	Binary odor detection, steer relative to wind, fixed heading upwind, no stopping rule	Wheeled robot	Does not account for signal loss, wind direction determined by maximum sensor readings when rotated full circle
Jayaram et al. (2022)	2 chemical sensors, flow direction	Biased upwind turning	Turns are biased upwind based in binary odor detection if intermittence or frequency above threshold	Simulation based on Gaussian odor packet dispersion	Investigated the role of intermittence and frequency detection
Kuwana and Shimoyama (1998)	2 chemical sensors	Bilateral signal contrast	Detection by sensor activates contralateral movement	Wheeled robot	Insect antennae used as sensors

(continued)

Table 9.2 (continued)

Source	Sensors	Strategies	Rules	Platform	Notes/findings
Li et al. (2001)	Single chemical sensor, flow direction presumed	Surge-cast, counterturning	Steer relative to wind, fixed turn counter turn angle, fixed heading upwind, fixed loss time, no stopping rule	Simulation using wind tunnel and flume data to model filament distribution in variable wind	Focused on examining various counterturning strategies, thresholds and wait times to reacquire; responses to local wind direction least effective
Liberzon et al. (2018)	Chemical and flow sensing	Surge-cast, counterturning	Steer relative to wind, cast if odor is not contacted within interval set by previous encounter at random angle, fixed odor detection threshold, no stopping rule	Simulation of pulsed plume using idealized advection-diffusion equation	Incorporating a simple dynamic rule based on odor contact time improves performance
Lo Iacono (2010)	1 chemical sensor, flow presumed	Surge-cast, gradient detection	Steer to wind, cast if odor is not detected, cast if instantaneous odor concentration decreases in time, no stopping rule	Simulation of wind tunnel data using second order stochastic modeling	Adding a concentration-based rule improves performance
Michaelis et al. (2020)	2 chemical sensors, flow direction presumed	Bilateral signal contrast for steering; upstream movement	Move at fixed angle to flow toward more strongly stimulated side, move directly upstream if equal, odor absence initiates stopping; fixed stopping rule	Simulation using CFD software	Signal strength defined as intermittence and concentration
Pyk et al. (2006)	6 chemical, 1 wind sensor	Surge-cast, counterturning	Binary detection, steer relative to wind, no source declaration	Areal drone in small wind tunnel	Incorporates vision-based collision avoidance, counter turn direction independent of behavior

Rutkowski et al. (2004)	1 chemical sensor; flow direct presumed	Surge-cast, counterturning	Concentration-dependent flight speeds and angles, counterturn timer resets at odor detection, no stopping rule	2D linear drive robotic gantry in wind tunnel	Discrete and preset concentration thresholds for modulating flight behavior
Webster et al. (2012)	3 chemical sensors, flow direction presumed	Surge-cast, steering based on bilateral odor signal contrast	Increased average odor spike concentration elicits upstream movement, turn toward more strongly stimulated side, frequency based stopping rule	2D linear drive robotic gantry in water tunnel	Sampling and movement rates, other parameters based on organism; performance compared with organism
Yuan et al. (2015)	2 chemical sensors	Move based on concentration change and bilateral contrast	Move upstream if signal increasing beyond fixed threshold; turn toward side of greater concentration if above fixed difference; no stopping rule	Wheeled robot wirelessly connected to computer	Petrinet method for computing movement

trying to design tracking agents based on biological mechanisms. Compensating for these undesirable properties of MOX sensors is possible using reverse estimation or fuzzy logic, but adds computational complexity (Shigaki et al. 2018, 2022).

More fundamentally, these and other similar studies often rely on determining the sensor output in test conditions in order to define an appropriate threshold; changing test conditions therefore requires defining a new threshold. A primary constraint here is the unpredictability of turbulent plumes, and this introduces another problem. Intermittent signal loss may occur even within the time-averaged plume boundaries, and odor filaments may impinge on sensors outside this boundary. Indeed, the need to define a criterion for, and response to, being outside the plume was motivated by earlier tests where the stage "followed" a spurious signal to move well outside beyond the plume boundary. Many of the strategies used by organisms are designed to reduce or compensate for this unpredictability.

Other efforts use a similar approach, differing in sensor properties and decision algorithms. Harvey et al. (2008) used a wheeled robot with a single chemical and flow sensor to test various biological and nonbiological strategies. Chemical sensing was provided by an ionized air detector that has rise and recovery times similar to animal sensors, which avoids the problems associated with slow MOX response and return speeds. The most important result here is that counterturning is by itself an ineffective strategy that often takes the tracker away from the plume. Note though that the counterturning strategy employed here was not dynamic. Pyk et al. (2006) used a wheeled robot with six MOX gas sensors and a wind vein, employing a cast-surge algorithm combined with counterturning to successfully locate an odor source in a 4 m wind tunnel. Again, binary odor presence or absence mediated the transition between upwind motion with counterturning and cross-wind casting. The robot was unable to locate the source in an odor field without wind, showing the importance of integrating flow and chemical sensing modalities. Yuan et al. (2015) use paired MOX sensors in a wheeled robot to track plumes based on both the change in chemical concentration and signal strength differences, replicating some presumed mechanisms used by aquatic creatures. Significantly, they do not incorporate wind sensing; movement direction is based on a Petrinet implemented by a computer that communicates wirelessly with the robot. It is unclear whether the translation of signal properties to movement rules incorporates biologically based strategies, and eschewing basic information provided by flow results in algorithms that currently cannot be implemented easily in a computationally limited autonomous vehicle.

Rutkowski et al. (2004) used a mobile gantry equipped with a fast ionization sensor sampling at 50 Hz (Robomoth). Upstream movement occurred in response to pheromone detection in a concentration-dependent manner (velocity and heading angles decrease with concentration), with counterturns being executed if the time between successive detection events was longer than 500 ms, otherwise Robomoth surged upstream if odor was detected. The incorporation of concentration-dependency and an internal counterturning timer was motivated by observations of tracking moths showing time between counterturns can be consistent, and moths fly slower and straighter paths as they approach the source. These algorithmic features show how state-dependent or dynamic rules can be implemented in the tracking scheme, at least in a basic fashion, even if they do not fully embody the biological mechanism. It is likely that the frequency of stimulation accounts for both the suppression of reduced flight speed (Shigaki et al. 2022) and counterturning, and that the critical interstimulus interval reflects ambient conditions (Shigaki et al. 2022). Edwards et al. (2005) presents an initial account of Robomoth testing counter-turning rules based on interstimulus intervals vs. odor spike concentration differences in time.

The effectiveness of cast-surge plus endogenous counter counterturning also has been examined in flying drones. Anderson et al. (2020) employed this approach using a biohybrid approach where odor sensing was provided by intact antennae from silk moths, which retain viability for several hours after being removed from the organism (see also Kuwana and Shimoyama 1998; Terutsuki et al. 2021 for examples of this biohybrid approach). A novel feature of this study is the use of wind vanes that passively stabilize the drone so that it points upwind, eliminating the need for additional sensors, computation, and/or an external field of reference. Upwind surge occurred upon odor detection and repeated detection elicits alternating counterturns. The drone transitions to casting in the absence of odor detection. Cast length increases if odor is not detected after a set distance (related to time by movement speed), which also triggers a counter turn. As is common, the interstimulus period, cast distances, and odor detection thresholds are predetermined (see also Shigaki et al. 2022). The drone was able to repeatedly track odor plumes in 2D, even in the presence of obstacles, which were avoided using laser range finders that trigger a counterturn when the drone closes to within 10 cm of an object. Although reminiscent of other studies, Anderson demonstrates the utility of rapid and discrete odorant sensing combined with a simple and largely passive method to orient to flow.

A few studies have examined BID chemosensory tracking in water, which is more challenging given the stringent design considerations for an underwater robot, as well as the need for more specialized sensors. Farrell et al. (2005) used a REMUS AUV equipped with flow sensors (acoustic doppler current profilers) and a fluorescence detector (10 Hz sampling) to track a source in an open-water shallow (4–8 m) operational area roughly 1000 × 400 m. The tracking strategies are the same as used in air, with slight variations. Chemical detection within a set time window triggers upcurrent movement at a fixed angle to the flow, with counterturns evoked if odor presence is detected within that time window. Detections slightly beyond that window evoke a modified counterturn, causing the robot to move a set distance upstream in the other direction and represents a compromise between cross-current casting and upstream movement. The transition to this action (reacquire) is predicated on the assumption that odor detection slightly outside the temporal window indicates the AUV has exited the plume, but not catastrophically. Longer periods of signal absence cause the AUV to transition to a modified casting behavior, which is repeated over a fixed series of points between the current position and the last position where odor was detected. The AUV returns to a grid searching routine to find the plume if this sequence ends unsuccessfully. Interstimulus periods, turn angle, and distance of the track legs for the casting maneuvers are all predetermined using prior data of odor dynamics at the test site. The AUV successfully located the odor source in 7 of 8 runs, with the bulk of the search time occurring close to the source where the AUV repeatedly loses contact with the narrowing odor plume. A primary difficulty is the fixed track distance traveled by the AUV in the reacquire mode can result in traveling far outside the plume. As noted by Farrell, fixed distances based on estimates of plume width are ineffective because of plume meander or eddy shedding and more dynamic rules could increase performance.

Webster et al. (2012) implemented tracking algorithms using a gantry approach in a water tunnel, similar to that of Rutkowski et al. (2004). The operational scenario was of a tracking agent on the substrate, and both algorithmic and other features are inspired by observations in crustaceans, particularly blue crabs. A major difference between this and moth-inspired studies is that steering based on difference in bilateral signal strength replaces counter-turning. The gantry utilized three conductivity sensors, one of which was located higher in the water than the other pair positioned slightly above the bed. Sensors operated at 2 Hz. The upper single sensor mediated upstream movement in response to odor detection, whereas lower sensor pair mediated cross-stream movement based on the difference in signal strength between left and right sensors. Odor detection was based on a dynamic rule of $C_{avg}/2$, where C_{avg} is the running average peak odor concentration of odor spikes in the preceding 10 s. Only spikes exceeding this threshold were considered detection events, and forward movement was initiated when spikes were equal to or greater than the previous values. Cross-stream motion occurs either with or without upstream movement and is mediated by the lower sensor pair. The agent moves cross-stream in the direction of greater signal strength and cross- and upstream based on the upstream movement rule noted above. Signal loss over 10 s, defined as an absence of peaks exceeding the previous average value, results in cross-stream sweeps. The agent stops (source is declared) when the above-threshold spike frequency is greater than 2 Hz. This agent, when tested in the same conditions in which trials were performed with blue crabs, resulted in performance and kinematics indistinguishable from that of the animal that inspired this design. Like similar studies with non-MOX sensors (Rutkowski et al. 2004; Edwards et al. 2005; Farrell et al. 2005; Anderson et al. 2020), this work shows the effectiveness of biologically inspired algorithms based on simple responses to iterative odor contact with transient odor filaments, and the utility of dynamic rules.

Grasso et al. (2000) used a wheeled robot in a 3.5 m water tunnel to examine tracking mechanisms potentially used by an animal walking on the seafloor, similar to that of Webster et al. (2012). The agent was equipped with two chemical (conductivity) sensors operating at 2.5 Hz. Forward movement occurred when the sensors register equal concentrations above a fixed threshold, otherwise turn toward the side with stronger signal. The agent moves backward if concentrations fall below fixed threshold. Tracking success was highly variable and dependent on the orientation relative to the flow. Agents without the "move backward" rule almost never closed to within 10 cm of the source because there was no mechanism to regain contact with the plume when forward motion resulted in signal loss. Backward movement improved performance to 0–30% success rate depending on distance and orientation. Thus, while purely chemical information can be useful on its own, tracking strategies based on both chemical and flow information are superior, as has been demonstrated in both aquatic and terrestrial environments. The authors also determined that a sensor separation of 3 cm resulted in closer approaches, likely because larger sensor separation resulted in asymmetrical patterns of detection that moved the agent away from the plume as it approached. This illustrates the utility of either mobile sensors or multiple sensors capable of resolving differences across a range of distances. Note that both of these strategies are used by organisms; lobsters, crayfish, and related animals have long and mobile sensors. The eight walking legs on decapods (crabs) allow for encoding signal contrast at multiple distances.

Increasing computational power has allowed investigators to examine bioinspired tracking algorithms in silico using a variety of methods to simulate odor plume dynamics, with varying degrees of accuracy and precision. Some efforts are explicitly based on physically derived models of plume structure in either air or water (e.g., Farrell et al. 2002; Lo Iacono 2010) or eddy simulation approaches that are specifically designed to simulate instantaneous plume structure (e.g., Michaelis et al. 2020), but this is not always the case (Li et al. 2001; Gao et al. 2015). Results of such studies must be interpreted in light of the ability of models to accurately capture odor plume dynamics, and there is a wide range of transparency, fidelity, and accuracy of reported computational approaches to simulating odor plumes in tracking studies.

Michaelis et al. (2020) used the CFD software package CFX (Ansys Inc., USA) to simulate turbulent transport of odorants in unidirectional flow over a sand bed with defined odor release conditions relevant to benthic marine invertebrates such as lobsters and crabs. The goal was to test concentration-dependent tracking using sensor pairs legislating movement based on comparisons of intermittence or intensity. The algorithms were guided by the extensive observations of lobsters and crabs, suggesting animals often track odors by successfully straddling the plume (edge) as they move upstream. Here, the comparison of signal properties between the paired sensors is a way to detect the edge. Sampling rate was set at 0.5 Hz, which approximates the abilities of crustacean chemosensors, with an integration time of 2 s and based on odor dynamics collected in a similar environment as the simulation. The searcher moved forward if there were equal positive signals from each sensor. If sensor outputs were different, the searcher moved toward the side of the sensor with stronger readings at 30° relative to flow. Absence of signal caused the searcher to remain in position. The performance of the intermittence-based algorithm exceeded that of the concentration-based algorithm, even when odor sampling frequency was low (0.5 Hz); the intermittence algorithm caused fictive movement along the edge of the plume to the source, whereas the concentration-dependent algorithm resulted in movement toward the plume center. These results show the utility of detection frequency as a tracking strategy even when sampling in time is coarse if sensors are rapid enough to acquire a discrete sample and the tracker is sufficiently large relative to the plume so as to be able to straddle the "edge." Jayaram et al. (2022) used the Gaussian packet dispersion approach (Farrell et al. 2002) to simulate the performance of agents capable of responding to both intermittence and frequency, where stimulation results in upwind turn bias demonstrated in flies. They showed that both intermittence and frequency can be used for tracking to a source, but performance (defined as the percent of agents that find the source) generally peaks with either intermittence or frequency sensing across a range of plume types and there are strong performance tradeoffs of one strategy the other such that one sensing mode generally performs best in a certain plume. Note that it is unclear the extent to which these two plume properties vary independently in real-world odor plumes.

Simulation approaches have been applied to 3D as well as 2D problems, perhaps because of the greater experimental difficulty of working with areal drones and the need to understand how the flight mechanism impacts flow, and therefore chemical signal structure, around the drone, as, for instance, in Eu (Eu and Yap 2018). Here, the goal was to examine a fuzzy logic-based system that surged upwind using course angles that integrated wind direction and vector information based on comparing readings from four sensors on a quadrotor. Sensors were placed so that they were within the region of self-generated flow of a particular rotor, which seems loosely inspired by observations of animals that use similar strategies (e.g., Fields and Yen 1993; Kanzaki 1998). The fuzzy logic algorithms had no biological basis and were designed to accommodate the constraints of MOX sensors. Aside from morphological filtering, bioinspired elements were the use of a 3D spiral search algorithms and zig-zag casting to find and regain contact with the plume. Not unexpectedly, 3D search is necessary and is most successful when combined with zig-zag casting; both algorithms have distinct advantages for locking on to the plume depending on the initial location relative to the center line. Gao et al. (2015) also use a computational approach to simulate tracking in a drone, with the bioinspired element being confined to counterturning and upwind surge behaviors. The utility of both of these studies is difficult to determine based on the lack of details.

Liberzon et al. (2018) used a simulation approach to examine the utility of dynamic rules for surge-cast behaviors based on the time it takes a flier to traverse a detectable odor "puff" (defined as a nonzero odor concentration) in a turbulent wind field. Using wind speeds typical for insect-tracking studies, the simulation modeled the spread of individual odor pulses using standard models of time-averaged dispersion of an odor cloud in a turbulent field, which also sets concentration. Turbulence intensity is used to vary the local instantaneous vector for wind direction to model the effect of wind gusts on the location of the odor puff, which simulates the meander of odor plumes in a turbulent field. The tracker surges upwind (i.e., 180° relative to the direction of the instantaneous local wind vector) as long as it is in an odor puff, and counterturns (with a random angle between 30° and 120°) if it does not encounter another odor puff within a period equal or less than the time taken to traverse the last puff. Increased turbulence intensity creates a more dispersed and wider plume, lengthening casting, cross-wind flight, and search time, while decreasing success rate. This approach successfully incorporates context-sensitive surge-cast and counterturning mechanisms, showing that binary odor presence–absence detection of discrete odor patches is sufficient for successful tracking when combined with simple dynamic rules. Certainly, the resultant behaviors reproduce well-known aspects of odor-mediated search in moths. Notably, there is no firm evidence that odor patch crossing time is an information-carrying parameter, although it is related to intermittence that declines with distance to the source (Webster and Weissburg 2001; Crimaldi et al. 2002; Rahman and Webster 2005) and may therefore have utility. Here, the use of a casting rule dependent on odor puff size, like intermittence-based rules, keeps the tracker from moving too far beyond the plume since odor puffs expand as they move downstream. This mechanism, however, must be reconciled with the tendency of moths to cease upwind flight in homogeneous odor clouds. It also is not clear that organisms are sensitive to the local instantaneous wind vector.

Nag and van Breugel (2024) employ a novel approach using data collected from field environments to determine the extent to which temporal patterns can be used to discern progress toward the source. Using an ionization-based sensor at a final frequency of 90 Hz, they determined whether parameters such as concentrations mean, standard deviation, and duration and frequency of odor encounters could be used to discern distance to the source. The results indicate high variance in properties of individual odor filaments but that averages are predictive if sampling time is long enough (10 s) and at a high enough frequency (20 Hz). These results parallel those of other studies suggesting backward-looking estimation of odor properties require time long time windows to resolve patterns (Webster and Weissburg 2001). Insect olfactory receptors can encode peak concentrations within several milliseconds, and ensemble averages of insect olfactory neurons can follow 100 Hz pulses (Szyszka et al. 2014), but it is unclear whether central nervous system processing integrates information on sufficiently long time scales to use such strategies (Cardé 2021; Steele et al. 2023).

Notably absent from this review are other simulation approaches to chemically guided navigation, particularly Bayesian methods such as infotaxis (Vergassola et al. 2007) and related models, which are not bioinspired at least as defined here. These typically high-level theoretical studies focus on information extracting and decision-making and incorporate biologically unrealistic requirements such as the need for agents to have an explicit spatial map, and large working memories. They also are computationally burdensome and are therefore not appropriate for computationally limited agents.

A few recent and novel studies use approaches that may nonetheless provide useful insights for developing algorithms for navigation. Verano et al. (2023) show that finite state controllers can be used to regulate upwind wind behavior in the plume vs. plume recontact transitions based on the frequency of recent above-threshold odor contacts. Simulations using data from lab-generated odor plumes show that in-plume (i.e., move upstream) vs. out-of-plume (i.e., local search) behaviors can be effectively governed using four memory states that together provide dynamic estimates of odor arrival rate and the resulting agent behavior resembles that displayed in olfactory-guided navigation. The use of stimulation frequency as the basis of the decision-making algorithm, combined with dynamic control requiring a small (<10) number of memory states, suggests a way forward for implementing more sophisticated bioinspired algorithms. Rando et al. (2025) use a very different computational approach to examine the utility of dynamic encoding of odor frequency over short time intervals as a way to regulate transitions between in-plume upstream movement and out-of-plume searching. The critical parameter is the "memory window" governing whether the agent transitions to out-of-plume search behavior. Odor contact is not continuous even within a turbulent plume, and the optimal memory window defines the period over which odor absence indicates the agent has moved beyond the plume and must transition to local search. This window may be determined by agents using a Q-learning paradigm and results in effective orientation in plumes simulated with turbulent plume DNS models. Given the training requirements, the approach as implemented is not feasible for agents where plume properties are a priori unknown. However, it represents a generalizable and adaptive rule that depends only on odor frequency, which is clearly behaviorally salient. A more heuristic approach to estimate this critical memory window offers a potentially valuable and biologically realistic way to implement context-dependent frequency-based rules seemingly used by animals (Vickers and Baker 1994; Page et al. 2011a).

9.4 Conclusions and Prospectus

Experiments and supporting simulation work clearly demonstrate that biologically inspired strategies can be used to track chemical plumes. Control algorithms have been implemented for autonomous trackers on land, in air, and in water (both substrate bound and free swimming), using a variety of strategies presumed to underly chemosensory tracking in animals. Simple counter-turning and surge-cast motor programs used by insects have been the most commonly examined and shown to be effective in the largest set of general conditions (i.e., on land, and in 2D and 3D in air). Tracking plumes in water has been less well studied, but the few studies in these environments provide some of the best evidence for how to design effective algorithms.

Despite these successes, the current state of these efforts is rather short of what is necessary to demonstrate bioinspired tracking strategies can be effective in real-world situations. The development of robust and efficient strategies for truly autonomous tracking is hampered by sensor technology, but also by partial implementation of the mechanisms used by animals, and test beds that are not well quantified and likely unrealistic.

Nearly all implementations of bioinspired strategies taking place on land or in air use MOX sensors with long rise and fall times that are inconsistent with the temporal sampling properties of animal chemical sensors. This hampers strategies based on sensing the instantaneous properties of discrete odor filaments such as instantaneous peak concentration that are the basis for animal-tracking strategies. It also reduces the ability to employ strategies that depend on intermittence and encoding odor encounter frequency that are crucial elements for effective tracking in animals. While signal conditioning algorithms can help improve the feature extraction of

MOX sensors, these are post hoc solutions that add considerable computational complexity and reduce response speed, both of which are undesirable features, particularly for airborne tracking agents. Complete and accurate assessment of biologically derived tracking algorithms cannot be obtained using MOX sensors.

Current studies have incompletely translated animal-tracking algorithms into robotic agents. It is clear that replicating the success of animals such as insects is contingent on counter-turning and surge-cast motor programs; nearly all successful studies use these two mechanisms in combination. When combined with amenotaxis, autonomous trackers can successfully find odor sources in a variety of conditions.

However, most of the current studies implement the counter-turn and surge-cast programs using fixed rules and thresholds that prevent dynamic responses to odor properties, and which compromise tracking speed and success. Typical problems include tracking beyond the plume and failure to regain plume contact if lost. Animal strategies that use short-term aspects of stimulus history, particularly frequency, to modulate turn frequency, cast length/movement speed, or surge speed reduce the possibility that a searcher will move sufficiently beyond the plume as to preclude regaining contact. These features have rarely been implemented in artificial tracking algorithms and would no doubt improve their performance.

Dynamic and adaptive algorithms seem to represent a major challenge in developing biologically inspired strategies on other fronts as well. Nearly all studies use fixed thresholds for detecting odor presence or a priori defined critical frequencies for evoking specific behavior. These parameters often are tuned to the experimental conditions using previous data or initial studies. It is difficult to imagine that animals employ such predetermined rules given the substantial variance in plume properties even under a particular physical regime, and at least some evidence indicates animals do not. The feasibility of implementing more dynamic rules is demonstrated by a few studies, and this must continue in order to develop agents capable of autonomous tracking. Real-world scenarios would be unlikely to present opportunities for parameter tuning that currently appear necessary. Similarly, many studies use fixed criteria for success, primarily based on agent distance from the source, as opposed to agents that can self-declare. Given that animals likely use either visual cues or those related to chemical stimulus properties to determine they have reached or are near the source, such features need to be integrated into existing strategies. Unfortunately, existing animal studies have focused on the tracking phase so the strategies that mediate switching to near-source behaviors remain poorly known.

Investigations of tracking in autonomous agents have not taken full advantage of the techniques used by biologists to study tracking. Understanding the strategies used by animals has been strongly enabled by comparative studies both across taxa and environments, and by the deliberate use of controlled experiments with specific flow or stimulus regimes. Animal investigations, particularly early on, were complemented with extensive independent measurements of odor plume stimulus dynamics and fluid dynamic parameters in a way that could be related easily to the environment in which animals lived. Doing so enabled understanding broad patterns such as the use of flow, the reliance on instantaneous odor filament properties, and the role of temporal vs. spatial information.

The application of these methods has been infrequent in studies of autonomous agents. Detailed measurements of odor plume properties are scant, and many studies supply no information at all beyond basic flow speeds. Investigations using poorly controlled fluid mechanical conditions are most common and difficult to reconcile with real-world scenarios. There are very few attempts to deliberately alter flow or stimulus properties. These factors make it exceedingly difficult to fully understand the strengths and limitations of particular approaches or define operational scenarios where a particular approach is effective.

Translating animal-tracking strategies remains a fascinating and useful endeavor that can both develop novel technology and also enhance biological understanding. Robots are in fact test beds to examine the utility of hypothesized tracking mechanisms, and strong collaborations between biologists and engineers have, and will continue to, yield new insights and technology. However, success in this research program will require closer collaboration between biologists and engineers with skills in robotics, signal processing, and fluid dynamics.

References

Alansari AF, Ucciferri CC, Stevens H, Youssef K, Wyeth RC (2025) Navigational behaviour of Lymnaea stagnalis in response to chemical and flow cues. J Exp Biol 228(16): jeb250746. https://doi.org/10.1242/jeb.250746

Álvarez-Salvado E, Licata AM, Connor EG, McHugh MK, King BMN, Stavropoulos N, Victor JD, Crimaldi JP, Nagel KI (2018) Elementary sensory-motor transformations underlying olfactory navigation in walking fruit-flies. elife 7:e37815

Anderson MJ, Sullivan JG, Horiuchi TK, Fuller SB, Daniel TL (2020) A bio-hybrid odor-guided autonomous palm-sized air vehicle. Bioinspir Biomim 16:026002

Baker TC, Haynes KF (1987) Manoeuvres used by flying male and oriental fruit moths to relocate a sex pheromone plume in an experimentally shifted wind field. Physiol Entomol 12:263–279

Baker TC, Willis MA, Haynes KF, L PP (1985) A pulsed cloud of sex pheromone elicits upwind flight in male moths. Physiol Entomol 10:257–265

Baker CF, Montgomery JC, Dennis TE (2002) The sensory basis of olfactory search beahvior in the banded kokopu (*Galaxias fasciatus*). J Comp Physiol A Neuroethol Sens Neural Behav Physiol 188:533–560

Belanger JH, Willis MA (1996) Adaptive control of odor-guided locomotion: Behavioral flexibility as an antidote to environmental unpredictability. Adaptive Behavior 4:217–253

Bell WJ, Tobin TR (1982) Chemo-orientation. Biol Rev 57:219–260

Cardé RT (2021) Navigation along windborne plumes of pheromone and resource-linked odors. Annu Rev. Entomol 66:317–336

Crimaldi JP, Wiley MB, Koseff JR (2002) The relationship between mean and instantaneous structure in turbulent passive scalar plumes. J Turbul 3:1–24

de Croon GCHE, Dupeyroux JJG, Fuller SB, Marshall JAR (2022) Insect-inspired AI for autonomous robots. Science. Robotics 7:eabl6334

Demir M, Kadakia N, Anderson HD, Clark DA, Emonet T (2020) Walking drosophila navigate complex plumes using stochastic decisions biased by the timing of odor encounters. elife 9:e57524

Derby CD, Weissburg MJ (2014) The chemical senses and chemosensory ecology of crustaceans. In: Derby C, Thiel M (eds) Nervous systems and control of behavior, Oxford, London

Dickman BD, Webster DR, Page JL, Weissburg MJ (2009) Three-dimensional odorant concentration measurements around actively tracking blue crabs. Limnol Oceangrap Meth 7:96–108

Dove ADM (2015) Foraging and Ingestive behaviors of whale sharks, Rhincodon typus, in response to chemical stimulus cues. Biol Bull 228:65–74

Duistermars JB, Chow DM, Frye MA (2009) Flies require bilateral sensory input to track odor gradients. Curr Biol 19:1310–1307

Dusenbery DB (1992) Sensory ecology: how organisms acquire and respond to information. W.H. Freeman and Co., New York

Dusenbery DB (1998) Spatial sensing of stimulus gradients can be superior to temporal sensing for free-swimming bacteria. Biophys J 74:2272–2277

Edwards S, Rutkowski AJ, Quinn RD, Willis MA (2005) Moth-inspired plume tracking strategies in three-dimensions. In: Proceedings of the 2005 IEEE international conference on robotics and automation, pp 1669–1674

Eu KS, Yap KM (2018) Chemical plume tracing: a three-dimensional technique for quadrotors by considering the altitude control of the robot in the casting stage. Int J Adv Robot Syst 15:1729881418755877

Farrell JA, Murlis J, Long X, Li W, Cardé RT (2002) Filament-based atmospheric dispersion model to achieve short time-scale structure of odor plumes. Environ Fluid Mech 2:143–169

Farrell JA, Pang S, Li W (2005) Chemical plume tracing via an autonomous underwater vehicle. IEEE J Ocean Eng 30:428–442

Ferner MC, Weissburg MJ (2005) Slow-moving predatory gastropods track prey odors in fast and turbulent flow. J Exp Biol 208:809–819

Fields DM, Yen J (1993) Outer limits and inner structure: the 3-dimensional flow field of *Pleuromamma xiphias* (Calanoida: Metridinidae). Bull Mar Sci 53:84–95

Gao B, Li H, Sun F (2015) 3D moth-inspired chemical plume tracking. In: 2015 IEEE international conference on Robotics and Biomimetics (ROBIO), pp 1786–1791

Gardiner JM, Atema J (2010) The function of bilateral odor arrival time differences in olfactory orientation of sharks. Curr Biol 20:1187–1191

Gomez G, Atema J (1996) Temporal resolution in olfaction: stimulus integration time of lobster chemoreceptor cells. J Exp Biol 199:1771–1779

Gomez G, Voigt R, Atema J (1994) Frequency filter properties of lobster chemoreceptor cells determined with high-resolution stimulus measurement. J Comp Physiol 174:803–811

Grasso FW, Consi TR, Mountain DC, Atema J (2000) Biomimetic robot lobster performs chemo-orientation in turbulence using a pair of spatially separated sensors: progress and challenges. Robot Auton Syst 30:115–131

Harvey DJ, Lu TF, Keller MA (2008) Comparing insect-inspired chemical plume tracking algorithms using a mobile robot. IEEE Trans Robot 24:307–317

Hodgson ES, Mathewson RF (1971) Chemosensory orientation of sharks. Ann New York Acad Sci 188:174–182

Ishida H, Suetsugu K, Nakamoto T, Moriizumi T (1994) Study of autonomous mobile sensing system for localization of odor source using gas sensors and anemometric sensors. Sensors Actuators A Phys 45:153–157

Ishida H, Kagawa Y, NakamotoT, Moriizumi T (1996) Odor-source localization in the clean room by an autonomous mobile sensing system. Sensors and Actuators B-Chemical 33(1–3):115–121

Jayaram V, Kadakia N, Emonet T (2022) Sensing complementary temporal features of odor signals enhances navigation of diverse turbulent plumes. elife 11:e72415

Kanzaki R (1998) Coordination of wing motion and walking suggests common control of zigzag motor program in a male silkworm moth. J Comp Physiol A 182:267–276

Kanzaki R, Sugi N, Shibuya T (1992) Self-generated zigzag turning of Bombyx mori males during pheromone-mediated upwind walking. Zool Sci 9(3):515–527

Keller T, Powell I, Weissburg MJ (2003) Role of appendages in chemically mediated orientation of blue crabs. Mar Ecol Prog Ser 261:217–231

Kennedy JS, Marsh D (1974) Pheromone-regulated Anemotaxis in flying moths. Science 184:999–1001

Kennedy JS, Ludlow AR, Sanders CJ (1980) Guidance system used in moth sex attraction. Nature 288:475–477

Koehl MAR, Koseff JR, Crimaldi JP, McCay MG, Cooper T, Wiley MB, Moore PA (2001) Lobster sniffing: antennule design and hydrodynamic filtering of information in an odor plume. Science 294:1948–1951

Kowadlo G, Russell RA (2008) Robot odor localization: a taxonomy and survey. Int J Robot Res 27:869–894

Kuwana Y, Shimoyama I (1998) A pheromone-guided mobile robot that behaves like a silkworm moth with living antennae as pheromone sensors. Int J Robot Res 17:924–933

Li W, Farrell JA, Card RT (2001) Tracking of fluid-advected odor plumes: strategies inspired by insect orientation to pheromone. Adapt Behav 9:143–170

Liberzon A, Harrington K, Daniel N, Gurka R, Harari A, Zilman G (2018) Moth-inspired navigation algorithm in a turbulent odor plume from a pulsating source. PLoS One 13(6):e0198422. https://doi.org/10.1371/journal.pone.0198422

Lo Iacono G (2010) A comparison of different searching strategies to locate sources of odor in turbulent flows. Adapt Behav 18:155–170

Loudon C, Koehl MAR (2000) Sniffing by a silkworm moth: wing fanning enhances air penetration through and pheromone interception by antennae. J Exp Biol 203:2977–2990

Mafra-Neto A, Carde RT (1994) Fine-scale structure of pheromone plumes modulates upwind orientation of flying moths. Nature 369:142–144

Mafra-Neto A, Carde RT (1995) Effect of the fine-scale structure of pheromone plumes – pulse frequency modulates activation and upwind flight of almond moth males. Physiol Entomol 20:229–242

McLeese DW (1973) Orientation of lobsters (*Homarus americanus*) to odor. J Fish Res Board Can 30:838–840

Mead KS, Wiley MB, Koehl MAR, Koseff JR (2003) Fine-scale patterns of odor encounter by the antenule of the mantis shrimp tracking turbulent plumes in wave-affected and unidirectional flow. J Exp Biol 206:181–193

Michaelis BT, Leathers KW, Bobkov YV, Ache BW, Principe JC, Baharloo R, Park IM, Reidenbach MA (2020) Odor tracking in aquatic organisms: the importance of temporal and spatial intermittency of the turbulent plume. Sci Rep 10:7961

Mishima T, Kanzaki R (1998) Coordination of flipflopping neural signals and head turning during pheromone-mediated walking in a male silkworm moth Bombyx mori. J Comp Physiol A 183:273–282

Montgomery J, Carton G, Voigt R, Baker C, Diebel C (2000) Sensory processing of water currents by fishes. Philos Trans R Soc Lond B Biol Sci 355:1325–1327

Moore PA, Grills JL (1999) Chemical orientation to food by the crayfish, *Orconectes rusticus,* influence by hydrodynamics. Anim Behav 58:953–963

Moore PA, Scholz N, Atema J (1991) Chemical orientation of lobsters, *Homarus americanus* in turbulent odor plumes. J Chem Ecol 17:1293–1307

Nag A, van Breugel F (2024) Odour source distance is predictable from a time history of odour statistics for large scale outdoor plumes. J R Soc Interface 21:20240169

Page JL, Dickman DB, Webster DR, Weissburg MJ (2011a) Getting ahead: context-dependent responses to odorant filaments drives along-stream progress during odor tracking in blue crabs. J Exp Biol 214:1498–1512

Page JL, Dickman DB, Webster DR, Weissburg MJ (2011b) Staying the course: chemical signal spatial properties and concentration mediate cross-stream motion in turbulent plumes. J Exp Biol 214:1513–1522

Prakash A, Nair AR, Arunav H, R RP, Akhil VM, Tawk C, Shankar KV (2024) Bioinspiration and biomimetics in marine robotics: a review on current applications and future trends. Bioinspir Biomim 19:031002

Pyk P, Bermúdez i Badia S, Bernardet U, Knüsel P, Carlsson M, Gu J, Chanie E, Hansson BS, Pearce TC, Verschure PFMJ (2006) An artificial moth: chemical source localization using a robot based neuronal model of moth optomotor anemotactic search. Auton Robot 20:197–213

Rahman S, Webster DR (2005) The effect of bed roughness on scalar fluctuations in turbulent boundary layers. Exp Fluids 38:372–384

Rando M, James M, Verri A, Rosasco L, Seminara A (2025) Q-learning with temporal memory to navigate turbulence. elife 13:RP102906

Reeder PB, Ache BW (1980) Chemotaxis in the Florida spiny lobster, *Panulirus argus*. Anim Behav 28:831–839

Rutkowski AJ, Edwards S, Willis MA, Quinn RD, Causey GC (2004) A robotic platform for testing moth-inspired plume tracking strategies. In: IEEE international conference on robotics and automation, 2004. Proceedings. ICRA '04, vol 3314, pp 3319–3324

Schuech R, Stacey MT, Barad MF, Koehl MAR (2011) Numerical simulations of odorant detection by biologically inspired sensor arrays. Bioinspir Biomim 7:016001

Sharpe T, Webb B (1998) Simulated and situated models of chemical trail following in ants

Shigaki S, Fikri MR, Kurabayashi D (2018) Design and experimental evaluation of an odor sensing method for a pocket-sized quadcopter. Sensors 18:3720

Shigaki S, Shiota Y, Kurabayashi D, Kanzaki R (2020) Modeling of the adaptive chemical plume tracing algorithm of an insect using fuzzy inference. IEEE Trans Fuzzy Syst 28:72–84

Shigaki S, Yoshimura Y, Kurabayashi D, Hosoda K (2022) Palm-sized quadcopter for three-dimensional chemical plume tracking. IEEE Trans Instrum Meas 71:1–12

Shine R, Phillips B, Waye H, LeMaster M, Mason RT (2003) Chemosensory cues allow courting male garter snakes to assess body length and body condition of potential mates. Behav Ecol Sociobiol 54:162–166

Shu LH, Ueda K, Chiu I, Cheong H (2011) Biologically inspired design. CIRP Ann 60:673–693

Steele TJ, Lanz AJ, Nagel KI (2023) Olfactory navigation in arthropods. J Comp Physiol A 209:467–488

Stupski SD, van Breugel F (2024) Wind gates olfaction-driven search states in free flight. Curr Biol 34:4397–4411.e4396

Swenson DP, McClintock JB (1998) A quantitative assessment of chemically-mediated rheotaxis in the asteroid Coscinasterias tenuispina. Mar Freshw Behav Physiol 31:63–80

Szyszka P, Gerkin RC, Galizia CG, Smith BH (2014) High-speed odor transduction and pulse tracking by insect olfactory receptor neurons. Proc Natl Acad Sci 111:16925–16930

Takasaki T, Namiki S, Kanzaki R (2012) Use of bilateral information to determine the walking direction during orientation to a pheromone source in the silkmoth Bombyx mori. J Comp Physiol A 198:295–307

Terutsuki D, Uchida T, Fukui C, Sukekawa Y, Okamoto Y, Kanzaki R (2021) Real-time odor concentration and direction recognition for efficient odor source localization using a small bio-hybrid drone. Sensors Actuators B Chem 339:129770

van Breugel F, Dickinson MH (2014) Plume-tracking behavior of flying Drosophila emerges from a set of distinct sensory-motor reflexes. Curr Biol 24:274–286

Verano KVB, Panizon E, Celani A (2023) Olfactory search with finite-state controllers. Proc Natl Acad Sci U S A 120:e2304230120

Vergassola M, Villermaux E, Shraiman BI (2007) 'Infotaxis' as a strategy for searching without gradients. Nature 445:406–409

Vickers NJ (2000) Mechanisms of animal navigation in odor plumes. Biol Bull 198:203–212

Vickers NJ, Baker TC (1994) Reiterative responses to single strands of odor promote sustained upwind flight and odor source location by moths. Proc Natl Acad Sci USA 91:5756–5760

Webster DR, Weissburg MJ (2001) Chemosensory guidance cues in a turbulent odor plume. Limnol Oceanogr 46:1048–1053

Webster DR, Weissburg MJ (2009) The hydrodynamics of chemical cues among aquatic organisms. Annu Rev. Fluid Mech 41:73–90

Webster DR, Rahman S, Dasi LP (2001) On the usefulness of bilateral comparison to tracking turbulent chemical odor plumes. Limnol Oceanogr 46:1048–1053

Webster DR, Volyanskyy KY, Weissburg MJ (2012) Bioinspired algorithm for autonomous sensor-driven guidance in turbulent chemical plumes. Bioinspir Biomim 7:036023
Wechsler SP, Bhandawat V (2023) Behavioral algorithms and neural mechanisms underlying odor-modulated locomotion in insects. J Exp Biol 226
Weissburg MJ (2000) The fluid dynamical context of chemosensory behavior. Biol Bull 198:188–202
Weissburg MJ, Zimmer-Faust RK (1993) Life and death in moving fluids: hydrodynamic effects on chemosensory-mediated predation. Ecology 74:1428–1443
Weissburg MJ, Zimmer-Faust RK (1994) Odor plumes and how blue crabs use them to find prey. J Exp Biol 197:349–375
Willis MA, Arbas EA (1991) Odor-modulated upwind flight of the sphinx moth, Manduca sexta L. J Comp Physiol A 169:427–440
Willis MA, Arbas EA (1998) Variability in odor-modulated flight by moths. J Comp Physiol A 182:191–202
Willis MA, Avondet JL (2005) Odor-modulated orientation in walking male cockroaches Periplaneta americana, and the effects of odor plumes of different structure. J Exp Biol 208:721–735
Willis M, Baker TC (1984) Effects of intermittent and continuous pheromone stimulation on the flight behaviour of the oriental fruit moth *Grapholita molesta*. Physiol Entomol 9:341–354
Willis MA, Baker TC (1994) Behavior of flying oriental fruit moth males during approach to sex-pheromone sources. Physiol Entomol 19:61–69
Wyeth RC (2019) Olfactory navigation in aquatic gastropods. J Exp Biol 222:jeb185843
Wyeth RC, Woodward OM, Willows AOD (2006) Orientation and navigation relative to water flow, prey, conspecifics, and predators by the nudibranch mollusc *Tritonia diomedea*. Biol Bull 210:97–108
Yen J, Weissburg MJ, Doall MH (1998) The fluid physics of signal perception by mate tracking copepods. Philos Trans R Soc Lond B 353:787–804
Yu Z-L, Hu N, Yang M-J, Song H, Hu Z, Wang X-L, Zhou C, Zhang Z-X, Zhang T (2019) Environmental water flow can boost foraging success of the juvenile rapa whelk Rapana venosa (Muricidae) in aquaculture tanks with still or flowing water: indication of chemosensory foraging. Aquaculture 513:734392
Yuan J, Oswald D, Li W (2015) Autonomous tracking of chemical plumes developed in both diffusive and turbulent airflow environments using Petri nets. Expert Syst Appl 42:527–538
Zimmer-Faust RK, Finelli CM, Pentcheff ND, Wethey DS (1995) Odor plumes and animal navigation in turbulent water flow. A field study. Biol Bull 188:111–116

Chapter 10
Head in Bilaterian Animals: A Robot Model?

Vincent Bels and Florence Kirchhoff

Abstract The presence of a head is a defining feature of bilaterian animals and a recurring design element in bioinspired robotics. While the structural forms differ considerably, the head serves analogous functions in animals and machines by concentrating sensory systems, neural processing, and motor control. This chapter examines the evolutionary origins and functional roles of the head, highlighting its contribution to voluntary movements, locomotor diversity, and adaptive behaviors. In bilaterians, cephalization has centralized sensory integration and motor coordination, enhancing survival through efficient predator–prey interactions, feeding strategies, and reproductive behaviors. The head position and mobility critically influence the biomechanical properties of locomotor patterns, across aquatic, terrestrial, and aerial environments. Sessile organisms also illustrate how sensory–motor integration contributes to survival strategies, emphasizing the broad ecological significance of head-related functions. By analyzing morphological designs, central pattern generators, and ecological constraints, we argue that the head represents an evolutionary innovation facilitating complex voluntary movements and behavioral plasticity. Comparative perspectives from robotics reveal that humanoid and animal-inspired machines often incorporate heads to improve navigation, interaction, and task performance. This chapter underscores the importance of integrating functional morphology, ethology, and biomimicry to better understand the role of the head in shaping performance and adaptability. Ultimately, this approach provides new insights to show how studies of biological functions can inspire the design of future autonomous robots and reversely.

Keywords Cephalization · Voluntary movements · Behavior · Locomotion · Sensory–motor integration

V. Bels (✉) · F. Kirchhoff
Muséum national d'Histoire naturelle, Sorbonne Université, Institut de Systématique, Évolution, Biodiversité (ISYEB), UMR 7205 CNRS/MNHN/EPHE/UA, Paris, France

Station de Biologie Marine MNHN/IFREMER, Dinard, France
e-mail: vincent.bels@mnhn.fr

V. Bels et al. (eds.), *Biomimicry – A Functional Approach*, Biologically-Inspired Systems 20, https://doi.org/10.1007/978-3-032-11949-0_10

10.1 Introduction

A large number of bioinspired robots have a head (Rojas-Quintero and Rodríguez-Liñán 2021). Why? Is this question similar to the question posed to Alan Turing at the police station in the film *Imitation Game* "*Can machines think*"? He replied: "*This is a stupid question ... machine is different.*" "*Why robots have a head*" is also a "*stupid question*"? While the head in animals and machines differs greatly in structure and shape, it fulfills comparable functional roles. The reason for installing a head on robots remains largely discussed since the early days of robotics (Bailly et al. 2018; Luria et al. 2018; McGinn 2020; Li et al. 2024a, b). To propose a discussion to the question of the head in robots, it is empirically necessary to functionally investigate the shape and the biological role of the head in living organisms compared with robots. Conversely, the investigation of robot functioning with or without a head can provide opportunities to understand some characteristics of traits in animals, including their functionality that refers to the way the traits of animals (i.e., biological structures, systems, and behaviors) operate to their fitness emphasizing their adaptive roles rather than their mere presence of shape (Gravish and Lauder 2018). Various robots are primarily designed to perform a specific, often repetitive task; however, multiform machines are increasingly being developed to carry out multiple tasks (Cham et al. 2002; Garcia et al. 2007; Šabanović 2010). Many robots have no distinct head or anterior part of the body (Fig. 10.1). Numerous animals used for mobile and fixed bioinspired robots only have an oral–aboral axis (Fig. 10.2; see, for example, (Owaki et al. 2024). It is widely acknowledged that many humanoid robots feature a head to enhance their interactions with humans and improve their social lives (Said et al. 2024; Yan et al. 2024). The survey of the literature demonstrates that almost all organisms with a head are used for bioinspired robots, but the shape of the head of robots is diverse compared to the head overall structures in animals (Mo et al. 2020) (i.e., Buschmann and Trimmer 2017; McGinn 2020; Mo et al. 2020; Bao et al. 2023).

10.2 Predator–Prey Interactions: A Key Point to Explain Animal Shape

Numerous animals with a head used for bioinspired aquatic, terrestrial, and arboreal mobile robots are bilaterian organisms (Ryczko et al. 2020; Schiebel et al. 2020; Kumar and Dutta 2021). One of the most significant outcomes of the Cambrian explosion was the emergence and rapid radiation of these bilaterian animals, which today encompass the vast majority of animal phyla (Erwin et al. 2011). Multiple interacting factors have been proposed to explain the rapid diversification of organisms, including predator bilaterian animals during the Cambrian (Fig. 10.3). The evolution of developmental control genes, especially the Hox gene clusters, probably enabled greater flexibility and modularity in Baupläne construction (Car-

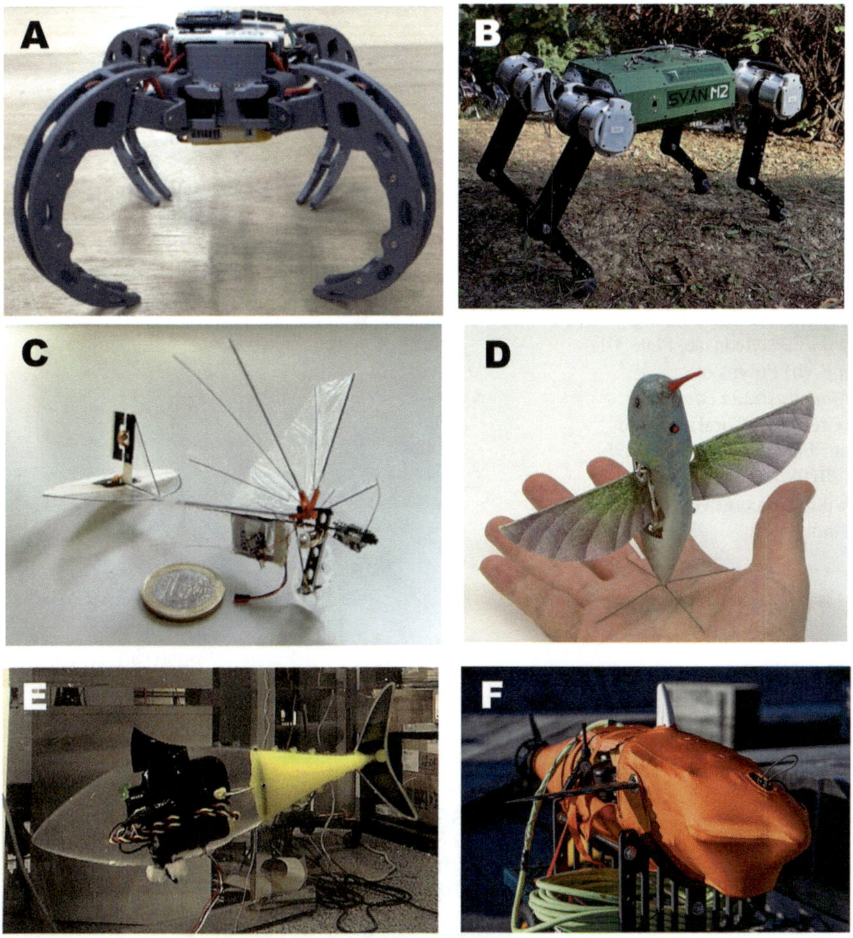

Fig. 10.1 Illustrations of bioinspired robots demonstrate that the anterior region is built on the comparison with the head in bilateral animals. (**a**, **b**) Terrestrial robots (https://commons.wikimedia.org/wiki/File%3AHuntsman-spider-inspired-quadruped-robot-a-walking-form-and-b-rolling-form.jpg?utm_source=chatgpt.com; https://commons.wikimedia.org/wiki/File:Svan_M2-_A_quadruped_robot.jpg). (**c**, **d**) Aerial robots (https://lh4.googleusercontent.com/_kIWY2DV0KnE/TV3-KlLEsOI/AAAAAAAAIro/osutcDWDqJ4/Nano%20Hummingbird.jpg; https://commons.wikimedia.org/wiki/File%3ANano_Hummingbird.jpg?utm_source=chatgpt.com). (**e**, **f**) Aquatic robots (https://commons.wikimedia.org/wiki/File:Soft_under-actuated_fish-like_robot.jpg; https://commons.wikimedia.org/wiki/File:The_Silent_NEmo.jpg)

roll 2000; Erwin 2000). The emergence of predation (interactions between predators and preys) as a major behavior influencing the niche diversification and trophic specialization (i.e., Conway 1998; Bicknell et al. 2023, 2025) has probably played a key role in the development of the behavioral repertoire of predators and its flexibility (i.e., mobility), both factors being impacted by numerous environmental changes in

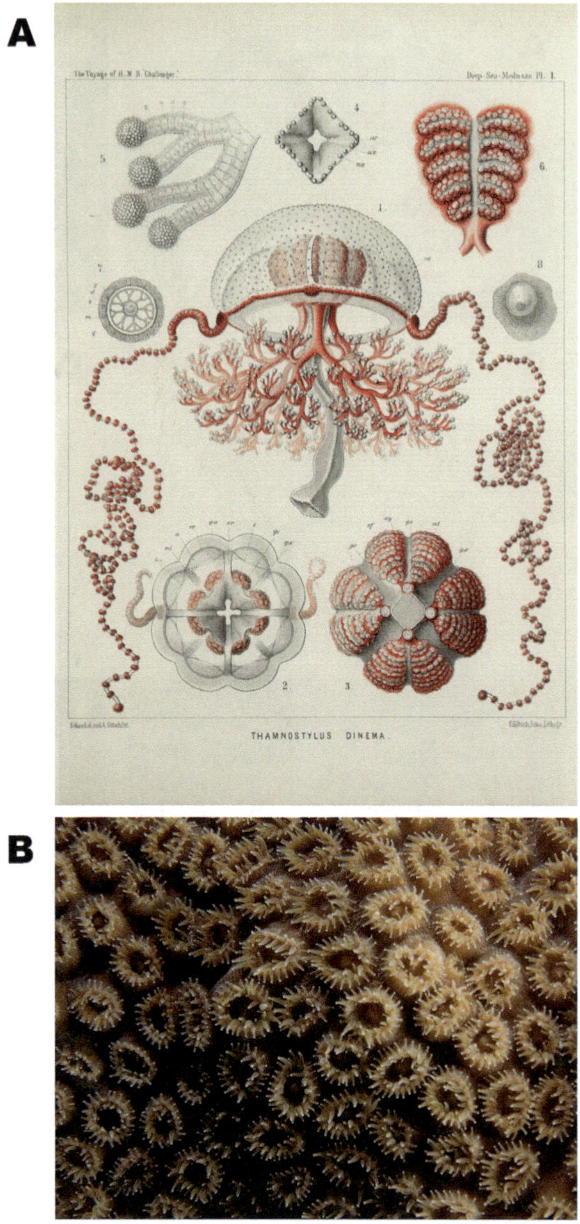

Fig. 10.2 Illustrations of soft animals with oral-aboral axis. (**a**) *Thamnostylus dinema*. From "*Report on Deep-sea Medusae dredged by H.M.S.Challenger during the Years 1873-1876*" (Haekel, 1882). These illustrations were published between 1885 and 1895 under the supervision of John Murray (1841–1914)—https://commons.wikimedia.org/wiki/File:Medusae_Plate_01.jpg. (**b**) Polyps of *Monstrastraea cavernosa* (Great Star Coral). (Image attributed to Nick Hobgood (2010)—https://commons.wikimedia.org/wiki/User:Nhobgood)

a short period (approximately 56 My). This behavioral repertoire involves a wide range of behaviors, including hunting strategies used individually or in groups to capture their prey that has been described in the literature. Some of these strategies fall into categories such as pursuit, ambush, and interception, with anticipating

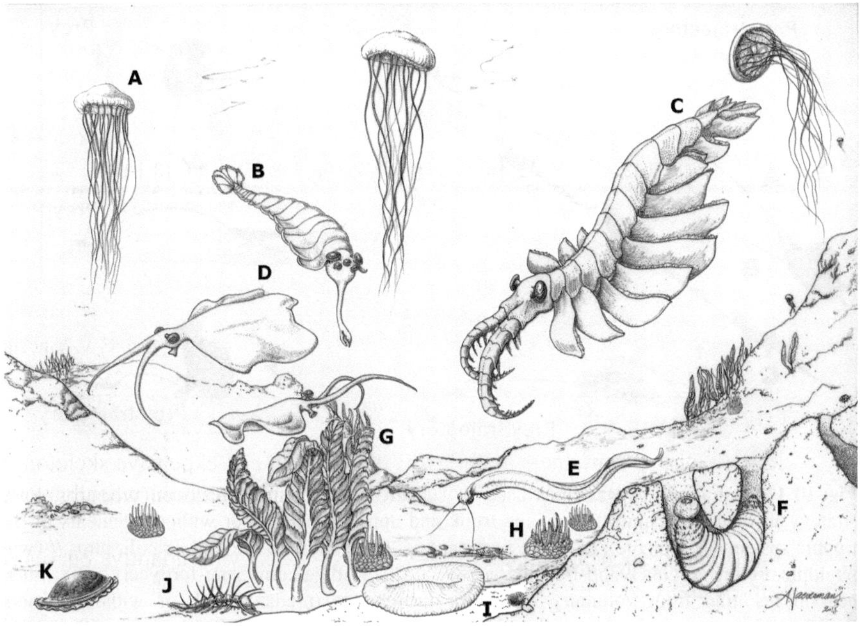

Fig. 10.3 Representative species of the late Edicarian ant Cambrian fauna demonstrate the presence of an oral-aboral axis (**a**), the presence of a head (**c–f**), and without an identified head based on the fossils (**g**, **k**). All specimens are drawn by Agathe Haevermans (Copyright: Agathe Haevermans, ISYEB MNHN 2025). (**a**) *Medusozoa*; (**b**) *Opabinia regalis*; (**c**) *Anomalocaris*; (**d**) *Nectocaris*; (**e**) *Pikaia gracilens*; (**f**) *Priapulid worm*; (**g**) Algae; (**h**) *Archaeocyatha*; (**i**) *Dickinsonia*; (**j**) *Hallucigenia sparsa*; (K) early Mollusca

movement of the prey (Fig. 10.4). These strategies rely on body and/or appendage mobilities (i.e., locomotor modes and maneuverability) associated with postural control under sensory control (Zhao et al. 2023). They combine these mobilities with movement of the trophic design (i.e., tongue and prehensile appendages). Such complex behavior associates different CPGs to produce performances that permit the realization the prey capture. The pursuit strategy involving chasing prey over various distances depends on their design (i.e., limbs), speed under the control of muscular actions, stamina related to energetic control, with often a combination of both. For instance, except for the lions, the majority of large felids (i.e., the cheetah) chase individually or in small temporary groups. These predators use short, explosive bursts of speed to catch fast-moving prey (i.e., large herbivorous prey). In contrast, canids (i.e., wolves and lycaons) rely on endurance for tracking and following weaker or injured animals over long distances in groups. Another common method is ambush hunting, where the predator waits and attacks by surprise, as defined for sit-and-wait predators. These hunters depend on various factors such as stealth, camouflage, and timing. Some predators employ specialized

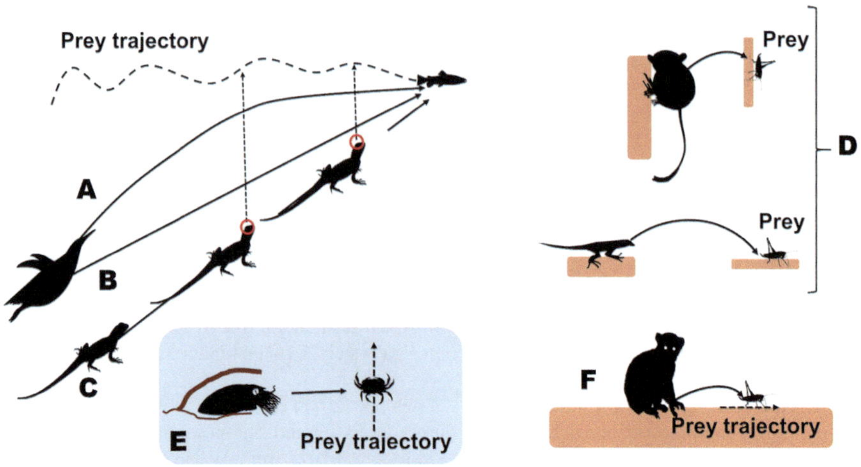

Fig. 10.4 Illustrations of strategies used by aquatic and terrestrial bilaterian predators. Each strategy involves the connection of the trunk and appendages with or without elements of the trophic system (i.e., tongue), ensuring prey capture. (**a**) Pursuit (@Michael Tripoli, https://www.phylopic.org/images/98a393bd-66cc-456a-a893-cc29133ff97a/kumimanu-fordycei). (**b**) Interception resulting from voluntary locomotor movements (predator with or without camouflage). (**c**) Linear optical strategy (https://www.phylopic.org/images/169b14e8-c89a-400b-8aa8-81b18fca7f99/amphibolurus-muricatus). The red circle indicates the position of the eye, providing visual anchorage on the prey throughout the predator's movement. (**d**) Two examples of ambush. The sit-and-wait posture of the predator is followed by a sudden, rapid movement of the entire organism toward the mobile (often poorly) prey. (**e, f**) Interception of a moving target by the predator. (**e**) Interception of a crab moving in the water column by a cuttlefish. (**f**) Interception of a slowly moving prey by a primate using rapid voluntary hand movements. Both predators anticipate the prey's position. (https://www.phylopic.org/images/24a5d2d8-3471-4600-bca6-7da162729989/arctocebus-calabarensis). Prey (**a**: https://www.phylopic.org/images/5034d9bf-a395-433d-b967-e335fd7185ca/salmo-trutta; **b**. https://www.phylopic.org/images/25aa3701-b086-49f2-9da7-0a0f574db185/norops-sagrei; **c**: @ Fritz Geller-Grimm, https://www.phylopic.org/images/28f5634b-b523-4542-978b-ddc1c0979260/bicolorana-bicolor). (**d**) Interception (https://www.phylopic.org/images/f598fb39-facf-43ea-a576-1861304b2fe4/carlito-syrichta). (**f**) Example of interception from a sit-and-wait (camouflage) position (https://www.phylopic.org/images/8f7c46e0-ed5e-455f-b5ae-68168151d01d/cassinoceras; @ Harold N. Eyster, https://www.phylopic.org/images/604d2670-7bf3-4543-b8e3-fbf8e370a530/metacarcinus-magister)

strategies or even use tools (not developed here because these strategies are often strongly related to the use of appendages in prehension).

10.3 Functionality in Animals: From Morphological Design to Behaviors

Functionally, the body of bilaterian organisms (Fig. 10.5) can be divided into a rigid anterior part called the head and a variably deforming posterior region or

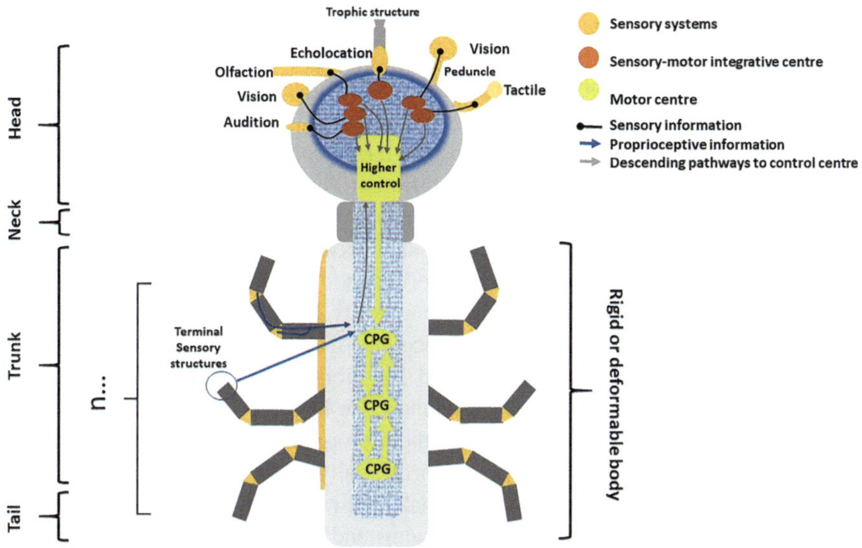

Fig. 10.5 Highly simplified body plan of bilateral organisms. The animal body is divided into a head with the majority of sensory systems collecting external information, a rigid or deformable body involving ventral or dorsal cord, and a series of multi-articulated appendicular systems. The head may be connected to the body directly or by a neck to the body that can terminate with a tail. External and internal sensory information is integrated in various centers before passing to higher centers that coordinate the CPG. Only locomotor-related CPG is represented here (@ Vincent Bels)

trunk with the mobile appendicular systems attached (i.e., appendices, fins, legs, flippers, limbs). The head is either strongly attached to the rest of the body (i.e., Crustacea as crabs) or mobile, thanks to a neck as in multiple Arthropoda (i.e., insects) and Vertebrata (Kuroda et al. 2021; Marek and Felice 2023; Büschges and Ache 2025). The appendicular system involves a diverse number of multisegmental elements that can produce a plethora of motions permitting the realization of highly different behaviors (i.e., locomotion, prehension, and displaying with stereotyped ritualized movements). As in animals, the combined axial and appendicular rhythmic movements can play a key role in the majority of behaviors, as demonstrated in numerous mobile robots (Semini et al. 2016; Ijspeert 2020; Baru et al. 2025). Locomotor modes involving such rhythmic movements are the basic behaviors of mobility in animals, regardless of their habitats (Blickhan and Full 1993; Full and Koditschek 1999; Grillner 2021; Gutarra and Rahman 2022; Suryanarayana et al. 2022; Malul et al. 2024; Pan et al. 2022). But the appendicular rhymic movements are also essential for a lot of behaviors, including food or object prehension and manipulation (Christel and Billard 2002; Reghem et al. 2011; Pouydebat et al. 2014, 2023; Böhmer et al. 2019; Pouydebat and Bardo 2019; Whishaw and Karl 2019).

Also, from an ecological point of view, aquatic animals can be fixed in their habitats, while terrestrial tripoblastic animals are always mobile (except some parasites). Some temporarily sessile animals (e.g., *Patella* sp.) are living at the interface between water and land (DeSalle and Schierwater 2007). However, they are capable of moving when submerged (Hill 2001). These animals can serve to approach the question not only of adhesion but also navigation (mobility) (Franz and Mallot 2000). In bilaterian animals, the head contains a major part of the central nervous system (CNS—brain or central ganglia) in continuity with a dorsal (spinal) or ventral cord or chain. Following the different Phyla, the head-like area and the true head support the entrance of the digestive tract or mouth, showing various specialized features (i.e., radula in mollusks, buccal appendices in arthropods, and tongue and in vertebrates) used for gathering food, display, defense (biting), and aggressive actions. The head supports the majority of the sensory organs related to the CNS, assuring the collection of exteroceptive information.

Bilateral organisms with a head used for bioinspired robots can be roughly divided into (i) organisms without an identifiable head in Echinodermata that can be mobile or fixed (Fig. 10.2), and (ii) mobile organisms with a head belonging to different Phyla (Table 1a,b). One major difficulty comes from the definition and description of the anterior region in animals and robots. In robots, this region can encompass the information-command systems and can play a key role as indicated by Bailly et al. (2018), who stated: "*Though the head is a key structure in animals able to perform voluntary spatial actions* (see Fig. 10.1) *and despite numerous computational arguments presented in this paper, today few robots are supplied with a head that plays a functional role in their navigation.*" But this presence of a structure that can be called a head depends on the conception and building of robots (Fig. 10.1) because the command system integrating exteroceptive information and also proprioceptive information from the motor systems itself (i.e., return from the appendicular and motor system) can be positioned along the body of robots and not particularly in an anterior region (Knüsel et al. 2020; Liu and Bagnall 2023; Li et al. 2024a, b) as in a few number of species (i.e., various Mollusca). Given the number of papers describing animal functioning and behaviors used as models for building robots, we propose here some empirical functional paradigms to open the comparative discussion about the presence of the head in bilaterian organisms and bioinspired robots.

The functional evolution of body organization in the Baupläne of hundreds of multicellular organisms is approached from diverse perspectives in robotics (i.e., Bao et al. 2023; Chu et al. 2024; Cornejo et al. 2024; Ishida et al. 2024). One primarily relates to the ability of organisms to move in their environment, regardless of their phylogeny, their shape, ecology, and the complexity of their behaviors (Lauder and Tangorra 2015; Gravish and Lauder 2018). For comparing behaviors in animals and robots, some authors suggest building an ethogram of robot behaviors (Karakaya et al. 2020; Korcsok and Korondi 2023; Abdai and Miklósi 2024) as it was proposed as a determined tool for comparing animal behaviors in ethology and behavioral ecology (Lorenz and Tinbergen 1938; Tinbergen 1963; MacNulty et al. 2007; Stanton et al. 2015; Kappel et al. 2024).

In any behaviors, all organisms utilize determined morphological traits (head, neck, trunk, tail, and appendages), here called "designs" to exhibit the wide diversity of their behaviors (Fig. 10.6), generating fitness for survival and reproduction.

Table 1a Brief summary of the characteristics of the anterior body region and the head of larvae/first stages in Protostomia and Deuterostomia

Clade	Subdivision	Phylum	Subphylum	Examples	Head in larvae/early stages	Major traits	References
Protostomia	Ecdysozoa	Arthropoda		Insects, arachnids, crustaceans	Present in embryos/ neonates	Early embryos show cephalization with brain primordia and sensory structures	Brusca et al. (2016)
		Nematoda			Present in larvae	Larvae with cephalic papillae and nerve ring around mouth region	De Ley and Blaxter (2002)
	Lophotrochozoa	Mollusca			Present in trochophore larvae	Trochophore larvae with apical sensory organ	Ruppert et al. (2004)
		Annelida		Segmented worms	Present in trochophore larvae	Trochophore larvae show apical sensory tuft and anterior ganglia	Brusca et al. (2016)
		Platyhelminthes		Flatworms	Present in free-living larvae	Free-living larvae with anterior eyespots and sensory cells	Ruppert et al. (2004)
Deuterostomia		Chordata	Cephalochordata	Lancelets	Present in embryos/ neonates	Embryos and neonates exhibit strong cephalization, with early brain and eyes	Kardong (2018), Ruppert et al. (2004)
			Tunicata	Sea squirts, salps	Present in embryos/ neonates	Larvae and embryos already possess a simple cerebral vesicle and sensory organs	Lacalli (2004), Ruppert et al. (2004)
			Vertebrata (Craniata)	Fishes, amphibians, reptiles, mammals	Present in neonates	Strong development of cephalization with brain vesicles and notochord	Delsuc et al. (2006), Tassy et al. (2010)
		Echinodermata		Sea urchins, sea stars, sea cucumbers	Present in bilaterally symmetrical larvae	Larvae are bilaterally symmetrical with an apical sensory organ and anterior nerve concentration. Cephalization disappears during metamorphosis	Mooi and David (2008), Brusca et al. (2016)

(continued)

Table 1b Brief summary of the characteristics of the anterior body region and the head of adult in Protostomia and Deuterostomia

Clade	Subdivision	Phylum	Subphylum	Examples	Cephalization in adults	Approximate percentage of species with a head	Major traits	References
Protostomia	Ecdysozoa	Arthropoda		Insects, arachnids, crustaceans	Strong	≈100%	Distinct cephalic capsule with sensory organs and cerebral ganglia	Brusca et al. (2016), Chapman et al. (2013)
		Nematoda			Moderate	≈80–90%	Anterior nerve ring and cephalic papillae; morphological head poorly differentiated	De Ley and Blaxter (2002)
	Lophotrochozoa	Mollusca			Variable	≈65–75%	Cephalopods and gastropods possess a well-developed head; bivalves, chitons, and scaphopods lack a true head	Ruppert et al. (2004)
		Annelida		Segmented worms	Moderate to strong	≈90–100%	Anterior prostomium with sensory structures and cerebral ganglia	Brusca et al. (2016)
		Platyhelminthes		Flatworms	Moderate	≈80–90%	Planarians show cephalization (eyespots, auricles); adult parasitic forms have reduced cephalization	Ruppert et al. (2004)
Deuterostomia		Chordata	Cephalochordata	Lancelets	Strong	≈100%	Adults retain a simple distinct anterior region with a cerebral vesicle and sensory organs	Lacalli (2004), Ruppert et al. (2004)
			Tunicata	Sea squirts, salps	Poor	≈0–10%	Adults undergo metamorphosis into sessile or pelagic forms with strongly reduced or absent cephalization	Delsuc et al. (2006), Tassy et al. (2010)
			Vertebrata (Craniata)	Fishes, amphibians, reptiles, mammals	Strong	≈100%	Major cephalization with protected brain; adult tunicates are sessile but larvae are cephalized	Kardong (2018), Ruppert et al. (2004)
		Echinodermata		Sea urchins, sea stars, sea cucumbers	Absent	0%	Adults are radially symmetrical without a head; larvae are bilaterally symmetrical	Brusca et al. (2016), Mooi and David (2008)

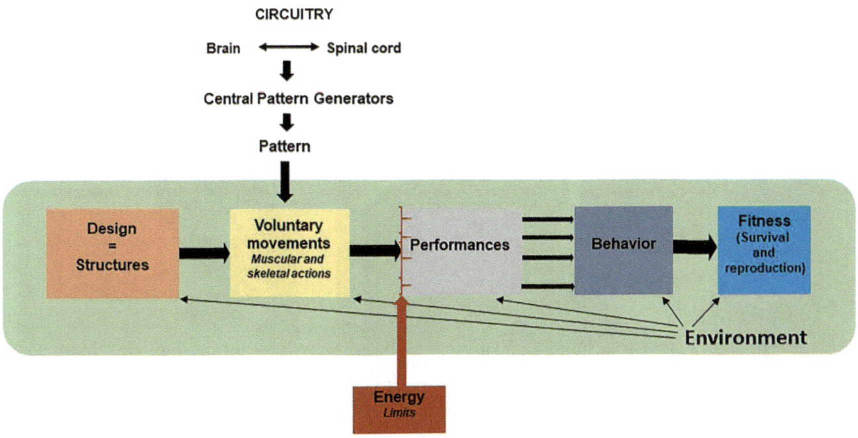

Fig. 10.6 Suggested paradigm associating the terms used in various disciplines to connect designs, voluntary movements, performances, and behaviors at the individual level. All performances are constrained by the ratio between cost and benefit, and the behaviors have been selected to play a biological role in the fitness submitted to natural selection (survival) and sexual selection (reproduction). All the links must be considered in robot designing and functioning

Although the difficulty to use the term "design" ("structural design") because it can simultaneously refer to a structure with its properties or "arrangement" and an underlying conscious or unconscious process as stated by Irschick and Higham (2016), we use this term to describe the morphological structures and their properties, as in a lot of ecomorphological studies (Arnold 1983; Garland Jr. and Losos 1994; Wainwright and Bellwood 2002; Wainwright et al. 2002; Betz 2006). Each design is characterized by a large number of properties at both macro (i.e., geometrical, material) and micro (i.e., molecular, tissular) levels. These properties are often shaped under a trade-off between several pressures related to their variable movements and biological roles in different environmental contexts (Gerald 2017), and have been used in bioinspired robots for a long time now (Coiffet 2005; Delcomyn 2004; Venkateswaran et al. 2019; Tamborini 2021; Acero et al. 2022; Lin et al. 2024). One major morphological design in the body plans used for bioinspired robots is the neck because the head is either immobile (i.e., fixed to the rest of the body) or mobile at various degrees. Although it plays a key role in the mobility of the head to the rest of the body, the neck structure and performances are not easily determined from kinematic studies, except in the case of cineradiographic analysis (Arnold 1983; Van Der Leeuw et al. 2001; Paulk and Gilbert 2006; Marek et al. 2021; Prusty and Sane 2024).

Thus, the role of the neck remains relatively undetermined in a lot of animals (i.e., birds) used as models for bioinspired locomotion, although it plays a key role in head mobility during or at the initiation of locomotor behaviors (Fig. 10.7). For example, the neck morphology becomes to be thoroughly studied in a lot of vertebrates living at the interface between land and water (Böhmer et al. 2019; Keller et al. 2025),

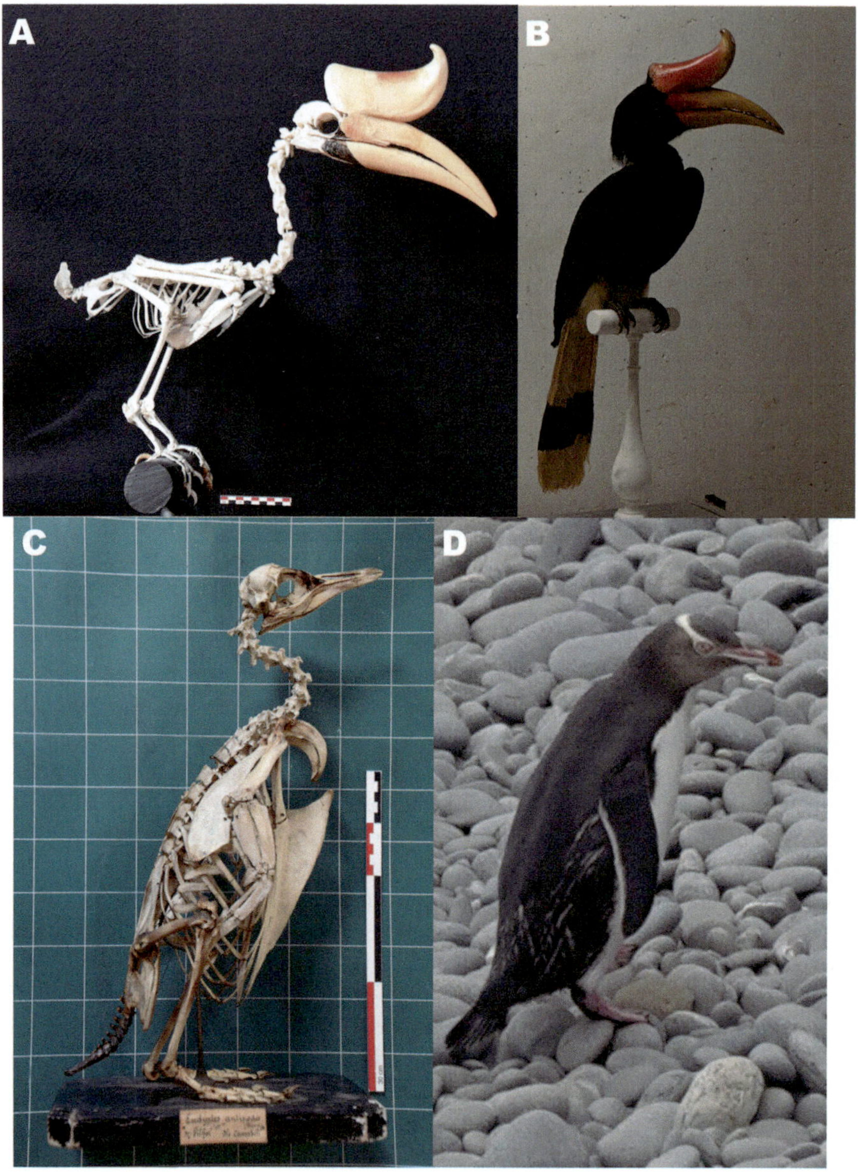

Fig. 10.7 Comparative neck structure and head posture in two species of birds using highly different locomotor modes in highly different habitats. (**a**, **b**) Skeleton showing neck shape and head posture in a preserved specimen (Muséum national d'Histoire naturelle, Paris, France; @ Florence Kirchhoff) in an arboreal hornbill *Buceros rhinoceros*. (**c**) Skeleton showing neck shape of *Megadyptes antipodes* (Muséum national d'Histoire naturelle, Paris, France; @ Eric Pellé). (**d**) Posture during bipedal locomotion of *Megadyptes antipodes* on the beach of Oamaru (New Zealand, @ Florence Kirchhoff)

Fig. 10.8 Head posture during various swimming modes in the Humboldt penguin (*Spheniscus humboldti*). (**a**) High-speed swimming in the water column. (**b**) Swimming under the water surface. (**c**) Swimming at low speed in connection with environmental monitoring. (**d**) a–d: Four images showing the head movements during a complex behavior associating swimming and environmental monitoring (@Florence Kirchhoff)

but the links between the morphological characteristics if this design and the head position assuring the plasticity of head movements in an animal able to pass from water to land and vice versa remain to be studied for improving inspired robots from the reptiles (Shen et al. 2020, 2025; Pan et al. 2022). Such animals cannot only be used for swimming at various speeds with head "fixed" in a determined posture, improving swimming behavior, but can show head-flexible movements in water and on land (Fig. 10.8). The head posture that can play a key role in the sensory flow during directional locomotion is either under the control of the muscular neck system (i.e., in a lot of birds and mammals) or a by-product of the head directly attached to the rest of the body without a neck (i.e., Crustacea).

During locomotion, optic flow (but not only) provides essential cues about direction, speed, and the structure of the surrounding environment. The dynamic visual cues continuously integrated with vestibular and proprioceptive inputs help

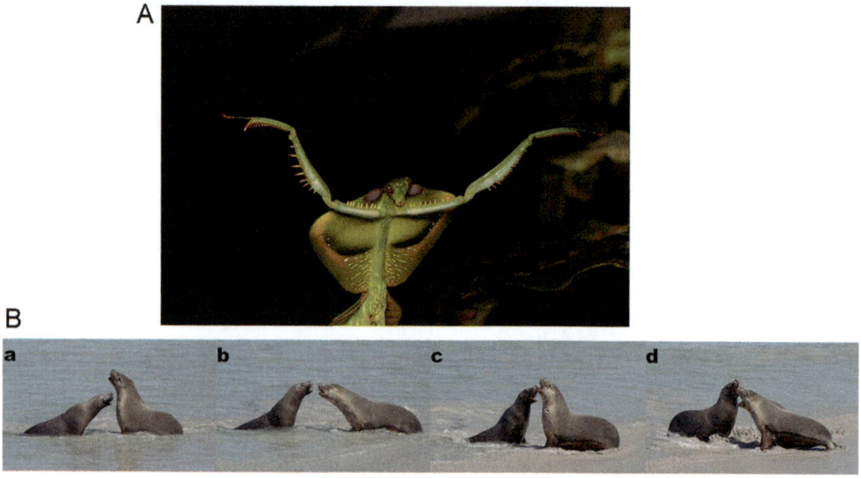

Fig. 10.9 (**a**) Anti-predator posture in *Choeradodis* sp. (@ Frederic Legendre, MNHN, France). (**b**) a–d: Four successive fighting movements involving the jaws and the anterior part of the body, while the Australian sea lions (*Neophoca cinerea*). The limbs are maintained in a frozen position. During fighting, the animal remains stabilized on the limbs in contact with the sand, permitting the seal to firmly maintain the posture during the fight (@Florence Kirchhoff). The limbs are maintained in a determined position to optimize the other behavioral pattern involving the trunk and the head

to maintain postural stability and enable effective movement through space with necessary bioinspired applications in robot movements (i.e., Ruffier et al. 2003; Altshuler et al. 2015; Egelhaaf 2023; Gutiérrez-Ibáñez et al. 2023). Also, a large number of studies demonstrate the eye of various species are either able to move in their orbit as for the visual system in chameleons (Pettigrew et al. 1999; Xu et al. 2023; Inoue et al. 2024) or being at the top of a peduncle (Wilkie and Wann 2003; Land 2015). In the last case, the sensory system can evolve independent mobility under a specialized ocular system for gaining information (Ye et al. 2022). The whole design (i.e., trunk and appendages) or one part of the design (i.e., appendage only) performs complex voluntary movements that are the result of muscular contractions identifying a determined pattern under the control of firing motor neurons organized in a circuitry forming the central-pattern-generators (CPGs) situated at different levels of the nervous system (Katz 2016; Barkan and Zornik 2019; Dubuc et al. 2023). These voluntary movements can terminate in postures such as resting in a frozen position of the body, facilitating other behaviors such as manipulation, displaying, fighting, and anti-predation (Fig. 10.9a, b).

It is widely known that cyclic movements of the four limbs are controlled by CPG to produce each mode of locomotion involved in a lot of circumstances, providing a key plasticity of these movements for achieving successful behaviors (van Baaren and Candolin 2018; Ducatez et al. 2020). Furthermore, they not only play a major role in the exploration and foraging behaviors, but also in capturing the food in association with the mode of prehension with limbs and hands (Montuelle and Kane 2019). Indeed, animals use diverse modes of locomotion based on limb or appendicular rhythmic cycles to move in various directions in aquatic, aerial, and terrestrial habitats for various tasks (Holmes and Phillips 1985; Blickhan and Full 1993; Full and Koditschek 1999; Mullins et al. 2011). Also, a plethora of robots are built for realizing one or several tasks (particularly locomotor tasks) as emphasized by: "*A real-life transforming robot designed to mimic the movement of animals that roam Planet Earth will eventually change the way other planets and celestial bodies will be explored*" (Jacobo 2023). But any movement (i.e., locomotor modes) selected by animals is more than a response of a CPG. It emerges from a continuous feedback loop involving the neural commands from central circuits (CPGs), the mechanical responses of the designs (i.e., bone articulations, muscles, and tendons), and the interactions of the morphological designs with the environment (e.g., friction, resistance) (i.e., Nishikawa et al. 2007).

The phenotypic characteristics of the organisms, from their designs to their functionings, are governed by physical principles and environmental constraints. Higham et al. (2021) proposed to determine for each organism realizing a task (i.e., locomotion, prehension) an *ecomechanical framework* that integrates functional organismal traits (e.g., morphology, material properties, biomechanics) with environmental variables (e.g., fluid flow, mechanical loads, habitat structure) to explain these characteristics. This approach enables predictive modeling of organismal voluntary movements and performances under environmental scenarios that organisms can meet. Ecomechanical models thus reveal key phenotypic constraints on our understanding of phenotypic diversity of the designs used in voluntary movements that have to be considered for bioinspired robots Each voluntary movement (i.e., locomotor behavior) is also associated with postural control depending on the proximal environmental conditions in a determined energetic context, as demonstrated for numerous animals (Zhang and Shu 2021; Stein 2022; Van Bijlert et al. 2024). They can also show adjustments at various levels (i.e., species-specific and individual levels) as shown in arboreal locomotion (Higurashi and Kumakura 2021; Pouydebat et al. 2023; Young 2023). The design and possible voluntary movements of the appendage system in locomotion are influenced by the fact the animals achieve other voluntary movements than locomotion with their appendage. These voluntary movements play a major role in diverse behaviors such as food acquisition, transport and manipulation during feeding, escaping and chasing behaviors (Christel and Billard 2002; Sustaita et al. 2013; Pouydebat et al. 2014; Pouydebat and Bardo 2019; Whishaw and Karl 2019; Song 2021). The motor control of limb/hand movement with the orofacial structures in the orofacial food manipulation of mammals is one example showing the role of the limbs in other behaviors with potential robotic application (An et al. 2022).

Forelimb cyclic movements play a key biological role in other behaviors of vertebrates, such as displaying (Brusatte et al. 2015; Clark et al. 2019; Fialko and Price 2024; MacGillavry et al. 2024). However, their CPGs remain unknown or must be more extensively studied (Bucher et al. 2015; Bels et al. 2019) to be used for determining robot characteristics and use. It is interesting to note that these CPGs can derive from CPGs of the locomotor behaviors under evolutionary processes such as behavioral ritualization (Bels 2021). Voluntary movements of appendicular systems frequently act as visual or tactile signals during courtship behaviors as demonstrated in whimbrel, *Numenius phaeopus* (Fig. 10.10a). In Iguanid lizards, dewlap extension is strongly related to the elevation of the forebody to convey information about individual conditions and fighting ability (Fig. 10.10b). Among birds, the wings are included as a major visual signal in the courtship behavioral sequences (McGillavry et al. 2024; Suzuki et al. 2024). Appendages also occupy a central role in agonistic interactions, where they are employed either as weapons or as ritualized displays that mediate conflict. Male fiddler crabs (*Uca* spp.) exemplify this dual function: their hypertrophied cheliped is used in visual signaling, thereby influencing territorial dynamics and mating success (Backwell et al. 2000; Murray and Backwell, 2006). Appendage forms and movements are also essential for parental care, facilitating the protection, grooming, and provisioning of offspring. For example, crustaceans use specialized appendages for egg brooding and grooming of developing young. In mammals, forelimbs are indispensable for infant transport, grooming, and defense, and primates exhibit extensive manual care that influences the integration of appendages into the social relationships (Maestripieri 1994).

Because the voluntary movements result in performances, the relationships between performances and behavior remain to be clarified. In their book *Animal Athletes*, Irschick and Higham (2016) built a heuristic diagram showing that performances are a central point influenced by morphological structure, physiology, and behavior (Fig. 1.2, p. 7). Several authors insert performances between designs and behaviors, considering behaviors as a "filter" between the performances realized by the organisms and their fitness as in Arnold's paradigm (Arnold 1983) (Garland et al. 1990; Garland Jr. and Losos 1994). This link relating design(s), including all kinds of structures, performances, and behavior(s), is also supported by Bels et al. (2019) in the paradigm called "*Neuro-ethological Morphology*" integrating approaches in ethology (including behavioral ecology and neuroethology already classically associated) and in functional morphology (including biomechanics and physics). Functional morphology mainly investigates (but not only) skeletal (internal in Chordates and external [exoskeleton] in Arthropods) and muscular traits of all kinds of Baupläne. In such a paradigm (Fig. 10.6), each behavior of all organisms is a sum of performances directly influenced by sensory-neuro-motor activities, not only depending on the properties of the designs but also on their abilities to integrate cognitive and sensory information with motor response. Performances depend on the plasticity of the voluntary movements due to environmental constraints as demonstrated in a large number of flexible behaviors such as locomotor behaviors (Fig. 10.11a, b).

Fig. 10.10 (**a**) a, b: Typical series of postures showing the voluntary movements of the wings during fighting in (**b**) typical display of *Iguana delicatissima*, wherein the forelimbs contribute critically both to the execution of push-up sequences and to the elevation of the body, a posture that facilitates optimal dewlap extension. Both behaviors provide visual signals in courtship, territorial, and threat displays

Fig. 10.10 (continued)

10.4 Complexity of Voluntary Movements

Each voluntary movement is variably complex in all organisms, depending on the structures and their properties, the performances needed for driving the behaviors, and the environmental contexts. The complexity of the behaviors depending on performances that result from a series of voluntary movements of determined designs are influenced by their physical properties such as the articulation between the elements of the appendicular system and proprioceptive information (Miki et al. 2022). One example of voluntary movements used by mobile organisms to be able to exploit their environment are their modalities of locomotion following the characteristics of the environment, that is, swimming and flying through fluid and walking, galloping, climbing, jumping, and hopping under gravity. The majority of locomotor behaviors are the result of performances achieved by limb voluntary movements resulting from muscular rhythmic activities controlled by neural activities produced by CPGs that form neural circuits in the spinal cord (Katz 2016; Grillner 2021; Bellegarda and Ijspeert 2022; Suryanarayana et al. 2022; Ramdya and Ijspeert 2023; Dubuc et al. 2023). What is observed, recorded, described, measured, and modeled in terms of behaviors of organisms is their responses integrating the links between the designs with all of their historical characteristics (i.e., morphological, physiological, biochemical) and their fitness in relationship with the properties of the environment where these organisms life and reproduce. These behaviors can also be influenced by the cognitive abilities (i.e., attention, learning, and memory) of the organisms (Lewis et al. 2021). Three types of voluntary movements can be arbitrarily observed in the plethora of behaviors: (i) voluntary movements concerning one design (i.e., cyclic limb movements in

Fig. 10.11 (**a**, **b**) Illustrations of the flexible body and limb movements during locomotor voluntary movements in *Arctocephalus forsteri* (für seals from New Zealand @ Florence Kirchhoff) associated with different performances constrained by the properties of the substrates. (**a**) A substrate formed by partially submerged algae, which constitutes an inherently unstable surface. (**b**) Stable substrate of rocks

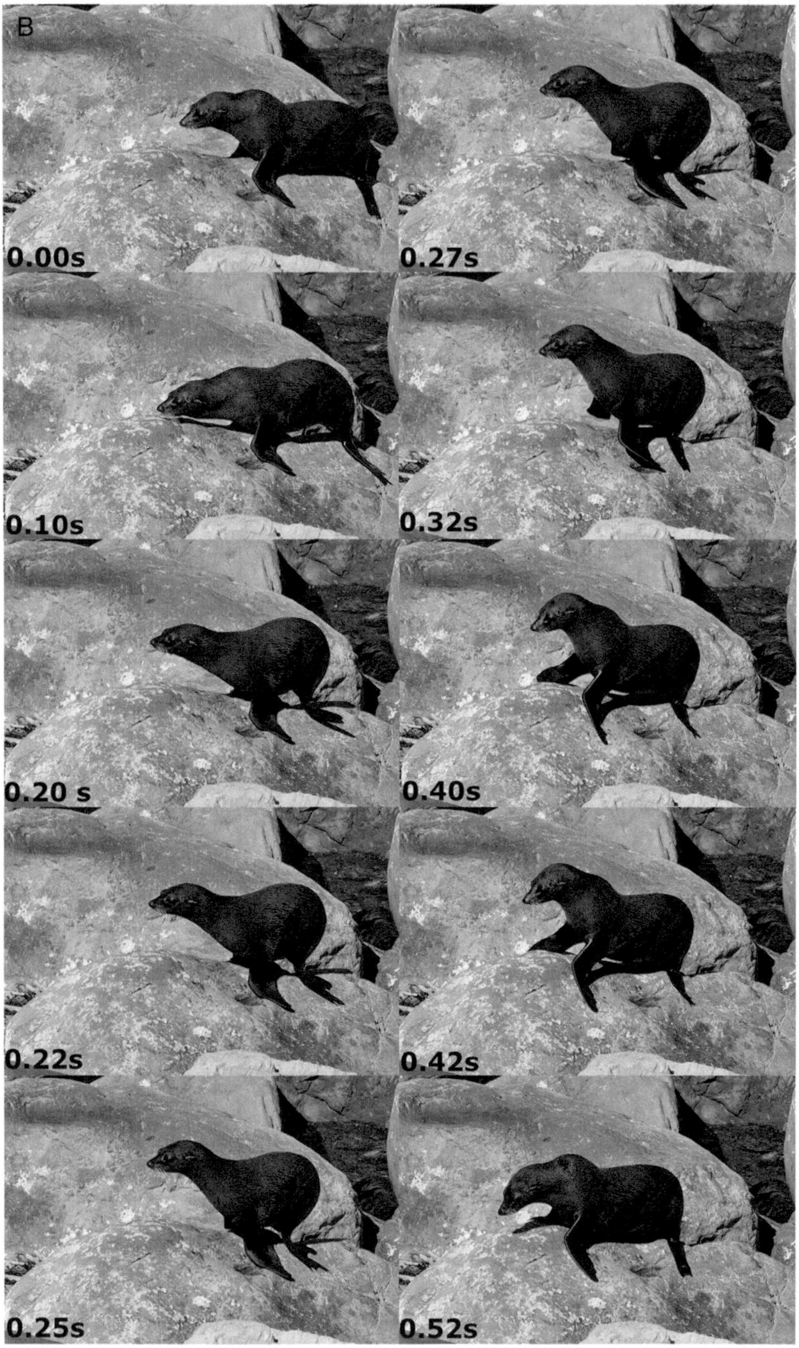

Fig. 10.11 (continued)

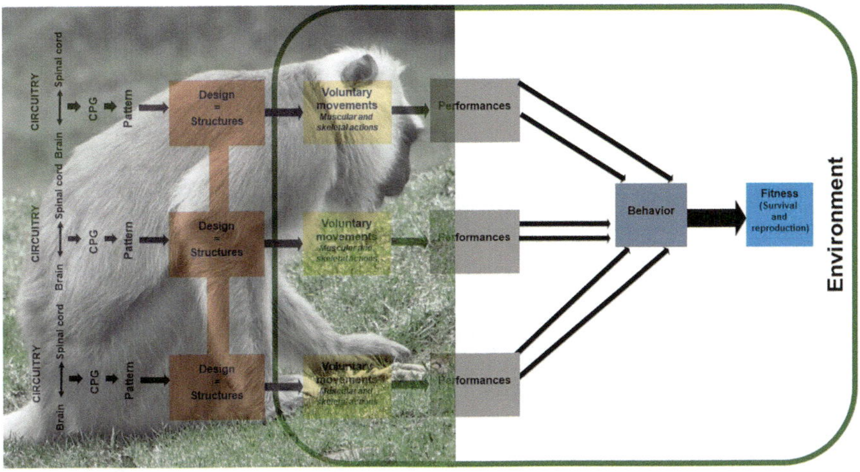

Fig. 10.12 Combined CPGs are used in animals exhibiting behaviors that ensure their fitness. For example, gray langur (*Semnopithecus entellus*) adjusts a typical posture of the head, the whole body, and one of the limbs to be able to sort and prehend small pieces of food on the ground

any mode of locomotion) as primarily used for primarily assuring the diverse mobilities of robots, (ii) series of voluntary movements related to one or several designs (i.e., succession of limb/body movements and postures during successive locomotor (Sun et al. 2021), and (iii) integrated voluntary movements involved in a complex behavior (i.e., integration of several movements of different designs for food prehension).

Some behaviors depend on multiple voluntary movements, each being under of one CPG. Thus, a majority of behaviors are under the control of the integration of several CPGs. In such behaviors, the animals realize various performances related to each of the designs implied in the voluntary movements selected by the animals (Hinde 1966; Hazlett 2012; Bels 2021; Golani and Kafkafi 2025). It can be illustrated by animals taking a determined body posture for optimizing the digit performances involved in food or object sorting and prehension (Laird et al. 2022; Lin et al. 2024). One classical example concerns the primates sitting and placing their body and forelimbs in a determined position to facilitate digit movements for selecting, and manipulating small food (Fig. 10.12). The animals must combine CPGs, adjusting the body and limb posture to facilitate digit voluntary movements. In the meantime, such posture can have a complementary biological role in facilitating the visual and auditory survey of the environment through the integration of other voluntary movements of another design governed by its proper CPG, that is, head rotation. The CPG of finger voluntary movements that play a major role in food/object prehension while the monkey keeps a relevant body and limb posture (Bels 2021) begins to be known (Xu et al. 2024). In animals, multiple behaviors are the result of the use the same design under different CPGs. For example, during foraging, different locomotor patterns selected by the animals

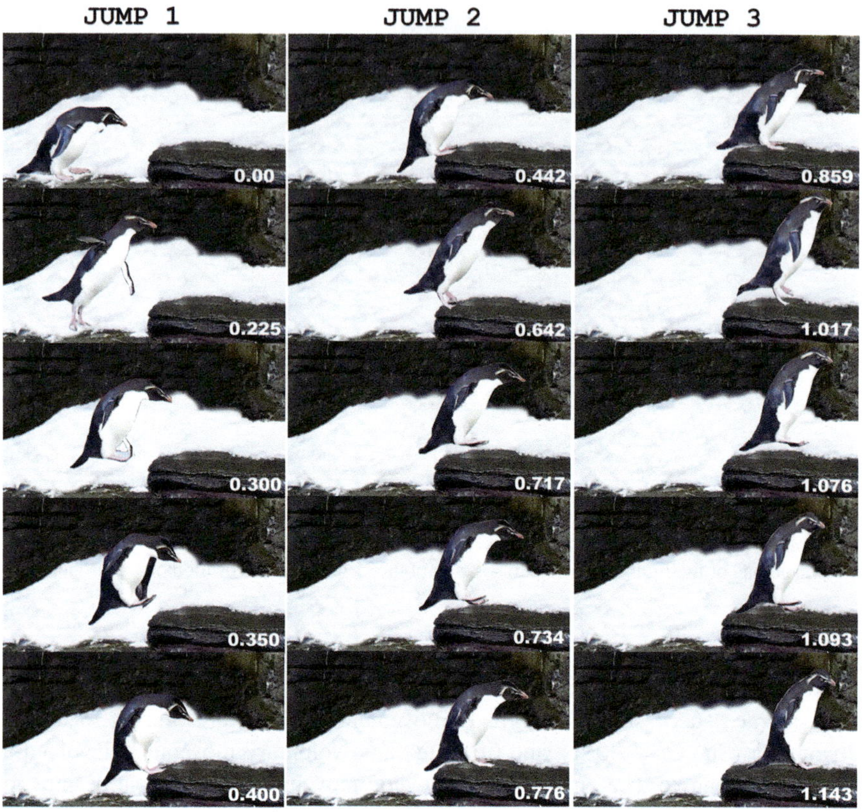

Fig. 10.13 A series of images demonstrating three successive jumps in a hopping sequence of *Eudyptes chrysocome*. The take-off first jump in the series always occurs with synchronized hind-feet positions. In the following jumps, feet can variably show synchronized or unsynchronized positions at the take-off (@ Florence Kirchhoff)

can succeed, although some are favored following the ecological circumstances as demonstrated in a lot of "everyday" activities" of vertebrates such as penguins. These aquatic birds are all able to swim and maneuver in water from the surface to deep water. Their swimming modes can be used for modeling aquatic robots. All of them are able to use bipedal classical bird-like bipedal locomotion to move on various substrates, including sand, ice, and rocks. Many species use jumps to move in rocky habitats, and some others called rockhoppers develop hopping as a series of short jumps to move on various habitats (Fig. 10.13). This simple example shows the diversity of animal locomotor behaviors in phylogenetically related species under relatively unknown environmental constraints (Matarić 1998; Ramdya and Ijspeert 2023). Animals are using very complex behavioral repertoire primarily depicted in ethograms built in diverse ecological and social contexts (Lehner 1987; Stanton et al. 2015; Casarrubea et al. 2022; Korcsok and Korondi 2023). Evidently, the voluntary movements are variably subjected to exteroceptive sensory control from various sources associated with proprioceptive information provided by the design or parts of the designs in movements (Ryu and Kuo 2021; Deng et al. 2022; Stein 2022). For example, the initiation of the take-off phase in jumping can be preceded by a time corresponding to preparation, where the animal integrates sensory information (i.e., distance to be jumped, height of the obstacle, distance to the food source to be captured) for optimizing the success of the jump.

10.5 Head in Voluntary Movements

The processes under the development of the head and its specializations remain to be clarified by combining various domains, including comparative embryology, paleontology, molecular biology, and functional morphology (Hombría et al. 2021; De Robertis and Tejeda-Muñoz 2022; Malakhov and Gantsevich 2022). Several mobile bilaterian larval or adult organisms have a head as illustrated in Crustacea (Fig. 10.14a, b). The head plays a key role in the achievement of voluntary movements acting in all the behaviors as demonstrated in a plethora of organisms (Lin et al. 2022; Virameteekul and Bhidayasiri 2022). In summary, the head characteristics (i.e., shape, morphological and mechanical properties, size, volume) are the results of the trade-off between completely different activities of the animals and, different biological roles. The origin of the head in animal Baupläne remains an intriguing domain, including combined approaches of various disciplines (i.e., evo-devo, comparative anatomy, and neurobiology) to understand how and why animals developed a distinct head region during the evolutionary history of animals. Empirically, the evolution of all the traits has become an evolutionary innovation that provides crucial advantages for the fitness of many animal species with different Bauplänes. In metazoan phyla, the evolution of the oral region and the head is a complex process, as summarized in Fig. 10.15. The trochophore larvae are ciliated organisms able to swim present in several lophotrochozoan lineages, including Annelida and Mollusca. These larvae exhibit bilateral symmetry with an anterior preoral ciliated band (the prototroch) and an apical tuft of cilia associated with a cluster of sensory and neural cells. This apical organ can be regarded as an ancestral precursor of cephalization (Nielsen 2004; Hejnol and Martín-Durán 2015). In ancestral trochophore-like forms, the apical region primarily mediated sensory perception and locomotor coordination. In more derived groups, this anterior sensory complex became progressively integrated into a head-like region. In Annelida, the larval apical organ seems to develop into the cerebral ganglia, which later differentiate into the brain, associated with anterior appendages and sensory structures such as antennae, palps, and eyes (Anderson 1973; Rouse and Pleijel 2001). By contrast, in Sipuncula and Echiura, phylogenetically related to annelids, the apical sensory region is secondarily reduced in adults, and no distinct head is present. These characteristics can be associated with their sedentary, burrowing, and deposit-feeding lifestyle (Nielsen 2004; Schulze and Rice 2002). In Mollusca, the trochophore develops into the veliger larva, where the apical region is elaborated into the velum and associated sensory structures, but do not show voluntary movements (Pechenik 1999). In cephalopods, this anterior specialization results in a highly centralized head with eyes and muscular feeding appendages (Shigeno et al. 2008). Thus, comparative studies of trochophore-bearing phyla demonstrate that the apical organ represents a conserved larval sensory structure that, depending on lineage-specific ecological pressures (particularly mobility and habitats), has either been elaborated into a complex head (i.e., annelids, and mollusks) or reduced in groups adopting a sedentary or burrowing lifestyle (i.e., sipunculans, and echiurans). Roughly, a lot of Bilateria improve their responses to the environment by facilitating and improving directional voluntary movements. But the question of when voluntary locomotor movements arose in bilateral Baupläne still remains debated.

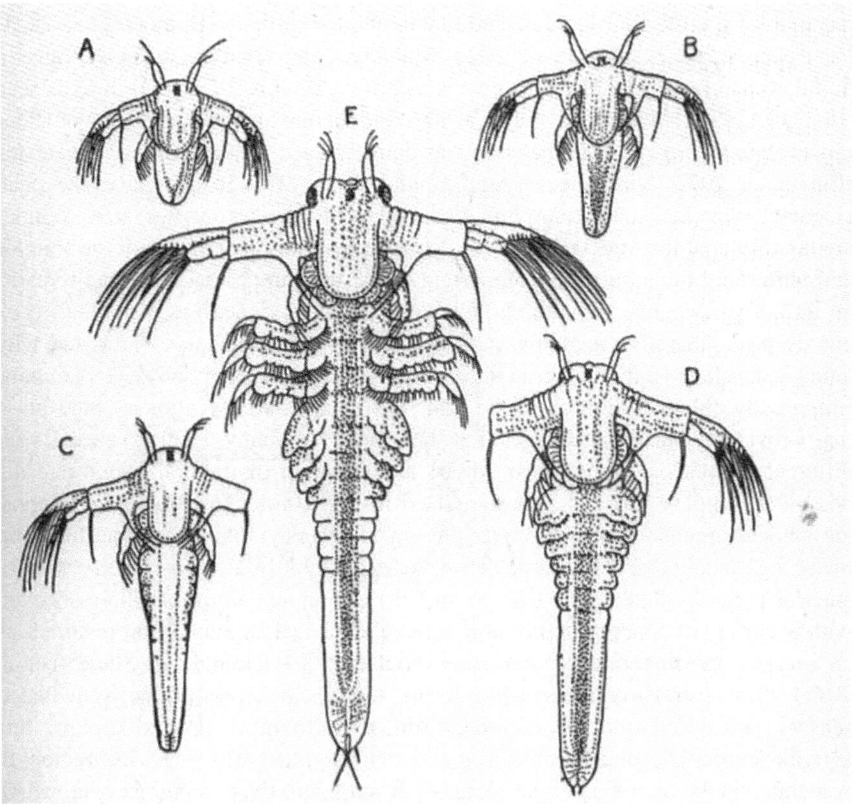

Fig. 10.14 (**a**) Larval stages of the brine shrimp (*Artemia salina*) demonstrating the development of the Bauplan (https://commons.wikimedia.org/w/index.php?search=larval+crustacea&title=Special%3AMediaSearch&type=image). (**b**) Illustrations of Crustacea with the typical Bauplan Arthropoda demonstrating various body and head shapes involving sensory systems

Fig. 10.14 (continued)

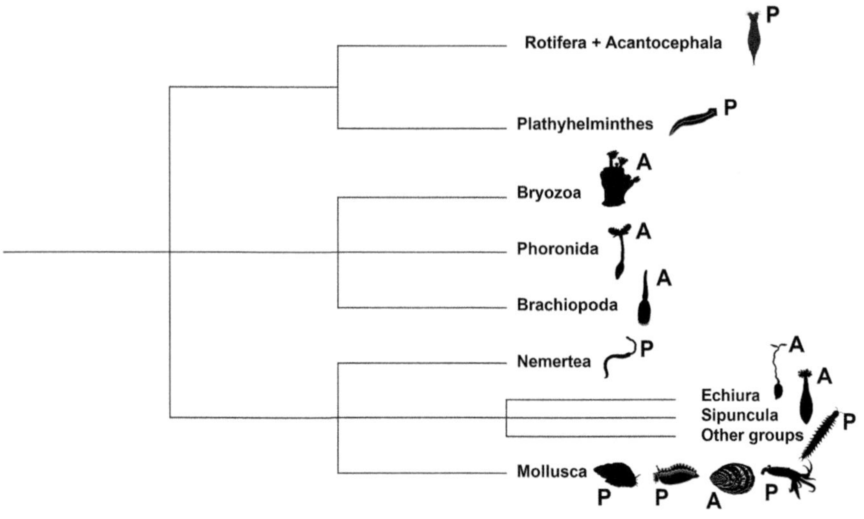

Fig. 10.15 Phylogenetic relationships among major lophotrochozoan lineages, showing the forms of the animals. The cladogram illustrates the evolutionary placement of Rotifera (+ Acanthocephala), Platyhelminthes, Bryozoa, Phoronida, Brachiopoda, Nemertea, Mollusca, and Annelida-related groups (Echiura, Sipuncula, and others). P, head present; A, head absent. Silhouettes representing representative body plans for each group are from phylopic (https://www.phylopic.org): .Rotifera—Fernando Carezzano, https://www.phylopic.org/nodes/b4596ecd-11f1-4c18-92ad-a983f5512846/rotifera%E2%80%91silhouettes, Platyhelminthes—Cristina Guijarro, https://www.phylopic.org/nodes/2852ad17-b8a4-4732-afa2-5a3291d39184/platyhelminthes%E2%80%91silhouettes; Bryozoa—Guillaume Dera, https://www.phylopic.org/nodes/1557a129-7de6-41ba-80eb-97fc9db3b606/ectoprocta-Phoronida - https://www.phylopic.org/images/e225d12f-acd5-4483-80c2-01a757bd4738/phoronopsis; Brachiopoda—Martin R. Smith, https://www.phylopic.org/images/e225d12f-acd5-4483-80c2-01a757bd4738/phoronopsis, Nemertea—Scott Hartman, https://www.phylopic.org/images/429429ed-d07d-463f-9316-62c7adde7e71/nemertea; Mollusca—Alexandra Hahn, https://www.phylopic.org/images/3244258c-e93a-497b-8f37-ac008e574c44/littorina-littorea; Diego Alexander Canque Sosa, https://www.phylopic.org/images/041878b2-0f61-4351-9b97-279ae453fa35/tryblidium-reticulatum; Margot Michaud—https://www.phylopic.org/images/938a7ea7-5262-4bbb-85ba-cb39181002f0/todarodes-pacificus; MABRI and FathomVerse, https://www.phylopic.org/images/2df44006-6939-4a83-aa4a-2bbff4438b33/tritonia-tetraquetra; Echiura, Guillame Dera, https://www.phylopic.org/images/c211dd21-1a06-408d-b1ee-e55b2821455f/bonellia-viridis; Sipuncula, Guillame Dera—https://www.phylopic.org/images/1d133e83-c034-45de-8037-330ce64c6ad5/themiste; other Annelida, Scott Hartman—https://www.phylopic.org/images/2dee030d-9f6d-4bab-87c4-c46869839b30/annelida

A lot of features related to the head can be associated with its role in protecting the brain and its functioning. For example, Stone et al. (2023) describe 12 mechanisms involved in brain protection with three shared with all body tissues (i.e., cranium in vertebrates) and four shared with tissues specialized for the brain. They show that these mechanisms permit the ability of the brain to function despite some potential damage that can accumulate during the life of the organism and to keep its cognitive function. Three (but not only) characteristics can be mainly considered relevant in relationship to the realization of voluntary movements, regardless of the ecological and motivational contexts in which they occur. Indeed, the first role recognized for the head concerns the concentration of neurons and the reorganization of the central nervous system (CNS) due to the various processes of cephalization and specialization (Garcia et al. 2007; Tzahor 2015; Grimaldi and Tajbakhsh 2021; Gainett et al. 2023; Ray and Medeiros 2023; Lofeu et al. 2024). This concentration can be present not only in the brain but also in the spinal cord and coordinated with structures of the brain in the head, such as the cerebellum. The concentration of major sensory systems (i.e., visual, olfactory, vomer olfactory, and auditory organs) with specialization often related to performances associated with specialized behavior of various behaviors involved in the survival (i.e., foraging, feeding, locomotion), but also in reproduction (i.e., locomotion, display), and parental care (Suryanarayana et al. 2022; Büschges and Ache 2025). This concentration provides information from the environment that are playing a key role in various voluntary movements in association with the proprioceptive effect on the voluntary movements.

A major role concerns the integration of sensory information with motor commands, arising first in feeding behaviors and most likely later in locomotion as shown in Fig. 10.15. The placement of the brain as an integrative command center in the "true" head provides numerous advantages, primarily due to the need to associate exteroceptive and interoceptive sensory information with the efficient coordination of responses. Recently, it has been suggested that much early sensing—such as gravity sensing, flow sensing, and proprioception—was predominantly reafferent, arising from translocational (whole-body movement) and deformational (body-shape change) actions of the organism. Such reafferent sensing, observed in sponges, ctenophores, placozoans, cnidarians, and bilaterians, did not require complex corollary discharge mechanisms and may have been central to the emergence of an integrated "body-self" in early nervous system evolution. This primary form of integration most likely preceded the more elaborate sensory-motor coordination required for locomotion (Jékely et al. 2021).

Evolution has favored Baupläne with a centralized brain in the head to integrate sensory information with cognitive possibilities and memory in order to produce efficient motor actions (Table 1a,b). Numerous studies have identified environmental factors explaining why the location of sensory organs and the brain in the head is crucial for the production and control of voluntary movements (Koontz et al. 2023; Nguyen and Duong 2023; Dorkenwald et al. 2024). The speed of signal transmission between sensory exteroceptors and voluntary movements involved in performing a specific task (e.g., modes of selected locomotion) can vary considerably and presumably depends on environmental constraints.

The involvement of the visual system and its neuromotor control provides a classical example of the biological relevance of associating sensory and motor signals in the head to enable rapid movements, such as those required to escape a predator (e.g., C-start locomotion or threat-evoked responses). In vertebrates, the first neural response in the visual cortex occurs approximately 40–100 ms after the onset of a visual stimulus, depending on the type of stimulus, its intensity, and the properties of the visual system. In humans, simple visual stimuli can reach the brain in approximately 50–70 ms, while subcortical pathways, such as the superior colliculus, respond even faster (around 20–30 ms), particularly in the processing of motion and threat-related cues (Carretié 2014; Tamietto and de Gelder 2010).

One rapid visuomotor response involving the visual system and motor control centers in the head is the C-start escape response in fish (Domenici and Kapoor 2010). Environmental changes are detected by photoreceptors in the retina within 5–10 ms, and the signal is then transmitted to the optic tectum—the primary visual processing center in fish—within approximately 10–20 ms. From there, it reaches the Mauthner cells, which initiate the coordinated muscular escape pattern (body deformation) in about 5–10 ms. This C-start response exemplifies a reflex-based architecture in which visual inputs bypass central processing and directly trigger pre-coded motor patterns (e.g., directional turns or thrusts) (Fotowat and Gabbiani 2011). Birds can also respond rapidly to visual threats, initiating flapping or jumping within 80–120 ms. In these cases, visual information is processed through both subcortical (tectal) and cortical pathways. In insects, visual signals are detected by compound eyes, where specialized photoreceptor cells convert light into electrical signals (phototransduction), initiating visual processing. The total transmission time from the eye to the brain is approximately 20–40 ms.

The evolution of the sensory and motor characteristics of the nervous system has likely been constrained by the need to control voluntary movements, not only for locomotion but also for executing diverse behaviors within the species-specific repertoire (ethogram). This evolutionary process is limited by the properties of the anatomical and physiological traits of the designs (e.g., the body, appendages, or limbs) that are shaped by multiple functional trade-offs. For instance, the forelimbs of many mammals are adapted to distinct functions, including various locomotor modes, object and food manipulation, display, and social communication. These anatomical designs operate through distinct motor patterns governed by central pattern generators located in the brain within the head. In bilaterians, this functionality is not only limited by the traits of the structures, and their constituent elements, but also by their interrelationships (Christel and Billard 2002; Reghem et al. 2011, 2013; Sustaita et al. 2013; Pouydebat et al. 2023).

During locomotion undertaken for different purposes (e.g., moving across habitats, pursuing prey, or escaping predators), animals employ a variety of patterns such as walking, trotting, running, galloping, jumping, and hopping, often at different speeds (Gambarian 2002). The head plays a pivotal role in integrating sensory input from external organs and internal proprioceptive systems to coordinate the neuromotor control of voluntary movements that underlie these locomotor performances. Furthermore, this key structure contributes to the biomechanical properties of different locomotor patterns by its positioning and movement (e.g., head movement in jumping of penguins).

One illustration of the role of perception in voluntary movement, closely tied to mobility, concerns the orientation of visual sensory organs during locomotion. Comparative studies demonstrate that this orientation can be achieved through three mechanisms, operating independently or in combination (Martinez-Morales and Wittbrodt 2009; Land 2011, 2015, 2018): (i) movement of the body while maintaining a fixed head (e.g., crabs), (ii) movement of the sensory organ itself (e.g., eyestalks in crabs; independent eye movement in chameleons), and (iii) head movement, both in aquatic and terrestrial contexts. Head orientation during behavior is therefore strongly tied to sensory information, which can play a decisive role in regulating decision-making processes.

A further dimension involves the evolution of trophic structures enabling exploitation of novel food resources as predators, and the development of novel anti-predator strategies as prey, reducing the risk of injury or death from predators or conspecifics (Drossel et al. 2001; Legreneur et al. 2012; Furness et al. 2025). Such trophic structures are central to voluntary movements within food webs, for instance, during prey capture. Elephants and other vertebrates that use trunks for object manipulation, or chameleons capturing prey with their tongues, exemplify the significance of trophic adaptations in both biological function and bioinspired robotics (Lou et al. 2022; Nguyen and Duong 2023; Li et al. 2024a; Leanza et al. 2024; Ma et al. 2025).

Therefore, a key evolutionary trend has been the specialization of the anterior body region into a head, which enhances survival in food webs both as predators and as prey, while also supporting reproductive success across diverse habitats. Evolution has favored head configurations that centralize the brain, allowing the integration of sensory information with cognition and memory to generate motor actions. Numerous studies highlight the importance of this centralized arrangement for survival (Koontz et al. 2023; Nguyen and Duong 2023; Dorkenwald et al. 2024). Integration of sensory-motor information required for voluntary movement depends on sensory organs capable of processing external and internal (proprioceptive) inputs, following the evolution of specialized sensory morphologies and functions. These organs collect and drive information to neural structures for integration. Jékely et al. (2021) proposed that much early sensing (e.g., gravity, flow, proprioception) was reafferent, reflecting translocational or deformational body actions in early metazoans such as sponges, ctenophores, placozoans, cnidarians, and bilaterians. Taken together, the morphology and role of the head underpin the organization of sensory-motor patterns and associated behaviors, as exemplified in the Baupläne that first diversified during the Cambrian (Fig. 10.3). Animals from this period presumably exhibited complex voluntary movements crucial for survival, such as foraging and defense. Baupläne thus represent adaptive outcomes shaping a wide diversity of traits embedded within food webs that originally arose in aquatic environments (Zhang et al. 2021; Zhang and Shu 2021; Mitchell and Pates 2025).

Sessile organisms have also evolved survival strategies requiring integrated sensory information to generate voluntary movements. Often exposed to predation, they display not only morphological defenses (e.g., protective armors) but also behavioral strategies (e.g., burrowing) that reduce predation risk (Legreneur et al. 2012; Bels et al. 2019; Rind and Bels 2024). The complexity of sensory-motor control is therefore fundamental in structuring organisms' ecological roles within any food webs, consistent with "arms race" theory. A crucial feature underlying success in food webs is the capacity for multimodal locomotor patterns that enable

Fig. 10.16 Fiddler crabs use unimodal locomotor sideward movements when they move out of or into their burrow

diverse behaviors such as foraging, pursuit, capture, feeding, and escape. These locomotor strategies are shaped by distinct constraints in aquatic and terrestrial habitats, as well as at their interface, and may involve (i) appendage or limb use, (ii) body deformation, or (iii) combinations of both. Numerous examples of uni- and multimodal locomotion within the same habitat illustrate the diversity of integrated sensory information required for behavioral transitions (Ferrario et al. 2021; Suryanarayana et al. 2022).

One example of unimodal locomotion is provided by Decapoda, whose arthropod Bauplan has made them models for examining "form–function" relationships in swimming, walking, and burrowing (Kwak and Bae 2018; Bao et al. 2023). Some species, such as sand crabs, have been studied as bioinspired models for burrowing, which involves specialized appendage movements (Wei et al. 2024), while walking is primarily unimodal. Fiddler crabs move across aquatic, terrestrial, and intertidal environments, using stereotyped sideways locomotion when entering and exiting burrows, which may displace 300–600 cm^3/m^2 of mudflat (Fig. 10.16). In contrast, other animals illustrate multimodal locomotion. For example, koalas (*Phascolarctos cinereus*) transition from quadrupedal locomotion to jumping, highlighting the complexity of integrating sensory information with voluntary motor control during successive locomotor modes (Fig. 10.17). These herbivorous marsupials navigate arboreal three-dimensional habitats as well as terrestrial environments to move between trees (Davies et al. 2013; Gaschk et al. 2019; Amanat et al. 2022; Crowther et al. 2022; Richards et al. 2023). Their locomotion involves moving on branches

Fig. 10.17 Example of complex movements in the koala (*Phascolarctos cinereus*) for moving in the 2D arboreal habitat. The animal uses multimodal locomotor behavior for moving between branches in the same tree. (**a**) Beginning of a rear locomotor mode to take a posture for take-off between two branches. (**b**) The animal takes the posture for initiating take-off. (**c**) Jumping. Each locomotor mode may be viewed as a task from the animal's point of view. The animal must change its posture between each mode of locomotion. For example, jumping needs to take a determined posture to improve the take-off and the landing. A lot of traits of the animal (i.e., knowledge of its structural morphology, experience, memory) influence and can improve the ability to move in this habitat. They also must produce head movements to gather needed sensory information before reaching one task (i.e., jumping between branches)

of varying diameters, which serve as both supports and obstacles. The structural complexity constrains locomotor behaviors, which include forward and backward walking and leaping between branches. Transitions between locomotor patterns require specific postures that optimize movement. For instance, when shifting from backward locomotion to jumping, koalas adopt preparatory postures for efficient takeoff relative to landing distance. Head and sensory system reorientation are also necessary during locomotor transitions, as shown in shifts from backward locomotion to jumping. The direction of the jump constrains sensory orientation and head posture, since these arboreal mammals must continuously reconfigure body alignment when switching locomotor modes within their three-dimensional habitat.

In aquatic and terrestrial animals, changing from one voluntary movement to another relies on the evolution and specialization of their designs (i.e., neck) and the sensory organs with their connections in the central nervous system. Regarding the locomotor behaviors, the motor abilities to move in liquid and on land are strongly different (i.e., gravity and friction in terrestrial locomotion), particularly when they pass from one environment to another (i.e., transition between terrestrial, aquatic, and aerial environments, transition between locomotion in and on the substratum as sediments), or when they have to move in a 3D habitat. A lot of organisms use unimodal motor patterns to move under the control of CPG in the spinal cord but are influenced by the functioning of the brainstem (Frigon et al. 2022; Ali and Benarroch 2022; Leiras et al. 2022; Dubuc et al. 2023). For example, some crab species (i.e., fiddler crab) use the same mode of side-way locomotion (Chen et al. 2017; Shen 2023) to move in and out their burrow or move on the land to find a feeding place (Fig. 10.13). In contrast, multimodal locomotor modes are largely described in animals moving in fluid (Tieck et al. 2018; Wang et al. 2024).

10.6 Head and Multimodal Locomotor Behavior in a Changing Environment

Bilaterian organisms with a fixed or mobile head can move at the interface between water and land (i.e., Tetrapoda such as amphibians, reptiles, including birds, and pinnipeds). Transitions between media require robust mechanical adaptability, as the propulsion forces, balance control, shape, and resistance vary significantly, that is, between land and water. The rapid development of actuators suggests that future amphibious robots will become increasingly autonomous, resilient, and capable of operating in harsh or variable environments (Mazouchova et al. 2013; van der Geest et al. 2023a; Aydin et al. 2024).

The case of marine vertebrates is particularly interesting (e.g., marine turtles, penguins, and pinnipeds). These animals are required to transition between fluid and terrestrial environments that differ in gravitational and frictional constraints and are characterized by diverse substrates such as ice, sand, and rock. On land, they display quadrupedal or bipedal locomotor patterns with varying degrees of body deformation, as shown in penguins (e.g., jumping) and quadrupedal locomotion in pinnipeds. In swimming, the body may or may not undergo deformation. For example, in turtles, the shell, even when modified, as in *Dermochelys* (Bang et al. 2016), constitutes a rigid element. Its shape and associated structures (e.g., the dorsal keels in *Dermochelys*) can optimize swimming performance without having implications for terrestrial locomotion. In water, these vertebrates employ various swimming modes and complex maneuvering, with or without body deformation (e.g., turtles). Their voluntary movements during transitions between water and land (e.g., in pinnipeds and penguins) may involve only the limbs, both limbs and body, or the body alone.

In marine turtles, locomotion on flat substrates such as beaches relies exclusively on synchronous or asynchronous limb cycles (Renous et al. 2007). In penguins, bipedal locomotion does not involve body deformation regardless of substrate type (e.g., ice versus granular or grassy surfaces). However, penguins deform the body when jumping, which enables them to avoid obstacles or negotiate changes in ground level by hopping. Empirical studies of pinniped locomotion in both aquatic and terrestrial contexts raise questions about the potential use of secondarily aquatic vertebrates as models for bioinspired robotics, particularly in relation to size and limb morphology. Despite their morphological specialization for aquatic life, these vertebrates retain the ability to move effectively in various terrestrial habitats. For example, penguins and pinnipeds can traverse soft beaches and rocky terrains with complex three-dimensional structures.

Pinniped locomotor patterns illustrate evolutionary trade-offs associated with a semi-aquatic lifestyle, balancing the competing demands of effective movement in both environments. In water, pinnipeds employ two primary modes of propulsion depending on the subgroup: phocids (true seals) rely on lateral undulation of the hindlimbs, whereas otariids (sea lions and fur seals) generate thrust through paddling with the fore flippers. Phocids swim using a pelvic oscillation mechanism in which the hind flippers drive thrust via side-to-side movements of the posterior body, while the forelimbs contribute minimally to propulsion, serving mainly as stabilizers that reduce rolling, aid maneuverability, and streamline the body to reduce drag (Fish et al. 2025). By contrast, otariids use a pectoral oscillation mode, similar to that of penguins and sea turtles, in which the fore flippers produce lift-based thrust, providing enhanced maneuverability and acceleration, particularly in shallow or near-surface waters (Feldkamp 1987).

At the water–land interface, these tetrapods variably employ their bodies and limbs depending on the substrate. Terrestrial locomotion differs markedly between groups. Otariids are capable of quadrupedal walking by rotating the hind flippers forward beneath the body, which allows a bounding gait. Phocids, in contrast, employ "galumphing" or undulatory crawling, in which trunk flexion and abdominal muscle contractions pull the body forward, assisted by limited forelimb use (Trites 1991; Supplementary Materials). Despite this seemingly cumbersome mode, some phocids (e.g., *Mirounga angustirostris*) can achieve rapid terrestrial movement over short distances. Walruses (*Odobenus rosmarus*), representing a distinct pinniped lineage, combine traits of both groups: they primarily use hindlimb propulsion in water, like phocids, but can also rotate their hind flippers forward to aid terrestrial locomotion, as in otariids, though in a slower and more cumbersome manner (Ray 1976). These species serve as models of locomotor flexibility at the water–land interface. For instance, a bioinspired robotic seal designed to replicate the seemingly awkward "bobbing gait" of pinnipeds uses two front limbs and a single hind limb to move efficiently across both aquatic and terrestrial environments (Arachchige et al. 2023; Fish et al. 2025).

Another fascinating example of multimodal locomotion is hatchling sea turtles (Fig. 10.18). The hatchlings have to dig to move out of their nest by using the flowing sand and the rigid bodies of the other surrounding turtles as potential contact to reach the horizontal surface of the sand and move. They also have to act against the actions of the other turtles (i.e., movement of the fore flippers) that can disturb their proximal environment (i.e., modifying the tricking of the sand, affecting the movements of their fore flippers). As soon as they move on sand, the turtles encounter various granular properties constraining their locomotion through the limb penetration in the substratum (Mazouchova et al. 2013; Guo et al. 2025). Some species have to change their locomotor modes in a very short time at the transition between terrestrial and aquatic habitats (Fig. 10.17) while others use a similar synchronized pattern of forelimb movements on land and in water for swimming (Renous et al. 1989a, b, 2007; Renous and Bels 1993). This short time is presumably related to their escaping performances under the selective pressure of the presence of terrestrial predators (Martins et al. 2021; Soeiro et al. 2022; Avenant et al. 2024a, b). Change from terrestrial locomotion to swimming that involves completely synchronized fore-flipper cycles for regular swimming and maneuvering as soon as they are in contact with water. This contact not only needs to change their locomotor patterns in a short time but also turtles face completely new physical constraints during this type, that is, changes of water contact by the waves. The use of two different locomotor modes to be able to move from the nest to swimming in the sea can give potential perspectives in bioinspired movements of appendages at the interface between terrestrial and aquatic environment (Low et al. 2007; Sun et al. 2021). By integrating limb compliance, gait modulation, and morphing appendages, a suite of bioinspired amphibious robots replicates marine turtle-like locomotor patterns (i.e., (Mazouchova et al. 2013; van der Geest 2023; van der Geest et al. 2023a, b; Aydin et al. 2024).

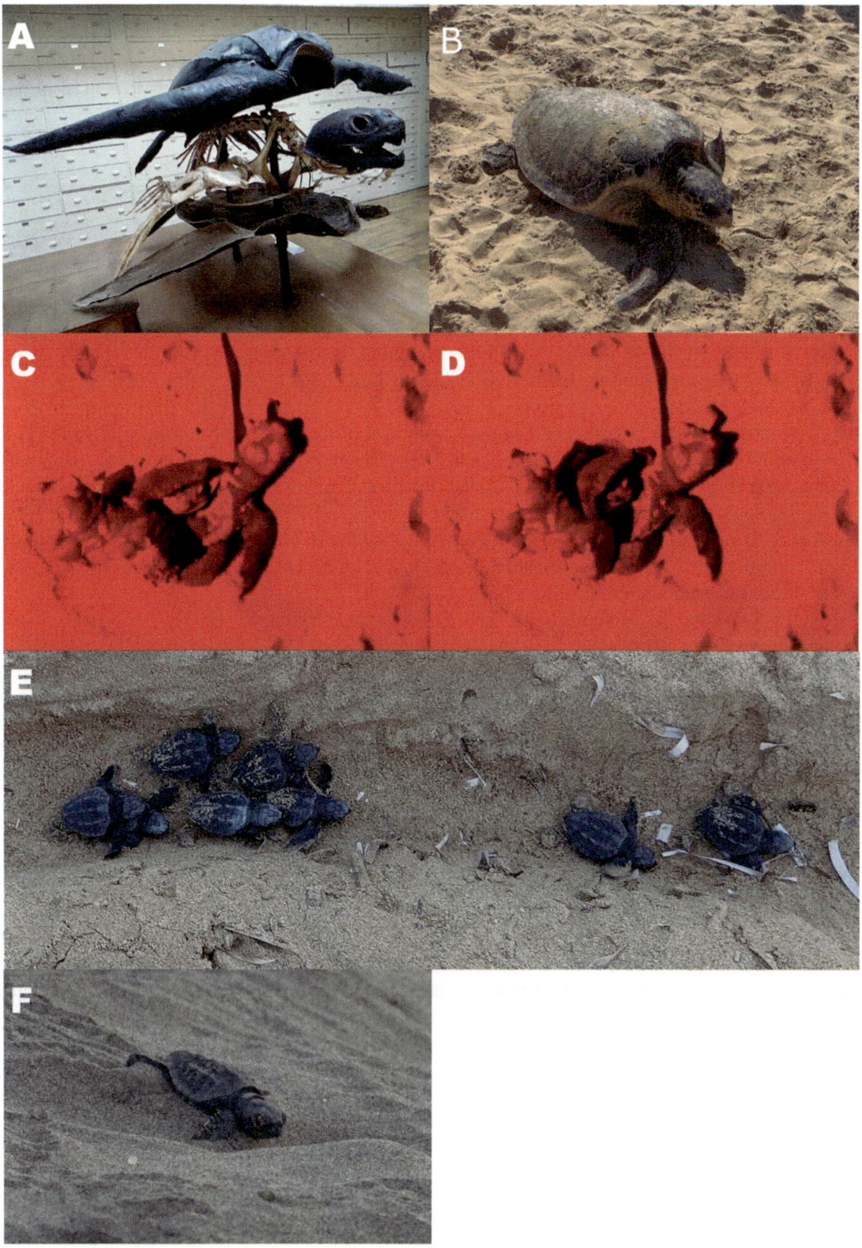

Fig. 10.18 (**a**) Body structure of *Dermochelys coriacea* demonstrating the specialized traits adapted to marine including transformations of the carapace and the limbs. (**b**) Frozen step of a terrestrial gait in *Caretta caretta* (dry sand of Ammes Beach in Kephalonia, Greece). (**c**, **d**) Two frames showing the limb voluntary movements in emerging *Caretta caretta* (@ Prune Poidevin). (**e**) Voluntary diagonal movements illustrating quadrupedal gaits in *Caretta caretta* (@ Florence Kirchhoff). (**f**) Voluntary synchronous forelimb movements of *Caretta caretta* moving from a hole in the beach surface in Zakynthos (@ Florence Kirchhoff)

Acknowledgments A lot of examples of behaviors illustrating this chapter is from field researches based on collaboration with various teams. We are particularly grateful to Chanel Comis and Nikos Vallianos from Wildlife Sense (https://wildlifesense.com/en/) for their help in the work on *Caretta caretta* in Kephalonia. We thank one volunteer Prune Poidevin for providing videos used in our study of nest emergence in *C. caretta*.

References

Abdai J, Miklósi Á (2024) An introduction to Ethorobotics: robotics and the study of animal behaviour. Taylor & Francis

Acero F, Yuan K, Li Z (2022) Learning perceptual locomotion on uneven terrains using sparse visual observations. IEEE Robot Autom Lett 7:8611–8618. https://doi.org/10.1109/LRA.2022.3188108

Ali F, Benarroch E (2022) What is the brainstem control of locomotion? Neurology 98:446–451

Altshuler DL, Bahlman JW, Dakin R, Gaede AH, Goller B, Lentink D, Segre PS, Skandalis DA (2015) The biophysics of bird flight: functional relationships integrate aerodynamics, morphology, kinematics, muscles, and sensors. Can J Zool 93:961–975

Amanat S, Srinivasan P, Mayer J, Bhavsar R, Ali Z, Paracha H, Granatosky MC (2022) Marsupial Locomotion. In: Vonk J, Shackelford TK (eds) Encyclopedia of animal cognition and behavior. Springer International Publishing, Cham, pp 4072–4077

An X, Li Y, Matho K, Mohan H, Xu XH, Whishaw IQ, Huang ZJ (2022) A cortical circuit for orchestrating oromanual food manipulation. bioRxiv, 2022–12. https://doi.org/10.1101/2022.12.12.520093

Anderson DT (1973) Embryology and phylogeny in annelids and arthropods. Pergamon Press, Oxford

Arachchige DD, Varshney T, Huzaifa U, Kanj I, Nanayakkara T, Chen Y, Gilbert HB, Godage IS (2023) Study on soft robotic pinniped locomotion. In: In 2023 IEEE/ASME international conference on advanced intelligent mechatronics (AIM). IEEE, pp 65–71

Arnold SJ (1983) Morphology, performance and fitness. Am Zool 23:347–361. Oxford University Press UK

Avenant C, Fossette S, Whiting S, Hyndes GA (2024a) Predation rates on flatback turtle Natator depressus eggs and hatchlings at an Island rookery. Mar Biol 171:228

Avenant C, Whiting S, Fossette S, Barnes P, Hyndes GA (2024b) Extreme predation of eggs and hatchlings for loggerhead turtles in eastern Indian Ocean. Biodivers Conserv 33:135–159

Aydin YO, Chikere N, Mcelroy J, Ozkan-Aydin Y (2024) Factors affecting flipper-based robotic locomotion in complex terrains. Authorea Preprints. Authorea

Backwell PRY, Christy JH, Telford SR, Jennions MD, Passmore NI (2000) Dishonest signalling in a fiddler crab. Proc R Soc B 267:719–724

Bailly F, Pouydebat E, Watier B, Bels V, Souères P (2018) Should Mobile robots have a head?: a rationale based on behavior, automatic control and signal processing. In: Vouloutsi V, Halloy J, Mura A, Mangan M, Lepora N, Prescott TJ, Verschure PFMJ (eds) Biomimetic and biohybrid systems. Springer International Publishing, Cham, pp 28–39

Bang K, Kim J, Lee S-I, Choi H (2016) Hydrodynamic role of longitudinal dorsal ridges in a leatherback turtle swimming. Sci Rep 6:34283. https://doi.org/10.1038/srep34283

Bao P, Shi L, Duan L, Guo S, Li Z (2023) A review: from aquatic lives locomotion to bio-inspired robot mechanical designations. J Bionic Eng 20:2487–2511

Barkan CL, Zornik E (2019) Feedback to the future: motor neuron contributions to central pattern generator function. J Exp Biol 222:jeb193318. The Company of Biologists Ltd

Baru A, Tomori M, Irizarry J (2025) Exploring the use of quadruped robots in construction environments: the case of small item transportation

Bels VL (2021) Biomechanics of display behaviour in tetrapods: throat display in squamates. In: Biomechanics in animal behaviour. Garland Science, pp 259–274
Bels VL, Pallandre J-P, Charlier S, Maillard A, Legreneur P, Russell AP, Paindavoine A-S, Zghikh L-N, Paulet E, Van Gysel E (2019) Predatory behavior in lizards: strategies and mechanisms for catching prey. In: Behavior of lizards. CRC Press, pp 107–141
Bellegarda G, Ijspeert AJ (2022) CPG-RL: learning central pattern generators for locomotion control in quadruped systems. IEEE Robot Autom Lett 99:1–8. https://doi.org/10.1109/LRA.2022.3218167
Betz O (2006) Biomechanics of locomotion: utilitarian structure and function in animal movement. J Zool 269(3):225–232
Bucher D, Haspel G, Golowasch J, Nadim F (2015) Central pattern generators. eLS 11:1–12. https://doi.org/10.1002/9780470015902.a0000032.pub2
Bicknell RDC, Schmidt M, Rahman IA, Edgecombe GD, Gutarra S, Daley AC, Melzer RR, Wroe S, Paterson JR (2023) Raptorial appendages of the Cambrian apex predator Anomalocaris canadensis are built for soft prey and speed. Proc R Soc B Biol Sci 290:20230638
Bicknell RD, Campione NE, Brock GA, Paterson JR (2025) Adaptive responses in Cambrian predator and prey highlight the arms race during the rise of animals. Curr Biol 35:882. Elsevier
Blickhan R, Full RJ (1993) Similarity in multilegged locomotion: bouncing like a monopode. J Comp Physiol A 173:509–517
Böhmer C, Fabre A-C, Taverne M, Herbin M, Peigné S, Herrel A (2019) Functional relationship between myology and ecology in carnivores: do forelimb muscles reflect adaptations to prehension? Biol J Linn Soc 127:661–680. Oxford University Press UK
Brusatte SL, Butler RJ, Barrett PM, Carrano MT, Evans DC, Lloyd GT et al (2015) The extinction of the dinosaurs. Biol Rev 90:628–642. https://doi.org/10.1111/brv.12128
Brusca RC, Moore W, Shuster SM (2016) Invertebrates, 3rd edn. Sinauer Associates (an imprint of Oxford University Press), Sunderland. ISBN 978-1-60535-375-3
Büschges A, Ache JM (2025) Motor control on the move: from insights in insects to general mechanisms. Physiol Rev 105:975–1031
Buschmann T, Trimmer B (2017) Bio-inspired robot locomotion. In: Hooper SL, Büschges A (eds) Neurobiology of motor control, 1st edn. Wiley, pp 443–472
Carretié L (2014) Exogenous (automatic) attention to emotional stimuli: a review. Cogn Affect Behav Neurosci 14(4):1228–1258. https://doi.org/10.3758/s13415-014-0270-2
Carroll RL (2000) Towards a new evolutionary synthesis. Trends Ecol Evol 15:27–32. Elsevier
Casarrubea M, Leca J-B, Gunst N, Jonsson GK, Portell M, Di Giovanni G, Aiello S, Crescimanno G (2022) Structural analyses in the study of behavior: from rodents to non-human primates. Front Psychol 13:1033561. Frontiers Media SA
Cham JG, Bailey SA, Clark JE, Full RJ, Cutkosky MR (2002) Fast and robust: Hexapedal robots via shape deposition manufacturing. Int J Robot Res 21:869–882
Chapman, R. F., Simpson, S. J., & Douglas, A. E. (Eds.). (2013). The insects: structure and function (5). Cambridge University Press. ISBN 978-0-521-11389-2
Chen Y, Wang H, Helbling EF, Jafferis NT, Zufferey R, Ong A, Ma K, Gravish N, Chirarattananon P, Kovac M, Wood RJ (2017) A biologically inspired, flapping-wing, hybrid aerial-aquatic microrobot. Sci Robot 2:eaao5619
Christel MI, Billard A (2002) Comparison between macaques' and humans' kinematics of prehension: the role of morphological differences and control mechanisms. Behav Brain Res 131:169–184. Elsevier
Chu D, Xiong C, Huang Z, Yang J, Ma J, Zhang J, Sun B, Cai J (2024) Human palm performance evaluation and the palm design of humanoid robotic hands. IEEE Robot Autom Lett 9(3):2463–2470
Clark DL, Macedonia JM, Rowe JW, Austin MR, Centurione IM, Valle CA (2019) Galápagos lava lizards (*Microlophus bivittatus*) respond dynamically to displays from interactive conspecific robots. Behav Ecol Sociobiol 73(10):136
Coiffet P (2005) Robotics: fundamentals and applications. Hermes Science Publications, Paris

Cornejo J, Sierra-Garcia JE, Gomez-Gil FJ, Weitzenfeld A, Acevedo FE, Escalante I, Recuero E, Wehrtmann IS (2024) Bio-inspired design of hard-bodied mobile robots based on arthropod morphologies: a 10 year systematic review and bibliometric analysis. Bioinspir Biomim 19(5):051001. https://doi.org/10.1088/1748-3190/ad5778
Conway Morris S (1998) The crucible of creation: the burgess shale and the rise of animals. (No Title)
Crowther MS, Rus AI, Mella VS, Krockenberger MB, Lindsay J, Moore BD, McArthur C (2022) Patch quality and habitat fragmentation shape the foraging patterns of a specialist folivore. Behav Ecol 33:1007–1017. Oxford University Press UK
Davies N, Gramotnev G, Seabrook L, Bradley A, Baxter G, Rhodes J, Lunney D, McAlpine C (2013) Movement patterns of an arboreal marsupial at the edge of its range: a case study of the koala. Mov Ecol 1:8
De Ley P, Blaxter ML (2002) Systematic position and phylogeny. In: Lee DL (ed) The biology of nematodes. Taylor & Francis, London, UK, pp 1–30
Delcomyn F (2004) Insect walking and robotics. Ann Rev Entomol 49(1):51–60
DeSalle R, Schierwater B (2007) Animal evolution and phylogeny: bridging the gap between molecules and morphology. BioEssays 29(8):701–712
De Robertis EM, Tejeda-Muñoz N (2022) Evo-Devo of Urbilateria and its larval forms. Dev Biol 487:10–20. Elsevier
Delsuc F, Brinkmann H, Chourrout D, Philippe H (2006) Tunicates and not cephalochordates are the closest living relatives of vertebrates. Nature 439(7079):965–968
Deng K, Hunt AJ, Szczecinski NS, Tresch MC, Chiel HJ, Heckman CJ, Quinn RD (2022) Biomechanical and sensory feedback regularize the behavior of different locomotor central pattern generators. Biomimetics 7:226. MDPI
Domenici P, Kapoor BG (2010) Escape responses in fish: kinematics, performance and behavior. In: Fish locomotion: an eco-ethological perspective. Science Publishers, Enfield, pp 123–170
Dorkenwald S, Matsliah A, Sterling AR, Schlegel P, Yu S-C, McKellar CE, Lin A, Costa M, Eichler K, Yin Y (2024) Neuronal wiring diagram of an adult brain. Nature 634:124–138. Nature Publishing Group UK London
Drossel B, Higgs PG, McKane AJ (2001) The influence of predator–prey population dynamics on the long-term evolution of food web structure. J Theor Biol 208:91–107. Elsevier
Dubuc R, Cabelguen J-M, Ryczko D (2023) Locomotor pattern generation and descending control: a historical perspective. J Neurophysiol 130:401–416
Ducatez S, Sol D, Sayol F, Lefebvre L (2020) Behavioural plasticity is associated with reduced extinction risk in birds. Nat Ecol Evol 4:788–793
Egelhaaf M (2023) Optic flow based spatial vision in insects. J Comp Physiol A 209:541–561
Erwin DH (2000) Macroevolution is more than repeated rounds of microevolution. Evol Dev 2:78–84
Erwin DH, Laflamme M, Tweedt SM, Sperling EA, Pisani D, Peterson KJ (2011) The Cambrian conundrum: early divergence and later ecological success in the early history of animals. Science 334:1091–1097
Feldkamp SD (1987) Swimming in the California Sea lion: morphometrics, drag and energetics. J Exp Biol 131:117–135. The Company of Biologists Ltd.
Ferrario A, Palyanov A, Koutsikou S, Li W, Soffe S, Roberts A, Borisyuk R (2021) From decision to action: detailed modelling of frog tadpoles reveals neuronal mechanisms of decision-making and reproduces unpredictable swimming movements in response to sensory signals. PLoS Comput Biol 17:e1009654. Public Library of Science San Francisco, CA, USA
Fialko K, Price TD (2024) Modularity in the evolution of visual signals associated with aggressive displays. bioRxiv. https://doi.org/10.1101/2024.02.22.581650
Fish FE, Kramer-Bottiglio RK, Leftwich MC, Tangorra JL, Kwatny HG (2025) Development of bio-inspired amphibious AUVs based on the morphological and swimming kinematics of secondarily aquatic vertebrates. In: Biologically inspired materials, processes, and systems (BIMPS). SPIE, pp 23–37

Fotowat H, Gabbiani F (2011) Collision detection as a model for sensory-motor integration. Annu Rev Neurosci 34:1–19
Franz MO, Mallot HA (2000) Biomimetic robot navigation. Robot Auton Syst 30:133–153. Elsevier
Frigon A, Akay T, Prilutsky BI (2022) Control of mammalian locomotion by somatosensory feedback. Compr Physiol 12:2877–2947
Full RJ, Koditschek DE (1999) Templates and anchors: neuromechanical hypotheses of legged locomotion on land. J Exp Biol 202:3325–3332. The Company of Biologists Ltd.
Furness EN, Speed JDM, Garwood RJ, Sutton MD (2025) The role of evolutionary processes in determining trophic structure. Oikos 2025:e10597
Gainett G, Klementz BC, Blaszczyk PO, Bruce HS, Patel NH, Sharma PP (2023) Dual functions of labial resolve the Hox logic of chelicerate head segments. Mol Biol Evol 40:msad037. Oxford University Press US
Gambarian PP (2002) Evolution of tetrapod locomotion. Zhurnal Obshchei Biologii 63:426–445
Garland T, Huey RB, Bennett AF (1990) Evolutionary physiology. Annu Rev Ecol Syst 21:541–566
Garland T Jr, Losos JB (1994) Ecological morphology of locomotor performance in squamate reptiles. In: Wainwright PC, Reilly TM (eds) Ecological morphology: integrative organismal biology. University of Chicago Press, pp 240–302
Garcia E, Jimenez MA, De Santos PG, Armada M (2007) The evolution of robotics research. IEEE Robot Autom Mag 14:90–103. IEEE
Gaschk JL, Frère CH, Clemente CJ (2019) Quantifying koala locomotion strategies: implications for the evolution of arborealism in marsupials. J Exp Biol 222:jeb207506. The Company of Biologists Ltd.
Gerald GW (2017) Effects of morphology, substrate use, and potential trade-offs on locomotor performance during multiple modes of snake locomotion. Evol Ecol Res 18(6):601–620
Golani I, Kafkafi N (2025) On growth and form of animal behavior. Front Integr Neurosci 18:1476233. Frontiers Media SA
Gravish N, Lauder GV (2018) Robotics-inspired biology. J Exp Biol 221:jeb138438. The Company of Biologists Ltd
Grillner S (2021) Evolution of the vertebrate motor system-from forebrain to spinal cord. Curr Opin Neurobiol 71:11–18. https://doi.org/10.1016/j.conb.2021.07.016
Grimaldi A, Tajbakhsh S (2021) Diversity in cranial muscles: origins and developmental programs. Curr Opin Cell Biol 73:110–116. Elsevier
Gutarra S, Rahman IA (2022) The locomotion of extinct secondarily aquatic tetrapods. Biol Rev 97(1):67–98. https://doi.org/10.1111/brv.12790
Guo Y, Zhu J, Wang Y, Zhang L, Wang S, Dou B, Yang S, Li B, Niu F, Hao C (2025) Sea turtle-inspired magnetic soft robot demonstrates versatile Land-to-submerged locomotion. J Bionic Eng 22:1039
Gutiérrez-Ibáñez C, Wylie DR, Altshuler DL (2023) From the eye to the wing: neural circuits for transforming optic flow into motor output in avian flight. J Comp Physiol A 209:839–854
Hejnol A, Martín-Durán JM (2015) Getting to the bottom of anal evolution. J Exp Zool B Mol Dev Evol 324:426–437. https://doi.org/10.1002/jez.b.22600
Hazlett B (2012) Ritualization in marine Crustacea. In: Behavior of marine Aniamals: current perspectives in research volume 1: invertebrates, p 97
Higham TE, Ferry LA, Schmitz L, Irschick DJ, Starko S, Anderson PSL et al (2021) Linking ecomechanical models and functional traits to understand phenotypic diversity. Trends Ecol Evol 36:860–873. https://doi.org/10.1016/j.tree.2021.05.009
Higurashi Y, Kumakura H (2021) A comparative anatomical analysis of the hindlimb muscles of Hylobates and Symphalangus with implications for brachiation. Primates 62:757–770. https://doi.org/10.1007/s10329-021-00937-3
Hill JM (2001) Patella vulgata common limpet. Marine life information network: biology and sensitivity key information reviews. Marine Biological Association of the United Kingdom, Plymouth. Disponible en ligne: https://www.marlin.ac.uk/species/detail/137

Hinde RA (1966) Animal behavior: a synthesis of ethology and comparative psychology. McGraw-Hill
Holmes WN, Phillips JG (1985) The avian salt gland. Biol Rev 60:213–256
Hombría JC-G, García-Ferrés M, Sánchez-Higueras C (2021) Anterior Hox genes and the process of cephalization. Front Cell Dev Biol 9:718175. Frontiers Media SA
Ijspeert AJ (2020) Amphibious and sprawling locomotion: from biology to robotics and back. Ann Rev Contr Robot Autonom Syst 3:173–193
Inoue Y, Katsumata Y, Kitazaki M (2024) Chameleon eyes: a visual augmentation system to present independent field of view to both eyes with single gaze control. The Eurographics Association
Irschick DJ, Higham TE (2016) Animal performance ecology: an integrative approach. J Exp Biol 219:2123–2130. https://doi.org/10.1242/jeb.124028
Ishida M, Berio F, Di Santo V, Shubin NH, Iida F (2024) Paleoinspired robotics as an experimental approach to the history of life. Sci Robot 9(95):eadn1125
Jacobo J (2023) The future of space exploration is a transforming, animal-like robot, scientists say. ABC News, 27 June 2023. https://abcnews.go.com/Technology/future-space-exploration-transforming-animal-robot-scientists/story?id=100386844
Jékely G, Godfrey-Smith P, Keijzer F (2021) Reafference and the origin of the self in early nervous systems. Philos Trans R Soc B 376:20190764. https://doi.org/10.1098/rstb.2019.0764
Kappel I, Riedel M-C, Becker F, Hicks S, Warlich-Zach N, Ganslosser U (2024) Ethogram of the domestic cat. Pets 1(3):284–314. https://doi.org/10.3390/pets1030021
Knüsel B et al (2020) Distributed sensor fusion for adaptive locomotion in modular robots. IEEE Int Conf Robot Autom. https://doi.org/10.3389/fnbot.2020.604426
Kuroda S, Adachi N, Kusakabe R, Kuratani S (2021) Developmental fates of shark head cavities reveal mesodermal contributions to the vertebrate head. Zool Lett 7:3. https://doi.org/10.1186/s40851-021-00177-8
Karakaya M, Macrì S, Porfiri M (2020) Behavioral teleporting of individual ethograms onto inanimate robots: experiments on social interactions in live zebrafish. iScience 23. Elsevier
Kardong KV (2018) Vertebrates: comparative anatomy, function, evolution, 8th edn. McGraw-Hill education. ISBN 978-1-260-09333-9
Katz PS (2016) Evolution of central pattern generators and rhythmic behaviours. Philos Trans R Soc B Biol Sci 371:20150057
Keller J, Berta A, Juhn M, Van Valkenburgh B (2025) Morphology and function of pinniped necks: the long and short of it. Anat Rec 308:ar.25642
Koontz A, Urrutia HA, Bronner ME (2023) Making a head: neural crest and ectodermal placodes in cranial sensory development. Semin Cell Dev Biol:15–27. Elsevier
Korcsok B, Korondi P (2023) How do you do the things that you do? Ethological approach to the description of robot behaviour. Biologia Futura 74:253–279
Kumar J, Dutta A (2021) Using bilateral symmetry of the biped robot mechanism for efficient and faster optimal gait learning on uneven terrain. Int J Intell Robot Appl 5:429–464
Kwak B, Bae J (2018) Locomotion of arthropods in aquatic environment and their applications in robotics. Bioinspir Biomim 13:041002. IOP Publishing
Lacalli TC (2004) Sensory systems in amphioxus: a window on the ancestral chordate condition. Brain Behav Evol 64:148–162
Laird MF, Punjani Z, Oshay RR, Wright BW, Fogaça MD, Van Casteren A, Izar P, Visalberghi E, Fragazy D, Strait DS, Ross CF, Wright KA (2022) Feeding postural behaviors and food geometric and material properties in bearded capuchin monkeys (Sapajus libidinosus). Am J Biol Anthropol 178:3–16
Land MF (2011) Oculomotor behaviour in vertebrates and invertebrates. In: The Oxford handbook of eye movements. Oxford University Press, New York, pp 121–132
Land MF (2015) Eye movements of vertebrates and their relation to eye form and function. J Comp Physiol A 201:195–214
Land MF (2018) The evolution of gaze shifting eye movements. In: Hodgson T (ed) Processes of visuospatial attention and working memory. Springer International Publishing, Cham, pp 3–11

Lauder GV, Tangorra JL (2015) Fish locomotion: biology and robotics. Annu Rev Mar Sci 7:521–545. https://doi.org/10.1146/annurev-marine-010814-015754

Lewis MA, Fagan WF, Auger-Méthé M, Frair J, Fryxell JM, Gros C, Gurarie E, Healy SD, Merkle JA (2021) Learning and animal movement. Front Ecol Evol 9:681704

Liu X, Bagnall MW (2023) Proprioceptive feedback control of locomotion. Robot Auton Syst 170:104521. https://doi.org/10.1016/j.robot.2023.104521

Lorenz K, Tinbergen N (1938) Taxis und Instinkthandlung in der Eirollbewegung der Graugans. Zeitschrift für Tierpsychologie 2:1–29

Leanza S, Lu-Yang J, Kaczmarski B, Wu S, Kuhl E, Zhao RR (2024) Elephant trunk inspired multimodal deformations and movements of soft robotic arms. Adv Funct Mater 34:2400396

Legreneur P, Laurin M, Bels V (2012) Predator–prey interactions paradigm: a new tool for artificial intelligence. Adapt Behav 20:3–9

Lehner PN (1987) Design and execution of animal behavior research: an overview. J Anim Sci 65:1213–1219. Oxford University Press

Leiras R, Cregg JM, Kiehn O (2022) Brainstem circuits for locomotion. Annu Rev Neurosci 45:63–85

Li C, Deng T, Wang Y, Sun F, Wolff B, Jiangzuo Q, Ma J, Xing L, Fu J, Zhang J (2024a) The trunk replaces the longer mandible as the main feeding organ in elephant evolution. Elife 12:RP90908. eLife Sciences Publications Limited

Li Y, Zhu L, Zhang Z, Guo M, Li Z, Li Y, Hashimoto M (2024b) Humanoid robot heads for human-robot interaction: a review. SCIENCE CHINA Technol Sci 67:357–379

Lin A, Witvliet D, Hernandez-Nunez L, Linderman SW, Samuel AD, Venkatachalam V (2022) Imaging whole-brain activity to understand behaviour. Nat Rev Phys 4:292–305. Nature Publishing Group UK London

Lin C, Liu X, Yang Y, Niu Y, Yu W, Zhang T, Tan J, Boots B, Zhao D (2024) Locoman: advancing versatile quadrupedal dexterity with lightweight loco-manipulators. In: 2024 IEEE/RSJ international conference on intelligent robots and systems (IROS). IEEE, pp 6877–6884

Lofeu L, Montefeltro F, Simon MN, Kohlsdorf T (2024) Functional modularity and mechanical stress shape plastic responses during fish development. Evolution 78:1568–1582. Oxford University Press US

Lou B, Zuo S, Liu J (2022) Bio-inspired vacuum driven soft robot for esophageal obstruction removal. In: 2022 5th international conference on intelligent robotics and control engineering (IRCE). IEEE, pp 32–38

Low K-H, Zhou C, Ong TW, Yu J (2007) Modular design and initial gait study of an amphibian robotic turtle. In: 2007 IEEE international conference on robotics and biomimetics (ROBIO). IEEE, pp 535–540

Luria M, Forlizzi J, Hodgins J (2018) The effects of eye design on the perception of social robots. In: 2018 27th IEEE international symposium on robot and human interactive communication (RO-MAN). IEEE, pp 1032–1037

Ma C, Song W, Zhao X, Yu H, Lu J, Qu Z, Wu W, Han Z, Mu Z, Yan J, Ren L (2025) Bioinspired soft robots with integrated biological motion mechanisms and rigid–flexible coupling systems. Small Methods 9:2402264

MacGillavry T, Pomeroy D, Lekuona JM, Langen TA, Uy JAC (2024) Courtship mechanics in a classic ornament: the wire-like tail feathers of the riflebird. Biol J Linn Soc blae077. https://doi.org/10.1093/biolinnean/blae077

Malakhov VV, Gantsevich MM (2022) The origin and Main trends in the evolution of bilaterally symmetrical animals. Paleontol J 56:887–937

Marek RD, Falkingham PL, Benson RBJ, Gardiner JD, Maddox TW, Bates KT (2021) Evolutionary versatility of the avian neck. Proc R Soc B Biol Sci 288:20203150

Martinez-Morales JR, Wittbrodt J (2009) Shaping the vertebrate eye. Curr Opin Genet Dev 19:511–517. Elsevier

Martins S, Sierra L, Rodrigues E, Oñate-Casado J, Galán IT, Clarke LJ, Marco A (2021) Ecological drivers of the high predation of sea turtle hatchlings during emergence. Mar Ecol Prog Ser 668:97–106

Matarić, M.J. (1998) Behavior-based robotics as a tool for synthesis of artificial behavior and analysis of natural behavior. Trends Cogn Sci 2, 82–86. Elsevier

Mazouchova N, Umbanovar PB, Goldman DI (2013) Flipper-driven terrestrial locomotion of a sea turtle-inspired robot. Bioinspir Biomim 8:026007. IOP Publishing

MacNulty DR, Mech LD, Smith DW (2007) A proposed ethogram of large-carnivore predatory behavior. J Mammal 88(3):595–605

Maestripieri D (1994) Primate psychology. Harvard University Press

Malul D, Berman H, Solodoch A, Tal O, Barak N, Mizrahi G, Berenshtein I, Toledo H, Lotan T, Sher D, Shavit U, Lehahn Y (2024) Directional swimming patterns of jellyfish emerge from simple behavioral rules. Curr Biol 34(17):4033–4038.e5. https://doi.org/10.1016/j.cub.2024.07.012

Marek RD, Felice RN (2023) A surrogate forelimb for reconstructing limb posture and function in extinct tetrapods. J Morphol 284(7):e21586. https://doi.org/10.1002/jmor.21586

Miki T, Lee J, Hwangbo J, Wellhausen L, Koltun V, Hutter M (2022) Learning robust perceptive locomotion for quadrupedal robots in the wild. Sci Robot 7(62):eabj6652. https://doi.org/10.1126/scirobotics.abj6652

Montuelle SJ, Kane EA (2019) Food capture in vertebrates: a complex integrative performance of the cranial and postcranial systems. In: Bels VL, Whishaw IQ (eds) Feeding in vertebrates: evolution, morphology, behavior, biomechanics. Springer, Cham, pp 71-137. 71–137. https://doi.org/10.1007/978-3-030-13739-7_4

Murray AJ, Backwell PRY (2006) Confrontation versus avoidance during aggressive interactions in fiddler crabs (*Uca annulipes*): a field test of game theory predictions. Behav Ecol Sociobiol 59:844–851. https://doi.org/10.1007/s00265-006-0195-5

McGinn C (2020) Why do robots need a head? The role of social interfaces on service robots. Int J Soc Robot 12:281–295

Mitchell EG, Pates S (2025) From organisms to biodiversity: the ecology of the Ediacaran/Cambrian transition. Paleobiology:1–24. Cambridge University Press

Mo X, Ge W, Miraglia M, Inglese F, Zhao D, Stefanini C, Romano D (2020) Jumping locomotion strategies: from animals to bioinspired robots. Appl Sci 10:8607. MDPI

Mooi R, David B (2008) Radial symmetry, the anterior/posterior axis, and echinoderm Hox genes. Annu Rev Ecol Evol Syst 39:43–62. https://doi.org/10.1146/annurev.ecolsys.39.110707.173521

Mullins OJ, Hackett JT, Buchanan JT, Friesen WO (2011) Neuronal control of swimming behavior: comparison of vertebrate and invertebrate model systems. Prog Neurobiol 93:244–269. Elsevier

Nguyen JD, Duong H (2023) Anatomy, head and neck: eye nerves. In StatPearls [Internet] p. StatPearls Publishing

Nielsen C (2004) Trochophora larvae: cell lineages, ciliary bands, and body regions. J Exp Zool B Mol Dev Evol 302B:35–68. https://doi.org/10.1002/jez.b.20005

Nishikawa K, Biewener AA, Aerts P, Ahn AN, Chiel HJ, Daley MA, Daniel TL, Full RJ, Hale ME, Hedrick TL (2007) Neuromechanics: an integrative approach for understanding motor control. Integr Comp Biol 47:16–54. Oxford University Press

Owaki D, Austin M, Ikeda S, Okuizumi K, Nakajima K (2024) A jellyfish cyborg: exploiting natural embodied intelligence as soft robots. arXiv. http://arxiv.org/abs/2408.01941. Accessed 22 June 2025

Pan J, Zhou Z, Wang J, Zhang P, Yu J (2022) Development of a penguin-inspired swimming robot with air lubrication system. IEEE Trans Ind Electron 70:2780–2789. IEEE

Paulk A, Gilbert C (2006) Proprioceptive encoding of head position in the black soldier fly, Hermetia illucens (L.)(Stratiomyidae). J Exp Biol 209:3913–3924. Company of Biologists

Pechenik JA (1999) On the advantages and disadvantages of larval stages in benthic marine invertebrate life cycles. Mar Ecol Prog Ser 177:269–297

Pettigrew JD, Collin SP, Ott M (1999) Convergence of specialised behaviour, eye movements and visual optics in the sandlance (Teleostei) and the chameleon (Reptilia). Curr Biol 9:421–424. Elsevier

Pouydebat E, Bardo A (2019) An interdisciplinary approach to the evolution of grasping and manipulation. Oxford University Press UK. Biol J Linnean Soc. https://academic.oup.com/biolinnean/article-abstract/127/3/535/5521438. Accessed 10 May 2025
Pouydebat E, Fragaszy D, Kivell TL (2014) Grasping in primates: for feeding, moving and human specificities. BMSAP 26:129–133
Pouydebat E, Boulinguez-Ambroise G, Manzano A, Abdala V, Sustaita D (2023) Convergent evolution of manual and pedal grasping capabilities in Tetrapods. In: Bels VL, Russell AP (eds) Convergent evolution. Springer International Publishing, Cham, pp 323–389
Prusty AD, Sane SP (2024) The motor apparatus of head movements in the Oleander hawkmoth (Daphnis nerii, Lepidoptera). J Comp Neurol 532:e25577
Ramdya P, Ijspeert AJ (2023) The neuromechanics of animal locomotion: from biology to robotics and back. Sci Robot 8:eadg0279
Ray CE (1976) Geography of phocid evolution. Syst Zool 25:391–406. Society of Systematic Zoology
Ray L, Medeiros D (2023) Linking vertebrate gene duplications to the new head hypothesis. Biology 12:1213. MDPI
Reghem E, Tia B, Bels V, Pouydebat E (2011) Food prehension and manipulation in Microcebus murinus (Prosimii, Cheirogaleidae). Folia Primatol 82:177–188
Reghem E, Chèze L, Coppens Y, Pouydebat E (2013) Unconstrained 3D-kinematics of prehension in five primates: lemur, capuchin, gorilla, chimpanzee, human. J Hum Evol 65:303–312. Elsevier
Renous S, Bels V (1993) Comparison between aquatic and terrestrial locomotions of the leatherback sea turle (Dermochelys coriacea). J Zool 230:357–378
Renous S, Lescure J, Gasc J-P, Bels V (1989a) Intervention des membres dans la locomotion et le creusement du nid chez la tortue luth. Amphibia-Reptilia 10:355–369. Brill
Renous S, Rimblot-Baly F, Fretey J, Pieau C (1989b) Caractéristiques du développement embryonnaire de la Tortue Luth, Dermochelys coriacea (Vandelli, 1761). Ann Sci Nat Zool Biol Anim:197–229
Renous S, de Broin F d L, Depecker M, Davenport J, Bels V (2007) Evolution of locomotion in aquatic turtles. In: Biology of turtles. CRC Press, pp 111–152
Richards HL, Adams JW, Evans AR (2023) Hanging on and digging deep: comparative forelimb myology of the koala (Phascolarctos cinereus) and common wombat (Vombatus ursinus). Zool J Linn Soc 199:60–82. Oxford University Press UK
Rind FC, Bels V (2024) Prey-predator interactions. Frontiers Media SA. Front Behav Neurosci. https://www.frontiersin.org/articles/10.3389/fnbeh.2024.1367484/full. Accessed 5 May 2025
Rouse GW, Pleijel F (2001) Polychaetes. Oxford University Press
Rojas-Quintero JA, Rodríguez-Liñán M d C (2021) A literature review of sensor heads for humanoid robots. Robot Auton Syst 143:103834. Elsevier
Ruffier F, Viollet S, Amic S, Franceschini N (2003) Bio-inspired optical flow circuits for the visual guidance of micro air vehicles. In: Proceedings of the 2003 international symposium on circuits and systems, 2003. ISCAS'03. IEEE, p III
Ruppert EE, Fox RS, Barnes RD (2004) Invertebrate zoology: a functional evolutionary approach, 7th edn. Thomson Brooks/Cole, Belmont
Ryczko D, Simon A, Ijspeert AJ (2020) Walking with salamanders: from molecules to biorobotics. Trends Neurosci 43:916–930. Elsevier
Ryu HX, Kuo AD (2021) An optimality principle for locomotor central pattern generators. Sci Rep 11:13140. Nature Publishing Group UK London
Šabanović S (2010) Robots in society, society in robots: mutual shaping of society and technology as a framework for social robot design. Int J Soc Robot 2:439–450
Said S, Youssef K, Prasad B, Alasfour G, Alkork S, Beyrouthy T (2024) Design and implementation of Adam: a humanoid robotic head with social interaction capabilities. Appl Syst Innovat 7:42. MDPI
Schiebel PE, Maisonneuve MC, Diaz K, Rieser JM, Goldman DI (2020) Robophysical modeling of bilaterally activated and soft limbless locomotors. In: Vouloutsi V, Mura A, Tauber F, Speck T,

Prescott TJ, Verschure PFMJ (eds) Biomimetic and biohybrid systems. Springer International Publishing, Cham, pp 300–311

Schulze A, Rice ME (2002) Molecular phylogeny of Sipuncula. Mar Biol 141:1–10.

Shigeno S, Tsuchiya K, O'Dor RK, Kuwasawa K (2008) Cephalopod brain development: conserved patterns and unique features. Zool Sci 25(2):123–134. https://doi.org/10.2108/zsj.25.123

Suzuki T, Nakamura M, Ueda K (2024) Dynamic wing displays facilitate intrasexual competition in birds. J Avian Biol 55(1):e03245. https://doi.org/10.1111/jav.03245

Semini C, Barasuol V, Goldsmith J, Frigerio M, Focchi M, Gao Y, Caldwell DG (2016) Design of the hydraulically actuated, torque-controlled quadruped robot HyQ2Max. IEEE ASME Trans Mechatron 22:635–646. IEEE

Shen N (2023) How crabs walk and what can robots learn from it?

Shen Y, Harada N, Katagiri S, Tanaka H (2020) Biomimetic realization of a robotic penguin wing: design and thrust characteristics. IEEE ASME Trans Mechatron 26:2350–2361. IEEE

Shen Y, Ding Z, Wang X, Mao Z, Huang Z, Chen B (2025) Biomimetic hydrofoil propulsion: harnessing the propulsive capabilities of sea turtles and penguins for robotics. Biomimetics 10:272. MDPI

Soeiro G, Mendes da Silva E, Leduc AO (2022) Sea turtle hatchlings can distinguish between coastal and oceanic seawaters. J Exp Biol 225:jeb244702. The Company of Biologists Ltd

Song A (2021) Neuromuscular control strategy during object transport while walking: adaptive integration of upper and lower limb movements. Louisiana State University and Agricultural & Mechanical College

Stanton LA, Sullivan MS, Fazio JM (2015) A standardized ethogram for the felidae: a tool for behavioral researchers. Appl Anim Behav Sci 173:3–16. Elsevier

Stein W (2022) Sensory input to central pattern generators. In: Jaeger D, Jung R (eds) Encyclopedia of computational neuroscience. Springer New York, New York, pp 3082–3090

Stone J, Mitrofanis J, Johnstone DM, Robinson SR (2023) Twelve protections evolved for the brain, and their roles in extending its functional life. Front Neuroanat 17:1280275. Frontiers Media SA

Sun X, Wang F, Xu J (2021) A novel dynamic stabilization and vibration isolation structure inspired by the role of avian neck. Int J Mech Sci 193:106166. Elsevier

Suryanarayana SM, Robertson B, Grillner S (2022) The neural bases of vertebrate motor behaviour through the lens of evolution. Philos Trans R Soc B Biol Sci 377:20200521

Sustaita D, Pouydebat E, Manzano A, Abdala V, Hertel F, Herrel A (2013) Getting a grip on tetrapod grasping: form, function, and evolution. Biol Rev 88:380–405

Tamietto M, de Gelder B (2010) Neural bases of the non-conscious perception of emotional signals. Nat Rev Neurosci 11(10):697–709. https://doi.org/10.1038/nrn2889

Tassy O, Dauga D, Daian F, Sobral D, Robin F, Khoueiry P, Salgado D et al (2010) The ANISEED database: digital representation, formalization, and elucidation of a chordate developmental program. Genome Res 20(10):1459–1468

Tamborini M (2021) The material turn in the study of form: from Goethe to morphology. Perspect Sci 29(5):643–665

Tieck JCV, Steffen L, Kaiser J, Roennau A, Dillmann R (2018) Multi-modal motion activation for robot control using spiking neurons. In: 2018 7th IEEE international conference on biomedical robotics and biomechatronics (biorob). IEEE, pp 291–298

Tinbergen N (1963) On aims and methods of ethology. Zeitschrift für Tierpsychologie 20:410–433

Trites AW (1991) Does tagging and handling affect the growth of northern fur seal pups (Callorhinus ursinus)? Can J Fish Aquat Sci 48:2436–2442

Tzahor E (2015) Head muscle development. In: Brand-Saberi B (ed) Vertebrate Myogenesis. Springer Berlin Heidelberg, Berlin/Heidelberg, pp 123–142

van Baaren J, Candolin U (2018) Plasticity in a changing world: behavioural responses to human perturbations. Curr Opin Insect Sci 27:21–25. Elsevier

van der Geest NJJ (2023) Unlocking the aquatic ballet of sea turtles: Harnessing Sea turtle biomechanics for the evolution of marine technologies. PhD Thesis. Auckland University of Technology

van der Geest, N., Garcia, L., Borret, F., Nates, R. & Gonzalez, A. (2023a) Soft-robotic green sea turtle (Chelonia mydas) developed to replace animal experimentation provides new insight into their propulsive strategies Sci Rep 13, 11983. Nature Publishing Group UK London
van der Geest N, Garcia L, Nates R, Borrett F (2023b) New insights into sea turtle propulsion and their cost of transport point to a potential new generation of high-efficient underwater drones for ocean exploration. J Mar Sci Eng 11:1944. MDPI
Van Der Leeuw AH, Bout RG, Zweers GA (2001) Evolutionary morphology of the neck system in ratites, fowl and waterfowl. Neth J Zool 51:243
van Bijlert PA, Geijtenbeek T, Smit IH, Schulp AS, Bates KT (2024) Muscle-driven predictive physics simulations of quadrupedal locomotion in the horse. Integr Comp Biol 64(3):694–714. https://doi.org/10.1093/icb/icae095
Virameteekul S, Bhidayasiri R (2022) We move or are we moved? Unpicking the origins of voluntary movements to better understand semivoluntary movements. Front Neurol 13:834217. Frontiers Media SA
Venkateswaran S, Chablat D, Boyer F (2019) Numerical and experimental validation of a robust design methodology for parallel manipulators. Robotics 8(2):32. https://doi.org/10.3390/robotics8020032
Wainwright PC, Bellwood DR (2002) Ecomorphology of feeding in coral reef fishes. In: Sale PF (ed) Coral reef fishes: dynamics and diversity in a complex ecosystem. Academic Press, San Diego, pp 33–55. https://fishlab.ucdavis.edu/wp-content/uploads/sites/397/2020/06/Wainwright-Bellwood-2002.pdf
Wainwright PC, Bellwood DR, Westneat MW (2002) Ecomorphology of locomotion in labrid fishes. Environ Biol Fishes 65:47–62. https://doi.org/10.1023/A:1019671131001
Wang X, Li S, Chang J-C, Liu J, Axinte D, Dong X (2024) Multimodal locomotion ultra-thin soft robots for exploration of narrow spaces. Nat Commun 15:6296. Nature Publishing Group UK London
Wei X, Yan T, Liu J, Liu S, He B, Sun T (2024) Study on the interaction characteristics of hull-propeller-appendages of autonomous underwater vehicle based on unsteady conditions. Ocean Eng 312:119087. https://doi.org/10.1016/j.oceaneng.2024.119087
Whishaw IQ, Karl JM (2019) The evolution of the hand as a tool in feeding behavior: the multiple motor channel theory of hand use. In: Bels V, Whishaw IQ (eds) Feeding in Vertebrates. Springer International Publishing, Cham, pp 159–186
Wilkie RM, Wann JP (2003) Eye-movements aid the control of locomotion. J Vis 3:3–3. The Association for Research in Vision and Ophthalmology
Xu Y, Liu C, Cui H, Song Y, Yue X, Feng L, Wu L (2023) Environment perception with chameleon-inspired active vision based on shifty behavior for WMRs. Appl Sci 13:6069. MDPI
Xu J, Mawase F, Schieber MH (2024) Evolution, biomechanics, and neurobiology converge to explain selective finger motor control. Physiol Rev 104:983–1020
Yan Z, Song Y, Zhou R, Wang L, Wang Z, Dai Z (2024) Facial expression realization of humanoid robot head and strain-based anthropomorphic evaluation of robot facial expressions. Biomimetics 9:122. MDPI
Ye S, Yu X, Chen H, Zhang Y, Wu Q, Tan H, Song J, Saqib HSA, Farhadi A, Ikhwanuddin M (2022) Full-length transcriptome reconstruction reveals the genetic mechanisms of eyestalk displacement and its potential implications on the interspecific hybrid crab (Scylla serrata♀\times S. Paramamosain♂). Biology 11:1026. MDPI
Young JW (2023) Convergence of arboreal locomotor specialization. In: Bels VL, Russell AP (eds) Convergent evolution: animal form and function. Springer International Publishing, Cham, pp 289–322. https://doi.org/10.1007/978-3-031-11441-0_13
Zhang X, Shu D (2021) Current understanding on the Cambrian explosion: questions and answers. PalZ 95:641–660. Springer
Zhang X, Chang C, Cui L, Qiao Y (2021) Ecosystem reconstruction during the Cambrian explosion. Paleontol Res 25:305–314. BioOne
Zhao Z-D, Zhang L, Xiang X, Kim D, Li H, Cao P, Shen WL (2023) Neurocircuitry of predatory hunting. Neurosci Bull 39:817–831

Chapter 11
How to Tame Analogies for Use in Biomimetics

Julian F. V. Vincent

Abstract The underlying theme is how, and at what level of complexity, can useful analogies be established between biology and engineering. The search is for a pattern that is common to both technology and biology. The belief is that such a pattern exists. The basic pattern of biology is established by cells. They assemble to form an organised hierarchy, separating into organs, specialising as functional units, semi-autonomous. Through evolution, organisms achieve great efficiency, reliability and adaptiveness. Some of the changes and development involved can be traced in trade-offs that commonly allow the mechanisms behind adaptiveness to be dissected and understood. For the classification of trade-offs, we turn to Genrich Altshuller who, in common with all Russians, was taught that dialectics (of which trade-offs are a form) expressed as thesis and antithesis are universal, so are their syntheses. Based on this, he invented TRIZ, a successful problem-solving technique primarily used in technology. Using his techniques, it is possible to express biological systems in the same terms as technology and fuse the two disciplines.

Keywords Trade-off · Dialectics · Analogy · Solving problems · Creativity

11.1 Introduction

> ... to understand what life is, we must view it at a variety of different levels, all interacting with each other in a complex web. It is that emergent web, full of feedback between levels, from the gene to the wider environment, that is life. (Noble 2008)

In an essay exploring possible ways of extracting technically relevant concepts and innovations from biology, it helps to have a list of the basic characteristics of the 'source book'. On which page might we look for an answer to this or that technical problem? This list of the (possible) working components of living systems

J. F. V. Vincent (✉)
Heriot-Watt University, Edinburgh, UK
e-mail: J.Vincent@hw.ac.uk

V. Bels et al. (eds.), *Biomimicry – A Functional Approach*, Biologically-Inspired Systems 20, https://doi.org/10.1007/978-3-032-11949-0_11

is adapted from Denis Noble (2021), one of the leading lights in our understanding of cellular biology, evolution and basic mechanisms of the cell. By 'cell' is meant, unless otherwise stated, the eukaryotic cell. That is, the type of cell that is the basis of animals, plants and fungi as commonly known.

1. *Functionality is multilevel.* Everything happens everywhere (not all at once, though probably along parallel paths) as a continuous process. Although it's possible to trace the course of a chemical process—say, the capture of energy from glucose—it isn't always possible to say what controls it. This is partly because ...
2. *Transmission of information is not one way.* There is an order of process which could be said to start with the formation of an enzyme using information provided by DNA, but this is not a hierarchy in terms of 'command'. In reality, the concept of ...
3. *The Central Dogma, that there's one molecule to rule them all (DNA), is compelling but naive.* As analogy, in any group of people you never know from where the next brilliant idea will come. So it is in the cell. The 'top-down' model forces a path through the levels of the system until it reaches a molecular limit. But work back up again and there's no guarantee you'll take the same route in reverse. In a large and complex network there can be many equivalent paths. What's more ...
4. *DNA is not the sole transmitter of inheritance*, since ...
5. *There is no privileged level of causality.* The cell is a co-operative system in which decisions can come from any level, in turn implying that ...
6. *There are no 'programs' at any level.* Without planning there can be no instructions. That doesn't mean to say that ...
7. *There may be 'pattern books' (e.g. genetic information)*, equivalent to a library of 'what works', but at present we have little idea of what controls their use. The generation of a sequence to template a particular protein can involve segments ('exons') of DNA from almost anywhere on the genome. This means that the central pattern of a protein such as collagen—a triple helix that induces collagen molecules to assemble into a structure—appears in some 30 different types of collagen in different tissues and membranes ranging from the tough 'mermaid's purse' of sharks to the transparent cornea of the eye. The sum of these observations means that, for the present at least ...
8. *Survival can be the only quality criterion.* Since conditions for survival can change from moment to moment ...
9. *There is no genuine 'theory of biology'.*

These observations and deductions are not presented as guidelines but as a warning. All sciences must go through this phase of alchemical pragmatism where data is accumulated with the essential patterns still obscure. Biology can be no different. This should not be seen as a constraint. In fact, it is the very opposite. There may be 100 or 1000 ways of assembling a biochemical pathway, but for various reasons such as cost or reliability, a selection of less productive pathways may be the biological survivors. Under technical conditions—a reaction vessel, say—the limiting factors are likely to be different and one or more of the other available pathways might be preferable.

For the ambition of transferring mechanisms from biology to technology, this does not seem very hopeful. Direct transfer is possible, given current technology, only at a basic level. This is quite sufficient to support a vibrant biochemicals industry. Developing methods and techniques for higher levels of biological complexity needs novel techniques that promise a technology of adaptive change driven by exchange of information at any and all levels of organisation.

11.2 Cells, Their Evolution and Development

The basic unit of life is the cell, a hierarchical composite structure, of which there are at least three types. This account will concentrate mostly on the eukaryotic (from the Greek meaning true nucleus) cell as being the basic unit of plants and animals. Since this type of cell is the building block of nearly all the multicellular organisms and functions modelled by biomimicry, it is described in sufficient detail to suggest areas of interest even at this nanometric level. We need to go back further to glimpse their possible origins amongst the prokaryotic cells such as bacteria, which have no defined nucleus (i.e. there is no defined or confining membrane around the molecules normally found to be in the nucleus of the eukaryote cell).

Nobody knows how or where cells emerged. Early evidence comes from molecules that may be associated with 'living' systems detected in a few early rocks of the order of 4 billion years ago. A slightly earlier time has been assessed using the genetic 'clock' of changes in DNA plus (informed) guesswork. Eukaryotes arose from prokaryotes when they acquired organelles (nucleus, mitochondria, Golgi, etc.) as invasive, independent prokaryotes that shared their metabolism with their new host. When proto-chloroplasts joined the fray, energy metabolism largely changed from the easy but low-energy and wasteful anoxic fermentation to photosynthesis which, with help from the other organelles, could turn readily available resources such as carbon dioxide, water and sunlight into sugar. This was about 3.2 billion years ago and set the tone for living things to function due to the symbiotic interaction of a number of specialist units. Between 3.2 and 2.2 billion years ago modern biological metabolism developed within single cells and the atmosphere changed to oxygenic. This is the basic chemistry of our modern world.

Eukaryotes are complex. DNA, mostly in the nucleus, is a pattern library that, in Man, is a single molecule, 2 m long, with some three billion base-pairs capable of templating the 100,000 or so types of protein that make up our body. Reading the pattern to generate those proteins is very controlled and non-linear: the 'one gene one protein' mantra is illusion. There is no such thing as a dedicated gene for this or that. DNA provides patterns, but the way those patterns are linked to generate a particular protein (which may need 100 or more separate patterns from the DNA) is obscure to say the least. This implies that a single pattern (an exon) might act for several different types of protein with different functions. Unravelling how this works—how the selections are made—is the goal of the Gene Ontology project, the underpinning of modern medicine. Just as interesting is the degree of accuracy achieved by the data-checking mechanism (details currently obscure) which proofreads newly formed 'raw' DNA with an error-correction improvement of two to three orders of magnitude (Bebenek and Ziuzia-Graczyk 2018). Detection of errors is largely due to the DNA existing as a mirrored pair of linear molecules—the double helix—though which of the pair, and when, acts as the reference is not known. The structure of the required protein is then read off the DNA sequence and the amino acids assembled in order.

Thus we have, in the eukaryotic cell 2.2 billion years ago, a tiny unit containing thousands of different proteins, their patterns and all the other components, such as lipids, sugars, phenols, that allow it to exist independently. It is a unit capable of living on simple inputs and can grow, repair and replicate itself. It can sense its surroundings, defend itself, control its shape and secrete the chemicals produced by its metabolism to make external structures. It needs no external control or assistance. But note that the traffic of information is not solely bottom-up from the gene factory to the functioning cell. It is also top-down as the biochemical network signals and controls cell can sense its surroundings and change as necessary. The cell not only adapts, but in the evolutionary quest for reliability has back-up systems for each attribute.

11.3 Biology Is Different from Technology

The independence of cells and their ability to adapt as their environment changes needs to be emphasised. These two graphs (Fig. 11.1a, b; courtesy of the Royal Society (Vincent et al. 2006)) were derived from a survey of the importance of six parameters in the solution of problems presented (mostly) by environmental factors. The parameters are Substance (raw and recycled materials—the basic resource), Structure (shape and manipulation of shape at all length scales), Energy (all sources—mainly sunlight and sugars though many types of cell live in anoxic darkness), Space (to live in and to rearrange according to expedience), Time (includes periodicity, recurring events, etc.) and Information (whatever change-making effects can be passed to a cell from the external environment or from other cells, by whatever medium). The vertical axis is the proportion (0–1) of problems solved by changing, in whatever way and by whatever means, the particular parameter. The horizontal axis is logarithmic and refers to the size of the object (organism—part or whole, object, environmental unit, etc.). The dividing lines are necessarily rather imprecise, since although the data sample was in the tens or more at each inflection point, there was no easy way of estimating confidence limits. But the differences between the two graphs are strong enough for them to be useful.

The graph of technology is fairly obvious—to solve the majority of the problems raised by our existence in the largely artificial world in which we live we mostly

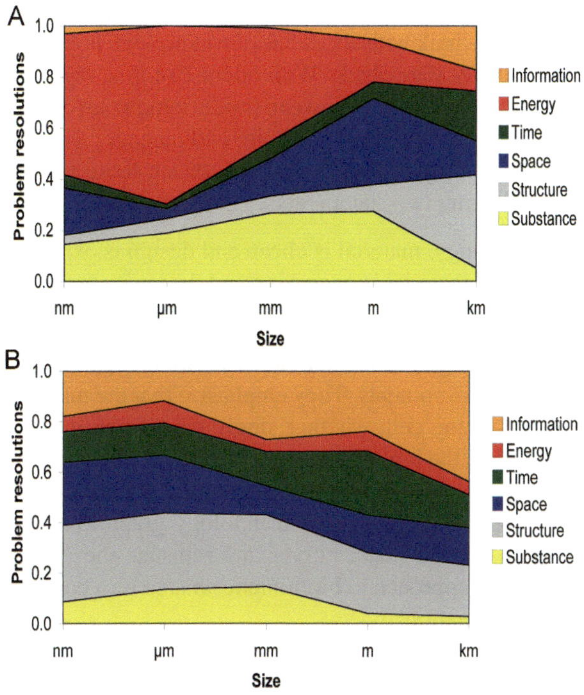

Fig. 11.1 (**a**) (Upper): Engineering solutions arranged according to size/hierarchy. (**b**) (Lower): Biological effects arranged according to size/hierarchy. (With permission from the Royal Society)

use Energy to process materials (at around the micrometre scale) such as steel and concrete and are very reliant on raw materials. Information entrained in the structure of raw materials is removed in order to obtain a homogeneous state. The mechanical performance of structures is primarily due to the basic properties of the component materials, rather than Structure at the molecular level or higher.

The biology graph is rather more subtle. Imagining the probable confidence limits on the raw data, most of the semi-horizontal lines are probably straight. The most rational interpretation is that not much changes over the entire size range. The main actors using and producing Information, the most prominent factor, are much the same size and importance, and their modes of generating, receiving and processing Information of all sorts are constant, starting at the smallest and increasing with the size of the organism. It suggests a single class of actor, most likely the cell, which are aggregated into tissues and organs and can still be independent effectors. This leads to the second most important factor, Structure, suggesting that organised shape, internal and external, natural and manufactured, is a product of behaviour consistently at all sizes. Materials, raw and (re)processed are less important, suggesting that the performance of biological 'materials' is controlled by their (manufactured) Structure. Necessary characteristics such as rigidity, durability and repairability are derived from Structure at all levels since there are very few (if any) organically produced 'materials' that are not composites, even down to the molecular level (Vincent 2013). Thus mother-of-pearl, made mainly of chalk with about 5% of wet protein, is orders of magnitude stiffer, stronger and tougher than chalk. With talent like this, organisms mostly do not need high-performance minerals and can produce high performance from locally available raw materials that are easy to acquire and manage.

The characteristic difference between these two graphs can be summarised by the following two statements:

In technology, material is cheap and design is expensive.
In nature, material is expensive and design is cheap.

This mantra, or aide memoire, has proved useful for architects and designers.

The two graphs make a rather brutal comparison between the manufactured and natural environments. They emphasise that the main actor in the natural world—our world—is the cell, whether singly or constituting animals and plants of all sizes. Over time, the biology graph has yielded to the technology graph. Although it is not possible to put history into reverse, we can pick out the major differences and see how we can transform the technology graph to look more like the biology one—smooth out the major curves and reinstate the overall left-to-right gradients. The analytical approach to biomimetics has to be a component in the realisation of such a transformation.

11.4 Cells Form Multicellular Organisms

Around 1.6 billion years ago, eukaryote cells were associating to form recognisable multicellular clusters. This happened in several lineages amongst plants, but only one lineage established animals—the fungus. Extant fungus such as slime moulds can assemble and develop reproductive organs, demonstrating differentiation and mobility. This seems to have happened around the era of the Great Oxygenation when the biochemistry of photosynthesis produced vast amounts of oxygen, a poison.

We can see the basic mechanisms of the transformation from uni- to multicellularity to this day (Bozdag et al. 2023). Experimentally yeast cells can form aggregations several millimetres across, which will grow and reproduce by budding. They are held together mechanically by cellular extensions. In some instances, adhesives are used, plus transmembrane type II collagens. Such stable aggregations, developed from fungi, became the sponges (Muller 2003). Immunity and apoptotic systems developed so that the individual cells knew they were part of a uniform community and could eject or destroy interlopers such as parasites. And so sponges transitioned from a colonial stage to integration and each aggregation effectively became an individual. Cells and structures specialised, moving the surrounding water along channels throughout the mass of cells, distributing food and oxygen, or reproducing, and so speciated.

Cellular differentiation within these individuals proceeded apace, and by 600 million years ago there was a huge expansion (probably not the first) of types of organism recognisable in the modern era of animals and plants. This expansion involved some more basic principles. The outcome was that eukaryote cells in a plant, animal or fungus exist quasi-independently and develop into organs with minimal central control, enabling them to optimise their disposition. Further expansion required mainly the development of organisational techniques based mainly on design and structure. This allowed organs to refine their activities independent from the rest of the body; expand on their specialisations; be switched on and off as required (hence the apparent disappearance of chosen functions 'out of season'); etc. The way became open for more sensitive control and integration of the organs and their activities.

The main message of this introduction is to show that the complexity of natural systems arose very early on in evolution. The cell is a system. Placing it in a hierarchy of systems allows it to specialise. But the basic biochemical pathways remain the same, often appearing more simple in evolutionarily older groups of organisms. Thus, if you want to base a technical idea on a biological organism or system, you might be advised to use the simplest organism to initiate your concept. Sophistication can still be your downstream target. A convincing case arises with *Lactococcus* cultured under various conditions (Bachmann et al. 2013).

Lactococcus is a genus of bacteria used in making fermented dairy products such as cheeses. Its main purpose in dairy production is the rapid acidification of milk which prevents the growth of spoilage bacteria. In a series of experiments, two environments were used to culture the bacteria: unstructured and structured. Growth in an unstructured suspension culture favours fast-growth and can be compared with a (properly) socialist society in which the availability and cost of energy is the same for all providing freely accessible public goods, whereas spatial structure is of importance for the evolution of high-yield strategies. Spatial structure was achieved with a water-in-oil emulsion: individual bacteria were thus isolated in droplets of water suspended in oil, creating a structured environment where the growth-limiting substrate is privatised for populations founded by individual cells. This leads to fast-growing and multiplying organisms in the unstructured culture which may have unused by-products which constitute a waste stream. Mutants that arise in the structured environment have a higher affinity for the substrate and use resources more efficiently under conditions of competition but grow more slowly. Thus, we have a trade-off between high growth (often associated with invasive organisms using a neglected but freely available resource) and high yield (per unit of resource). This demonstrates the effect of privatising a public good on the evolutionary outcome between competing metabolic strategies. It allows the investigation of fundamental questions in biology such as the evolution of cooperation, cell–cell interactions and the relationships between environmental and metabolic constraints. Not only does this example illustrate a simple analogy but shows that analogy can be scale-free. It conforms to basic interactions shown by (simple) examples of economics (though there may be oligarchs amongst the bacteria) comparing a co-operative structure with privatisation.

Perhaps this opens up an area of experimental biomimetics where unicellular organisms are cultured in the laboratory under conditions relevant to a social or economic model, manipulated to resemble real-world conditions. This is analogous to Bill Phillips' water-computer (Bissell 2007) in the UK Treasury.

11.5 We Have a Problem . . .

> If I were given one hour to save the planet, I would spend 59 minutes defining the problem and one minute resolving it. (Attrib. Einstein)

The significant omission in most methods of biomimetics is a definition of the problem in question. And since the main user of the technique is going to be an engineer, probably with little knowledge of biology, the problem itself will be an engineering problem rather than a biological one; much effort will be saved if the method is familiar to engineers. Better still, if the problem can be couched in neutral terms, it will serve the bridge between biology and engineering. Even so, as Einstein (or one of his many avatars) was aware, this can be the most difficult aspect of any investigation, and it is crucial.

There are at least two criteria before we start to attempt resolution:

1. We need the technique to be sure that we have identified the root of the problem.
2. The problem must be expressed neutrally with no reference to context, wet, dry, large or microscopic.

This ideal starting point is only rarely seen in the literature. When the researcher has a novel problem to solve, one of their first reactions is to investigate and organise any available information. This commonly starts with the construction of a database that can range from a simple list of words—a vocabulary or thesaurus—to a linked list or to a Simple Knowledge Organisation System (SKOS). A database enables acquisition, storage, management and sharing of data and information and is the simplest and shortest path to a measurable outcome, but not necessarily to a valuable outcome. But a database is a 'closed world' in which if a topic is not listed, it doesn't exist and there is no mechanism to question an apparent omission (Glick 2013).

Unfortunately the database is the most common method purporting to describe and summarise aspects of a biological system, often with little or no mention of the effect or change provided by individual components, nor with the integration of those 'components' into the workings of the organism. The hierarchy of the biological system is thus provided and tagged, but biological anatomy and hierarchy are complicated and subtle and a word list is not sufficient for understanding. Without information of 'services rendered'—disposition and integration—the list is *useless*. Worse than that, the technologist who wishes to extract a principle direct from a comparatively novel and otherwise alien system, in a form that can be incorporated into apparently unrelated engineering, finds little or no help in the language of biology. The biologist is then retired from the project team of engineers, who continue as engineers, steering the project down paths familiar to them. A source of possible novelty is in danger of being discarded with the comment that it's unrelated and irrelevant: 'Interesting but trivial'. This is a very common—almost the normal—state of affairs.

The first step should be to isolate processes and establish connections with a view to abstracting disposition. A SKOS can provide help in this direction, but is not sufficiently detailed without more lexical analysis (Miranda et al. 2016). Similarly lacking in sufficient detail, large language models and other AI techniques may help, but they must be used with skill to make them reliable. Biological texts are often complex; important detail is glossed over or, more probably, not recognised. For the moment such texts may be better analysed individually by a domain expert. The case for developing a semantic tool for a more complete and integrated system—an ontology—is not helped by databases that are presented in the form of an ontology but lack dynamics. To be of use the items listed in the database must be linked in such a way that one can ask of them questions such as 'What does it do?'; 'Is it unique?'; 'What else will do the same?'; 'How will it react under ... conditions?' and so on. A half-way step is provided by lexical analysis (Vakili and Shu 2001) and natural language processing (NLP) (Cheong and Shu 2014).

An ontology will necessarily require information from one or more ordered sources, but is often more concerned with the definition of concepts and inter-relationships, concentrating on the adequacy and effectiveness of the methods of abstraction, thus establishing a reliable, standardised and precise vocabulary, revealing its reliance on philosophy (Yargan and Jansen 2024). Certainly precision is important for reliability. With care and structural uniformity an ontology can be interoperable with other ontologies using the same basic structure. Unfortunately, the creation of a conventional ontology is not easy, especially if it is to be linked to a series of definitions and solutions to problems in biomimetics. One answer is to break ontological convention and to include something broader—perhaps SKOS.

11.5.1 ... and Can Solve It

> A creative mind ...
> loves contradictions and non sequiturs;
> searches for connections between remote objects and processes that seem to be unconnected;
> uses deduction by analogy, moving between systems using induction, deduction and traduction;
> combines logic and intuition, inventing rules for thinking.
>
> *Remember these.* (Terninko et al. 1998)

The philosophers still have a part to play.

In the history of human thinking there have been two conceptions concerning the development of the universe—the idealistic and the materialistic. The union of these two is the dialectic. Materialist dialectics is based on contradiction, or as Lenin put it: 'The splitting of a single whole and the cognition [analysis] of its contradictory parts'. Lenin is referring to Hegel's 'thesis' and 'antithesis'. In simple terms, if you have a problem to solve (thesis) its solution can be retrieved by recognising a contradiction (antithesis). The resolution (synthesis) eliminates the contradiction to give the result you need. If your definition of the Contradiction is ideal, then there is a good chance that the synthesis will also be ideal (Savransky 2000). But as far as I know the ideality of this approach has not been proved analytically, although it survives as a much-defended article of faith. An intriguing aspect of this approach to solving problems is that the antithesis does not always have to be apparent.

Conceptually the model is a triangle with the third side defined by the length and relationship between the other two. Thus, it's quite acceptable to invent the antithesis, in consideration of the thesis and the target of an ideal result, and to proceed with the analysis. Whether this will work with biological models has not been explored.

One of the main characteristics of the successful—that is, fully resolved—Contradiction is that the pre-existing properties are maintained or improved. For this to happen, it's necessary that the two halves of the Contradiction are not, or only weakly, linked. Indeed, one of the characteristics of the Contradiction is that either half of the dialectic can be identified as the thesis, so that the Contradiction so defined will have two resolutions. Finally, a supporting analytical framework is provided by TRIZ (a dialectic method for the inventive solution of problems). That this technique should be Russian comes as no surprise when it is realised that dialectical materialism is at the heart of Russian thinking and is taught in primary schools. TRIZ was adapted from Hegel's dialectic by Genrich Altshuller (1999). From the examination of many 'strong' (i.e. inventive and often counter-intuitive) patents the main characteristics of the problems, already classified in dialectic form, were sorted into groups according to their main characteristics. Traditionally there are 39 of these groups, usually called the technical parameters. Since each Parameter in a dialectic pair can be the thesis or the antithesis, up to 39×39 problems can be tabbed, each providing a theoretically unique Contradiction. Resolution of the Contradiction is achieved by implementing one or more changes based on those suggested by the appropriate strong patent. These changes (the Inventive Principles) are very general and neutral in terms of techniques. I have listed them at the end of this chapter showing technical examples of each Principle together with biological equivalents that I have developed. Conventionally the Parameters are presented in a 39×39 matrix; the vertical axis is thesis, the horizontal one antithesis.

Since thesis and antithesis are interchangeable the entire matrix is populated. Where each combination of thesis and antithesis intersects, a maximum of four relevant Inventive Principles is identified, doubling when both versions of the Contradiction are identified. Thus, the matrix is a tabbed database of problems and their solution in rather general terms. The primary purpose of the matrix is not to solve a problem so much as to restrict the area most likely to provide a strong solution.

However, if the same Contradiction emerges when thesis and antithesis are swapped, it implies a much closer relation between them, and that they are linked. This is a *routine* design solution in TRIZ, and results in an interaction between the two Parameters so that improvement in one Parameter is accompanied by deterioration (or at least, change) in the other, implying a balance between the two factors. This is what one typically observes in a trade-off, where changing the balance allows the action of the trade-off to be modulated according to circumstance. Thus the biological trade-off is adaptive, whereas the dialectic Contradiction is not. The tradition of TRIZ regards the Contradiction as strong and the trade-off as weak, without robust evidence to support this opinion. A TRIZ master will reject talk of a trade-off out of hand. Yet there is no evidence that a Contradiction is 'strong' other than the strength of any related patent. And the Contradiction matrix has some 25% of routine (i.e. symmetrical) solutions (Vincent 2017), so that it's already 'infected'!

The trade-off has many useful characteristics. Its action can be modelled mathematically and a line or front calculated that follows the theoretical optimum generated by the changing balance between the two halves of the trade-off. This is the Pareto front and is rather more than a map of the range of performance. It traces the range and limit behaviour achieved by the organism (Shoval et al. 2012). The opinion that biology fudges its systems so that their performance is 'good enough' is wrong. Biological systems work to the highest efficiency within limits of safety and reliability (Ng et al. 2019). The Pareto Criterion thus replaces the criterion of classic TRIZ that the Contradiction is an ideal solution. Trade-offs interact with each other and continually adjust. The more trade-offs that interact in a specific case, the less decidable the outcome will be and the less likely that a simple model of two or three factors will be able to predict the behaviour of a particular trade-off embedded within a network of trade-offs (Garland Jr et al. 2022). Indeed, one of the skills necessary for successful experimental biology is to recognise any controlling trade-offs before choosing the organism to be investigated.

This is illustrated by the transport network in plants, comprising the xylem and phloem vessels that connect the photosynthetic system in the leaf with the rest of the plant (Conn et al. 2017).

Data on the geometry of the system were collected using laser scanning and used to calculate the Pareto Front taking the two limit cases of the cost-efficiency trade-off in the morphology of the network: the maximum total distance (highest construction cost) is given by each photosynthetic source having its own route to the plant stem; the minimum distance (lowest construction cost) requires all the sources to sit on a single, rather circuitous, route. The actual network parameters measured on the plant could then be placed in the Cartesian space occupied by the calculated Pareto Front. In nearly all instances (using data from tomato, sorghum and tobacco plants), the plant datum was on the Pareto Front, indicating that the transport network was at a performance maximum. The adaptability of the trade-off was shown by the different positions of the plant datum along the Front when the plants were raised in different regimes of temperature, light and hydration (Fig. 11.2). The physiological significance and causality of this variation along the Front

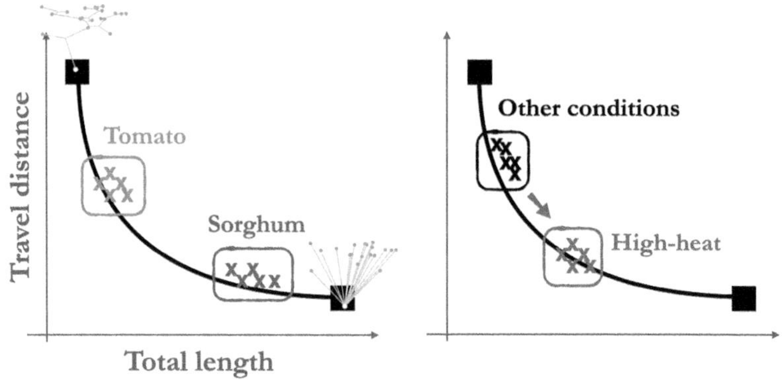

Fig. 11.2 Pareto Front: the optimal trade-off between two common network design measures: minimising construction costs (top left of each figure) and minimising transport distance (bottom right of each figure). The only architectures that will remain in the population are those that lie on the Pareto front, as these architectures cannot be improved along both construction and transport objectives at the same time. Labelled boxes indicate different behaviour of species (left) and a single species subjected to different environmental conditions (right). (Adapted with permission from a diagram by Saket Navlakha)

are yet to be determined, but it illustrates another possible advantage of a biomimetic approach to design: biology can be inherently and predictably adaptive. Moreover, the Pareto Criterion confirms that the trade-off can present an ideal solution even in varying conditions. Can we build such adaptability and performance rating into our technology as a standard design factor?

In the limit, this adaptability has to be delivered by the cells constructing the organism since to be adaptive it must be constructed with reference to conditions external to the cell. This brings up the problem of function (Yargan and Jansen 2024). Since the basis of our interpretation of living organisms requires that a cell cannot have intention, there can be no such phenomenon as a cellular 'function' since this implies knowledge or expectation of a certain outcome resulting from an action by the cell. The result of an action by a cell (secretion of a hormone, for instance) has to be due to natural selection in some way. This is the way technology works. But it means that when describing a chain of actions it is necessary to abstain from 'function' and resort to 'disposition' in order to retain clarity of thought.

However one approaches the definition and resolution of a problem, the trade-off/Contradiction approach seems by far the best way to define biological effects for the benefit of bioinspired design. It has a solid foundation in history, philosophy and mathematics (e.g. multiobjective analysis) and is the core of a method that is used in both biology and technology. Transfer of biological effects to technology becomes a formality. It remains to support it with a procedure that can decide which biological trade-off will provide the synthesis. The answer, of course, is TRIZ.

11.6 Biology Invades Technical TRIZ

Population of a matrix for biomimetics is a long and hard job. One advantage of using published data is that the trade-off Parameters and, quite often, the Inventive Principles that solve the trade-off, have been identified and published by the scientist(s) who did the research, leaving a verifiable paper trail. The equivalent data are not available for the standard TRIZ matrix. The main problem is assessing the quality of these observations and the quality of the interpretations. An even bigger problem is the relative dearth of publications that provide such information.

From a collection of over 16,000 selected biological references in my database, I identified nearly 3000 that had the word 'trade' somewhere in the text. This selection covered all of biology, including *Homo*. Of course, the term 'trade-off' is not the only means of recognition in a biological text. Negative relationship; optimisation; inverse; compromise and several others will do so, but 'trade-off' is by far the most common. These data were reviewed several times with (I hope) increasing skill in distilling the meaning from the words (biology papers are much more difficult that technical papers in this respect). This yielded some 270 'solved' trade-offs. However, biology seems not to generate a wide enough range of trade-offs to populate the entire half-matrix—out of 780 available squares in the matrix—each representing a possible trade-off—only about 500 are achieved and fewer than that number are solved. The remaining trade-offs may be rare or not invoked. It's quite possible that a trade-off can be partially side-stepped by a behavioural mechanism. Some of these are observable such as decision-making by social insects moving to a new site (Marshall et al. 2006), or choice of a nest site that will be less likely to be invaded by predators. More trade-offs must have left their mark on the cellular control systems, but we are aware of only a few. However, the mathematics of the Pareto Front can be used to identify otherwise unseen trade-offs (Hart et al. 2015), so further analysis of this model may increase the range of solutions that nature can offer technology.

11.6.1 Parameters and Principles

For convenience and cross-comparison, the biomimetic versions of the Technical Parameters and the Inventive Principles are organised following the TRIZ headings that have been carefully chosen over the years by Altshuller and TRIZ experts. There is no conceivable reason to change them, and it would only cause confusion. In practical terms, which will be less open to dispute and totally open to modification, the text-book explanations and examples of the individual parameters and inventive principles need to be expanded in order to make the interface between biology and technology clearer. The required degree of expansion is uneven, reflecting and amplifying the differences between the two disciplines. Thus, amongst the

Parameters it is the two which identify threats (disease, predation, starvation, etc.), both internal and external, which need the most biological subcategories—about twice as many as the other Parameters. Unfortunately, the comparable statistics from TRIZ are unavailable because there is no record of the total number or identity of the Parameters from which those in the TRIZ matrix were winnowed, so any data will be biased. A comparison of the inventive principles can be made, with a certain amount of caution, using data from the commercial company Ideation, quoted in Terninko et al. (1998). The top 10 in this list are as follows:

- Change a parameter (change state, concentration, consistency, compliance, temperature)
- Prepare for action (ready to perform or react)
- Segmentation (divide the object or process into similar or different segments)
- Exchange a mechanical effect for non-mechanical (chemical, sensory, light, colour, etc.)
- Extract and separate (isolate, remove, specialise)
- Convert to dynamics (replace stiff materials and concepts with flexible ones)
- Periodic action (replace continuous action with impulses)
- Vibration or pattern (vibrate, oscillate, resonate)
- Colour (interaction with light, transparency, change colour, camouflage)
- Inversion (other way around, change parameter to its opposite).

The comparable list drawn up from the current work is

- Convert to dynamics (related to adaptation and locomotion)
- Feedback (obviously needed in any adaptive control system)
- Prior counteraction (defence, which relates to externally originating harm)
- Convert harm to benefit (including adaptive trade-offs)
- Segmentation (divide the object or process into similar or different units)
- Local quality (make each part of an object fulfil a different and useful function)
- Safety (sometimes called 'prior cushioning'—be prepared for any eventuality)
- Consolidate (merge identical or similar objects to perform parallel operations)
- Composite material (fibres or particles in a matrix, layers, etc.)
- Intermediary (merge two or more objects temporarily, perhaps transferring information or one object changing the other)

Comparing all 40 inventive principles in the two lists gives an overall similarity of their order of significance of about 60%. It appears that biology is not only suggesting different methods for delivering solutions compared to technology, but the class of solution can be very different. Take the first inventive principle in each list as an example. The technical TRIZ recommendation is to change the characteristics or identity of one or more of the constants forming the solution to the problem, whereas biology suggests an adaptive change. Perhaps this list is a focal point for the conversion of Fig. 11.1a to look more like Fig. 11.1b, as suggested at the end of Sect. 11.3.

11.6.2 TRIZ in Biomimetics

In use, the TRIZ method for bioinspired design is no different from standard TRIZ and the relevant inventive principles that can suggest its synthesis looked up in the biological matrix. This works because the biological trade-off, if different from the technical one in the details of its resolution, will point to different inventive principles. The different sources of information and reasoning are not visible to the engineer/technician using the method, only the fact that the suggested solution(s) may go down a different pathway. However, setting up the biological trade-off matrix is a very different exercise. All the Parameters and all the Inventive Principles have to be revised so that they conform to the general description of traditional TRIZ but present a biological vision. It's this transition or translation between the technological language of TRIZ and the biological version that is at the heart of this method.

The following example has its origin in a radio programme about electric vehicles. The problem was, as ever, the batteries, which are always too heavy and don't get lighter as the battery discharges. The limit of weight has been reached with current technology, so the battery needs greater charge density, but that will make charging time longer. But there's also a limit on the minimum charging time since charging heats the battery, threatening its structure, reducing the charging rate and the charge density. The contradiction/trade-off has to be defined. The factors can be broken down like this:

A. Nett energy content too low (it always is!)
B. (1) Thermal condition; (2) external source of harm

This yields two contradictions: [A, B1]; [A, B2]. For both cases, A is classed as parameter [20], energy available to a fixed object. Thermal condition is parameter [17], and external source of harm is Parameter [30].

The standard TRIZ matrix has a blank square where [17] and [20] cross and gives the following IPs for [20]:[30]

Extraction (extract only the necessary part or property)
Prepare for action (arrange objects so that they can go into action)
Convert harm into benefit (increase harm until it becomes beneficial)
Thermal expansion (use metals with different coefficients of expansion)

Only four inventive principles are available here since this combination of Parameters turns out to define a *routine* solution. The suggestions in brackets are the detailed suggestions deduced from the research paper, interpreted to conform with the definitions of the various inventive principles. Extraction could refer to removing excess heat or introducing a cooling system, which would be a preparation for action. This would allow a higher rate of charging at constant temperature, eliminating the need to take thermal expansion into consideration.

Turning to the biomimetic solution, there is no solution to the pair [17]:[20], but luckily one of the 270 trade-off cases so far parsed responds to the Parameter pair [20]:[30]. It concerns a green alga, a microscopical plant (Zhu et al. 2016). The analogy with the battery arises because the alga gets its energy from sunlight but suffers from animals (fish) eating it, which represents the external source of harm.

So the problem is to maximise survival of the alga.

The IPs that the biological matrix recommends are

Consolidate (assemble identical objects, form a swarm or group)
Defence (chemical defence—introduce a toxin; mechanical defence—change size)
Convert to Dynamics (change design in some active way)
Feedback (allow active feedback and control)

Note that three of the above four recommended design changes occur in the top 10 of IPs listed above.

For this bridge to work, there must be some basic analogies between the biology and the technology. Unravelling them suggests that further understanding of the biology would lead to technical (and probably biological) advances. Energy from the sun, captured by photosynthesis, charges the alga with energy, in this instance chemical which has a high energy density. Each algal cell is therefore a store for energy. It can either retain its store (be a miniature battery) or deliver it to a larger storage device. The algal energy is consumed by predators (in this example a water flea) who typically achieve absorption of only 10% of what's available. That's where the toxin comes in—it protects the algal cells from being eaten. In the battery, it could also represent a mechanism to aid transfer of the energy to a larger electrode, rather like changing pH facilitating release of bound oxygen from haemoglobin. This may require less energy and militate against the battery overheating. These processes presumably require control, so there have to be active feedback loops. It is highly likely that chemistry will reveal better, cheaper, less polluting and resource-intensive methods storing electricity. At the time of writing, sodium has many advantages (cost, availability, stability, charging rate, safety, weight) but also disadvantages (lifetime, energy density). As with biological systems, the decider may be more to do with cost and reliability than with extreme performance.

In this instance, there was only one example that conformed with the defining trade-off. However, the biology is a specific example of a primary step in the food chain of herbivory. This analogy suggests more trade-offs or specific questions of biological mechanisms will exist, where the established analogies can be refined.

If you are interested in this approach, a complete ontology, populated with a starter 270 trade-offs and relevant background biology to stimulate analogy, is available online. It's in a closed area reserved for ontologies and can be accessed by application to . It's interactive and supervised, so you can play with it, add trade-offs and use it with advice and discussion. The ontology is licensed under Creative Commons 4.0 which allows you to have your own copy but not to use it for commercial purposes. There is also free tutorial material showing how to establish a database of biomimetic trade-offs; again apply to .

11.7 The Completed Picture

The biological version of the contradiction matrix is not merely presenting a way in which a biological organism can solve its trade-off and be more likely to survive; it is presenting a solution for a particular trade-off, independent of the bearer of the trade-off and its resolution. Trade-offs can be derived from subcellular mechanisms (Mori et al. 2019) or whole organisms. The organism is deemed to be a natural experiment in how to solve the particular trade-off. This establishes a relationship between Parameters and Inventive Principles that has a truth and identity of its own, independent of its origin. Living organisms are effectively examples of strong patents, many repeated in different applications across the range of products. This idea has been a strong mantra in the biomimicry area, though as far as I know nobody has tried to make it as widely applicable as this. The reason is that the dialectic approach is firmly grounded in both theory and practise, so it has a power that an advertising slogan lacks. This was Altshuller's original idea, and users of TRIZ worldwide have proved that the strategy works, with strong technical patents clarifying the problem and pointing towards its solution. Biology has a different way of doing things but that different way may also be of value in solving a range of apparently disconnected problems using biomimetics. TRIZ has given biomimetics a new meaning beyond the copying of biological systems. It's true that in simple, basic, one-dimensional copying-as mimicry new ideas and methods can be revealed and implemented, but with TRIZ analysis, biomimetics achieves a new dimension. This is probably the strongest argument for studying biomimetics—as an analytical system it offers unfamiliar ways of solving problems that, because they originate in the most successful invention on Earth—Life—may offer a step change for Man's technology to be more sustainable and adaptable. This makes it easier to derive strong analogies.

Analogical thinking is therefore an important part of solving a problem. If you believe Hofstadter and Sander (2013), it's probably the *only* way of thinking. Telling a joke, making a pun, writing a poem, doing a crossword puzzle (or almost any puzzle), creating a model (real or conceptual), explaining a complex idea (such as defining analogical thinking), doing an experiment ... all these activities, and more, can act as stimuli for creating useful analogies. The best analogist will be an Autolycus—a magnet for unconsidered trifles. Be interested in everything. The good artist borrows; the great artist steals.

11.8 Inventive Principles of TRIZ Comparing Biology and Technology

The numbered headings repeat those defined by TRIZ and are normally listed in books on TRIZ.

Biological examples are stated first, separated from technical ones by a double slash (//). Occasionally the same IP works both in biology and technology.

1. Segmentation
   - Groups of ants, bees or wasps act as a coherent unit when they defend the nest or swarm // multi-wire cables are safer and more durable; a large computer is replaced by a network of PCs.
   - Parts of plants and animals can be made to fall off in a controlled manner, e.g. autumnal leaf fall; loss of limbs or tail by many animals when challenged // crash protection in a car by a crumple zone, clips and plugs for temporary connection.
   - Many animals are simply segmented and most plants are modular (leaves, flowers, stem, root) // LEGO bricks can be made into a variety of structures and models.
   - In any structure, compressive and tensile elements should be minimised and kept separate.
2. Extract and separate
   - A large and strong animal has a deep voice // fibre optic cable allows a single light source to illuminate a number of objects separately.
   - A bat uses sonics to identify prey items // radar provides a local map
   - Many flies secrete digestive enzymes over food // a tank track distributes weight more evenly.
3. Local quality—introduce heterogeneity
   - Bees use different areas of the honeycomb for storing nectar, pollen, raising larvae and queen cells // communities have separate areas for housing and shops, a front door can have separate flap openings for dog and letters.
   - An insect can detect vibration, odours, flow and touch with its antennae // a Swiss penknife can have a knife, screwdriver, scissors, toothpick, etc.
4. Asymmetry
   - Make a hedgehog spine/quill slightly curved to encourage bending and allow shock-absorption // the chain wheel of a bicycle is oval so that the pedal pressure can be constant.
   - The woodpecker converts the sinusoidal movement of its body into an asymmetrical sawtooth profile that gives power amplification // the cruck timbers of an A-frame timber building are bent so they deform safely as the building is completed.
5. Merging
   - Plants have many identical, separate, leaves // PCs in a network, transistors on a single chip.
   - Cells assemble into a slime mould plug // ball- and roller-bearings reduce rotary friction.

6. Universality
   - The cuticle of arthropods is waterproof, the skeleton, a food store, a signalling system (sound and vision) and a sensing material // van seat adjusts to accommodate seating, sleeping and cargo.
7. Nested structures
   - In the female aphid, unborn aphids have their offspring developing inside them, even several generations further on // Russian nesting doll (Matrioshka).
   - Deployable structures in insects such as ovipositor, mouthparts // a mechanical pencil contains lead, telescope and aerial (antenna) work on the same principle.
8. Anti-weight
   - Plankton go up and down on a daily cycle either by active swimming or by changing their buoyancy, shark's heterocercal tail // aircraft and animals flying.
   - Swim bladder in fish gives them neutral buoyancy // dandelion seed blown by the wind
9. Preliminary counter-action
   - Penguin feathers give insulation, cork bark gives protection against fire // house insulation saves heating energy
   - Non-woody plants are prestressed which reduces the amount of solid material needed // reinforced concrete with pretensioned rebar.
10. Preliminary action
    - Diurnal rhythms and cycles, annual rhythms for reproduction ensures young are born when there is plenty of food // self-adhesive tape, stamps, etc.
11. Cushioning in advance
    - Hormones prepare for combat (e.g. adrenalin), birds moult before migration flight // armour.
    - Nervous system detects patterns in behaviour and surroundings // crashbarriers on fast roads.
12. Equipotentiality
    - Elastin provides elastic energy storage that evens out pulses of blood and keeps it just below turbulence level // spring counterbalances weight of an up-and-over garage door.
13. Reverse or invert
    - Exodigestion: e.g. house-fly exodigests food, starfish everts gut, carnivorous plants have no gut // moving pathway, escalator, instead of walking or climbing.

14. Spheroidality
    - Membranes or containers at all levels don't require an expensive skeleton // Dyson's ball replaces a wheel for movement in any direction (wheelbarrow, vacuum cleaner).
    - Joints and branches eliminate stress concentrations with adaptive curves // curved archway.
15. Dynamism
    - Flexible membranes such as skin, insect wings, deform rather than break // flexible endoscope.
    - Adaptive prestrain by muscles to stabilise movement // moveable controls in an aircraft or wingwarping.
16. Partial or excessive action
    - Muscles attain high contraction rates by using fewer cross-bridges at each contraction // ensure penetration of a reactant by over-filling and removing excess.
17. Another dimension
    - Plant apical meristem forms primordia by buckling the constrained membrane, archer fish shoots down prey items sitting on exposed plants // a mirror allows re-orientation of a sight-line.
18. Mechanical vibration
    - Communication by any form of sound production, bee waggle-dance // control or damp resonant vibration for sound over a wide range of frequencies.
    - Wing beat frequency of flying animals // tuning a musical instrument.
    - Sonic 'radar' of bats, dolphins, etc. // radar, ASDIC and other range finders.
19. Periodic action
    - Flashing of a firefly is more noticeable, aestivation, hibernation // Warning light flashes, lighthouse lights have identifying flashing sequences.
    - Plankton migrate vertically avoiding predation // shops have displayed opening times.
20. Continuity of useful action
    - 'Higher' organisms function at constant temperature, filter-feeders (e.g. basking shark, mussels) capture food continuously // a flywheel stores energy when the vehicle stops so that the engine can keep running optimally.
21. Perform a process at high speed
    - Male of a spider, mantis, etc., copulates quickly then escapes before it can be eaten, basilisk lizard must run very fast over water // cut a material at high speed before heat can propagate and deform it.

22. Convert harm to benefit
    - Autotomy enables an animal to shed a limb and escape a predator, animals and plants recycle dung, micro-cracks can stop a larger crack // learn from mistakes.
23. Feedback
    - Any control system needs feedback—locomotion, speech, nutrition, hearing ... // any system retaining a null point.
24. Intermediary
    - Beetle antifreeze, sandworm glue, camouflage // carrier bag, wheelbarrow, temporary glue.
25. Autonomous response
    - Control is passed to a subsystem, as in the primary feathers of an eagle's wing that deflect in response to airflow, much of locomotion and bodily movement becomes autonomous with training // halogen lamps regenerate during use re-depositing evaporated material, software can have error-checking built in.
26. Copy
    - Mimicry is common, as camouflage or as dangerous animals or plants // copies are usually cheaper versions of the real thing, and can be quickly made with a mirror.
    - A cell produces a copy of its genetic and related systems before dividing, with outstanding accuracy // a copy can be electronically analysed and displayed on a screen.
27. Obsolescence
    - There are many examples of 'living fossils' surviving in unique habitats that would be otherwise outcompeted // many disposable devices and materials are sterile and are to be used once.
28. Substitution of mechanical objects
    - This is a great energy saver for marking out territory and possessions // use energy fields (electric, magnetic, etc.) to interact with objects and organise them, use a puff of air to redirect an object down a different path.
29. Pneumatics and hydraulics
    - Innumerable animals and plants use water under pressure as a skeletal substitute (worms, nonwoody plants, snails, fish) and as an actuator, insect wings expand hydraulically using self-generated pressure // liquid is virtually incompressible and is used in control systems in h.

30. Flexible shells and thin films

    - Ammonite shells are thin and change shape with high pressure at depth, but complex sutures keep them together.

31. Cellular materials

    - All animals, plants, bacteria, etc., are made of one or more cells and can be lightweight though stiff and tough (e.g. wood, feathers, bone, quills, plastron . . .) // Store fuels in a 'sponge', honeycomb core in a sandwich structure.

32. Colour change

    - Octopus and cuttlefish and some teleosts and crustaceans can change colour and blend into the background // colour is used for masking for printing and for blocking specific wavelengths.

33. Homogeneity

    - Insect cuticle has soft intersegmental areas between stiffer sclerites, with uniform chemistry but different amounts of the constituents (chitin, protein, water, phenolics, etc.) // plastics produce similar gradients with varying cross-linking, fillers and plasticisers.

34. Discard and recover

    - Nearly all natural materials can be recycled. Locust mandibles and front teeth of rodents are self-sharpening // most technical materials can be recycled but there may be problems with alloys, a sharpener can be built in to a machine.

35. Change properties of a material (the 'Get Out of Jail Free card' of TRIZ)

    - Since natural materials are composites, this usually involves adding a component to the mix, or changing the volume fractions or controlling temperature or . . . // Change any of the material parameters—don't forget pressure, volume and temperature (PVT).

36. Phase transition

    - Many, possibly all natural materials are liquid crystalline and so can form structures autonomously // PVT.

37. Internally driven dimensional change

    - The maths defining osmotic pressure in controlling the expansion of a plant cell is identical to the maths defining the effect of temperature on a metal. Therefore, surprisingly, this is an important Inventive Principle for biological materials! // thermal effects on metals and elastomers can effect actuation.

38. Strong oxidants

    - Hydrogen peroxide and phenolase generate hot oxygen as a propellant in the Bombardier beetle, otherwise this is a good description of many enzyme systems // The Nazi rocket plane—Komet or Me 163—used chemistry similar to that of the Bombardier beetle to achieve speeds of 750 mph in 1944.

39. Inert medium

    - Radiation blockers such as melanin protect desert insects at high temperatures, other insects and plants can dehydrate and remain inert until rehydrated years later // oxygen is highly reactive, so removing it can prevent degradation; insulators such as foam can exclude oxygen from a fire.

40. Composite material

    - Most, if not all, biological materials are composite. They can achieve a wide range of properties by changing components, volume fractions, etc. // Composites are being adopted in technology for similar reasons.

References

Altshuller G (1999) The innovation algorithm, TRIZ, systematic innovation and technical creativity. Technical Innovation Center Inc, Worcester

Bachmann H, Fischlechner M, Rabbers I, Barfa N, Branco dos Santos F, Molenaar D et al (2013) Availability of public goods shapes the evolution of competing metabolic strategies. Proc Natl Acad Sci USA 110(35):14302–14307

Bebenek A, Ziuzia-Graczyk I (2018) Fidelity of DNA replication—a matter of proof reading. Curr Genet 64(5):985–996

Bissell C (2007) Historical perspectives. The Moniac, a hydromechanical analog computer of the 1950s (pdf). IEEE Control Syst Mag 27(1):69–74

Bozdag GO, Zamani-Dahaj SA, Day TC, Kahn PC, Burnetti AJ, Lac DT et al (2023) De novo evolution of macroscopic multicellularity. Nature 617(7962):747–754

Cheong H, Shu LH (2014) Retrieving causally related functions from natural-language text for biomimetic design. J Mech Des 136(8):081008

Conn A, Pedmale UV, Chory J, Navlakha S (2017) High-resolution laser scanning reveals plant architectures that reflect universal network design principles. Cell Syst 5(1):53–62 e3

Garland T Jr, Downs CJ, Ives AR (2022) Trade-offs (and constraints) in organismal biology. Physiol Biochem Zool 95(1):82–112

Glick J (2013) Ontologies and databases—knowledge engineering for materials informatics. In: Rajan K (ed) Informatics for materials science and engineering. Butterworth-Heinemann, Oxford, pp 147–187

Hart Y, Sheftel H, Hausser J, Szekely P, Ben-Moshe NB, Korem Y et al (2015) Inferring biological tasks using Pareto analysis of high-dimensional data. Nat Methods 12(3):233–235

Hofstadter D, Sander E (2013) Surfaces and essences: analogy as the fuel and fire of thinking. Basic Books, Philadelphia

Marshall JA, Dornhaus A, Franks NR, Kovacs T (2006) Noise, cost and speed-accuracy tradeoffs: decision-making in a decentralized system. J R Soc Interface 3(7):243–254

Miranda S, Orciuoli F, Sampson DG (2016) A SKOS-based framework for Subject Ontologies to improve learning experiences. Comput Hum Behav 61:609–621

Mori M, Marinari E, De Martino A (2019) A yield-cost tradeoff governs *Escherichia coli*'s decision between fermentation and respiration in carbon-limited growth. npj Syst Biol Appl 5:16

Muller WEG (2003) The origin of metazoan complexity: Porifera as integrated animals. Invert Comp Biol 43:3–10
Ng CY, Wang L, Chowdhury A, Maranas CD (2019) Pareto optimality explanation of the glycolytic alternatives in nature. Sci Rep 9(1):2633
Noble D (2008) The music of life: biology beyond genes. OUP, p xiii, 151
Noble D (2021) The illusions of the modern synthesis. Biosemiotics 14(1):5–24
Savransky SD (2000) Engineering of creativity: introduction to TRIZ methodology of inventive problem solving. CRC Press LLC, Boca Raton
Shoval O, Sheftel H, Shinar G, Hart Y, Ramote O, Mayo A et al (2012) Evolutionary trade-offs, Pareto optimality, and the geometry of phenotype space. Science 336(6085):1157–1160
Terninko J, Zusman A, Zlotin B (1998) Systematic innovation. An introduction to TRIZ. CRC Press
Vakili V, Shu LH (2001) Towards biomimetic concept generation. In: Proceedings of DETC'01 ASME 2001 design engineering technical conferences design theory and methodology
Vincent JFV (2013) Structural biomaterials. Princeton University Press
Vincent JFV (2017) The trade-off: a central concept for biomimetics. Bioinspired Biomimetic Nanobiomater 6(2):67–76
Vincent JFV, Bogatyreva OA, Bogatyrev NR, Bowyer A, Pahl A-K (2006) Biomimetics: its practice and theory. JRS Interface 3(9):471–482
Yargan D, Jansen L (2024) Does biomimetics require a unified account of function? In: Proceedings of the Joint Ontology Workshops, Enschede, The Netherlands
Zhu X, Wang J, Chen Q, Chen G, Huang Y, Yang Z (2016) Costs and trade-offs of grazer-induced defenses in *Scenedesmus* under deficient resource. Sci Rep 6:22594

Conclusion: Functional Biomimetics as a Paradigm for Product Development, Sustainability Solutions, and Biodiversity Protection

This book is written in the spirit of the form–function relationship concept within organisms related to bioinspired technical systems. It is apparent that in biology and engineering, functionality plays a decisive role. However, in biomimetics/biomimicry, biological models can also be used corresponding to the other criteria, such as beauty and aesthetics. Thus, we can explore functional biomimetics and aesthetic biomimetics.

The first one emphasizes the use of biological principles to solve engineering or design problems, such as designing more efficient airplane wings inspired by bird wings or creating self-cleaning surfaces based on the lotus effect. The priority is given to functionality, efficiency, and engineering-based problem-solving. The second one is focused on drawing inspirations from the visual and structural beauty of nature, though not directly related to specific functional demands. Examples of such an approach would be mimicking seashell patterns in decorative architectural elements or using organic forms of plants in furniture design. Priority is given to the aesthetic appeal and visual harmony with nature.

In biomimetics, the relationship between functionality and beauty is complex. While biomimicry often prioritizes functional solutions inspired by nature, it can also simultaneously lead to aesthetic designs, but not necessarily vice versa. For example, the intricate, lightweight self-supporting structures in nature can be both structurally efficient and visually stunning. Balancing these aspects is crucial for creating successful and desirable products and systems. In the balancing of both functionality and beauty, a synergetic approach can be of importance, however, with an emphasis on the bioinspired functionality. Striving for both functional and aesthetic goals can be challenging, requiring a deep understanding of biological systems, design principles, and engineering technologies, capable of transferring these principles into prototypes, and finally into novel successful products in the market.

The biomimicry's role should be mainly an inspiration and not necessarily an imitation. Biomimicry should go beyond copying nature's appearance for aesthetics.

V. Bels et al. (eds.), *Biomimicry – A Functional Approach*, Biologically-Inspired Systems 20, https://doi.org/10.1007/978-3-032-11949-0

Understanding the underlying principles and processes that lead to primarily functional and secondarily aesthetic outcomes is critical. The holistic approach that considers the whole system rather than focusing on various parts or aspects is the key to success. In biomimetics, one should research the interconnectedness and interdependencies of various components, recognizing that the whole is more than the sum of its parts. Therefore, an effective biomimetic design considers the interplay between form and function and their interaction with the environment. Thus, more experimental studies on organisms are needed to study their structures and physical properties and the optimization path chosen in their natural environment. With the exception of a few model organisms, this remains on the "to do" list of science.

The ultimate hope of the authors and editors of this book is that by learning from nature's efficient and resilient systems, biomimicry can lead to more sustainable and environmentally friendly solutions for human life. In addition, by studying other aspects of organisms' biology, biomimicry will contribute to a better understanding of all biological traits of animals and plants, to the development of a greater respect for every organism on Earth, and thus to a stronger intrinsic motivation of every human being to protect the biodiversity of our planet—an infinite source of ideas.

Paris, France — Vincent Bels
Kiel, Germany — Stanislav N. Gorb
Conway, SC, USA — Roi Gurka

GPSR Compliance

The European Union's (EU) General Product Safety Regulation (GPSR) is a set of rules that requires consumer products to be safe and our obligations to ensure this.

If you have any concerns about our products, you can contact us on ProductSafety@springernature.com

In case Publisher is established outside the EU, the EU authorized representative is:

Springer Nature Customer Service Center GmbH
Europaplatz 3
69115 Heidelberg, Germany

Batch number: 10726133

Printed by Printforce, the Netherlands